AF685129

Energy Efficiency and Carbon Neutrality in Buildings

Energy Efficiency and Carbon Neutrality in Buildings

Guest Editor

Amos Darko

MDPI

Basel • Beijing • Wuhan • Barcelona • Belgrade • Novi Sad • Cluj • Manchester

Guest Editor
Amos Darko
Department of Construction
Management
University of Washington
Seattle, WA
USA

Editorial Office
MDPI AG
Grosspeteranlage 5
4052 Basel, Switzerland

This is a reprint of the Special Issue, published open access by the journal *Buildings* (ISSN 2075-5309), freely accessible at: https://www.mdpi.com/journal/buildings/special_issues/MP476QK560.

For citation purposes, cite each article independently as indicated on the article page online and as indicated below:

Lastname, A.A.; Lastname, B.B. Article Title. *Journal Name* **Year**, *Volume Number*, Page Range.

ISBN 978-3-7258-5059-4 (Hbk)
ISBN 978-3-7258-5060-0 (PDF)
https://doi.org/10.3390/books978-3-7258-5060-0

Cover image courtesy of Amos Darko

© 2025 by the authors. Articles in this book are Open Access and distributed under the Creative Commons Attribution (CC BY) license. The book as a whole is distributed by MDPI under the terms and conditions of the Creative Commons Attribution-NonCommercial-NoDerivs (CC BY-NC-ND) license (https://creativecommons.org/licenses/by-nc-nd/4.0/).

Contents

About the Editor

Amos Darko

Amos Darko, Ph.D., M.ASCE is a John E. Schaufelberger Endowed Assistant Professor in the Department of Construction Management at the University of Washington. He is a highly cited and highly ranked researcher in green building and sustainability. He has completed several sustainability (energy and carbon)-related projects and has been awarded numerous research grants and awards. He won the prestigious Sustainability 2023 Young Investigator Award from the Sustainability journal in 2024 and the Green Talents Award from the Cabinet-Level German Federal Ministry of Education and Research in 2020. He completed his Ph.D. research project in 2019 at The Hong Kong Polytechnic University, developing strategies and models for promoting the adoption of green building technologies. He is ranked among the world's top 2% and top 0.05% of scientists by Stanford University and ScholarGPS, respectively. He is also ranked as the world's lifetime #1 highly ranked scholar in the field of green building by ScholarGPS. His current research interests lie in the sustainability (energy and carbon) and resiliency retrofits of existing buildings.

Preface

The world is in a period of sustainability breakdown, with climate change causing life-threatening floods, wildfires, storms, droughts, precipitation, and heat and cold waves in every corner of the world. With the buildings and construction sector being a major contributor to this challenge, we need rapid systemic changes in this sector if we are to avoid catastrophic climate hazards in the years ahead.

When I was approached by the Buildings editorial office to put together a Special Issue on enhancing energy and carbon efficiency in buildings, I asked myself one important question: Can this Special Issue contribute to achieving the above goal of making positive changes in the buildings and construction sector toward sustainability? My answer to that question was a resounding "yes"; therefore, I did not hesitate to take up the initiative to work on this Special Issue.

Energy Efficiency and Carbon Neutrality in Buildings is for those who must have a broad understanding of strategies, procedures, measures, methods, and tools to successfully plan, design, build, operate, and manage buildings in ways that reduce their energy use and carbon emissions. Those who conduct research on and/or study energy and carbon efficiency in buildings will also find this Reprint useful. Another group of individuals who will benefit from this Reprint are those who formulate and evaluate policies and regulations on energy and carbon efficiency in the buildings and construction sector.

The 13 papers in this Reprint are intended to provide a comprehensive understanding of innovative strategies, solutions, and practices to promote energy efficiency and carbon neutrality in the buildings and construction sector. They present innovative technologies to manage and reduce energy use and carbon emissions in buildings. They show how project management strategies and key competencies can be modified to successfully deliver energy efficiency and carbon neutrality goals in building projects. They also examine the impact of policies and regulations on building sector decarbonization. These dimensions have been examined from the perspectives of different building types, including residential, educational, and other public buildings. The audience of this Reprint will likely be policymakers, industry practitioners, researchers, and students interested in the dynamics of energy efficiency and carbon reduction in the buildings and construction sector.

A Reprint like this is needed because those in the buildings and construction sector must understand how to reduce or eliminate the environmental footprint of the sector. I hope that this Reprint will be an extremely valuable resource and companion in this knowledge-building journey. This Reprint provides useful insights into promoting energy efficiency and carbon neutrality in buildings.

So, roll up your sleeves (or perhaps grab a sweater!) and get ready to explore the many fascinating ideas and innovative solutions and strategies to reduce energy use and carbon emissions in buildings presented in this Reprint.

Amos Darko
Guest Editor

buildings

MDPI

Article

Building Energy Performance Modeling through Regression Analysis: A Case of Tyree Energy Technologies Building at UNSW Sydney

Faham Tahmasebinia [1,*], Ruihan He [1], Jiayang Chen [1], Shang Wang [1] and Samad M. E. Sepasgozar [2]

1 School of Civil Engineering, The University of Sydney, Sydney, NSW 2008, Australia
2 School of Built Environment, The University of New South Wales, Sydney, NSW 2052, Australia
* Correspondence: faham.tahmasebinia@sydney.edu.au

Abstract: Addressing clients' demands, designers have become increasingly concerned about the operation phases of buildings and, more specifically, energy consumption. This issue has become more prominent as people realize that the Earth's resources are limited and depleted, and buildings are major energy consumers. Building Information Modelling (BIM) has gained popularity in recent years and is now widely used by architects, engineers, and construction teams to collaborate and provide a comprehensive design that follows a sustainable strategy. The objective of this research is to examine how building variables are linked to energy consumption in various building shapes, achieved by building prototypes. The accuracy of the regression models is evaluated by undergoing a validation process. Consequently, this study created building information models of selected education facility office rooms and used Autodesk Insight 360 and Green Building Studio (GBS) to perform energy simulations. A 6 Green Star education building in Australia is chosen as the case study of this paper. Thirteen variables related to building internal design were examined, and five were found to endure a substantial effect on building energy consumption. The study also looked at the window-to-wall ratio (WWR), which was analyzed by multi-linear regression; however, the results showed that the model did not fit well, and the error obtained during the validation process ranged from 1.0% to 26.0%, which is unacceptable for this research. These findings highlight some limitations in using BIM tools and linear regression methods and discuss some potential improvements that can be achieved in future studies.

Keywords: green building; building information modelling; BIM; regression analysis; EUI; energy optimization

1. Introduction

1.1. Introduction

BIM provides a comprehensive platform for designers to gain a deeper understanding of the various building parameters to design buildings that are low in energy consumption [1,2].There are many tools on the market that use and are compatible with BIM technology, which facilitates designers from different fields. In addition, as an important part of statistics, regression models can provide a more significant indication of the dispersion and correlation of data. Combining the two allows designers to understand the parameters most critical to energy consumption. BIM is widely used by many companies, and a technical review of BIM applications suggests that two main considerations in BIM utilization are its compatibility with needs and its interoperability with other tools [3]. These two drivers can make BIM more useful to the community or industry. BIM integration with various tools and sources of data, such as building shape, geometry, materials, and equipment types, allows the creation of accurate energy modelling. Energy modelling practices are important since advanced models can be used to predict the energy consumption of various buildings. This outcome of simulations and estimations of energy consumption need to be compared with actual data acquired from energy meters or submeters recording the electrical, heating, ventilation, or air conditioning usage or the data collected and managed through building management systems in real time at smart buildings [4]. The validation of models using read data will eventually help to apply artificial intelligence (AI) algorithms to predict energy consumption and optimize energy efficiency based on changes in design variants, construction, or usage during the operation phase of the building [5]. Thus, the current practices of energy modelling are important to test or cross-verify the estimation methods for various design alternatives and shapes. This type of practice will eventually support advanced AI-based optimizations and predictions. Based on lessons from the current practices, AI will be further developed to automate the modelling process using big data created by numerous smart homes and buildings [4]. These modelling exercises, optimizations, and predictions are significant because they help facility managers and the building sectors to improve the energy performance, reduce greenhouse gas emissions, and contribute to sustainable development goals (SDGs), in particular, SDG 7, 11, and 13. These SDGs are respectively connected to the provision of modern and sustainable energy, the development of resilient and sustainable cities, and the implementation of urgent actions to combat climate change. The present paper on building energy performance modelling through regression analysis is very significant and relevant to achieving the SDGs. The study provides a means of evaluating and predicting building energy performance, which can contribute to reducing energy consumption, mitigating greenhouse gas emissions, and promoting green buildings or sustainable building design and operation practices. By investigating and improving energy efficiency in buildings, this study aligns with SDG 7 on affordable and clean energy, SDG 11 on sustainable cities and communities, and SDG 13 on climate action [6]; therefore, the current exercise is an important contribution to advancing sustainable development outcomes and addressing global challenges.

1.2. Problem Definition

As concerns about energy consumption have increased, standards have been specified to limit the energy consumption of buildings, such as the famous United States commercial building energy codes Standard 90.1 [7]. Furthermore, challenges such as the 2030 challenge have been developed to motivate designers to focus on the energy consumption of buildings [8]. Statistical methods have been popular since researchers are attracted to finding an easy way to evaluate building energy consumption. The statistical method provides a simple and accurate way to predict building energy consumption [9]. This paper aims to combine BIM technology and statistical methods to perform a further study of the 'building's internal variables that affect the 'building's energy consumption.

1.3. Objectives

There are four objectives that this paper aims to achieve. Firstly, the BIM models of six different shapes of offices in the building are created in Autodesk Revit. Using Revit allows users to develop a 3D solid model and modify building data efficiently [10]. Secondly, the energy simulation is developed in Autodesk Insight 360 and GBS. Thirdly, the statistical method is then used to investigate the relationship between selected building variables and energy consumption of six shapes of rooms separately. Furthermore, a validation is developed to obtain the error between simulation results and the one from a real-world case study. The Tyree Energy Technologies Building (TETB) of The University of New South Wales (UNSW) is chosen as the case study for this paper.

2. Literature Review

2.1. Implementing BIM in Green Building Design and Construction

2.1.1. Background of BIM Development in Green Construction

BIM was developed in the 1970s, but the promising concepts and values of BIM were agreed upon in 1999. Since this time, continuous efforts have been made to implement BIM in green building design. Succar revealed that BIM is comprised of interacting processes and technology, which can generate and manage building data throughout the entire project life cycle [11]. Abanda et al. then provided a list of 122 BIM software systems and indicated that the various functions of BIM can be competent in various aspects of green construction, such as building energy performance assessment, lighting analysis, and construction and demolition waste analysis [12]. Wong and Zhou presented a summary of literature reviews on BIM green construction but lacked the detail of specific frameworks to link BIM with green construction [13]. However, Lu et al. presented an essential assessment of the relationship between BIM and green buildings and proposed a Green BIM Triangle organization to conceptualize the interactions between BIM and green buildings.

2.1.2. Energy Simulation by BIM

BIM stores massive energy-related building information and has a strong capability for information exchange. It can extract all building geometry information from the architectural building model and forecast total energy consumption and make recommendations for energy optimization [14]. With these results, designers can build more sustainable buildings.

Kamel and Memari summarized previous case studies about the use of various BIM tools in energy analysis. These case studies indicate that the building models are mostly developed based on Revit or ArchiCAD [15]. Revit and ArchiCAD have strong conceptual massing capabilities, which can use basic shapes to model building form and orientation in the early design stage. The building energy model can be automatically generated from the building model through the prototyping of an Application Programming Interface (API). The user can update the design concept anytime in Revit or ArchiCAD and the energy model is also updated automatically. The two most prevalent BIM file schemas are gbXML and IFC, which allow data transfer. Feasible energy simulation tools are various, including EnergyPlus, OpenStudio, and GBS. They take either IFC or gbXML as the input to generate the building energy performance results.

2.2. Research Gap and Challenges for BIM-Based Energy Simulation

2.2.1. Limitations of Using BIM Tools

Kamel and Memari also investigated the feasibility of data exchange from BIM to BEM and found that the lack of interoperability was a major issue in implementing BIM tools [15]. Table 1 summarizes their findings in data transfer between different tools based on three case studies. Direct transfer from Revit to GBS is the most feasible and convenient solution, which is also applied in this study. Other BIM tools either have missing data or require manual input, which is not user-friendly. Secondly, today's industry is lacking clear construction standards and codes for various aspects of the BIM application. Thirty-six existing standards of Green BIM application have been studied by Chong et al., but those standards do not pay much attention to the specific execution of BIM application in green building design, such as refurbishment and demolition of green buildings [16]. Lastly, low industrial acceptance of BIM technology is also an important issue. BIM is an innovative technology that requires collaboration among various software tools. The lack of knowledge and expertise in the industry makes workers rely on conventional construction methods [17].

2.2.2. Future Work Directions of BIM

According to the issues identified in Section 2.2.1, the process of data extraction and exchange needs to be optimized by enhancing the functionality of gbXML and IFC Schema. An efficient algorithm also needs to be developed to enable the automatic extraction of specific types of energy data from the BIM model. More compatible middleware is required to develop if direct data transfer is prohibited. Moreover, future research should pay more attention to the technical details of implementing BIM in energy simulation and all relevant green building designs rather than just making recommendations to use BIM. A complete industry standard of BIM application and a more effective framework for the energy simulation process is expected to be developed.

Table 1. Data transfer issues between different BIM tools [15].

Data Type	Case Study 1		Case Study 2		Case Study 3
	Revit-GBS	**Revit-gbXML**	**gbXML-OpenStudio**	**OpenStudio-IDF**	**GBS-IDE**
Weather File Schedules	Automatically obtained through the location button in Revit Transfers	N/A	Requires manual input	N/A	Requires manual input
Schedule	Transfers	Transfers	Does not transfer and manually added using GUI	Transfers	N/A
Constructions	Complete transfer, GBS divides surfaces to sub-surfaces based on the energy analytical model	N/A	Complete and correct transfer	N/A	Transfers
Loads	Transfers	Does not Transfer	Does not transfer and manually added using GUI	Transfers	Transfers
Space Types	Transfers	N/A	Transfers	N/A	Transfers
Building Information	Transfers	Transfers	Partial Transfers For example, typical floor height is not transferred	Transfers	Transfers
Thermal Zones	Transfers thermostat set points based on default values	Does not transfer	Does not transfer and manually added using GUI	Transfers	N/A
HVAC	Transfers	Does not transfer	Does not transfer and manually added using GUI	Transfers	Transfers

N/A represents there is no information about this field.

2.3. Multiple Linear Regression (MLR) Models of Energy Simulation

2.3.1. Background

As the building sector is the largest energy and CO_2 emitter, energy prediction models have been researched and innovated over the past decades. In the early research, physical models with static energy analysis were first used to predict energy consumption, which is quite time-consuming and labor-intensive. Some mathematical models and energy forecasting software were then developed, which improve the prediction speed but cannot perfectly match the simulation data. Moreover, the involved mathematical formulas or programming language requires expertise [18]. Compared with various energy prediction tools, the multiple linear regression (MLR) models are the most appropriate prediction tool to be implemented in this thesis study due to its simplicity and high accuracy [19]. MLR can express the simulated energy data in a linear equation to predict the energy consumption of given material properties.

2.3.2. Application in Building Energy Simulation

Multiple regression models have been applied widely in different areas of energy analysis, such as residential building retrofit, heat load prediction and building energy usage estimation, etc. Walter and Sohn used a multivariate linear regression model to quantify the contribution of each building variable to the total energy usage. However, the model is not validated by the post-retrofit energy data [20]. Martinez et al. applied the same technique to assess the energy performance results of a residential house after retrofit and found that upgrading building envelopes increase energy savings [21]. Jaffa et al. also developed an accurate polynomial model summarizing the influence of heat consumption [22]. Lam et al. studied the energy consumption of office buildings in different climates in China. They imported 12 design variables, including building loads, HVAC system, etc., and surprisingly found that the difference between regression prediction and DOE simulation is within 10%. The result implies that the regression models developed can be used to estimate the likely energy savings or consumption during the initial design stage when different building schemes and design concepts are being considered [23].

Mottahedi et al. obtained a more accurate prediction with a maximum error of 5% on total energy consumption. The researchers made a multiple linear regression model to predict the energy consumption of an office building with seven different shapes [24]. The model includes the regression coefficient for each design parameter, and the format is also adopted in this study to generate an MLR model.

2.3.3. Limitations of Multiple Linear Regression Models (MLR)

The main issue of using a multiple linear regression model is the multicollinearity, which is caused by the high correlation among internal factors. Multicollinearity makes it hard to interpret the coefficients and thus reduces the capability of the model to present the effect of independent variables. Another issue is that MLR cannot perform accurate predictions as non-linear models produced by machine learning models, such as Artificial Neural Networks (ANNs) and Support Vector Machines (SVMs) [25]. This is because linearity cannot entirely represent reality, and the building model is so complicated that not every detail of the building can be modelled.

2.3.4. Future Work Direction of MLR

The MLR needs to be improved to detect the multicollinearity issue and illustrate the residuals from its best-fit regression equation. Moreover, the sensitivity analysis should be applied to aid in the regression approach as it can measure the relative standing of the effect limitations, making it possible to replace the design parameters that may influence an insignificant effect on the dependent variable. Moreover, the stepwise function can be used to determine whether to eliminate or add influence variables based on the level of correlation. However, the future trend should focus more on non-linear modelling to obtain more accurate forecast results. More machine-learning models should be introduced to the industries [25].

2.4. *Energy Regression Analysis Combine with BIM Tools*

2.4.1. Necessity for MLR Combine with BIM

Many case studies have indicated the applicability of BIM in energy simulation in Section 2.1.2, but few papers have verified the simulation results and tested their validity. The accuracy of the simulation is confronted by the challenge of interoperability issues among BIM tools. Interoperability refers to the ability of two or more systems to exchange information that is needed and available to use the information [16]. Several studies have summarized the potential issues of the two most proposed Data exchange Schemes (IFC, and gbXML). Choi et al. found that data loss frequently occurs when transferring IFC model data into the energy model, leading to incomplete or incorrect IFC files. Moreover, the definitions of IFC can vary with different BIM authoring tools, and thus unexpected errors were often created, reducing the reliability of results [26]. Pinherio et al. pointed out that the gbXML-enabled data exchange is inappropriate for large complex building shapes since the calculated surface areas and space volume differences might exceed the standard engineering tolerance and cause an overestimation of building energy consumption [27]. Moreover, the lack of the geometric representation of heating, ventilation, and air conditioning systems for gbXML might affect the simulation results [28].

These findings above indicate that the energy simulation process produces errors and thus, the model validation is necessary to decide whether the errors are within the acceptable range. Model validation can be achieved by constructing MLR with a series of statistical analyses, such as *T*-test, F-test, and ANOVA. MLR can predict the energy simulation results produced from the BIM energy analysis software and represent them in linear equations. The comparison between regression prediction and BIM simulation results can be visualized by generating bar plots on MATLAB.

2.4.2. Summary of Relevant Case Study

There is a lot of separate literature on regression analysis to predict energy use and BIM to simulate energy models, but few studies have mentioned how the two technologies work together. Table 2 below is a summary of case studies about the combination of these two techniques.

Table 2. Summary of case studies about regression analysis combined with BIM tools.

Author	Case Study	Modelling Technique	Findings
Mottahedia, M Mohammadpour, A Amiri, S, S Riley, D Asadi, S	Annual energy consumption of a typical office building with seven different building shapes in two different climates [24]	e-Quest and DOE-2: used to generate building energy model. MLR: used to predict total energy consumption in a linear equation. Coefficient determination (R2), F-test, and root mean squared error (RMSE): used to discuss the accuracy of regression models.	Space heating is the main source of energy consumption in the polar climate zone. T-shape buildings consume the most energy in both cold–dry and warm–marine climates.
Mottahedia, M. Asadi, S. Amiri, S.	Development of a Multiple Linear Regression Model to assess energy consumption in the early stage of commercial buildings design [29]	Monte Carlo Analysis: used to estimate the probability of getting acceptable results. eQuest and DOE-2: used to create and simulate modelling. MLR: used to identify the relationship between independent variables and the dependent variable. F-test, and RMSE: used to tell whether the design variables are significant	1. A basic framework is developed to link the regression model with the BIM simulation tool. 2. The errors between DOE simulation and regression prediction are within 5%.
Aghdaei, N Kokogiannakis, G Mccarthy, T	Prediction of annual heating and cooling demand in three types of Australian dwellings [30]	MLR with ANOVA approach: used to predict annual heating and cooling demand. Building Energy simulation method (not specified): used to verify the prediction results. Sensitive analysis: used to identify the most influential parameters.	The error between energy simulation and prediction results is less than 15%. R^2 was over 0.90, indicating a good agreement between the simulation and regression model.
González, J. Alberto, P. Soares, C, A, P. Najjar, M. Haddad, A, N.	BIM and BEM methodologies integration in energy-efficient buildings using experimental design [31]	AutoCAD Revit: used to define a physical model and perform model integration. GBS: used to run the generated model 42 times. Regression Models developed by Minitab software with *p*-value: used to determine the representativeness of the buildings	The higher the efficiency of lighting and applications, the lower the electric demand. The lack of information about thermal material properties affects the accuracy of the simulation.

2.5. *Comparative Study between Existing Building Energy Studies and IDEA Model*

The traditional methods to explore building energy consumption mainly focus on fundamental data and lack interaction of influencing elements from both macro and micro levels [32]. However, an innovative energy simulation model was introduced by Huo et al. to investigate the carbon emission of Chinese commercial buildings toward 2060, which is the integrated dynamic emission assessment mode (IDEAM) [33]. The IDEAM consists of the system dynamics (SDs) model and a bottom-up end-use decomposition model, which can not only reflect the influential parameters but also provide an interactive feedback mechanism between different levels of parameters [33]. The IDEAM has been successfully utilized in exploring the interaction among different influencing elements and predicting the urban residential building's carbon emissions [34].

2.6. Literature Summary and Critical Review

In this literature review, three articles provided the definition and wide application of BIM in green building constructions. One article identified that the energy simulation results obtained by BIM are comprehensive due to its rich database and information exchange capability. One article made a summary of case studies using BIM tools in energy analysis and found the two reliable modelling tools and file schemas used in data transfer. Three articles demonstrated three key limitations of using BIM, including the lack of interoperability, industry standards, and acceptance. It is thus recommended to develop an efficient algorithm or more compatible middleware to enable more efficient and accurate data transfer in the future. One article then summarized the traditional energy prediction methods and emphasized the ease of operation and precision for MLR through comparison. Five articles have applied regression techniques in various fields of energy measurements and found that most energy data can be verified by MLR. However, the multicollinearity issue is an inevitable challenge for MLR, and thus, one article suggests focusing more on non-linear model prediction and machine learning models in future work. The limitations mentioned above and findings from four articles indicated that BIM simulations would produce unexpected errors. Thus, it is necessary to combine MLR with BIM to validate the results.

Additionally, the innovation research method (IDEAM) that considers both macro and micro variables is compared with the traditional method; however, this article aims to explore the interaction between energy consumption and building internal variables, which are factors belonging to the micro level. The ambient influencing elements are ignored to simplify the whole process. Hence, the traditional fundamental study is more compatible with this research.

These articles have helped to deepen our understanding of the BIM concept, the breadth of its capabilities, and its distinct advantages over other analysis software. The articles on the application of BIM in the energy field provide a good theoretical basis for the methodology of this thesis. Although these articles also exposed some flaws and limitations of BIM, this thesis does not aim to solve these issues but summarizes the limitations during the energy simulation process.

3. Materials and Methods

3.1. Tools

Table 3 below illustrates the tools used in building BIM models and performing energy simulations.

Table 3. List of selected software.

Name	Function
Autodesk AutoCAD [35]	Drawing an approximate 2D plan of the TETB based on photos taken during the site investigation
Autodesk Revit [36]	Creating 3D solid building models based on the 2D CAD drawing created in the above stage
Autodesk Insight 360 [37]	Providing visual data and 3D analytical models
Autodesk GBS [38]	Acting as the back engine of Insight 360, providing detailed numerical data

3.2. Building Elements, Properties, and Energy Settings

Due to the limitation of information, only three building elements were used in the room models, and some types defaulted in Revit. Additionally, each room model has the same room area of 20 square meters to better prevent results from being influenced by too many independent variables.

Table 4 shows the energy setting used in this study. Due to the limited information available, most settings are set to default, while several of them were chosen accordingly.

Table 4. Summary of energy simulation settings.

	Energy Settings	
Energy modelling mode	Use Conceptual Masses and Building Elements	Default
Building Service	VAV—Single Duct	Default
Building Type	School or University	TETB is located in UNSW, which is a university facility
Building Operating Schedule	Default	Default
HVAC System	Central VAV, HW Heat, Chiller 5.96 COP, Boilers 84.5 Eff	Default
Export Category	Rooms	Each shape of the room has the same room area of 20 m^2

Table 5 shows the material properties used in energy simulation, and these include the major building components such as roofs, walls, floors, and glass type. Different types are measured by U-Value, which indicates the thermal transmittance of the material [39].

Table 5. Summary of energy simulation override settings.

	Analysis Properties
Roofs	4 in lightweight concrete (U = 1.275 W/(m^2·K))
Exterior Walls	8 in lightweight concrete block (U = 0.8108 W/(m^2·K))
Interior Walls	Frame partition with $\frac{3}{4}$ in gypsum board (U = 1.4733 W/(m^2·K))
Ceilings	8 in the lightweight concrete ceiling (U = 1.3610 W/(m^2·K))
Floors	Passive floor, no insulation, tile, or vinyl (U = 2.9582 W/(m^2·K))
Slabs	Un-insulated solid (U = 0.7059 W/(m^2·K))
Doors	Metal (U = 3.7021 W/(m^2·K))
Exterior Windows	Large double-glazed windows (reflective coating), industry (U = 2.9214 W/(m^2·K))
Interior Windows	Large single-glazed windows (U = 3.6898 W/(m^2·K), SHGC = 0.86)
Skylights	Large double-glazed windows (reflective coating), industry (U = 3.1956 W/(m^2·K))

3.3. *Autodesk Insight 360*

Autodesk Insight 360 is a cloud-based service that allows users to create different scenarios by changing the selected building parameters, such as building orientation, window shading, wall and roof construction type, and find out the best performance selection so that the most energy-efficient design can be generated [40]. In this study, Insight 360 was used as a preliminary analysis tool to provide the 3D analytical model based on the BIM models.

Insight 360 also provides the energy consumption required by the standard and the idealized design to better help users understand the parts that can be improved and realize the importance of designing energy-efficient buildings.

The six models of office rooms are displayed in the form of several surfaces with different colors. The interactive selection box allows the users to adjust the different building internal variables, including building orientation, Window-to-Wall ratio, window glass type, window shading, wall construction, roof construction, building infiltration rate, lighting efficiency, daylighting and occupancy controls, plug load efficiency, HVAC system, operating schedule, PV-Panel efficiency, PV-Payback limit, and PV-surface coverage. Figure 1 displays the analytical models of six different shapes developed in Insight 360. Each model is assigned a number, which is used to identify them in the following contents.

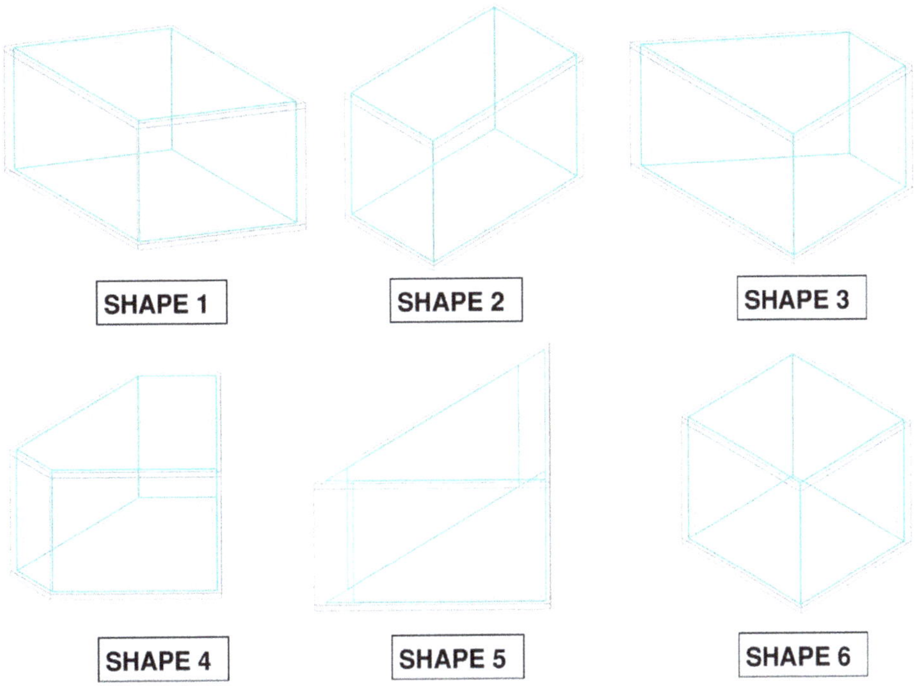

Figure 1. Revit analytical models.

3.4. *Autodesk Green Building Studio (GBS)*

Similar to Insight 360, GBS is also a cloud-based energy simulation tool. GSB can develop and evaluate several building variables and run alternatives simulation automatically by changing building variables [38]. Additionally, GBS is the back engine of Insight 360. It provides computational support for Insight 360. Therefore, the building variables in both tools are highly matched. Furthermore, the energy simulation data can be exported as a CSV file, which allows users to develop statistical analysis in other programming tools.

3.5. *Case Study*

The Tyree Energy Technologies Building (TETB) is the case study of this paper. TETB is a 6 Green Star laboratory building in Australia. Figure 2 presents some images from two internal areas along with thermal images. The photos also show the students' activities in those areas. This building is also used for teaching, meeting room, student group meetings, and studies. Thus, comfort, Co2, and temperature are critical factors for this building during operation.

Based on the introduction of Taylor Thomson Whitting (TTW), the Structural, Façade and civil engineering consultant of this project, several energy-saving technologies, such as the highly efficient Façade fabricated of glass and terracotta tiles, the featured steel roof span across the cantilevers over the entrances, makes this building not only full of technology and sustainable but also injected with research significance. In this research, six office rooms of TETB with typical shapes were chosen as the objects. Six shapes are extracted to allow researchers to find out the importance of different building variables in building energy simulation. The findings are compared with the original models to perform error analysis to corroborate the findings. Figures 3 and 4 display the BIM model of TETB and the office rooms selected for this research.

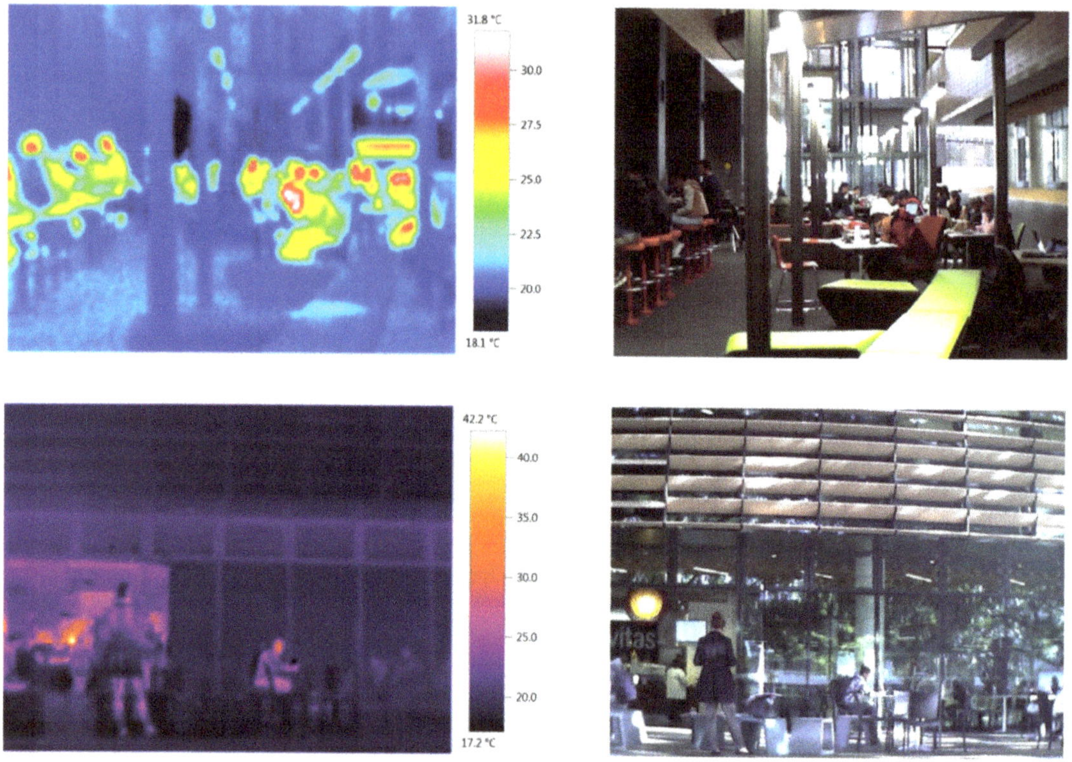

Figure 2. TETB images showing the thermal state, students' activities, and the façade design.

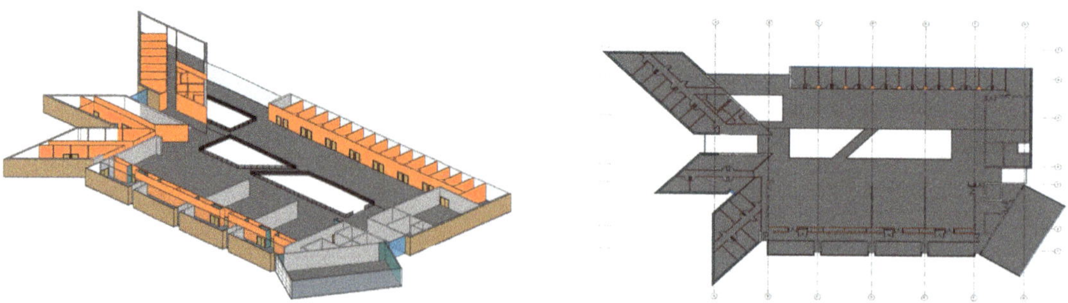

Figure 3. TETB BIM model and typical floor plan.

This paper aims to investigate the relationship between building internal variables and energy consumption. The BIM models of seven different shapes were built in Revit. The energy simulations were performed in Insight 360, which allows users to compare the original design with various benchmarks. The alternative designs were developed by GBS automatically, and the detailed calculation data were exported as a CSV file for the statistical analysis discussed in the next section. Additionally, six real-world rooms in TETB were prepared in Revit to prove the statistical analysis developed in the next step.

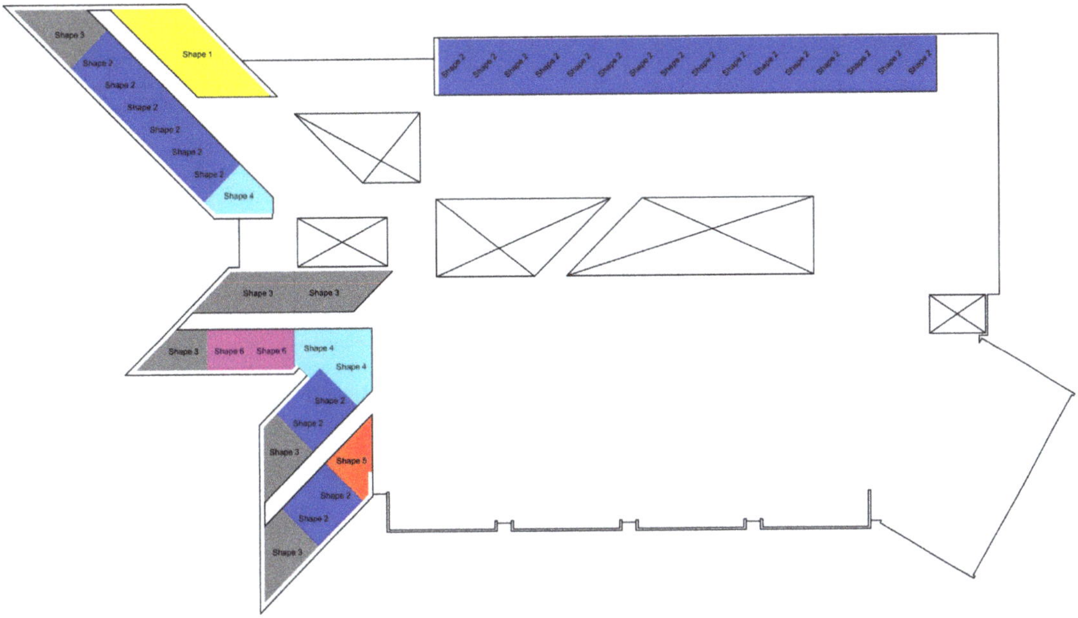

Figure 4. Selected shapes.

4. Obtaining, Filtering Data, and Performing Statistical Analysis

4.1. Introduction

Statistical analysis was used to investigate how each building variable can affect energy consumption.

The six initial data sets were generated by GBS and exported as CSV files. The appropriate data filtering work is necessary for large data analysis. Dropping bad and unnecessary data during the filtering process can balance the trade-off between the filtered data size and the information preserved and attenuate the effect of excessively abnormal data on the sample [41].

4.2. Tools

The statistical analysis is performed in MathWorks MATLAB. MATLAB is one of the most popular technical computing tools in engineering and scientific areas [42]. MATLAB is a comprehensive and powerful data computing platform that is compatible with a wide range of file formats, including the CSV file format. Moreover, it comes with multiple toolboxes to allow users to handle large volumes of data and calculations. In this study, several built-in functions, such as "fitlm" and "corrcoef" are called to calculate the statistical indices and to help researchers understand the data distribution and linear relationships.

4.3. Building Variables

The building variables play important roles in determining building energy consumption. Only the internal building variables were researched because internal variables, such as wall construction type and infiltration rate, can be easily controlled by using different materials during the construction stage, and they are more stable once the building is built. GBS output files and previous studies from others determine the selection of building variables. Several papers by others have already illustrated the importance of several building internal variables on energy consumption, which are shown in Table 6.

Table 6. Selected building variables from GBS.

Building Internal Variables	Description
Wall Construction Type (WCT)	• Indicates the material used in wall construction • Determined by wall U-Value
Roof Construction Type (RCT)	• Indicates the material used in roof construction • Determined by roof U-Value
Window-to-Wall Ratio (WWR)	• Contains several internal elements, which are the ratio of windows to walls, the type of window glass, window orientation, and the widow shading area • The ratio of windows to walls is a percentage • The type of window glass is determined by the glass U-Value • Window orientation is determined by the rotated degrees from the building North (0°) in the clockwise direction • Window shading area is a percentage
Building Orientation (ORI)	• Indicates the rotated degrees from the building North (0°) in the clockwise direction • This study only focuses on the selected internal rooms, solar and lighting analysis are not included in this paper
Infiltration (INF)	• Indicates the uncontrolled flow of air into the building through gaps • Determined by a certain value
Lighting Efficiency (LIG)	• Indicated the maximum circuit wattage consumed by lighting per unit floor area of an illuminated space • Determined by a certain value
Daylighting and Occupancy Control (DOC)	• Indicates how people control and manage the lighting system • Determined by a certain type
Plug Load Efficiency (PLE)	• Indicates the nominal equipment power density in Watts per unit area • Determined by a certain value
HVAC System (HVAC)	• Indicates the type of HVAC system • Determined by several HVAC system configurations • Affected by Ambient temperature, humidity • HVAC system has too many internal parameters and configurations, which are not fully listed in GBS
Operating Schedule (OPS)	• Determined by a certain type of operating schedule
Min/Max Internal Loads	• Only two runs are available from GBS • No relative information is available
Min/Max Envelop	
Min/Max Form	

According to other studies, building internal variables, including wall construction type, roof construction type, window-to-wall ratio, building orientation, infiltration, lighting efficiency, daylighting, and occupancy control, plug load efficiency, HVAC types, operating schedule, Min/Max internal loads, Min/Max envelope, and Min/Max form, were chosen as the independent variables in this study. Additionally, these parameters are also the design variables in GBS. GBS allows users to create alternative designs, which provides users with a convenient and straightforward way to analyze the relationships among them by changing these variables individually. Table 6 lists the selected variables and descriptions.

It is worth noting that not all the variables from GBS are within the scope of this paper. Based on the description in Table 6 and the box chart analysis in Section 5.2.1, the ORI has a very concentrated data distribution, suggesting that the ORI does not have a significant effect on energy consumption in this study, and it is discarded. In addition, although the HVAC system was considered a significant variable affecting energy consumption in other studies, the limited information available and that TETB, as a 6-star energy efficient building, makes extensive use of natural ventilation systems to reduce energy consumption, as well as the inability of GBS to fully output all internal variables of the HVAC system, led to the HVAC system being discarded in this study. Min/Max Internal Loads, Min/Max Envelop, and Min/Max Form are also discarded due to each with only two sets of available data and the inability to manually add design alternatives. Additionally, DOC and OPS are categorical variables and have no numerical meanings, which are discarded in this study.

There are several indicators provided by GBS to help designers evaluate building performance. These indicators include energy use intensity (EUI), electric cost, fuel cost, total annual cost, and total annual energy consumption. However, these variables, except EUI, are highly limited by building size and local currency and utility rates. Therefore, energy use intensity (EUI) is chosen as the dependent variable and indicator to evaluate building energy consumption.

4.4. Linear Regression Model

The regression method is a central part of many research projects and has been utilized in almost every field, including economics, biological science, and engineering [43]. The regression method provides users with a simple, efficient way to study the dependence, which can be understood if a result depends on other characteristics and how strong it is. Linear regression is the essential and most commonly used method in regression [44]. Previous research by others has validated the effectiveness of using linear regression methods to analysis building energy consumption [45–47]. Other researchers, such as Aydinalp-Kiksal et al., agreed that the regression method is easier to use without the specific expertise required [48,49]. This study aims to find the preliminary relationships between building variables and building energy consumption. These relationships can be explained as the dependence of energy consumption on multiple building variables. Therefore, the linear regression can summarize obtained data as simply as possible [44], which is highly compatible with this research.

Simple linear regression (SLR) can estimate the relationship between a single independent variable and a dependent variable by fitting the data with a straight line. The general formula is listed below:

$$y = \alpha + \beta_1 x + \epsilon \quad (1)$$

- y is the predicted value of the dependent variable;
- x is the independent variable;
- α is the intercept, which is the predicted value of y when $x = 0$;
- β_1 is the estimated regression coefficient, it decides the trend and the correlation strength between y and x;
- ϵ is the error of the estimate.

Apart from simple linear regression, multi-linear regression (MLR) is used to estimate the relationship between two or more independent variables and one dependent variable. The general formula is listed below:

$$y = \alpha + \beta_0 x_0 + \beta_1 x_1 + \beta_2 x_2 + \ldots + \beta_n x_n + \epsilon \quad (2)$$

- y is the predicted value of the dependent variable;
- $x_1, x_2 \ldots x_n$ are the independent variables;

- α is the intercept, which is the predicted value of y when $x = 0$;
- $\beta_1, \beta_2 \ldots \beta_n$ are the estimated regression coefficients, each of them decides the trend and the correlation strength between y and x individually;
- ϵ is the error of the estimate.

4.5. Statistical Indicators

To understand the distribution and meaning of data, performing the statistical test is an essential step. The indicators help to understand the distribution, validity, and significance of data. Four statistical indicators: correlation ecoefficiency, root mean square error (RMSE), adjusted R-square, and p-value are selected in this study. These indicators are generated by the MATLAB functions and are used to evaluate the reliability of results.

5. Simulation Results

5.1. Introduction

The corresponding statistical analysis was developed based on the research above. The energy simulation results, as well as statistical analysis values and diagrams, are demonstrated in the following parts. The distribution and trends of data are also discussed. Seven BIM models of realistic rooms of TETB are used to verify the proposed regression models. The error analysis is performed to find out the accuracy of the proposed model regression models based on the Monte Carlo method. The relationship between several independent variables and dependent variables is discussed and summarized in the following sections.

5.2. Energy Simulation Results

This section displays data related to selected variables from GBS in graphical form. Both data filtering and statistical analysis were completed in MATLAB.

5.2.1. EUI vs. Selected Building Variables

Figure 5 demonstrates the distribution of each design variable directly by using a boxplot.

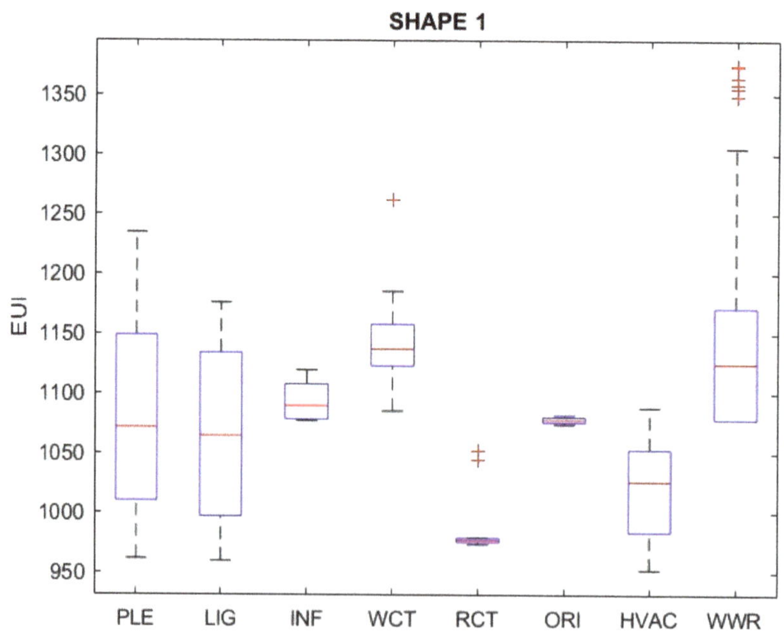

Figure 5. Boxplot of EUI vs. selected building variables.

The data distribution of PLE is observed to be skewed towards the lower side, which is consistent with the trend of the selected values of PLE. Similarly, the EUI values corresponding to LIG are in line with the distribution of the independent variable values, and the data distribution is visually more centered. The whisker for LIG is shorter, which means that the data for LIG are more concentrated than for PLE. The data distribution for INF is significantly lower, with the lower quartile almost coinciding with the lowest quartile and the median skewed towards the lower side. For WCT, the data are skewed towards the low side overall, and there is an outliner; this is because the lower values are more concentrated when the WCT variables are selected, and there is a larger independent variable, which makes the data distribution sparse as the independent variable increases. For the RCT, the graph shows that the data are very concentrated, and there are two outliners, which occur for similar reasons to the outliners in the WCT.

It is worth noting that the WCT and RCT are the main variables determining the building envelope. The distribution of WCT has a larger variation than RCT, which means changing the WCT (wall U-Value) has a greater impact on the EUI. The ORI also has a very concentrated data distribution, with the median in the middle of the box. The HVAC has a skewed high-side distribution with a moderate degree of variation. The WWR has many significant outliers and a high degree of variation, which is considered non-normal.

5.2.2. Correlation Analysis for WWR

As mentioned earlier, WWR has several internal influencing elements: window shading, window orientation, window glass type, and window-to-wall ratio. EUI is determined by these elements together, leading this paper to use multiple linear analysis rather than simple linear regression. The outliers and wide variations are considered the results of changing these four elements.

It is noteworthy that window glass type and orientation are categorical variables. Therefore, this paper uses window U-Values and the rotation angle from the north to represent these two variables, respectively. Meanwhile, both window shading and the window-to-wall ratio are percentages. Table 7 shows these elements and their numerical values.

Table 7. Summary of WWR internal influencing elements.

WWR Internal Influencing Elements	
Window Shading	100.0% (No change); 66.7% (2/3 Window); 33.3% (1/3 Window)
Window Orientation (degrees)	0° (North); 90° (East); 180° (South); 270° (West)
Window Glass Type (U-Value)	2.98 (No change); 6.17 (Single Clr); 2.74 (Double Clr); 1.99 (Double LoE); 1.55(Triple LoE)
Window-to-Wall Ratio	95%; 65%; 30%; 0%

Figure 6 displays the correlation coefficient (r) among WWR internal independent elements and dependent variables, which is EUI. The range of r should be from −1 to 1, while a negative r value represents two variables that have a negative correlation, and a positive r value means two variables have a positive correlation. This paper shows the correlation coefficients among variables from −0.2227 to 1.

According to Figure 6, the correlation between window glass type, window shading, and EUI is negative, which represents that EUI increases with the decreasing window glass type (U-Value) and window shading, while the window-to-wall percentage, building orientation, and EUI have a positive correlation. Additionally, the correlation between window-to-wall percentage and building orientation is considered as strong, while window glass type (U-Value) has a moderate correlation to EUI.

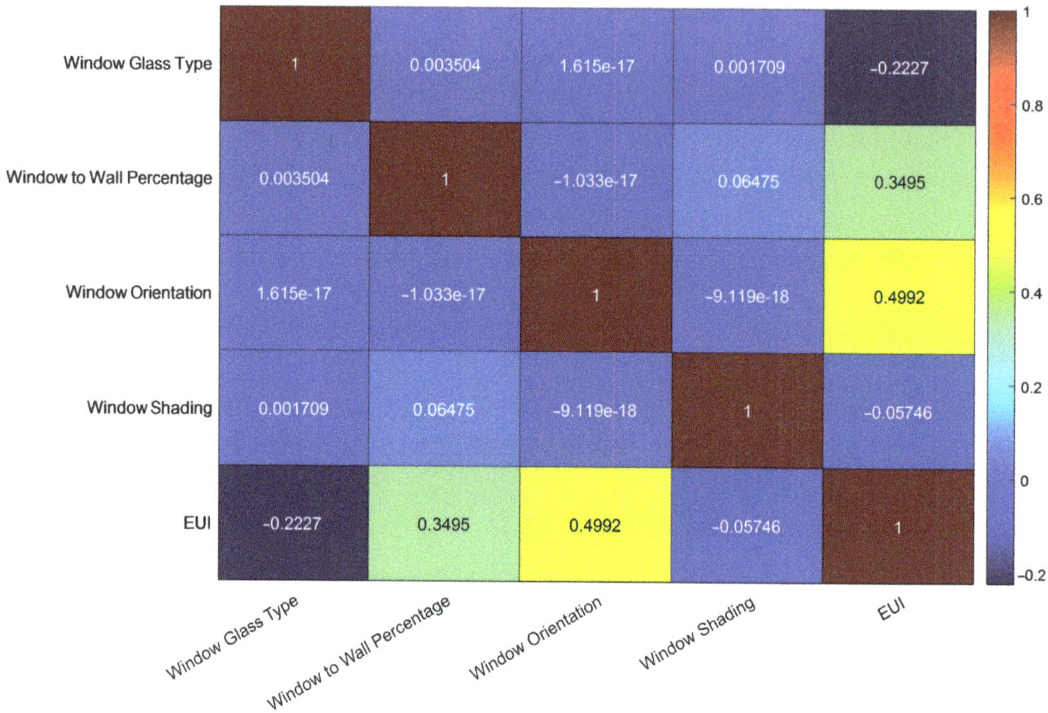

Figure 6. Correlation matrix for WWR internal influencing elements.

It is also worth noting that window shading has an extremely weak correlation to EUI, which is less than −0.1. Additionally, Table 8 summarizes the change in adjusted R-square when dropping different elements. It can be found that the adjusted R-square only decreases by 0.71% when dropping WS. By considering the above statement, window shading has an extremely weak correlation to EUI and a very limited contribution to the regression model. Therefore, window shading is dropped during the multi-linear regression analysis. For the adjusted R-square, as mentioned in Section 4.3, the EUI is chosen as the dependent variable to represent the building energy consumption. Therefore, the remaining three elements, which are window glass type (U-Value), window-to-wall percentage, and window orientation, are considered the independent variables.

Table 8. Change in adjusted R-square when dropping an element.

Internal Elements	Window Glass Type	Window-to-Wall Percentage	Window Orientation	Window Shading
Adjusted R-square when dropping an element	0.198	0.0903	0.268	0.279
Change (%) Datum: 0.298	91.9	71.2	4.66	0.71

The multi-linear regression model is developed in MATLAB. Table 9 shows the coefficients for three independent variables and intercept values for the equations of seven different shapes. The *p*-values of each shape are less than 0.05 (5%), which means the significance of the regression model is above 95%, and the model is significant.

Table 9. WWR multi-linear regression equation.

$y = \alpha + \beta_0 x_0 + \beta_1 x_1 + \beta_2 x_2$						
	Shape-1	**Shape-2**	**Shape-3**	**Shape-4**	**Shape-5**	**Shape-6**
α	1130.7	1115.6	1150.1	1192.9	1177.8	1133.9
β_0 (Glass Type)	−13.135	−12.018	−12.97	−9.1958	−8.6853	−11.574
β_1 (Window-to-Wall Percentage)	120.57	110.6	117.87	102.57	101.38	104.27
β_2 (Building Orientation)	0.09186	0.24142	−0.089786	−0.45074	−0.44391	0.036913
RMSE	60.5	58.2	66.8	57.1	58.7	49.4
Adjusted R-Square	0.285	0.34	0.237	0.475	0.452	0.303
F-test p-value	1×10^{-13}	8.88×10^{-17}	3.27×10^{-11}	1.09×10^{-25}	4.8×10^{-24}	1.13×10^{-14}

According to the multi-linear regression equations listed above, the regression models have a high level of RMSE, and the adjusted R-Square values are low. These indicators represent that the accuracy of models is low. The best-fitted line cannot represent the most data points. Figure 7 also verifies this statement.

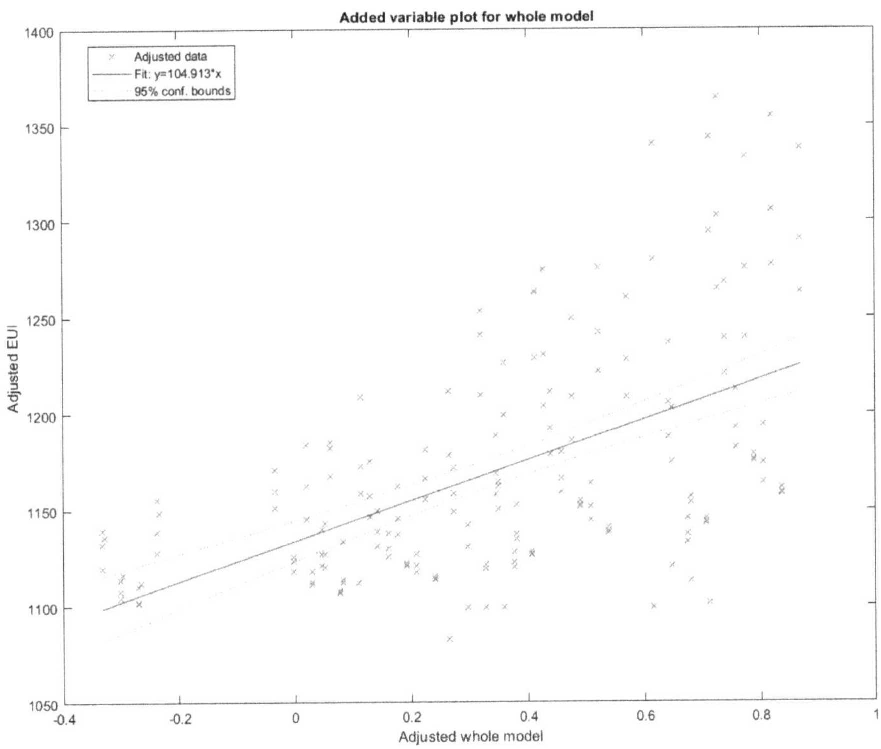

Figure 7. General data distribution and best-fitted line.

5.2.3. Interactions between Shapes and Building Variables

This part aims to discuss the results of the interaction between building shapes and building variables on EUI.

The effects of changing the wall construction type (U-Value) on EUI are demonstrated in Figure 8. EUI and wall construction type (U-Value) have a positive relationship. It is noted that the EUI increases rapidly when the U-Value is less than 0.5, and the trend becomes gentle when U-Value is higher than 0.5. These trends indicate that changing with a range of U-Value less than 0.5 significantly reduces EUI, and with the increasing in wall U-Value, it has less importance on EUI.

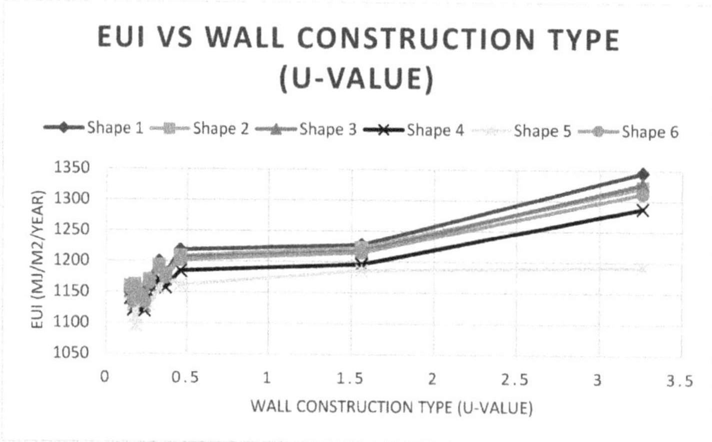

Figure 8. EUI of shapes vs. wall construction type (U-Value).

The obvious differences in EUI are observed among the six shapes. Shape 1 has the highest level of EUI value, while Shape 5 has the lowest one. It also shows using wall material with poor thermal insulation has a minor impact on EUI in Shape 5, when the wall U-Value is higher than 0.5.

Figure 9 displays the effect of roof construction type (U-Value) on RUI among different shapes. As demonstrated in the figure, EUI has a positive relationship with roof U-Value. EUI increases steeply when the roof U-Value is less than 0.5, and the slope increases slightly when U-Value is higher than 0.5. These trends also indicate that controlling EUI by changing the roof U-Value is less effective when the U-Value is higher than 0.5.

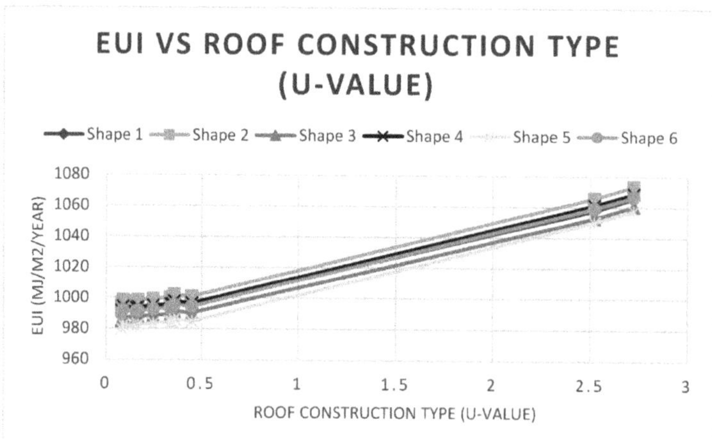

Figure 9. EUI of shapes vs. roof construction type (U-Value).

According to Figure 9, Shape 2 has the highest value of EUI among all roof U-Values, while Shape 5 has the lowest one.

The trends of EUI vs. infiltration rate are displayed in Figure 10. The infiltration rate and EUI are correlated positively. Generally, the slope is considered steady. The trends of all shapes show common linear relations with positive trends.

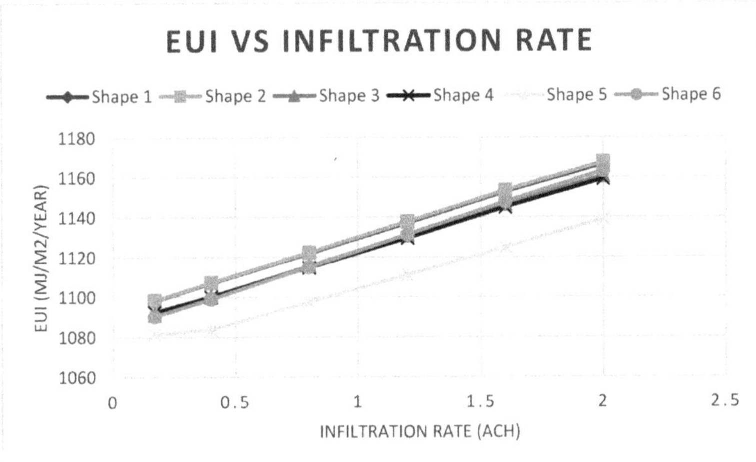

Figure 10. EUI of shapes vs. infiltration rate.

The figure also shows Shape 2 has the highest level of EUI in all infiltration types, while Shape 5 has the significantly lowest one.

Figure 11 demonstrates the trends between lighting efficiency and EUI among different shapes. The lighting efficiency has positive trends with EUI, and the trends are considered general linear relations for all shapes as all slopes are steady.

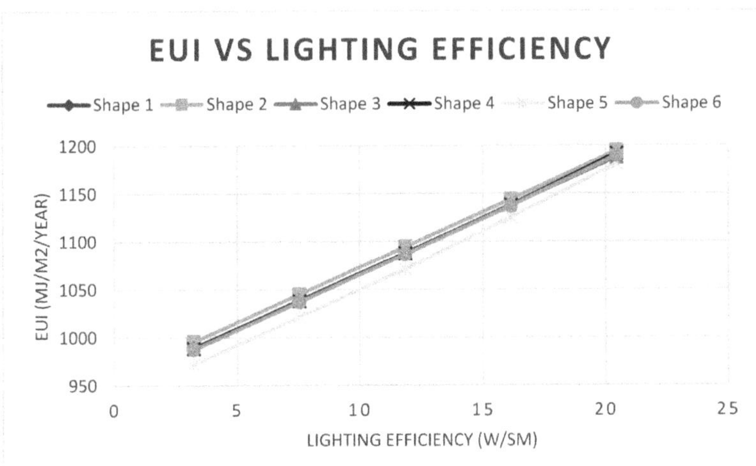

Figure 11. EUI of shapes vs. lighting efficiency.

Shape 2 has the highest level of EUI, while Shape 5 has the lowest one. Moreover, the difference in EUI is not significant among all shapes except five, as the lines of these five shapes are close to each other.

The relationships between EUI and plug load efficiency for different shapes are demonstrated in Figure 12. Similar to trends in lighting efficiency, plug load efficiency has a positive linear relationship with EUI, and different building shapes have minor effects on EUI except Shape 5.

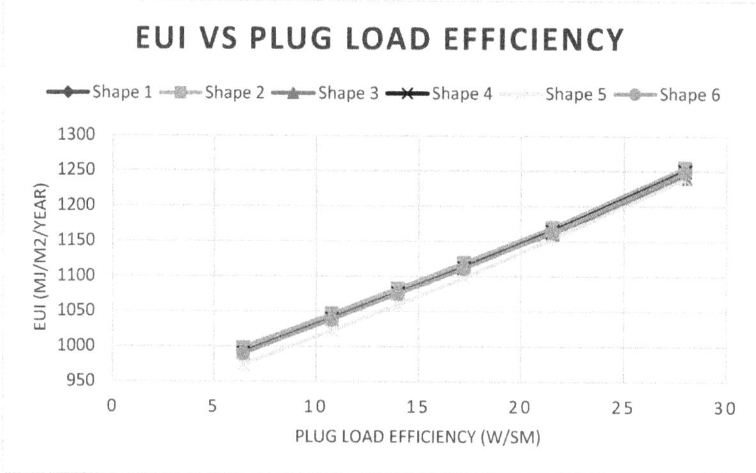

Figure 12. EUI of shapes vs. plug load efficiency.

To summarize the figures in Section 5.2.3, Shape 1 has the highest EUI of WCT, and Shape 2 has the highest EUI of all other variables. Shape 5 has the lowest EUI of all variables. This means Shape 5 is the most energy-saving one among all shapes. No significant difference in EUI of different shapes is observed, except in Shape 5.

5.3. *Simply Linear Regression*

The results of linear regression analysis are displayed and illustrated in the following content. The best-fit lines are generated. The correlation and significance are also discussed. The best-fitted line and 95% confidence bounds/intervals, root mean square error (RMSE), as well as adjusted R-square, are provided to help explain the results.

5.3.1. Wall Construction Type

From Figure 13, the relationship between WCT about EUI has the same trends in all six shapes, and they are positively correlated. EUI value increases with the value of WCT rising. The best-fit lines are also shown for six cases, and their equations are listed in Table 10.

Table 10. Fitted lines of wall construction type (U-Value).

$y = \alpha + \beta_1 x$						
	Shape 1	**Shape 2**	**Shape 3**	**Shape 4**	**Shape 5**	**Shape 6**
α	1145.9	1147.7	1138.3	1143.1	1128.7	1143.3
β_1	61.176	53.709	58.039	51.088	48.889	51.941
RMSE	20.3	17.0	19.1	16.8	15.8	16.7
Adjusted R-Squared	0.861	0.872	0.863	0.863	0.868	0.868
F-test p-value	3.89×10^{-7}	2.24×10^{-7}	3.47×10^{-7}	3.43×10^{-7}	2.76×10^{-7}	2.74×10^{-7}

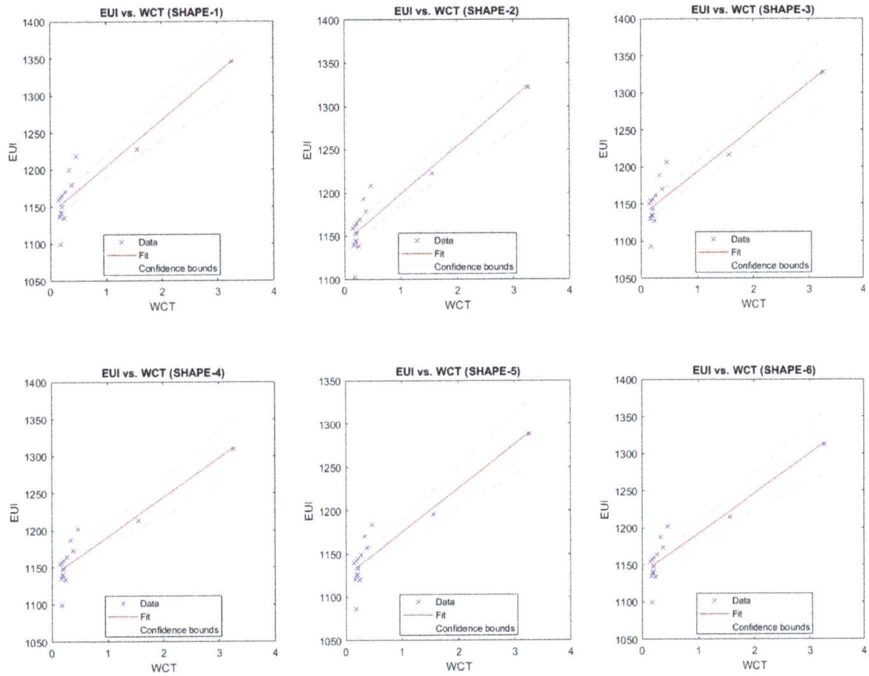

Figure 13. Fitted lines of wall construction type (U-Value).

As can be seen from the table, the *p*-values for all six cases are small, which means that the six cases have high significance for the fitness equation. The RMSE is low for all six cases, with Shape 5 having the smallest RMSE, which means that Shape 5 has the least dispersion of observations relative to the fit line. The adjusted R-squared is close to 0.80 for all six cases, which means that the fitted equations can explain 80% of the variation in the observations, which represents a good fit.

Further, Figure 13 also shows the 95% confidence interval for the data, and they are shown as dashed lines. The confidence intervals for the six cases follow the same trend and are narrower where the sample is dense and wider where the sample is sparse. This means that denser sampling reduces the distribution error of the data and thus improves the accuracy of the data. As GBS only offers a limited selection of WCTs, it leads to uneven sampling in the range of WCT values greater than 1.0. This results in wider confidence intervals and higher errors between the observed and true values of the objects.

5.3.2. Roof Construction Type

As can be seen from Table 11, the *p*-values for all six cases are much smaller than that of WCT, which means that the six cases have extremely high significance for the fitness equation. The RMSE is low for all six cases, with Shape 1 having the smallest RMSE, which means that Shape 1 has the least dispersion of observations relative to the fit line. The adjusted R-squared is close to 1.000 for all six cases, which means that the fitted equations can explain most of the variations in the observations. The results represent the observations are almost identical to the regression model, with positively linear correlations.

Table 11. Fitted lines of roof construction type (U-Value).

$y = \alpha + \beta_1 x$						
	Shape 1	**Shape 2**	**Shape 3**	**Shape 4**	**Shape 5**	**Shape 6**
α	986.24	992.23	981.43	987.67	975.97	985.55
β_1	28.809	29.465	28.7	29.422	29.625	29.474
RMSE	1.7	2.01	1.89	1.96	1.98	1.95
Adjusted R-Squared	0.996	0.995	0.996	0.995	0.995	0.995
F-test *p*-value	9.66×10^{-19}	7.39×10^{-18}	4.42×10^{-18}	5.46×10^{-18}	5.43×10^{-18}	4.86×10^{-18}

Additionally, similar to the WCT, the sampling data for the RCT is sparse in intervals greater than 0.5. Interestingly, it can be observed from the Figure 14 that the confidence interval for the RCT, even though gradually widening, remains relatively narrow when the RCT equals around 2.5 since the last two sample value lines of the RCT are closer than those in WCT, which characterizes the above statement that the larger the sample size, the narrower the confidence interval and the observations are more representative of the true values.

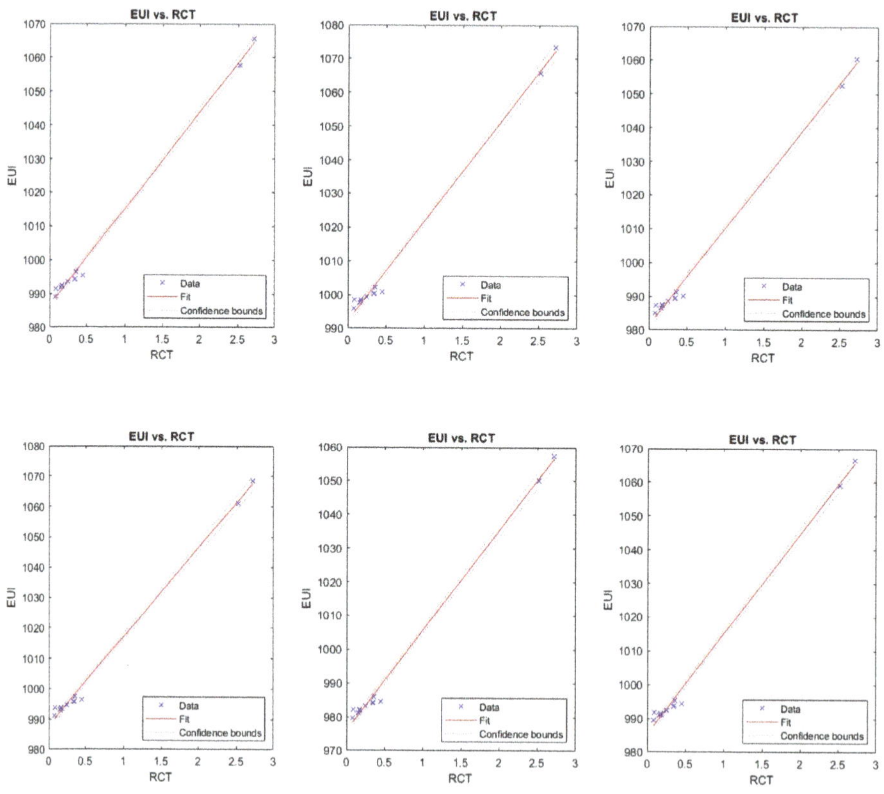

Figure 14. Fitted lines of roof construction type (U-Value).

5.3.3. Plug Load Efficiency

Figure 15 demonstrates the trends in PLE and EUI among six shapes. The EUI is positively correlated with the PLE, and the six shapes of the PLE adaptation lines are in general agreement with the observed data. The more evenly sampled values and the fact that one value was taken for each 0.4 resulted in a more consistent confidence interval, and the wider ends shown in WCT and RCT do not appear here.

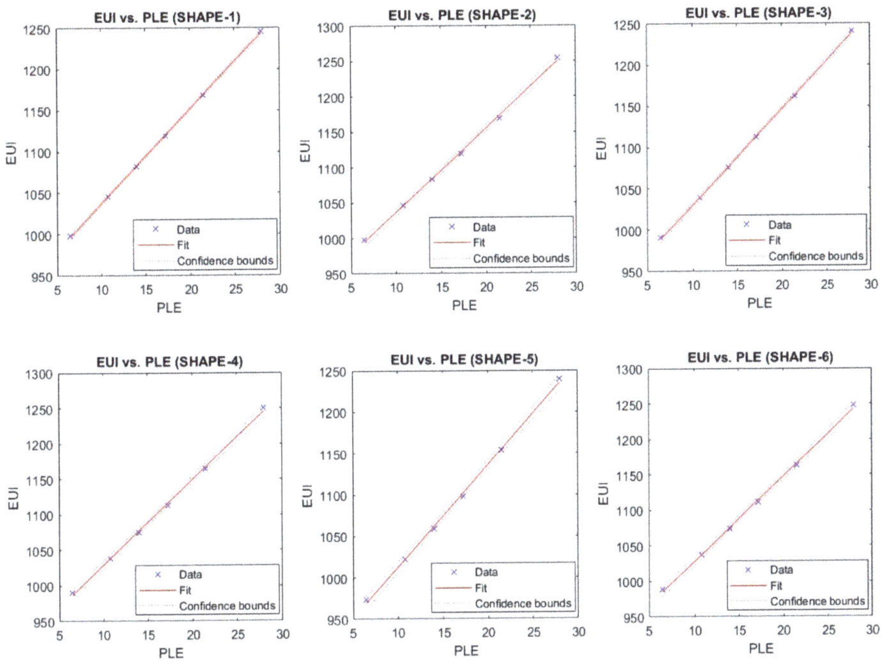

Figure 15. Fitted lines of plug load efficiency.

Table 12 shows the fit line equations and the results of statistical indices. Comparing the magnitude of the fit line coefficients, it can be concluded that the PLE has a more significant effect on the EUI in Shape 5. The regression models on all six shapes have very small RMSE and *p*-values, which means that the values predicted by the regression models have very small errors compared with the observed values and that the regression models are highly significant. It is worth noting that the adjusted R-squared values are consistent, and the value of Shape 1 is equal to 1, which means that the regression model explains all the observed variations completely and that there is a certain positive linear relationship exists between these two variables.

Table 12. Fitted lines of plug load efficiency.

$y = \alpha + \beta_1 x$						
	Shape 1	**Shape 2**	**Shape 3**	**Shape 4**	**Shape 5**	**Shape 6**
α	922.22	917.8	914.39	909.17	889.26	908.71
β_1	11.502	11.880	11.593	12.015	12.353	11.963
RMSE	1.53	3.78	2.08	3.89	4.16	3.82
Adjusted R-Squared	1.0	0.998	0.999	0.998	0.998	0.998
F-test *p*-value	2.13×10^{-8}	6.96×10^{-7}	7.04×10^{-8}	7.41×10^{-7}	8.7×10^{-7}	7.02×10^{-7}

5.3.4. Lighting Efficiency

The trends of EUI about LIG are shown in Figure 16. Similar to the PLE, the observations of the LIG are almost in a straight line, and the regression model is close to the observations. The uniform sampling allows for a narrow range of confidence intervals and small errors in the observations relative to the true values.

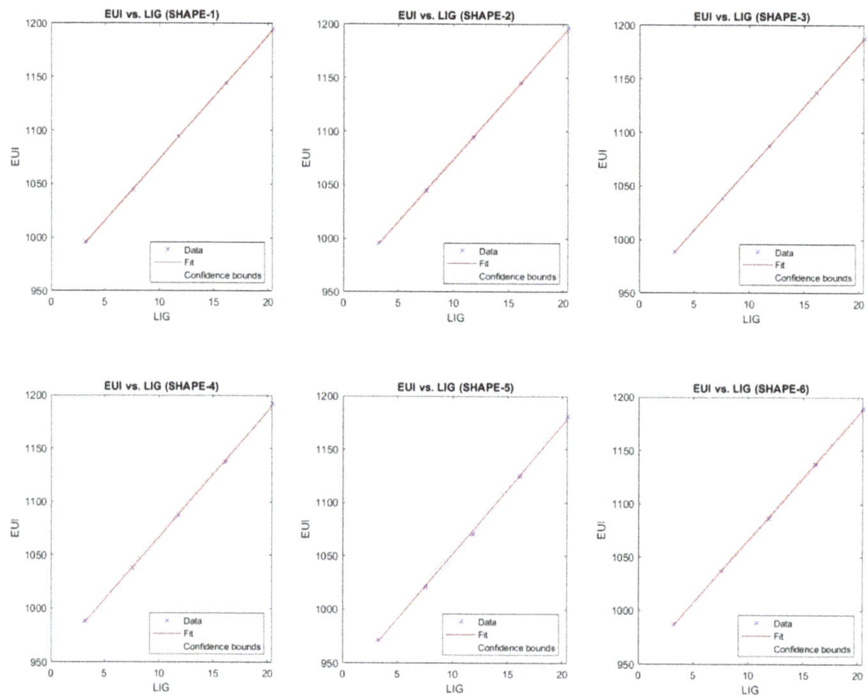

Figure 16. Fitted lines of lighting efficiency.

Table 13 shows that the effect of LIG on EUI is greatest for Shape 1, with small RMSE and *p*-values for all six shapes, meaning that the observations are evenly distributed, and the null hypothesis is less likely to be true, the null hypothesis can be rejected, and the regression model results are significant. Five of the six models have an adjusted R-squared equal to 1, which means that the observations are perfectly explained by the regression model and that LIG is positive and linearly associated with EUI.

Table 13. Fitted lines of lighting efficiency.

$y = \alpha + \beta_1 x$						
	Shape 1	**Shape 2**	**Shape 3**	**Shape 4**	**Shape 5**	**Shape 6**
α	958.11	957.68	951.51	949.48	930.77	948.9
β_1	11.531	11.630	11.532	11.751	12.102	11.706
RMSE	0.389	0.842	0.308	1.59	2.49	1.17
Adjusted R-Squared	1.0	1.0	1.0	1.0	0.999	1.0
F-test *p*-value	3.37×10^{-8}	3.32×10^{-7}	1.68×10^{-8}	2.15×10^{-6}	7.65×10^{-6}	8.82×10^{-7}

5.3.5. Infiltration Rate

The trends of EUI about INF are shown in Figure 17. The observations of the INF are almost in a straight line, and the regression model is close to the observations. The uniform sampling allows for a narrow range of confidence intervals and extremely small errors in the observations relative to the true values. However, in Shape 5, the confidence interval is larger at the beginning (in the range of 0.0 to 0.5) because the observed values are not distributed linearly. The different slopes resulting from the non-uniform distribution of the first point and the remaining make the confidence interval larger, which demonstrates that there is some error between the observed and true values.

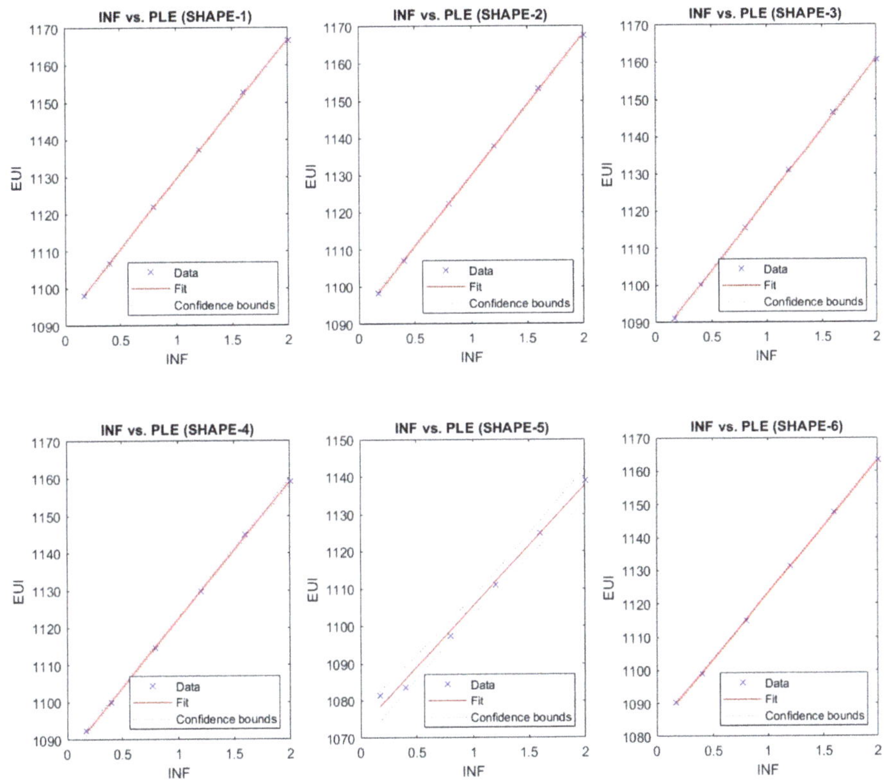

Figure 17. Fitted lines of infiltration rate.

Table 14 shows that the effect of INF on EUI is greatest for Shape 6, with small RMSE and p-values for all six shapes, meaning that the observations are evenly distributed, and the null hypothesis is less likely to be true, the null hypothesis can be rejected, and the regression model results are significant. Five of the six models have an adjusted R-squared equal to 1, which means that the regression model perfectly explains the observations and that LIG is positive and linearly associated with EUI. Only Shape 5 has a smaller adjusted R-square and higher RMSE due to the anomalies in observed values.

In summary, the RCT, PLE, LIG, and INF regression models fit the observed values well; the observed values have a small margin of error and can reflect the true value well. It is worth mentioning that WCT has a large confidence interval due to the uneven distribution of the sample values, leading to a large error with the true values. The regression model does not fit the observations very well compared with other models. However, WCT has a relatively high level of adjusted R-square; these minor dispersions of the observed values are attributed to a reasonable and acceptable range.

Table 14. Fitted lines of infiltration rate.

$y = \alpha + \beta_1 x$						
	Shape 1	**Shape 2**	**Shape 3**	**Shape 4**	**Shape 5**	**Shape 6**
α	1091.9	1092	1085	1085.7	1072.9	1083.2
β_1	37.635	38.035	38.089	36.804	32.465	40.094
RMSE	0.433	0.398	0.458	0.441	2.18	0.319
Adjusted R-Squared	1.0	1.0	1.0	1.0	0.991	1.0
F-test p-value	1.7×10^{-8}	1.16×10^{-8}	2.03×10^{-8}	2×10^{-8}	1.96×10^{-5}	3.91×10^{-9}

5.4. Parametric Analysis

Figures 18 and 19 demonstrate a comparison of the constants and coefficients of the regression models for all variables selected to perform a simple linear regression analysis for all six shapes. The effect of changing each variable on EUI is presented intuitively in the following content.

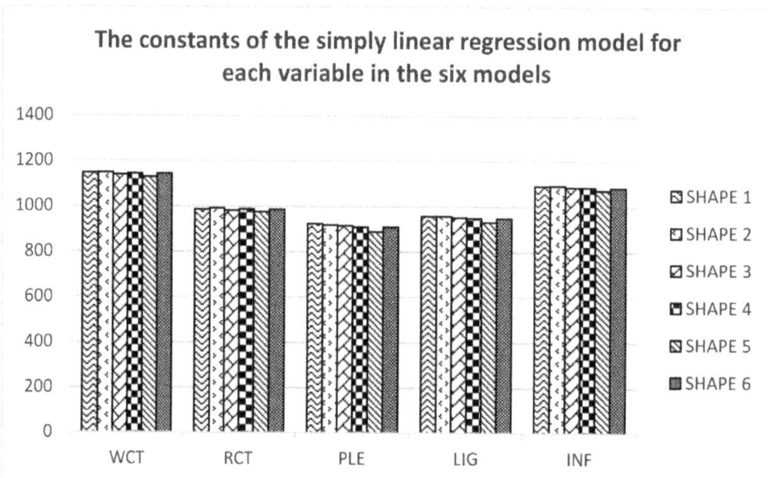

Figure 18. Constants plot of variables for 6 Shapes.

The parameter terms for each of the six shape variables are shown in Figure 18, with WCT and INF having the most significant constant terms, followed by RCT, LIG, and PLE. The larger the constant term, the larger the base of energy consumption and the higher the lower limit of consumption. According to the image, the constant terms of the regression models are close to each other. It is worth noting that Shape 1 has relatively high constant terms for PLE, LIG, and INF, and Shape 6 has higher constant terms for the WCT and ECT variables.

Figure 19 shows the coefficients for each variable of the six shape regression models. The coefficients in a regression model determine the degree of influence of that independent variable on the dependent variable, with larger absolute values representing the greater importance of that independent variable on the dependent variable. According to the images, WCT has the most significant coefficient, and WCT has the most effect on energy consumption, followed by INF and RCT. PLE and LIG have similar-sized coefficients and have the least effect on the dependent variable. Shape 1 has the largest coefficient in WCT, which means that changing WCT in Shape 1 has the largest influence on EUI. Conversely, Shape 1 has a non-significant effect on the other variables. RCT, PLE, LIG, and INF have a relatively significant effect on EUI in Shape 6. The RCT, PLE, and LIG have a relatively equivalent effect on EUI.

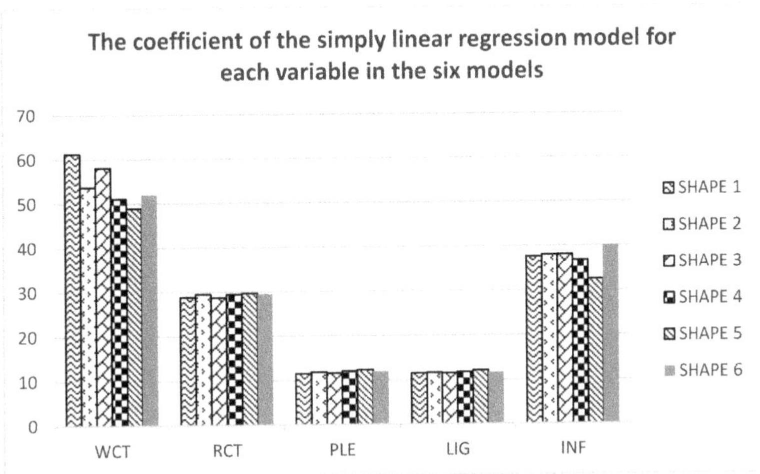

Figure 19. Coefficient plot of variables for 6 Shapes.

5.5. Validation and Error

The regression model obtained in the previous steps needs to be compared with real TETB office rooms to validate the regression models. Six validation models were built based on real word dimensions, and material properties were all referenced from papers related to TETB or similar projects and standards. Table 15 lists the real and assumed material information of validations.

Table 15. Material properties chosen for validation models.

Façade and Roof Type	**U-Value (W/M^2·K)**	**Reference**
External Wall Façade—1	0.500	Burgun, Bilbao [50]
External Wall Façade—2	0.333	
Lighting and Plug Load Efficiency	Intensity (W/sqm)	Reference
Lighting Efficiency (Power density)	10	UNSW [51]
Plug Load Efficiency (Power Load)	17.5	Dunn and Knight [52]
Infiltration Rate	**Air Change Per Hour (ACH)**	**Reference**
Infiltration Rate (Air Leakage Rate)	0.4	Speert and Legge [53]

The following diagrams compare regression models and the models built based on real dimensions. Errors are shown as absolute percentages (Figures 20–25).

The above figures show the comparison between the regression models and validation models. The error indicates the accuracy of the regression equations. The maximum error was observed at the lighting efficiency of Shape 6, while minor errors were found in the roof construction of Shapes 2 and 4.

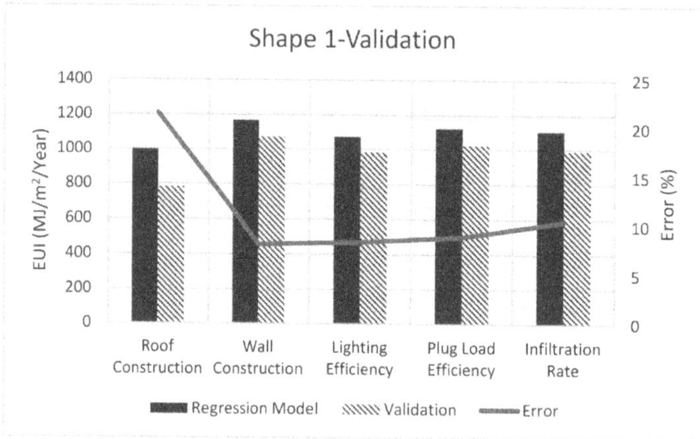

Figure 20. TETB Shape 1 validation.

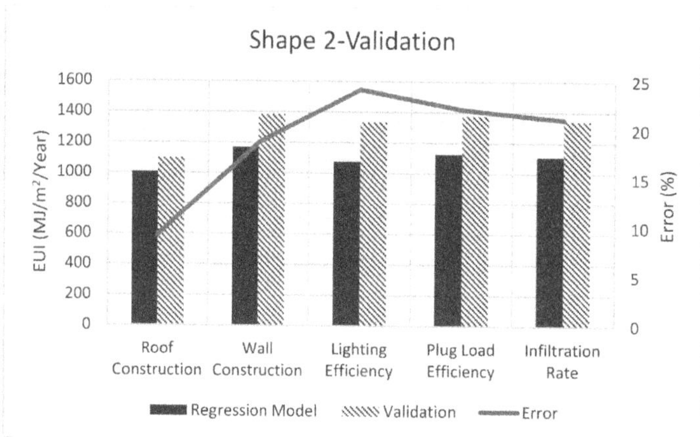

Figure 21. TETB Shape 2 validation.

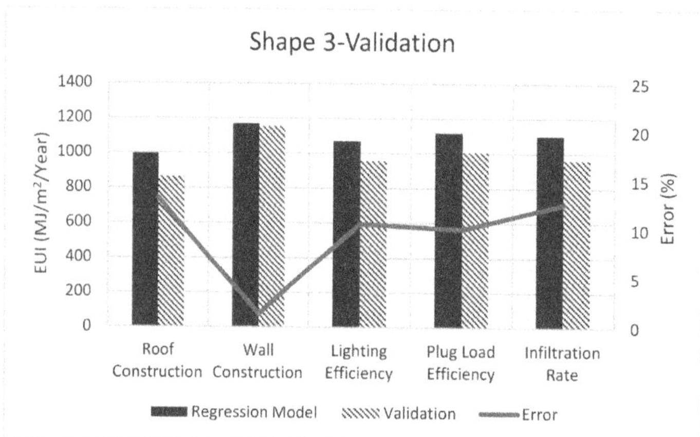

Figure 22. TETB Shape 3 validation.

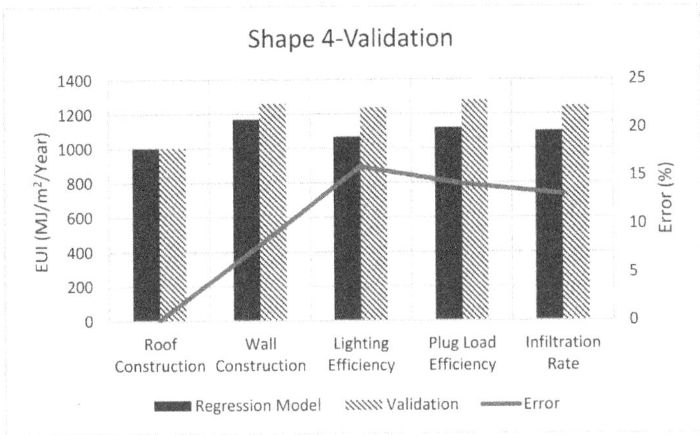

Figure 23. TETB Shape 4 validation.

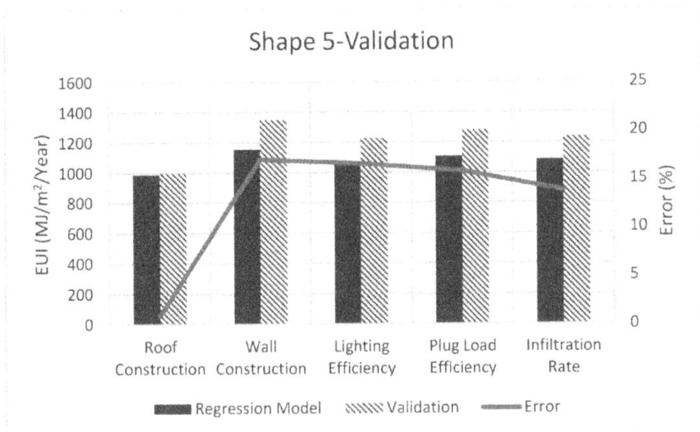

Figure 24. TETB Shape 5 validation.

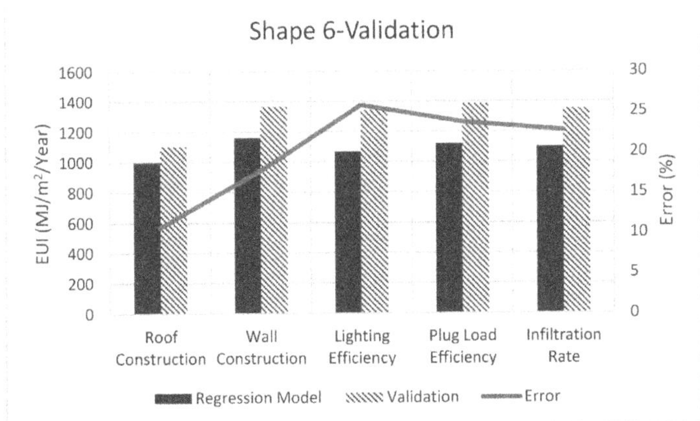

Figure 25. TETB Shape 6 validation.

It is worth noting that TETB has several office shapes with sharp angles. There were some gaps between two wall connections with the default analytical space and surface resolution because Revit ignored building components smaller than the resolution.

Generally, the error ranges between 10% and 25%, which is not an acceptable range. The regression equations do not pass the validation process. Hence, the accuracy of the regression models is low. The possible reasons and limitations are discussed below.

6. Conclusions, Recommendations, and Future Research

This study investigates the impact of essential building variables on energy consumption for different building shapes. The analysis utilizes two different linear regression methods to simplify the process and examine the different types of variables.

In recent years, designers have creates buildings that use energy efficiently. This research shows that building energy consumption is a complex issue affected by many variables that require careful investigation. Results indicate that roof construction, infiltration rate, lighting efficiency, and plug load efficiency have a strong linear relationship with EUI, while wall construction has a weaker correlation. The study used the multiple linear regression method to analyze the window-to-wall ratio (WWR) and found that window shading had little impact on energy consumption. Furthermore, the research found no significant differences in energy consumption between different building shapes.

This study aimed to investigate the relationships between building variables and energy consumption in different building shapes by creating prototypes. The accuracy of the regression models was measured through a validation process. However, the errors in the models were found to be outside of an acceptable range, and as a result, the models were deemed unqualified. The study acknowledges several potential reasons and limitations discussed earlier in the content.

The topic of this research paper is considered complex and has already been studied by many other researchers. This study utilized a simplified research method to achieve preliminary results. However, more extensive and detailed research is necessary to explore the relationship between building variables and energy performance fully.

Based on the research process and results of this paper, the author makes several recommendations and points out several directions for future studies. Firstly, a more reliable and accurate energy analysis software, such as EnergyPlus can be used to validate the results obtained from GBS. Similar to Garcia and Zhu's study, different energy analysis tools can be used to verify the calculations, and a real energy bill can be used to validate the results [54]. Secondly, as discussed in the limitation part, the geometry of the shape, including the aspect ratio and floor area, should be considered in future studies. The aspect ratio and floor area should be controlled to create a consistent simulation environment. Thirdly, more variables should be researched. Alothman pointed out that the HVAC system is an important energy consumption element [55]. Asadi also said that ceiling insulation type has a significant effect on energy consumption [29]. Fourthly, the authors found that TETB is surrounded by education facilities during the site inspection, and most of them have the same roof level as TETB. Adequate sunlight can enter the building and reduce the use of artificial light sources. Hence, the solar analysis and lighting analysis can be performed in future studies based on the preliminary BIM model of a typical floor in TETB as shown previously. Fifthly, an innovative method to explore more complex and realistic relationships between energy consumption and influencing factors of macro and micro levels has been introduced in the previous part, which is the IDEA model. More interactive research can be performed by considering building elements, occupant behaviors, ambient environment, and even socioeconomic factors.

This paper only considers interior elements and idealized scenarios with some technical limitations. Many default settings are selected during the analysis. Hence, real external parameters should be measured using a wide range of sensing technologies to offer accurate and detailed results in future studies. The sensing technologies enable digital twining and produce big data over time. Thus, artificial intelligence should be utilized for the optimization of energy efficiency and offer real-time recommendations on energy saving to the facility or building managers. Some algorithms valuable to be combined and tested are artificial neural networks to identify complex patterns in the building data; genetic algorithms to find an optimal combination of building parameters for consumption minimization, decision trees to develop decision rules for optimizations, and random forest, support vectors machines, and gradient boosting machines.

This requires the author to have more in-depth knowledge of the heating, cooling, ventilation, energy, and weather data. Realistic constructions that comply with Australian Standards and Building Codes can be applied during modelling.

Author Contributions: Conceptualization, F.T. and S.M.E.S.; Methodology, F.T., R.H., J.C., S.W. and S.M.E.S.; Software, F.T., R.H., J.C. and S.W.; Validation, F.T., R.H., J.C., S.W. and S.M.E.S.; Formal analysis, F.T., R.H., J.C., S.W. and S.M.E.S.; Investigation, F.T., R.H., J.C. and S.W.; Resources, F.T. and S.M.E.S.; Data curation, F.T.; Writing—original draft, F.T., R.H., J.C. and S.W.; Writing—review & editing, F.T. and S.M.E.S.; Supervision, F.T. and S.M.E.S.; Project administration, S.M.E.S. All authors have read and agreed to the published version of the manuscript.

Funding: This research received no external funding.

Data Availability Statement: Not applicable.

Conflicts of Interest: The authors declare no conflict of interest.

Abbreviations

BCA	Building Codes of Australia
BIM	Building Information Modelling
BRI	Building Orientation
DOC	Daylighting and Occupancy control
GBS	Autodesk Green Building Studio
INF	Infiltration Rate
LIG	Lighting Efficiency
MLR	Multi-Linear Regression
OPS	Operating Schedule
PLE	Plug Load Efficiency
RCT	Roof Construction Type
RMSE	Root Mean Square Error
TETB	Tyree Energy and Technologies Building
UNSW	University of New South Wales
WCT	Wall Construction Rate
WWR	Window-to-Wall Ratio

References

1. Abanda, F.; Byers, L. An investigation of the impact of building orientation on energy consumption in a domestic building using emerging BIM (Building Information Modelling). *Energy* **2016**, *97*, 517–527. [CrossRef]
2. Zhang, C.; Nizam, R.S.; Tian, L. BIM-based investigation of total energy consumption in delivering building products. *Adv. Eng. Informatics* **2018**, *38*, 370–380. [CrossRef]
3. Shirowzhan, S.; Sepasgozar, S.M.; Edwards, D.J.; Li, H.; Wang, C. BIM compatibility and its differentiation with interoperability challenges as an innovation factor. *Autom. Constr.* **2020**, *112*, 103086. [CrossRef]
4. Sepasgozar, S.; Karimi, R.; Farahzadi, L.; Moezzi, F.; Shirowzhan, S.; Ebrahimzadeh, S.M.; Hui, F.; Aye, L. A Systematic Content Review of Artificial Intelligence and the Internet of Things Applications in Smart Home. *Appl. Sci.* **2020**, *10*, 3074. [CrossRef]
5. López-Pérez, L.A.; Flores-Prieto, J.J. Adaptive thermal comfort approach to save energy in tropical climate educational building by artificial intelligence. *Energy* **2023**, *263*, 125706. [CrossRef]
6. Balcilar, M.; Uzuner, G.; Nwani, C.; Bekun, F.V. Boosting Energy Efficiency in Turkey: The Role of Public–Private Partnership Investment. *Sustainability* **2023**, *15*, 2273. [CrossRef]
7. Liu, B.; Rosenberg, M.; Athalye, R. National Impact Of ANSI/ASHRAE/IES Standard 90.1-2016. In Proceedings of the 2018 Building Performance Analysis Conference and SimBuild, Chicago, IL, USA, 26–28 September 2018.
8. Mazria, E.; Kershner, K. *Meeting the 2030 Challenge through Building Codes*; Architecture 2030: Santa Fe, NM, USA, 2008; Available online: https://sallan.org/pdf-docs/2030Challenge_Codes_WP-1.pdf (accessed on 19 October 2022).

9. Fu, H.; Baltazar, J.-C.; Claridge, D.E. Review of developments in whole-building statistical energy consumption models for commercial buildings. *Renew. Sustain. Energy Rev.* **2021**, *147*, 111248. [CrossRef]
10. Ferrandiz, J.; Banawi, A.; Peña, E. Evaluating the benefits of introducing "BIM" based on Revit in construction courses, without changing the course schedule. *Univers. Access Inf. Soc.* **2018**, *17*, 491–501. [CrossRef]
11. Succar, B. Building information modelling framework: A research and delivery foundation for industry stakeholders. *Autom. Constr.* **2009**, *18*, 357–375. [CrossRef]
12. Abanda, F.H.; Vidalakis, C.; Oti, A.H.; Tah, J.H. A critical analysis of Building Information Modelling systems used in construction projects. *Adv. Eng. Softw.* **2015**, *90*, 183–201. [CrossRef]
13. Wong, J.K.W.; Zhou, J. Enhancing environmental sustainability over building life cycles through green BIM: A review. *Autom. Constr.* **2015**, *57*, 156–165. [CrossRef]
14. Gao, H.; Zhang, L.; Koch, C.; Wu, Y. BIM-based real time building energy simulation and optimization in early design stage. *IOP Conf. Ser. Mater. Sci. Eng.* **2019**, *556*, 012064. [CrossRef]
15. Kamel, E.; Memari, A.M. Review of BIM's application in energy simulation: Tools, issues, and solutions. *Autom. Constr.* **2019**, *97*, 164–180. [CrossRef]
16. Chong, H.-Y.; Lee, C.-Y.; Wang, X. A mixed review of the adoption of Building Information Modelling (BIM) for sustainability. *J. Clean. Prod.* **2017**, *142*, 4114–4126. [CrossRef]
17. Lu, Y.; Wu, Z.; Chang, R.; Li, Y. Building Information Modeling (BIM) for green buildings: A critical review and future directions. *Autom. Constr.* **2017**, *83*, 134–148. [CrossRef]
18. Catalina, T.; Iordache, V.; Caracaleanu, B. Multiple regression model for fast prediction of the heating energy demand. *Energy Build.* **2013**, *57*, 302–312. [CrossRef]
19. Braun, M.; Altan, H.; Beck, S. Using regression analysis to predict the future energy consumption of a supermarket in the UK. *Appl. Energy* **2014**, *130*, 305–313. [CrossRef]
20. Walter, T.; Sohn, M.D. A regression-based approach to estimating retrofit savings using the Building Performance Database. *Appl. Energy* **2016**, *179*, 996–1005. [CrossRef]
21. Martinez, A.; Choi, J.-H. Analysis of energy impacts of facade-inclusive retrofit strategies, compared to system-only retrofits using regression models. *Energy Build.* **2018**, *158*, 261–267. [CrossRef]
22. Jaffal, I.; Inard, C.; Bozonnet, E. Toward integrated building design: A parametric method for evaluating heating demand. *Appl. Therm. Eng.* **2012**, *40*, 267–274. [CrossRef]
23. Lam, J.C.; Wan, K.K.; Liu, D.; Tsang, C. Multiple regression models for energy use in air-conditioned office buildings in different climates. *Energy Convers. Manag.* **2010**, *51*, 2692–2697. [CrossRef]
24. Mottahedi, M.; Mohammadpour, A.; Amiri, S.S.; Riley, D.; Asadi, S. Multi-linear Regression Models to Predict the Annual Energy Consumption of an Office Building with Different Shapes. *Procedia Eng.* **2015**, *118*, 622–629. [CrossRef]
25. Yildiz, B.; Bilbao, J.; Sproul, A. A review and analysis of regression and machine learning models on commercial building electricity load forecasting. *Renew. Sustain. Energy Rev.* **2017**, *73*, 1104–1122. [CrossRef]
26. Choi, J.; Shin, J.; Kim, M.; Kim, I. Development of openBIM-based energy analysis software to improve the interoperability of energy performance assessment. *Autom. Constr.* **2016**, *72*, 52–64. [CrossRef]
27. Pinheiro, S.; Wimmer, R.; O'Donnell, J.; Muhic, S.; Bazjanac, V.; Maile, T.; Frisch, J.; van Treeck, C. MVD based information exchange between BIM and building energy performance simulation. *Autom. Constr.* **2018**, *90*, 91–103. [CrossRef]
28. Liu, X.; Akinci, B.; Bergés, M.; Garrett, J.H. Extending the information delivery manual approach to identify information requirements for performance analysis of HVAC systems. *Adv. Eng. Inform.* **2013**, *27*, 496–505. [CrossRef]
29. Asadi, S.; Amiri, S.S.; Mottahedi, M. On the development of multi-linear regression analysis to assess energy consumption in the early stages of building design. *Energy Build.* **2014**, *85*, 246–255. [CrossRef]
30. Aghdaei, N.; Kokogiannakis, G.; Daly, D.; McCarthy, T. Linear regression models for prediction of annual heating and cooling demand in representative Australian residential dwellings. *Energy Procedia* **2017**, *121*, 79–86. [CrossRef]
31. González, J.; Soares, C.A.P.; Najjar, M.; Haddad, A.N. BIM and BEM Methodologies Integration in Energy-Efficient Buildings Using Experimental Design. *Buildings* **2021**, *11*, 491. [CrossRef]
32. Huo, T.; Ren, H.; Zhang, X.; Cai, W.; Feng, W.; Zhou, N.; Wang, X. China's energy consumption in the building sector: A Statistical Yearbook-Energy Balance Sheet based splitting method. *J. Clean. Prod.* **2018**, *185*, 665–679. [CrossRef]
33. Huo, T.; Xu, L.; Liu, B.; Cai, W.; Feng, W. China's commercial building carbon emissions toward 2060: An integrated dynamic emission assessment model. *Appl. Energy* **2022**, *325*, 119828. [CrossRef]
34. Huo, T.; Ma, Y.; Xu, L.; Feng, W.; Cai, W. Carbon emissions in China's urban residential building sector through 2060: A dynamic scenario simulation. *Energy* **2022**, *254*, 124395. [CrossRef]
35. Onstott, S. *AutoCAD 2015 and AutoCAD LT 2015 Essentials: Autodesk Official Press*; John Wiley & Sons: Hoboken, NJ, USA, 2014.
36. Kirby, L.; Krygiel, E.; Kim, M. *Mastering Autodesk Revit 2018*; John Wiley & Sons: Hoboken, NJ, USA, 2017.
37. Maglad, A.M.; Houda, M.; Alrowais, R.; Khan, A.M.; Jameel, M.; Rehman, S.K.U.; Khan, H.; Javed, M.F.; Rehman, M.F. Bim-based energy analysis and optimization using insight 360 (case study). *Case Stud. Constr. Mater.* **2023**, *18*, e01755. [CrossRef]
38. STUDIO, Green Building. Autodesk Green Building Studio. 2008. Available online: https://damassets.autodesk.net/content/dam/autodesk/www/industries/education/docs/Access-GBS-2014-New.pdf (accessed on 19 October 2022).

39. Nardi, I.; Paoletti, D.; Ambrosini, D.; de Rubeis, T.; Sfarra, S. U-value assessment by infrared thermography: A comparison of different calculation methods in a Guarded Hot Box. *Energy Build.* **2016**, *122*, 211–221. [CrossRef]
40. Islam, M.H.; Safayet, M.A.; Al Mamun, A. Building performance analysis for optimizing the energy consumption of an educational building. *Int. J. Build. Pathol. Adapt.* **2022**, *ahead of print*. [CrossRef]
41. Li, Y.; Deng, X.; Ba, S.; Myers, W.R.; Brenneman, W.A.; Lange, S.J.; Zink, R.; Jin, R. Cluster-based data filtering for manufacturing big data systems. *J. Qual. Technol.* **2021**, *54*, 290–302. [CrossRef]
42. Etter, D.M.; Kuncicky, D.C.; Hull, D.W. *Introduction to MATLAB*; Prentice Hall: Hoboken, NJ, USA, 2002; Volume 4.
43. Montgomery, D.C.; Peck, E.A.; Vining, G.G. *Introduction to Linear Regression Analysis*; John Wiley & Sons: Hoboken, NJ, USA, 2021.
44. Weisberg, S. *Applied Linear Regression*; John Wiley & Sons: Hoboken, NJ, USA, 2005; Volume 528. [CrossRef]
45. Ansari, F.A.; Mokhtar, A.S.; Abbas, K.A.; Adam, N.M. A Simple Approach for Building Cooling Load Estimation. *Am. J. Environ. Sci.* **2005**, *1*, 209–212. [CrossRef]
46. Dhar, A.; Reddy, T.A.; Claridge, D.E. A Fourier Series Model to Predict Hourly Heating and Cooling Energy Use in Commercial Buildings With Outdoor Temperature as the Only Weather Variable. *J. Sol. Energy Eng.* **1999**, *121*, 47–53. [CrossRef]
47. Kialashaki, A.; Reisel, J.R. Modeling of the energy demand of the residential sector in the United States using regression models and artificial neural networks. *Appl. Energy* **2013**, *108*, 271–280. [CrossRef]
48. Aydinalp-Koksal, M.; Ugursal, V.I. Comparison of neural network, conditional demand analysis, and engineering approaches for modeling end-use energy consumption in the residential sector. *Appl. Energy* **2008**, *85*, 271–296. [CrossRef]
49. Ciulla, G.; D'Amico, A. Building energy performance forecasting: A multiple linear regression approach. *Appl. Energy* **2019**, *253*, 113500. [CrossRef]
50. Burgun, F.; Bilbao, J.; Sproul, A.; Partridge, L.; Lowndes, P.; Pardon, J. First energy performance results of a university building and comparison to environmental rating simulation data. *Innovation* **2013**, *5*, 1.
51. UNSW. *UNSW Design and Construction Requirements 08-Part 3.2 Lighting (Rev 6)*; UNSW: Kensington, Australia, 2018.
52. Dunn, G.; Knight, I. Small power equipment loads in UK office environments. *Energy Build.* **2005**, *37*, 87–91. [CrossRef]
53. Speert, J.; Legge, C. Informed Mechanical Design Through Tested Air Leakage Rates. In Proceedings of the Building Enclosure Science & Technology (BEST3) Conference, Atlanta, GA, USA, 2–4 April 2012.
54. Garcia, E.G.; Zhu, Z. Interoperability from building design to building energy modeling. *J. Build. Eng.* **2015**, *1*, 33–41. [CrossRef]
55. Alothman, A.; Ashour, S.; Krishnaraj, L. Energy Performance Analysis of Building for Sustainable Design Using Bim: A Case Study on Institute Building. *Int. J. Renew. Energy Res.* **2021**, *11*, 556–565. [CrossRef]

Disclaimer/Publisher's Note: The statements, opinions and data contained in all publications are solely those of the individual author(s) and contributor(s) and not of MDPI and/or the editor(s). MDPI and/or the editor(s) disclaim responsibility for any injury to people or property resulting from any ideas, methods, instructions or products referred to in the content.

buildings

MDPI

Article

CO_2 Emission Allocation for Urban Public Buildings Considering Efficiency and Equity: An Application at the Provincial Level in China

Zhidong Zhang *, Yisheng Liu and Zhuoqun Du

School of Economics and Management, Beijing Jiaotong University, Beijing 100044, China; yshliu1@bjtu.edu.cn (Y.L.); duzhu0qun@163.com (Z.D.)
* Correspondence: davidzhang7@163.com

Abstract: China is currently recognized as the leading global energy consumer and CO_2 emitter. A significant amount of carbon emissions can be attributed to urban public buildings. Establishing an equitable and efficient carbon emission allocation mechanism is a crucial step to meeting the ambitious targets in China's 2030 carbon peak plan. In this study, we estimate the total amount of CO_2 emissions from urban public buildings by 2030 and propose a preliminary scheme of carbon quota assignment for each province. By means of applying the zero-sum gains data envelopment analysis (ZSG-DEA) model, the carbon emission quotas allocation of urban public buildings in China's 30 provinces is proposed, and the corresponding pressure to reduce provincial carbon emissions is analyzed. The results indicate that Qinghai has the lowest carbon emission rate (0.01%) for urban public buildings, while Guangdong has the highest (9.06%). Among the provinces, Jiangsu, Jiangxi, and Tianjin face the least pressure in reducing carbon emissions from urban public buildings. On the other hand, Hebei, Beijing, and Anhui are under great pressure to decrease carbon emissions. Notably, Hebei is predicted to have the highest emission reduction requirement of 95.66 million tons. In terms of pressures on carbon emissions reduction for urban public buildings, Jiangsu, Jiangxi, and Tianjin exhibit the least pressure. Hebei, Beijing, and Anhui are facing intense pressure to decrease carbon emissions. These findings offer policymakers valuable insights into developing a fair and efficient carbon allowance allocation strategy, while also contributing to China's efforts to mitigate carbon emissions and combat climate change.

Keywords: carbon peak; China; urban public building; carbon emission allocation; zero sum gains-data envelopment analysis (ZSG-DEA)

1. Introduction

Greenhouse gases are widely recognized as the primary contributors to climate change [1]. The control of greenhouse gas emissions and mitigation of global warming are not only a concern for developed countries, but also for developing countries. In 2015, over 190 parties signed the United Nations Framework Convention on Climate Change in Paris (UNFCCC). This international agreement stipulates that the average global temperature growth should not exceed preindustrial levels by 2 °C, and preferably 1.5 °C [2].

As the world's largest developing country, China has grown into the world's top consumer of energy and carbon emissions producer [3]. China released 28% of the world's total carbon emissions in 2017, which is significantly more than the United States, the second-largest emitter, which contributed only 15% [4]. China has taken significant measures to address climate change by presenting remarkably ambitious Intended Nationally Determined Contributions (INDCs) as part of the Paris Agreement [5]. By 2030, China has pledged to reduce carbon intensity by 60% to 65% from 2005 levels [6] and achieve carbon neutrality no later than 2060 [7]. The Action Plan for Peaking Carbon Dioxide by 2030 was released by China in 2021 in order to provide policy direction for the execution of various peaking carbon efforts and promote the achievement of the peaking carbon objective [8].

Buildings are a major source of both energy consumption and CO_2 emissions. In 2019, buildings' energy consumption in China amounted to approximately 1.11 billion tons of standard coal, leading to total carbon emissions of around 2.2 billion tons, which accounted for approximately 22% of the entire carbon emissions in China [9]. In 2018, the overall building area in China encompassed about 60.1 billion square meters, with approximately 12.8 billion square meters of urban public buildings accounting for 21.3% of the total building area [10]. The average annual building completion area increased by about 2.5 billion square meters from 2014 to 2018, with urban public areas accounting for about 33%. The urban public building area per capita rose from 9 square meters in 2001 to 15 square meters in 2018 [11]. Nevertheless, there remains ample room for future growth as it still lags behind the levels observed in developed countries.

Urban public buildings play a pivotal role as crucial venues for various activities, and their significant energy consumption throughout their lifecycle, including construction, operation, maintenance, and dismantling, has made carbon dioxide emissions from these buildings a focal point for emission reduction targets. It is imperative to reduce carbon dioxide emissions from urban public buildings to achieve China's carbon reduction goals. Among the array of strategies for emission reduction, carbon trading is widely acknowledged as one of the most effective methods [12]. Therefore, it is of paramount importance to research methods that incentivize the participation of urban public buildings in carbon trading, particularly through the equitable allocation of carbon dioxide emission quotas.

However, many studies have focused solely on the fairness or efficiency of carbon emission allocation, overlooking the importance of considering both aspects. For instance, Serrao et al. [13] employed a data envelopment analysis approach to redistribute agricultural greenhouse gas emissions among EU15 countries. Park et al. [14] employed the Boltzmann distribution, which takes efficiency into account, to allocate permits in carbon emissions trading. Meng et al. [15] considered the fairness principle and employed a three-indicator allocation model to measure the provincial quota allocation for carbon emissions in China's electric power sector. Jin et al. [16] investigated the allocation plan of carbon emission allowances among each province in China based on a two-stage DEA model. However, given the varying technological development levels and different stages of development across different regions, as well as the diverse emission reduction targets set for different regions, a research approach that solely emphasizes fairness or effectiveness appears insufficiently comprehensive and integrated. To establish a rational carbon emission allocation mechanism, we contend that both fairness and effectiveness principles should be simultaneously considered.

Furthermore, scholars have extensively researched energy consumption analysis and carbon emission measurement in the context of public buildings. Huang et al. [17] employed the life cycle assessment approach to evaluate the achievements of public buildings in Shenzhen by quantifying the reduction of CO_2 emissions over the past decades. Qian et al. [18] empirically investigated the carbon emissions of public infrastructure and their influence on the global greenhouse effect. You et al. [19] conducted a study on the carbon dioxide emissions from public and commercial buildings in China, analyzing the temporal and spatial changes as well as the driving factors of CO_2 emissions. Li et al. [20] examined the interoperability between energy simulation and building information modeling (BIM) tools, focusing on simulating the total energy consumption during the operational phase of buildings. The study compared the predicted total energy consumption with the actual monitored data to evaluate the accuracy of the simulation results. Liu et al. [21] and Huang et al. [22] developed a support vector machine (SVM) model for predicting and diagnosing the total energy consumption in public buildings. Yue et al. [23] utilized a dynamic life cycle model to accurately calculate the carbon footprint of hospital buildings, which provides a basis for developing energy conservation and emission reduction plans.

It is evident that only a limited number of studies have examined the allocation of carbon emission quotas at the provincial level, with a specific focus on urban public buildings. Gan et al. [9] conducted a comparison between provincial allocation results derived from the ZSG-DEA model and the fixed cost allocation model. They further utilized these models to develop an allocation scheme for carbon emission quotas in China's public buildings. Zhang et al. [24] employed the SBM-DEA model to allocate carbon emission quotas for public buildings across different provinces, with the aim of achieving carbon reduction targets. Therefore, there exists a research gap in the current literature regarding the allocation of carbon emission quotas at the provincial level, particularly in the context of urban public buildings.

This paper addresses the research gap by examining how to allocate carbon quotas among urban public buildings in provinces. The objective is to develop a scientifically rigorous allocation mechanism with the objective of determining the optimal interprovincial distribution of carbon quotas for urban public buildings in China. By doing so, we aim to provide informed guidance to policymakers and promote the sustainable and low-carbon development of urban public buildings. Firstly, we predicted the overall carbon emission of urban public buildings in 2030 and formulated an initial scheme for carbon quota distribution for each province. Then, we transformed the traditional model referred to by Lins et al. [25] into a novel zero-sum gain data envelopment analysis (ZSG-DEA) model, which maintains the constant total amount concept while simultaneously achieving the best input and output efficiency. Subsequently, in accordance with principles of both equality and efficiency, we utilized the ZSG-DEA approach to appraise the comprehensive efficiency of the above-mentioned allocation outcomes. In the adjustment of carbon emission allowances, the quota allocation among provinces is typically considered a zero-sum game, implying that an increase in the quota allowances for one province would lead to a corresponding decrease in the allocation for another province. The state of Pareto efficiency is achieved and the final allocation results are obtained.

Urban public buildings' carbon emission quota allocation can serve as an incentive for provinces to collaborate in achieving carbon emissions peak targets, effectively controlling carbon emissions in public buildings, and attaining energy conservation and emission reduction objectives. Additionally, it can provide policy guidance and support for carbon trading market quota allocation.

This study has made the following contributions.

1. Many earlier studies in the field of carbon emission allocation have concentrated on the distribution of CO_2 emissions among provinces primarily. However, we have directed our attention to carbon emission allocation in provincial public buildings, and fill the gap created by the lack of studies shedding light on carbon emission allocation in the public building sector. We developed an optimal scheme for carbon emission allocations in provincial urban public buildings in China which provides a clear and effective strategy for reducing carbon emissions in these buildings.
2. We have developed a ZSG-DEA model which takes into account the actual performance of each region in achieving its carbon emission goals. Compared to other models of previous research, this model gives more attention to the overall regional efficiency and allows for the reward of regions that achieve optimal performance in reducing their carbon emissions while penalizing those that do not meet the fixed target.
3. This study not only proposes an innovative scheme for carbon emission allocations in provincial urban public buildings but also provides significant policy recommendations for provincial-level allocation quotas for urban public buildings. It is significant in addressing China's serious challenge of decreasing carbon emissions.

The rest of this paper is structured as follows. Section 2 examines significant research on the allocation of carbon emission allowances. Section 3 describes the methods and data source. Section 4 shows the results and discussion. Finally, the summary of the key findings and the policy recommendations are presented in Section 5.

2. Literature Review

2.1. Equity and Efficiency in Carbon Emission Allocation

In recent years, there has been a considerable body of research on low-carbon development, with scholars focusing their attention on carbon emission allowances and showing a particular interest in the choosing of appropriate principles and approaches. In order to tackle climate change, the UNFCCC [26] emphasizes equality principles and proposes a framework of "shared but distinct responsibilities". The principle of equity is a broad concept that encompasses several criteria, including equality, historical responsibility, carbon emission abatement capability, and grandfathering [27].

In the recent literature, the equity principle has been explored through various indicators. Pan et al. [28] have proposed an allocation plan based on equal cumulative carbon emissions per person, emphasizing the notion of equality. Similarly, Ye et al. [29] utilize the population indicator to reflect the concept of equality. Fang et al. [30] employ a multi-criteria allocation approach for 30 provinces using an enhanced ZSG-DEA model, incorporating GDP and historical carbon emissions as measures of equity. Zhang et al. [31] put forth a regional CO_2 allocation scheme that aligns with the principle of equality. Furthermore, Han et al. [32] and Qin et al. [33] have considered indicators such as per capita GDP, historical CO_2 emissions, and CO_2 emissions per unit of industrial added value as indicators to represent the principle of equality. It can be seen that when it comes to the issue of fairness in carbon emission allocation, researchers frequently take population size, economic status, and historical carbon emissions into account.

As research advances and our knowledge deepens, these studies continue to enhance our comprehension of the equity in carbon emission allocation. However, it becomes increasingly apparent that the concept of absolute fairness in the allocation of carbon emissions does not necessarily serve the best interests of all nations and regions [4]. While the equity principle considers all regions, it lacks efficiency as it fails to incentivize provinces with low carbon emissions or constrain those with high emissions, thereby limiting overall efficiency. Therefore, efficiency should be a crucial principle to consider when allocating carbon emission allowances [34]. The objective of this principle is to optimize society's overall carbon emission efficiency, where regions that produce greater output with lower emissions ought to be granted a greater number of allowances [35]. The allocation of carbon emissions is primarily based on DEA-based efficiency measurements that reflect the production efficiency of countries or regions. This evaluation method offers a more comprehensive reflection of production factors and enables a detailed assessment of a country's or region's efficiency [36].

2.2. Data Envelopment Analysis (DEA) Model

The data envelopment analysis (DEA) model, which was first developed by Charnes [37], has gained extensive application in the distribution of resources, particularly when the input or output remains constant [38]. This method is categorized into two types: input-oriented models and output-oriented models. It is primarily utilized to evaluate the effectiveness of decision-making units (DMUs) that contain numerous inputs and outputs. The DEA model has been widely adopted by numerous scholars in their research on carbon emission allocation, serving as a valuable tool for evaluating the efficiency of carbon emission quotas. Lozano et al. [39] proposed a three-stage DEA method which involves maximizing the total desirable outputs, minimizing the total undesirable emissions, and minimizing the consumption of input resources, in order to reallocate emission permits.

The overall quantity of CO_2 emissions must stay within certain limits due to the establishment of carbon emission goals by countries and regions. Consequently, determining how to achieve maximum efficiency in carbon emission allocation remains a significant challenge. The ZSG-DEA model, introduced by Lins et al. [25], is a suitable approach to address this issue. The ZSG-DEA follows procedures that resemble a game with a zero-sum outcome, where a gain for one side corresponds to a loss for the other, but the total allotment of resources remains constant [40]. This model is easy to implement and enables the maximization of overall efficiency by adjusting the quotas to their maximum potential [41]. Gomes and Lins et al. [42] employed the ZSG-DEA model to analyze a case study on carbon emissions. Chiu et al. [43] investigated the reallocation of carbon emission allowances among the European Union's 24 members by applying the ZSG-DEA model. Xiong et al. [44] developed a weighted ZSG-DEA model to assess the allocation of energy consumption for every province. Furthermore, The appraisal of carbon emission allowances at a provincial level has been carried out by various scholars using the ZSG-DEA model. For instance, Cai et al. [45] utilized the ZSG-DEA model to allocate emission allowances throughout China's 30 provinces. Yang et al. [46] assessed carbon emission allocation schemes in China through the ZSG-DEA model. Cui et al. [47] and Fang et al. [30] redistributed carbon emission quotas among China's 30 provinces using the ZSG-DEA model. Song et al. [48] integrated the fairness principle into the ZSG-DEA model with the aim of improving the efficiency of carbon emission quotas and creating a more equitable allocation scheme among provinces. However, it is worth noting that the existing literature in this field predominantly focuses on either fairness or efficiency, with limited consideration given to the integrated principles of fairness and efficiency. By combining these two principles, not only can the equitable allocation of resources be ensured, but also the enhancement of carbon emission reduction efficiency can be promoted. Therefore, this study adopts a holistic approach that incorporates both fairness and efficiency considerations. By taking into account indicators such as population, economy, and historical carbon emissions, the ZSG-DEA model is employed to explore the allocation of carbon emission quotas for urban public buildings in 30 provinces in China.

3. Methodology and Data Sources

Due to significant disparities in development among different regions in China, the allocation of carbon emission quotas for urban public buildings should be coordinated and harmonized. The allocation process is illustrated in Figure 1. First, the research data were collected from China Statistical Yearbook and the China Energy Statistical Yearbook. Next, the carbon emissions from urban public buildings in China for the year 2030 were predicted using the GM (1,1) model. Subsequently, adhering to the principles of fairness and efficiency, and considering factors such as population, economic development, and energy consumption in each province, the iteration and reallocation results were carried out using the ZSG-DEA model. Lastly, analysis and discussion were conducted on the allocation results of carbon emission quotas for urban public buildings and the pressure to reduce emissions.

3.1. Methodology

DEA is a linear programming technique that was first introduced in 1978 by Charnes [37]. It is widely utilized to assess the effectiveness of homogenous collections of decision-making units (DMUs). The conventional DEA model makes the assumption that all inputs and outputs of all decision-making units are independent of one another and have no impact on other activities. However, the entire quantity of a certain input or output is constant in the allocation field. If a specific DMU's input or output increases, it will result in a decrease in other DMUs' input or output; this renders the typical DEA model ineffective. Consequently, Lins et al. [42] proposed the ZSG-DEA model. The ZSG-DEA model reconfigures the inputs and outputs of DMUs to achieve new targets for all DMUs while keeping the sum of the changing variables constant [4]. This feature enables the ZSG-DEA model to be more suitable for resource-constrained research, such as carbon emission allocation.

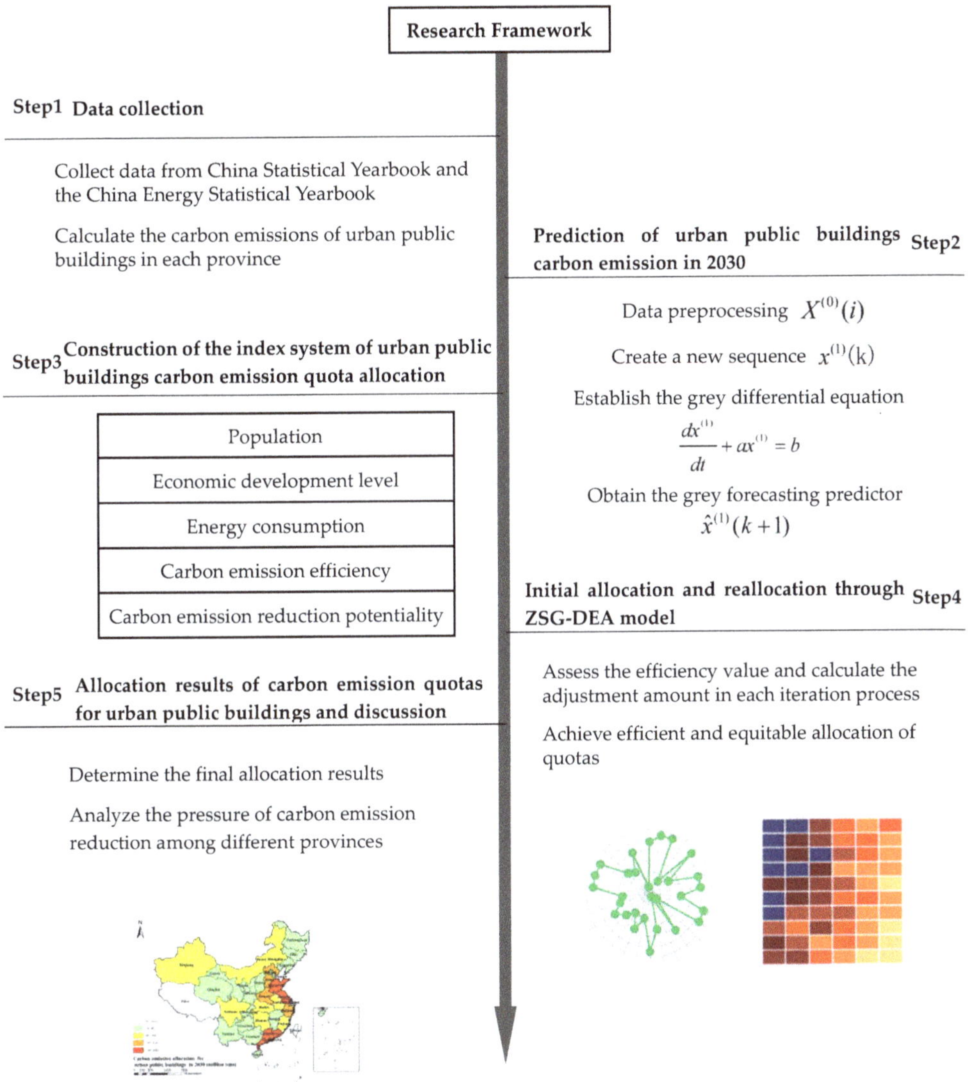

Figure 1. Flowchart of the method.

Input-oriented models are used to measure the efficiency of decision-making units. If DMU_0 is not technically efficient, it is assigned an efficiency value φ_0 under the ZSG-DEA model. To become DEA efficient, the DMU_0 must decrease its input k by a certain amount $u_0 = x_{0k}(1-\varphi_0)$. This decreased amount is then allocated proportionally to the other DMU. The value k of allocation that DMU_i obtained from DMU_0 is

$$\frac{x_{ik}}{\sum\limits_{i\neq 0} x_{ik}} \cdot x_{0k}(1-\varphi_0) \quad (1)$$

Since the percentage of input is being decreased simultaneously by each DMU, after all adjustments are completed, the redistribution of input k to DMU_i is

$$x'_{ik} = \sum_{i \neq 0} X_i \left[\frac{x_{ik}}{\sum\limits_{i \neq 0} x_{ik}} \cdot x_{0k}(1-\varphi_0) \right] - x_{ik}(1-\varphi_i), i = 1,2,3,\cdots,N \tag{2}$$

The input-oriented BCC model employs the ZSG-DEA approach to assess the relative efficiency of DMU_0. The result [49] is shown in Equation (3).

In the original model, all non-DEA-efficient DMUs allocate their excess inputs proportionally to achieve DEA efficiency. However, the consequence of doing so for all DMUs is that some DMUs still fail to achieve DEA efficiency after abating their inputs, even according to Equation (3). There are two approaches to address this problem: the proportional abatement formula method [25] and the iterative method [50]. In this study, the iterative method is used, and multiple redistributions of the input k can be achieved through multiple iterations. Eventually, all DMUs will attain the effective boundary, that is, all DMUs will be 100% effective. The result of the reallocation of inputs at this point is the allocation scheme that makes the best efficiency.

$$\begin{aligned} &\min \varphi_0 \\ &s.t. \begin{cases} \sum\limits_{i=1}^{N} \lambda_i y_{ij} \geq y_{0j}, & j = 1,2,3,\cdots,M \\ \sum\limits_{i=1}^{N} \lambda_i x_{ik} \left[1 + \frac{x_{0k}(1-\varphi_0)}{\sum\limits_{i \neq 0} x_{ik}} \right] \leq \varphi_0 x_{0k}, & k = 1,2,3,\cdots,R \\ \sum\limits_{i=1}^{N} \lambda_i = 1, & i = 1,2,3,\cdots,N \\ \lambda_i \geq 0, & i = 1,2,3,\cdots,N \end{cases} \end{aligned} \tag{3}$$

In the economic-environmental system, carbon emissions are non-desired outputs. The DEA-based environmental efficiency evaluation model offers several methods to handle non-desired outputs, including the main idea of "treating non-desired outputs as inputs". This approach aligns with the DEA model's requirements for input indicators, where the lower the input and the higher the desired output, the more efficient the technology is deemed to be. Similarly, in this study, we employed this approach and served carbon emissions as the input variable of the model.

Gomes et al. [42] have utilized GDP, population, and energy consumption as output variables. Choosing GDP as the desired output associated with carbon emissions is more comprehensible. Considering population as the desired output suggests that regions with larger populations are more efficient in terms of GDP when they maintain the same level of carbon emissions between regions. Moreover, selecting energy consumption as the desired output implies that regions with lower levels of carbon emissions are more efficient when they maintain comparable levels of energy consumption between regions.

The GM (1,1) model, which was first introduced by Deng in 1982, is a time series prediction model [51]. The GM (1,1) model employs a single input variable and differential equations distinct from first-order ones [52], allowing it to generate predictions for unavailable or incomplete time series data by analyzing existing information. It is frequently utilized to estimate population and has been widely applied in predicting resources and emissions [53]. Due to a lack of officially ordered and uniformly spaced statistics on GDP, population, and energy consumption for 2030, there is a strong demand for reliable prediction models to forecast these crucial variables. Therefore, this study adopts the GM (1,1) model to predict these indicators.

The formulas are as follows:

Step 1: Data preprocessing. This step involves the collection and preprocessing of the input data. Let the original data sequence be $X^{(0)}(i)$, where i denotes the i th sample. Additionally, the raw sequence of n samples is defined as

$$X^{(0)} = \left(x^{(0)}(1), x^{(0)}(2), \cdots, x(0)(n)\right), x^{(0)}(k) > 0,\ k = 1, 2, 3, \cdots, n \tag{4}$$

Step 2: Create a new sequence from the original data sequence. Using the accumulated generating operation (AGO), generate a new sequence $X^{(1)}$.

$$X^{(1)} = \left(x^{(1)}(1), x^{(1)}(2), \cdots, x^{(1)}(n)\right) \tag{5}$$

The $x^{(1)}(k)$ is calculated in the following way:

$$x^{(1)}(k) = \sum_{i=1}^{n} x^{(0)}(i), k = 1, 2, 3, \cdots, n \tag{6}$$

Step 3: Establish the grey differential equation. Transform the original data sequence into a first-order differential equation to approximate the time series using an exponential function. This results [54] in the grey differential model, which is represented as

$$\frac{dx^{(1)}}{dt} + ax^{(1)} = b \tag{7}$$

where a and b are the development coefficient and grey input, respectively, obtained through least squares estimation using Equations (8) and (10). Parameter $\hat{a}$ is represented

$$\hat{a} = (B^T B)^{-1} B^T Y \tag{8}$$

where

$$B = \begin{bmatrix} -z^{(1)}(2) & 1 \\ -z^{(1)}(3) & 1 \\ \vdots & \vdots \\ -z^{(1)}(n) & 1 \end{bmatrix}, z^{(1)} = (z^{(1)}(2), z^{(1)}(3), \cdots, z^{(1)}(n)) \tag{9}$$

In Equations (8) and (9), B is the accumulation matrix of the interval-type accumulated generating sequence, where the element in B is

$$z^{(1)}(k) = \frac{1}{2}\left(x^{(1)}(k) + x^{(1)}(k-1)\right), k = 2, 3, \cdots, n \tag{10}$$

and

$$Y = \begin{bmatrix} x^{(0)}(2) \\ x^{(0)}(3) \\ \vdots \\ x^{(0)}(n) \end{bmatrix} \tag{11}$$

Step 4: Obtain the grey forecasting predictor by substituting the estimated parameters obtained from Equations (8) to (11) into Equation (7). The resulting equation [55] is

$$\hat{x}^{(1)}(k+1) = (x^{(1)}(0) - \frac{b}{a})e^{-ak} + \frac{b}{a},\ k = 1, 2, 3, \cdots, n \tag{12}$$

where $\hat{x}^{(1)}(k+1)$ represents the predicted value of the future $k+1$ time point of the time series. a is the development coefficient of the GM (1,1) model, b is the gray input data, k is the prediction step, and $x^{(1)}(0)$ is the first element of the original data.

Step 5: Check the prediction process. Repeat Steps 1–4 to forecast process behavior after obtaining a new sample until abnormal circumstances emerge. If a forecast point falls between the upper and lower control boundaries, the process is under control, and monitoring can proceed. However, If the predicted point goes outside of these boundaries, this suggests the process has been out of control, necessitating further analysis and corrective action.

3.2. Data Sources

This study employs the ZSG-DEA approach to allocate 2030 CO_2 emission quotas for interprovincial urban public buildings across China. Hong Kong, Macao Taiwan, and Tibet are excluded owing to lacking data. The data period covers from 2005 to 2020, with the latest published data available until 2020. Additionally, the GM (1,1) model is employed in this research to forecast the CO_2 emissions of provincial public buildings in 2030.

Carbon emissions were selected as the undesirable input, with GDP, population, and energy consumption as the three output variables. Table 1 provides a description of these input and output variables. The China Energy Statistical Yearbook and China Statistical Yearbook were the sources of data for the years spanning from 2005 to 2020.

Table 1. Input–output variables.

Variable Classification	Specific Variable	Variable Explanation
Input variable	Population GDP Energy consumption	Population of each province in 2030 GDP of each province energy consumption of each province in 2030
Output variable	Carbon emissions	Estimated CO_2 emissions by regions of urban public buildings

4. Results

4.1. The Initial Carbon Quotas Allocation

The initial carbon quotas allocation for urban public buildings among Chinese provinces is a crucial step in managing carbon emissions and addressing climate change. As depicted in Figure 2, the initial carbon quota allocation results show that the provinces with the highest allocations are primarily located in the east and southeast regions of China, while those with the lowest allocations are primarily located in the western regions. The provinces of Hebei, Jiangsu, and Guangdong received the highest carbon quotas, accounting for 25.32% of the total initial allocation. This can be attributed to their significant economic development, large population, and high levels of carbon emissions. In contrast, Qinghai, Ningxia, and Gansu received the lowest carbon quotas, amounting to only 2.08% of the total allocation. These provinces exhibit low levels of economic development, small populations, and relatively low carbon emissions, which may explain the lower allocation.

Table 2 presents the energy consumption, GDP, population, and allocation results of urban public buildings' initial CO_2 emission quotas in 2030. The following are the reasons for allocating CO_2 emission allowances to urban public buildings. Provinces with lower CO_2 reduction potential or obligation, such as Hebei and Guangdong, were assigned higher quotas. Provinces with less CO_2 reduction responsibility but higher emission efficiency, such as Shanghai, Zhejiang, and Jiangsu, were given more quotas. Provinces with greater CO_2 sequestration potential, such as Fujian, Yunnan, and Sichuan, were considered to have higher emission feasibility and were given more quotas. Conversely, provinces with higher CO_2 reduction responsibility but lower economic feasibility or environmental sustainability were given fewer quotas. In conclusion, the initial carbon quota allocation for urban public buildings is a significant step toward carbon emission reduction in China. The allocation results indicate that provinces with the highest carbon emissions and greater economic development have been assigned higher quotas.

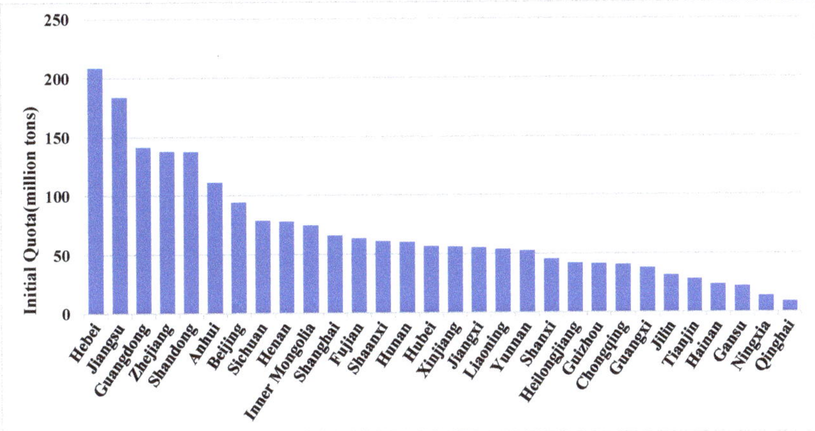

Figure 2. Initial carbon emission allocations for urban public buildings among thirty Chinese provinces in 2030.

Table 2. Initial carbon emission allocations for urban public buildings among thirty Chinese provinces in 2030.

Province	Carbon Quota (Million Tons)	Energy Consumption (Million Tce)	GDP (Billion Yuan)	Population (Million)
Beijing	94.30	78.07	9454.55	24.70
Tianjin	28.30	93.16	3166.36	14.80
Hebei	208.77	388.43	7575.23	77.81
Shanxi	45.61	245.37	3586.52	34.77
Inner Mongolia	74.69	354.29	4027.39	23.56
Liaoning	54.11	249.38	4585.15	41.57
Jilin	31.84	54.05	2397.51	21.31
Heilongjiang	42.03	117.17	2255.15	26.84
Shanghai	66.05	125.82	9230.37	27.03
Jiangsu	183.81	413.61	27,641.29	90.95
Zhejiang	137.66	319.70	16,398.45	76.59
Anhui	111.31	200.38	11,932.85	61.75
Fujian	63.57	196.03	13,902.60	46.86
Jiangxi	55.34	156.28	7415.71	45.94
Shandong	137.41	500.11	16,645.72	107.78
Henan	78.16	231.83	14,505.83	103.84
Hubei	56.55	189.89	13,680.23	60.13
Hunan	60.44	166.87	11,812.72	67.71
Guangdong	141.23	461.97	29,017.68	151.85
Guangxi	37.97	165.91	6077.94	53.25
Hainan	23.82	38.82	1612.12	11.62
Chongqing	40.84	100.34	8053.32	35.64
Sichuan	78.93	232.33	14,404.98	86.44
Guizhou	41.51	130.36	7308.79	41.29
Yunnan	52.61	173.13	7792.94	48.29
Shaanxi	61.01	211.84	7436.71	41.68
Gansu	21.75	101.15	2077.54	24.51
Qinghai	8.70	67.89	859.95	6.24
Ningxia	13.61	177.15	1048.71	8.17
Xinjiang	56.33	409.45	3771.64	30.43

4.2. Efficiency and Final Carbon Quotas Allocation

The initial carbon quotas allocation for urban public buildings across 30 Chinese provinces was found to be largely inefficient. As shown in Table 3, only 8 of the 30 provinces achieved an optimal efficiency score of 1, achieving the DEA frontier. This group includes Henan, Hubei, Guangdong, Guangxi, Qinghai, Ningxia, and Xinjiang. Following closely was Hunan, with an efficiency assessment of 0.932. The remaining 22 provinces had efficiency scores ranging from 0.4 to 0.9, indicating that the initial quota allocation in these regions was not entirely efficient. Specifically, Shanghai, Zhejiang, Jilin, Heilongjiang, Hainan, Anhui, Beijing, and Hebei were among the eight provinces with efficiency scores below 0.62, significantly below the DEA frontier. Hebei received the lowest efficiency score at 0.42.

To attain efficient and equitable quota allocation across provinces, an iterative process was applied to the adjustment of the initial quotas. In each iteration, the carbon quota allocation was recalibrated based on the ZSG-DEA model to ensure that the efficiency score of each province approached 1 while keeping the total carbon emissions constant throughout the iteration [45].

Figure 3 depicts the results of the iterative process in optimizing the carbon emission quota allocation plan. The graph presents the efficiency values for each province in each iteration. During the first iteration, there were numerous provinces with low efficiency values, while some provinces started to approach the efficient frontier. During the second iteration, the efficiency values of many provinces improved, and eight provinces achieved high efficiency values approaching 0.99. In the third and fourth iterations, many provinces with high efficiency values increased, and 26 and 29 provinces achieved near-perfect efficiency values above 0.99. Finally, in the fifth iteration, almost every province achieved an efficiency value of 1, indicating that the carbon emission quota allocation plan had been optimized. Therefore, Figure 3 demonstrates the effectiveness of the iterative process in improving provinces' efficiency values and ultimately achieving an optimal carbon emission quota allocation plan.

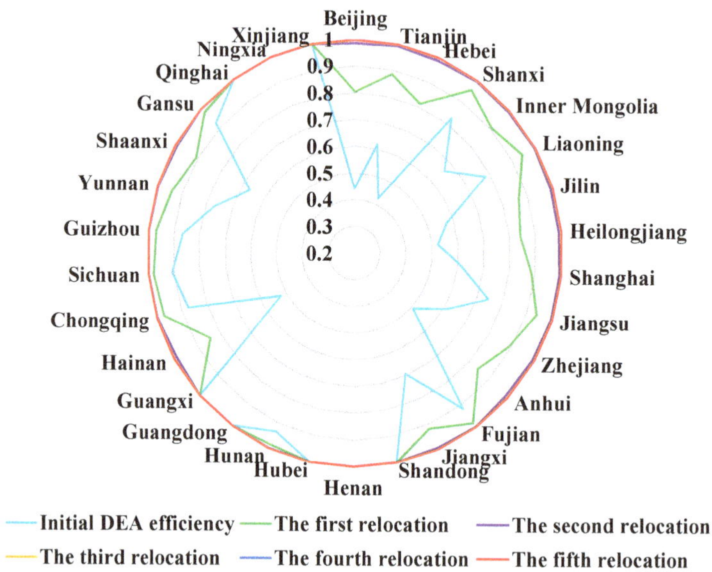

Figure 3. The DEA efficiency of initial allocation and reallocation.

In Figure 4, the carbon emission quotas for urban public buildings for 30 different provinces are shown as they are adjusted following every iteration. The overall carbon emission quota for provinces combined remains constant at 2108 million tons throughout the iterations. However, the amount of carbon emissions allowed for each province's urban public buildings changes with each iteration. It is crucial to emphasize that the total carbon emission quota adjustment for all urban public buildings in all provinces amounts to zero after each iteration. This means that any increase in the carbon emission quota for one province's urban public buildings must be offset by a decrease in another province's carbon emission quota.

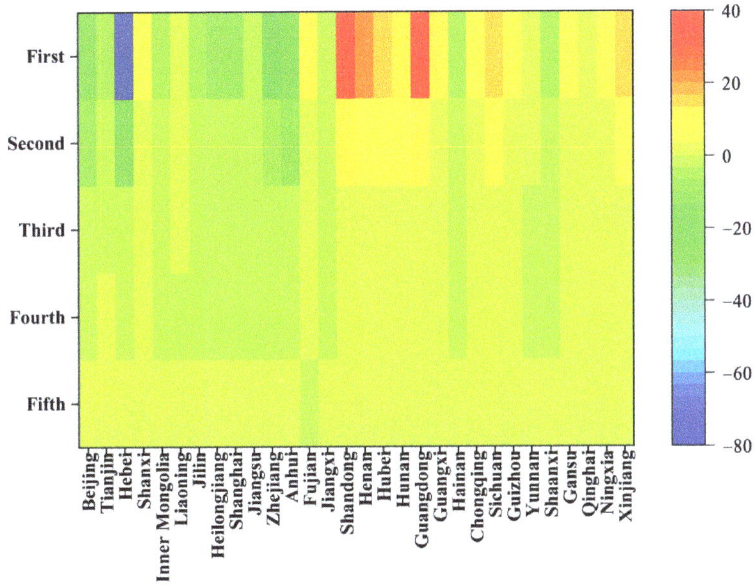

Figure 4. Quota adjustments for each provincial urban public building after each iteration.

As a result of these adjustments, some provinces will experience an increase in their carbon emission quotas, while others will see a decline. This suggests that there may be differences in urban public building carbon emissions across the provinces. By adjusting the carbon emission quotas after each iteration, the overall goal of reducing carbon emissions can be achieved while also accounting for provincial differences.

The first two iterations exhibit substantial changes in carbon quotas, with certain provinces experiencing much larger adjustments compared to others. The provinces with significant increases in carbon emission allowances include Guangdong, Shandong, Henan, Hubei, etc. Guangdong increased by 37.85 million tons, and Shandong increased by 36.82 million tons. Conversely, Hebei experienced a significant decrease in carbon emission allowances, primarily due to its lower efficiency value, resulting in a reduction of 77.10 million tons during the redistribution process. In contrast, the third, fourth, and fifth iterations exhibit minimal deviations in quotas, which suggests that quotas have stabilized over time. Furthermore, in the fifth iteration, the majority of provinces display an increase in carbon quotas, whereas Hebei and Beijing observe a decrease due to their initially low efficiency values. Overall, these findings demonstrate the potential for policy interventions to improve energy efficiency and carbon reduction in public buildings, which can be sustained over time with appropriate measures. The stabilization of carbon quotas over several iterations also implies that policymakers may refine and optimize their strategies for desirable outcomes.

Figure 5 illustrates the CO_2 emission quota allocation for urban public buildings. The data indicate that Guangdong, Shandong, and Jiangsu provinces had the highest carbon quotas, exceeding 150 million tons of carbon emissions per province. Notably, Guangdong province had the most substantial amount of carbon emissions at 191.05 million tons, representing 9.06% of the total quota for the country. The high carbon emission quota for this province can be attributed to its dense population and rapid industrialization. In contrast, the provinces of Qinghai, Hainan, and Ningxia received the smallest carbon quotas, with Qinghai and Hainan getting only 11.77 and 16.74 million tons each, respectively. These provinces have low population densities and are primarily focused on agriculture and tourism, explaining their reduced carbon quotas. Figure 5 demonstrates the wide variation of carbon quotas across different regions of China and underscores the need for a targeted and strategic approach to carbon emission reduction.

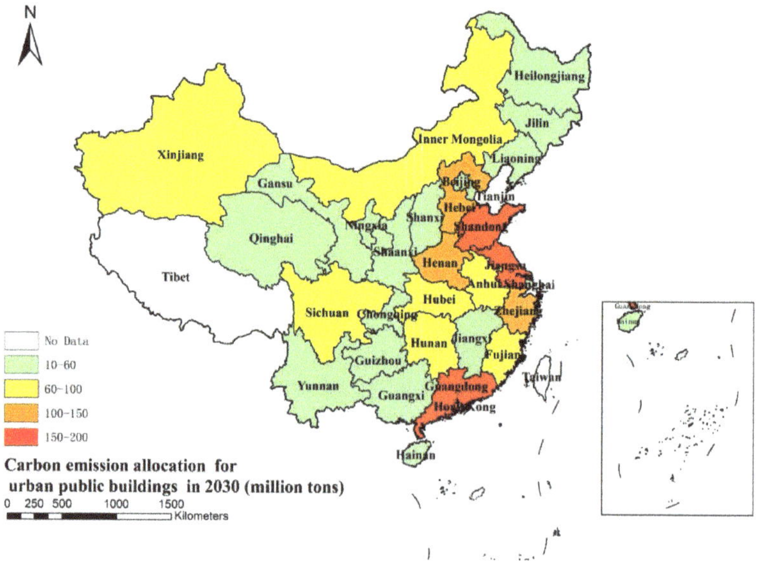

Figure 5. Allocation of final carbon emission quotas for urban public buildings at provincial level in China.

Table 3. Initial and final carbon emission quotas and allocation efficiency.

Province	Initial Efficiency	Final Efficiency	Adjustment Amount (Million Tons)	Final Quota (Million Tons)
Hebei	0.42576	1	95.67	113.1
Beijing	0.44369	1	39.14	55.16
Anhui	0.50604	1	37.16	74.15
Hainan	0.52254	1	7.08	16.74
Heilongjiang	0.52267	1	12.61	29.42
Jilin	0.57153	1	7.38	24.46
Zhejiang	0.61115	1	26.87	110.79
Shanghai	0.61239	1	12.02	54.03
Tianjin	0.61939	1	4.71	23.59
Inner Mongolia	0.66326	1	8.49	66.2
Shaanxi	0.66807	1	6.42	54.59
Jiangxi	0.69239	1	3.93	51.41
Jiangsu	0.74473	1	3.04	180.77
Yunnan	0.76735	1	−1.68	54.29
Liaoning	0.78163	1	−2.78	56.89
Shanxi	0.82855	1	−5.31	50.92
Guizhou	0.86662	1	−7.01	48.52

Table 3. *Cont.*

Province	Initial Efficiency	Final Efficiency	Adjustment Amount (Million Tons)	Final Quota (Million Tons)
Chongqing	0.87262	1	−7.25	48.09
Sichuan	0.90653	1	−17.51	96.44
Fujian	0.91565	1	−14.97	78.54
Gansu	0.92357	1	−5.4	27.15
Hunan	0.93201	1	−15.6	76.04
Shandong	1	1	−48.47	185.88
Henan	1	1	−27.56	105.72
Hubei	1	1	−19.95	76.5
Guangdong	1	1	−49.82	191.05
Guangxi	1	1	−13.39	51.36
Qinghai	1	1	−3.07	11.77
Ningxia	1	1	−4.79	18.4
Xinjiang	1	1	−19.86	76.19

5. Discussion

5.1. Pressure of Provincial Carbon Emission Reduction for Urban Public Buildings

This paper utilizes historical data on carbon emissions to forecast the expected CO_2 emissions for every province in China by 2030. Subsequently, we utilized the principles of equality and efficiency to distribute carbon quotas to each province, which provide a strong foundation for measuring the pressure of carbon emission reduction for provinces. Reduction pressure is a metric that quantifies each province's carbon emission reduction goal towards the country's target. If the projected carbon emissions for a province were to surpass its allocated quota, then it incurred a positive reduction pressure, indicating the need to reduce emissions further. Conversely, if the projected carbon emissions were lower than the allocated quota, then it had a quota surplus, resulting in negative reduction pressure. The carbon emission reduction pressure was estimated by the gap between the allocated quota and the projected carbon emissions of provinces, measured in millions of tons. The higher the reduction pressure value, the greater the mitigation efforts required by the province to meet the 2030 carbon emission reduction targets. Moreover, we utilized the percentage of carbon emission reduction pressure to provide a relative indication of the effort required for each province, calculated as the ratio of the carbon emission reduction pressure to the allocated carbon emission quota. Figure 6 provides a visual representation of the pressure on each province in China to reduce its carbon emissions.

The carbon reduction pressure varied significantly among provinces, and each province had different reasons behind its reduction pressure. In terms of the number of provinces under pressure to reduce their emissions, 17 provinces showed a surplus in their carbon emission quotas, which means that they are emitting less than their permitted limits. The other 13 provinces have experienced positive reduction pressure, which means that they need to put more effort into reducing their carbon emissions. By analyzing the provinces which possess a quota surplus or experience reduction pressure, it is clear that there is a fair distribution, demonstrating the effectiveness and equity of the 2030 carbon emission quota allocation scheme.

Guangdong, Shandong, Henan, Hubei, Xinjiang, Sichuan, Hunan, Fujian, and Guangxi had a significant quota surplus. The province of Guangdong received the highest excess allocation, which amounts to 49.82 million tons of carbon emissions. The majority of these provinces were in the beginning stages of carbon quotas and were expected to have comparatively decreased carbon emissions in 2030. Thus, they had a surplus in their quota. In contrast, Qinghai and Ningxia, positioned in the end stages of carbon quotas, enjoyed a large quota surplus due to the lowest predicted carbon emissions in China. Additionally, a total of nine other provinces, including Fujian, Guangxi, Chongqing, Guizhou, Gansu,

Shanxi, Liaoning, and Yunnan, had some quota surplus but not as much as the previously mentioned provinces.

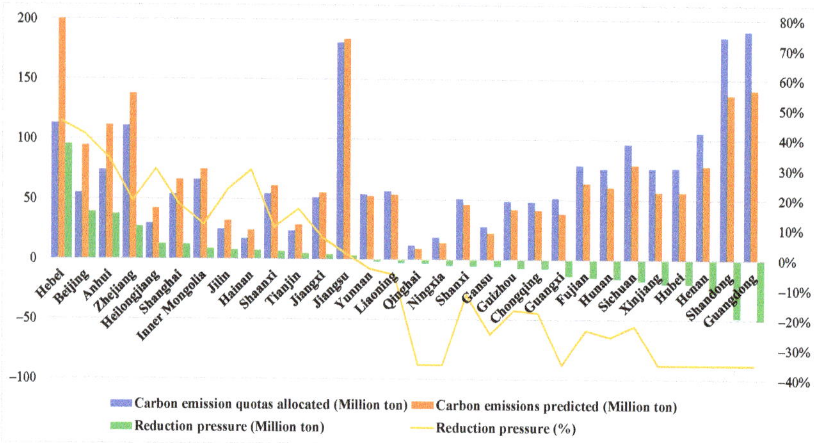

Figure 6. Carbon emission reduction pressure of provincial urban public buildings in 2030.

In the provinces where the carbon reduction pressure is positive, Jiangsu, Jiangxi, and Tianjin have less than 5 million tons of reduction pressure value, and they have the least reduction pressure. Conversely, Hebei had the highest reduction pressure at 95.66 million tons. In addition, reduction pressure in Shannxi, Hainan, Jilin, Inner Mongolia, Shanghai, Zhejiang, Anhui, and Beijing gradually increases, ranging from 6.41 million tons to 39.14 million tons. These reduction pressures reflected their responsibilities to mitigate carbon emissions to reach their allocated quotas. Overall, the reduction pressure values provided insight into the measures each province needed to take in order to meet their emission reduction goals and contribute towards China's carbon emission targets.

5.2. Comparison with the Existing Literature

This study proposes an allocation scheme for carbon quotas in urban public buildings that aims to balance fairness and efficiency principles. Our quota allocation scheme prioritizes provinces with greater potential for output increase while ensuring that the overall carbon emissions from Chinese urban public buildings do not increase. To evaluate the rationale and accuracy of our proposed scheme, we compare it with relevant published studies.

Figure 7 presents the comparison results between this study and the published literature regarding the allocation of carbon quotas in Chinese urban public buildings by 2030 derived from the existing literature [9,24]. Due to differences in allocation principles and measurement methods, the research findings on the allocation of carbon quotas for urban public buildings are inconsistent. By observing Figure 7a, we can note significant variations in the allocation results of carbon quotas for urban public buildings. Figure 7b illustrates the proportion of carbon quota allocations for urban public buildings among provinces. Further analyzing Figure 7b, we find that the provinces with the highest carbon quota allocations, exceeding 4%, include Jiangsu, Guangdong, Zhejiang, Shandong, Beijing, and Shanghai. These regions are located in the economically developed eastern coastal areas. In contrast, the provinces with the lowest carbon quota allocations, less than 1%, consist of Hainan, Gansu, Ningxia, and Qinghai. These areas belong to the less economically developed western regions.

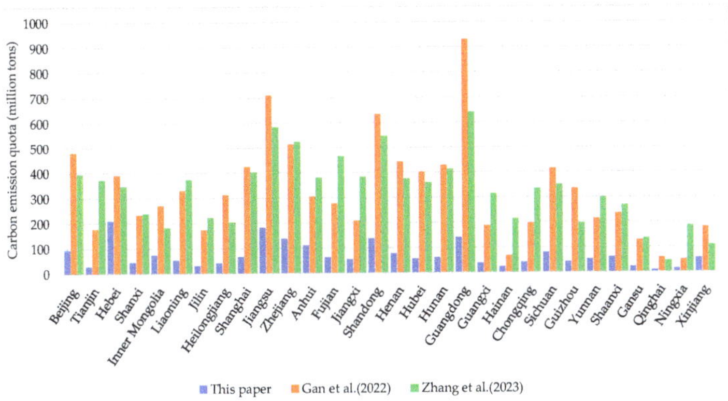

(a)

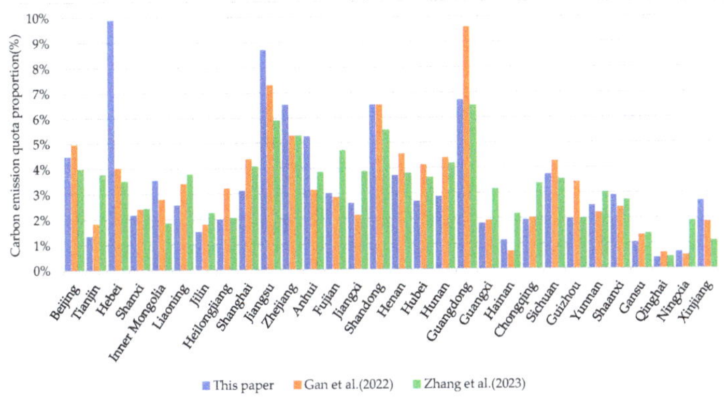

(b)

Figure 7. The comparison of allocation results for urban public buildings in 2030 between this paper and the published literature. (**a**) Carbon emission quota allocation results among provinces; (**b**) proportion of carbon emission quota allocation results among provinces (Gan et al. (2022) [9] and Zhang et al. (2023) [24]).

These observations indicate that the economically lagging western regions bear a smaller responsibility for CO_2 emissions reduction, providing greater space for future socioeconomic development. Conversely, the economically advanced eastern regions shoulder a larger burden of CO_2 emissions reduction, aligning with China's goal of balanced regional development to narrow the economic development gap.

Although there are some differences in detail between our study and others, the overall trends in our results align with existing research. Our study provides a suitable allocation scheme for carbon quotas in urban public buildings, promoting sustainable development and achieving carbon reduction goals.

6. Conclusions

6.1. Main Findings

The paper proposes a scheme for allocating carbon emission quotas to urban public buildings in China. It is founded on the principles of equality and efficiency and uses the ZSG-DEA model to ensure that each province receives an equitable and appropriate amount of carbon quotas. We first measure the overall CO_2 emissions of urban public buildings in China by 2030 using a number of factors, including economic development, historical carbon emissions, and population and, provide a preliminary scheme of carbon emission quotas allotment for all of the provinces at the beginning of the allocation process. Employing the ZSG-DEA model, carbon emission quotas for urban public buildings in 30 different Chinese provinces are determined. Lastly, CO_2 emission reduction pressure is investigated.

The research findings can be summarized in three main points:

1. Qinghai, Hainan, Ningxia, Tianjin, and Jilin had small carbon quotas for urban public buildings. These provinces, characterized by relatively low population density and abundant resources, have achieved a balance between economic and environmental development, facilitating the attainment of sustainable development objectives. Consequently, these provinces are encouraged to develop their economies while minimizing the negative impact on the environment. Provinces such as Guangdong, Shandon, Jiangsu, Hebei, Zhejiang, and Henan were assigned substantial carbon quotas due to their high population densities and superior input–output efficiencies. In essence, these provinces exhibit a superior ability to generate higher economic output with lower input, positioning them as significant drivers of economic growth. Nonetheless, this advantage often corresponds to increased carbon emissions. In order to balance growth and environmental protection in these provinces, the allocation scheme has awarded higher carbon quotas to enable them to continue developing their economies while minimizing the carbon footprint of their urban public buildings. This not only promotes sustainable development in these provinces but also contributes to achieving the national goal of reducing carbon emissions.
2. There is variation in carbon emission quotas across different regions of China, with quotas gradually increasing from the northwest to the southeast. This is because the southeast coastal areas are more developed economically, have larger populations, and also have higher carbon emission efficiency. As a result, provinces such as Guangdong, Jiangsu, Shandong, and Zhejiang have been assigned higher carbon quotas. The allocation of higher carbon quotas to these provinces is intended to incentivize them to mitigate their carbon emissions and also improve sustainable development. In contrast, the northwest regions have fewer economic advantages, smaller populations, and lower carbon emission efficiency. Consequently, provinces such as Qinghai, Hainan, Ningxia, and Gansu have been assigned relatively lower carbon emission quotas, which may provide incentives to develop cleaner and more sustainable industries and lead to long-term economic and environmental benefits.
3. According to the results of carbon emission reduction pressure, it has been observed that many of the 30 provinces in China are facing pressures to decrease carbon emissions. Out of these provinces, 17 have a surplus in their carbon emission quotas, including Guangdong, Shandong, Hubei, Hunan, and others. Guangdong and Shandong have the highest proportion of quota surplus among all the provinces. Both Guangdong and Shandong are considered highly industrialized and developed regions in China, which could contribute to their better performance in reducing carbon emissions. While some provinces are carbon emissions quota surplus, others are experiencing significant pressure to reduce emissions. Provinces such as Hebei, Beijing, Anhui, and Zhejiang are facing huge emission reduction pressure, indicating that they are struggling to bring their carbon emissions down to more sustainable levels. Provinces with abundant resources, such as Heilongjiang and Inner Mongolia, face significantly higher pressure to reduce emissions in urban public buildings. Conversely, provinces such as Sichuan, Fujian, and Hunan, which have low historical CO_2 emissions and strong economic development, experience relatively lower reduction pressure.

6.2. Policy Implications

The findings carry significant implications for policymakers who need to implement measures to mitigate carbon emissions and foster a sustainable future. Based on the findings, three main policy suggestions can be proposed.

Established differentiated goals and strategies: It is imperative for all provinces to contribute to carbon emission reduction by gaining a comprehensive understanding of the current state of greenhouse gas emissions. However, it is not feasible to hold all provinces to the same standards or targets. A one-size-fits-all approach is not practical, and therefore governments need to establish differential benchmarks that take into consideration economic development, energy-use efficiency, and carbon emissions. As a result, establishing differentiated goals and strategies will be key in ensuring that all countries are responding realistically to climate change.

Looser reduction targets for developed and energy-efficient provinces and stricter targets for developing and energy-dependent provinces: Some provinces may be more efficient than others in terms of their energy usage and might have more efficient policies in place that help reduce carbon emissions. In contrast, others might be transitioning away from carbon-intensive energy sources more slowly due to their reliance on them. Owing to these differences, a blanket target for carbon emissions reduction may not be feasible for every province. In such cases, developing and energy-dependent provinces need stricter reduction targets, whereas developed and energy-efficient provinces may have looser targets assigned.

Upgrading industrial structures and increasing clean energy use: Upgrading to cleaner production processes, reducing energy waste, and adopting emissions reduction technologies all contribute to cutting carbon emissions for less efficient provinces. Additionally, using cleaner renewable energy sources such as solar, wind, and hydroelectric power, rather than carbon-intensive sources such as coal, will be important in decreasing emissions. In less efficient provinces, particularly those that depend on coal and other fossil fuels for energy, increasing the use of clean energy should be a priority. It is a long-term process; however, it is necessary to transition towards carbon neutrality in the future.

6.3. Limitations and Future Research

Although this study provides valuable insights into the carbon quota allocation for urban public buildings, it is important to acknowledge the presence of certain limitations. Firstly, the study was carried out with limited data, which may not be generalized to a larger scope, and the results may not be entirely accurate. In future research, more data could be collected, and the sample size could be increased to improve the study's validity. Secondly, the application of the ZSG-DEA model in this study could be improved to refine the allocation process of CO_2 quotas. Researchers can investigate how to set constraints on intensity variables for each region to ensure fair and reasonable allocation of quotas. Thirdly, city-level CO_2 quota allocation can be prioritized in future research to enable sustainable development and effective carbon emission reductions. Research on this could consider how different cities can have unique carbon allocations that align with their respective circumstances.

Author Contributions: Z.Z. Conceptualization, Methodology, Writing—original draft, Writing—review and editing. Z.D. Data curation, Software, Visualization. Y.L. Conceptualization, Supervision, Project administration, Funding acquisition. All authors have read and agreed to the published version of the manuscript.

Funding: This research was funded by National Natural Science Foundation of China, grant number 71871014.

Data Availability Statement: The data used to support the results are available from the corresponding author upon request.

Conflicts of Interest: All the authors declare no conflict of interest.

References

1. Adnan, A.; Khan, S. *A Comparative Study of Various Strategies Used for the Mitigation of Global Warming*; 2021. Available online: https://scholar.uwindsor.ca/ttccap/2021/online/5/ (accessed on 25 June 2021).
2. Raupach, M.R.; Davis, S.J.; Peters, G.P.; Andrew, R.M.; Canadell, J.G.; Ciais, P.; Friedlingstein, P.; Jotzo, F.; van Vuuren, D.P.; Le Quere, C. Sharing a quota on cumulative carbon emissions. *Nat. Clim. Chang.* **2014**, *4*, 873–879. [CrossRef]
3. Zhang, M.W.; Gao, F.F.; Huang, B.; Yin, B. Provincial Carbon Emission Allocation and Efficiency in China Based on Carbon Peak Targets. *Energies* **2022**, *15*, 9181. [CrossRef]
4. Ma, D.L.; Xiao, Y.P.; Zhao, N. Optimization and Spatiotemporal Differentiation of Carbon Emission Rights Allocation in the Power Industry in the Yangtze River Economic Belt. *Sustainability* **2022**, *14*, 5201. [CrossRef]
5. Ye, B.; Jiang, J.; Li, C.; Miao, L.; Tang, J. Quantification and driving force analysis of provincial-level carbon emissions in China. *Appl. Energy* **2017**, *198*, 223–238. [CrossRef]
6. Enforcing Climate Change Response Initiative–Chinese National Determined Contribution. Available online: http://www.gov.cn/xinwen/2015-06/30/content_2887330.htm (accessed on 30 June 2015).
7. Normile, D. China's bold climate pledge earns praise-but is it feasible? *Science* **2020**, *370*, 17–18. [CrossRef]
8. Cheng, Y.L.; Gu, B.H.; Tan, X.C.; Yan, H.S.; Sheng, Y.H. Allocation of provincial carbon emission allowances under China's 2030 carbon peak target: A dynamic multi-criteria decision analysis method. *Sci. Total Environ.* **2022**, *837*, 155798. [CrossRef]
9. Gan, L.; Ren, H.; Cai, W.G.; Wu, K.; Liu, Y.; Liu, Y. Allocation of carbon emission quotas for China's provincial public buildings based on principles of equity and efficiency. *Build. Environ.* **2022**, *216*, 108994. [CrossRef]
10. Building Energy Research Centre of Tsinghua University. *Annual Report on China Building Energy Efficiency*; China Architecture and Building Press: Beijing, China, 2020.
11. Du, Z.Q.; Liu, Y.S.; Zhang, Z.D. Spatiotemporal Analysis of Influencing Factors of Carbon Emission in Public Buildings in China. *Buildings* **2022**, *12*, 424. [CrossRef]
12. Yu, N.; Chen, J.H.; Cheng, L. Evolutionary Game Analysis of Carbon Emission Reduction between Government and Enterprises under Carbon Quota Trading Policy. *Int. J. Environ. Res. Public Health* **2022**, *19*, 8565. [CrossRef] [PubMed]
13. Serrao, A. *Reallocating Agricultural Greenhouse Gas Emission in EU15 Countries*; Agricultural and Applied Economics Association: San Antonio, TX, USA, 2010.
14. Park, J.-W.; Kim, C.U.; Isard, W. Permit allocation in emissions trading using the Boltzmann distribution. *Phys. A-Stat. Mech. Its Appl.* **2012**, *391*, 4883–4890. [CrossRef]
15. Meng, M.; Wang, L.X.; Chen, Q. Quota Allocation for Carbon Emissions in China's Electric Power Industry Based upon the Fairness Principle. *Energies* **2018**, *11*, 2256. [CrossRef]
16. Jin, X.; Zou, B.; Wang, C.; Rao, K.; Tang, X. Carbon Emission Allocation in a Chinese Province-Level Region Based on Two-Stage Network Structures. *Sustainability* **2019**, *11*, 1369. [CrossRef]
17. Huang, Y.; Duan, H.; Dong, D.; Song, Q.; Zuo, J.; Jiang, W. How to evaluate the efforts on reducing CO2 emissions for megacities? Public building practices in Shenzhen city. *Resour. Conserv. Recycl.* **2019**, *149*, 427–434. [CrossRef]
18. Qian, Y.; Sun, R.; Zhao, J.; Song, S.; Wang, P. How does public infrastructure construction influence CO_2 emissions? A global case. *Front. Environ. Sci.* **2022**, *10*, 1048297. [CrossRef]
19. You, K.; Yu, Y.; Cai, W.; Liu, Z. The change in temporal trend and spatial distribution of CO_2 emissions of China's public and commercial buildings. *Build. Environ.* **2023**, *229*, 109956. [CrossRef]
20. Li, H.X.; Ma, Z.; Liu, H.; Wang, J.; Al-Hussein, M.; Mills, A. Exploring and verifying BIM-based energy simulation for building operations. *Eng. Constr. Archit. Manag.* **2020**, *27*, 1679–1702. [CrossRef]
21. Liu, Y.; Chen, H.; Zhang, L.; Wu, X.; Wang, X.-J. Energy consumption prediction and diagnosis of public buildings based on support vector machine learning: A case study in China. *J. Clean. Prod.* **2020**, *272*, 122542. [CrossRef]
22. Huang, J.; Kaewunruen, S. Forecasting Energy Consumption of a Public Building Using Transformer and Support Vector Regression. *Energies* **2023**, *16*, 966. [CrossRef]
23. Yue, J.; Li, W.; Cheng, J.; Xiong, H.; Xue, Y.; Deng, X.; Zheng, T. A dynamic calculation model of the carbon footprint in the life cycle of hospital building: A case study in China. *Eng. Constr. Archit. Manag.* **2022**. [CrossRef]
24. Zhang, L.; Liu, Y.; Du, Z. *Research on Carbon Allowance Allocation for Public Buildings in China Based on the Target of Carbon Emission Intensity Reduction in 2030*; Research Square: Durham, NC, USA, 2023. [CrossRef]
25. Lins, M.P.E.; Gomes, E.G.; de Mello, J.; de Mello, A. Olympic ranking based on a zero sum gains DEA model. *Eur. J. Oper. Res.* **2003**, *148*, 312–322. [CrossRef]
26. Raju, N.S. United Nations Framework Convention on Climate Change. *Int. J. Empir. Educ. Res.* **2019**, 13–25. [CrossRef]
27. Wu, F.; Huang, N.; Liu, G.; Niu, L.; Qiao, Z. Pathway optimization of China's carbon emission reduction and its provincial allocation under temperature control threshold. *J. Environ. Manag.* **2020**, *271*, 111034. [CrossRef] [PubMed]
28. Pan, X.Z.; Teng, F.H.; Wang, G. Sharing emission space at an equitable basis: Allocation scheme based on the equal cumulative emission per capita principle. *Appl. Energy* **2014**, *113*, 1810–1818. [CrossRef]
29. Li, L.X.; Li, Y.N.; Ye, F.; Zhang, L. Carbon dioxide emissions quotas allocation in the Pearl River Delta region: Evidence from the maximum deviation method. *J. Clean. Prod.* **2018**, *177*, 207–217. [CrossRef]

30. Fang, K.; Zhang, Q.F.; Long, Y.; Yoshida, Y.; Sun, L.; Zhang, H.R.; Dou, Y.; Li, S. How can China achieve its Intended Nationally Determined Contributions by 2030? A multi-criteria allocation of China's carbon emission allowance. *Appl. Energy* **2019**, *241*, 380–389. [CrossRef]
31. Zhang, Y.J.; Wang, A.D.; Da, Y.B. Regional allocation of carbon emission quotas in China: Evidence from the Shapley value method. *Energy Policy* **2014**, *74*, 454–464. [CrossRef]
32. Han, R.; Tang, B.J.; Fan, J.L.; Liu, L.C.; Wei, Y.M. Integrated weighting approach to carbon emission quotas: An application case of Beijing-Tianjin-Hebei region. *J. Clean. Prod.* **2016**, *131*, 448–459. [CrossRef]
33. Qin, Q.D.; Liu, Y.; Li, X.; Li, H.A. A multi-criteria decision analysis model for carbon emission quota allocation in China's east coastal areas: Efficiency and equity. *J. Clean. Prod.* **2017**, *168*, 410–419. [CrossRef]
34. Du, J.; Pan, M.; Chen, Y.; Duan, Y.R. An efficiency-based allocation of carbon emissions allowance: A case study in China. *J. Clean. Prod.* **2020**, *251*, 119346. [CrossRef]
35. Phylipsen, G.; Bode, J.W.; Blok, K.; Merkus, H.; Metz, B. A Triptych sectoral approach to burden differentiation; GHG emissions in the European bubble. *Energy Policy* **1998**, *26*, 929–943. [CrossRef]
36. Zhang, N.; Wei, X. Dynamic total factor carbon emissions performance changes in the Chinese transportation industry—ScienceDirect. *Appl. Energy* **2015**, *146*, 409–420. [CrossRef]
37. Charnes, A.; Cooper, W.W.; Rhodes, E. Measuring the efficiency of decision making units. *Eur. J. Oper. Res.* **1978**, *2*, 429–444. [CrossRef]
38. Wu, F.; Fan, L.W.; Zhou, P.; Zhou, D.Q. Industrial energy efficiency with CO2 emissions in China: A nonparametric analysis. *Energy Policy* **2012**, *49*, 164–172. [CrossRef]
39. Lozano, S.; Villa, G.; Brännlund, R. Centralised reallocation of emission permits using DEA. *Eur. J. Oper. Res.* **2009**, *193*, 752–760. [CrossRef]
40. Ma, C.-Q.; Ren, Y.-S.; Zhang, Y.-J.; Sharp, B. The allocation of carbon emission quotas to five major power generation corporations in China. *J. Clean. Prod.* **2018**, *189*, 1–12. [CrossRef]
41. Yang, F.; Lee, H. An innovative provincial CO_2 emission quota allocation scheme for Chinese low-carbon transition. *Technol. Forecast. Soc. Change* **2022**, *182*, 121823. [CrossRef]
42. Gomes, E.G.; Lins, M. Modelling undesirable outputs with zero sum gains data envelopment analysis models. *Taylor Fr.* **2008**, *59*, 616–623. [CrossRef]
43. Chiu, Y.H.; Lin, J.C.; Su, W.N.; Liu, J.K. An Efficiency Evaluation of the EU's Allocation of Carbon Emission Allowances. *Energy Sources Part B-Econ. Plan. Policy* **2015**, *10*, 192–200. [CrossRef]
44. Xiong, S.; Tian, Y.; Ji, J.; Ma, X. Allocation of Energy Consumption among Provinces in China: A Weighted ZSG-DEA Model. *Sustainability* **2017**, *9*, 2115. [CrossRef]
45. Cai, W.G.; Ye, P.Y. A more scientific allocation scheme of carbon dioxide emissions allowances: The case from China. *J. Clean. Prod.* **2019**, *215*, 903–912. [CrossRef]
46. Yang, M.; Hou, Y.R.; Ji, Q.; Zhang, D.Y. Assessment and optimization of provincial CO_2 emission reduction scheme in China: An improved ZSG-DEA approach. *Energy Econ.* **2020**, *91*, 104931. [CrossRef]
47. Cui, X.Y.; Zhao, T.; Wang, J. Allocation of carbon emission quotas in China's provincial power sector based on entropy method and ZSG-DEA. *J. Clean. Prod.* **2021**, *284*, 124683. [CrossRef]
48. Song, J.; Chen, R.; Ma, X. Provincial Allocation of Energy Consumption, Air Pollutant and CO2 Emission Quotas in China: Based on a Weighted Environment ZSG-DEA Model. *Sustainability* **2022**, *14*, 2243. [CrossRef]
49. Zheng, L. Sharing the Carbon Emission Reduction Responsibility across Chinese Provinces: A Zero Sum Gains DEA Model. *Resour. Sci.* **2012**, *34*, 2087–2096.
50. Lin, T.; Ning, J.F. Study on allocation efficiency of carbon emission permit in EU ETS based on ZSG-DEA model. *J. Quant. Tech. Econ.* **2011**, *3*, 36–50.
51. Deng, J.L. Control Problems of Grey Systems. *Syst. Control Lett.* **1982**, *1*, 288–294.
52. Chang, A.-Y. Prioritising the types of manufacturing flexibility in an uncertain environment. *Int. J. Prod. Res.* **2012**, *50*, 2133–2149. [CrossRef]
53. Hamzacebi, C.; Es, H.A. Forecasting the annual electricity consumption of Turkey using an optimized grey model. *Energy* **2014**, *70*, 165–171. [CrossRef]
54. Abbas, S.; Yousaf, H.; Khan, S.; Rehman, M.Z.; Blueschke, D. Analysis and Projection of Transport Sector Demand for Energy and Carbon Emission: An Application of the Grey Model in Pakistan. *Mathematics* **2023**, *11*, 1443. [CrossRef]
55. Akay, D.; Atak, M. Grey prediction with rolling mechanism for electricity demand forecasting of Turkey. *Energy* **2007**, *32*, 1670–1675. [CrossRef]

Disclaimer/Publisher's Note: The statements, opinions and data contained in all publications are solely those of the individual author(s) and contributor(s) and not of MDPI and/or the editor(s). MDPI and/or the editor(s) disclaim responsibility for any injury to people or property resulting from any ideas, methods, instructions or products referred to in the content.

buildings

MDPI

Article

Demand-Side Management Method for Households with Self-Generation and Storage of Electricity

Vicente León-Martínez *, Clara Andrada-Monrós, Elisa Peñalvo-López and Juan Ángel Saiz-Jiménez

Department of Electrical Engineering, Universitat Politècnica de València, Camino de Vera 14, 46022 Valencia, Spain; claanmon@etsii.upv.es (C.A.-M.); elpealpe@upvnet.upv.es (E.P.-L.); jasaiz@die.upv.es (J.Á.S.-J.)
* Correspondence: vleon@die.upv.es; Tel.: +34-96-387-95-92

Abstract: The main objective is to propose a calculation method for assessing the benefits of individual domestic prosumers in self-consumption and economic savings when managing their own energy resources. The paper applies the demand-side management concept in the residential sector from the individual domestic perspective so that customers can understand the value of their own sustainable energy resources, conducting self-generation and demand management. The novelty lies in allowing the prosumer to manage their own energy resources to their benefit at a reasonable cost, instead of participating in automated large residential demand-side-management programmes that respond to the means of the grid system operator or other energy service companies, such as aggregators. A methodology for calculating the self-consumption rate and the economic benefit for the consumer is proposed, including three different cases: consumer demand is higher than self-generation, and consumer demand is equal to self-generation, and consumer demand is lower than self-generation. The methodology is validated with actual data from a household in Valencia (Spain) during a complete year, obtaining an average reduction in the annual electricity bill of 70% and a demand coverage with the self-renewable system reaching values of 80% throughout the year. The significance of this methodology goes beyond the economic revenue of the individual consumer; it also aims to guide consumers towards efficient practices in the use of their available energy resources and raise awareness on their energy behaviour.

Keywords: energy transition; active demand-side management; residential load management; renewable energy; load prioritisation

1. Introduction

According to the current energy model, cities are responsible for 75% of total primary energy consumption and 80% of greenhouse gases [1], mainly due to their concentrated populations, commercial and industrial activities, and transport systems. Urban areas often have a higher density of residential buildings, leading to increased energy consumption for heating, cooling, lighting, and appliances. Based on the Building Stock Observatory of the European Commission [2], there are approximately 196 million buildings within the Member States of the European Union, and most of them are residential [2]. Specifically, the floor area of the residential building stock accounts for approximately 75% of the total, while the remaining 25% is non-residential [2]. Regarding energy consumption, buildings represent 20 to 40% of the global final energy consumption in Europe [3]. In 2022, final energy consumption in European residential buildings represented 21% of the overall building stock, including the categories: Residential, Non-residential, Building construction industry, Other construction industry, and Other [4].

During the last year, the average household electricity prices in Europe were 28.9€ per 100 kWh, including all components in the electricity bill (the cost of electricity, power capacity, distribution, and taxes). The largest increase was in the Netherlands (953.2%), followed by Lithuania (87.8%) and Romania (77.3%) [5]. On other continents like Asia, Africa, or South America, the average cost of electricity is approximately 0.06 €/kWh, 0.1 €/kWh, and 0.11 €/kWh, respectively [6]. However, low electricity prices may represent the use of pollutant fossil fuels, which are very detrimental to the planet in the medium to long term.

These high electricity prices and preoccupations about electricity produced with polluting energy sources have persuaded residential consumers to invest in small renewable home systems for self-consumption and modify their energy consumption patterns, translating loads to low-price periods of time-of-use tariffs.

Assistance to the consumer has been provided through expensive Home Energy Management Systems, which have been introduced in recent years to facilitate automatic energy management in the home, reducing energy demand by shifting load from peak to off-peak hours [7], but requiring major investments. These systems have evolved to disaggregate domestic loads for better management and have integrated new loads existing at the residential level, such as electric vehicles [8]. However, these systems must be integrated into households, whose users must be able to afford the economic and technological resources involved. In this context, it is important to emphasise that the main motivation of individual domestic consumers is to reduce electricity bills by adapting to their own generation profile while maintaining comfort by modifying their energy behaviour and implementing simple demand-side management actions [9]. In this sense, it is important to understand the factors that drive behaviours and encourage actions to be taken to achieve greater efficiency [10]. Studies show that limited knowledge about energy management and the amount of energy used in their household affect motivation to act. Providing residential consumers with comprehensive information on the direct impact their actions have on reducing energy consumption, as well as other indirect actions (such as reducing showering time or using warm water for washing up) results in a reduction of energy consumption [11,12].

The aim of this work is to develop a methodology to empower domestic consumers in managing their own energy resources. The novelty lies in locating the domestic consumer at the centre of the energy transition paradigm [13,14], engaging them in first-hand management of their own energy resources, such as their home renewable generation system, modifying their energy behaviour (demand flexibility), or using a storage system.

It represents a procedure to actively involve domestic consumers in the energy transition for the building sector, minimising the greenhouse emissions, and enhancing the use of renewable energy sources. Alongside, this approach raises energy conservation awareness in society, organically motivating changes in lifestyle with a lower price than automated systems and favouring the energy transition in the residential sector.

2. State-of-the-Art

The origin of Electricity Demand Management dates back to the 1970s in the United States (USA) with the issuance of the Public Utility Regulatory Policies Act (PURPA), which was part of the Energy Policy and Conservation Act [15]. Later, in 1985, Gellings et al. [16] presented Demand-Side-Management (DSM) as the cornerstone of utility planning, implementing a series of demand response actions to influence the time and amount of energy used by customers to achieve desired changes in the utility's system load shape [17–21].

Demand Response (DR) Programmes for utilities [22,23] are classified into price-based demand response and incentives offered:

- Time-based DR programmes are based on offering consumers different tariffs for different time periods. In this context, consumers tend to transfer their manageable loads to periods with a lower price, resulting in a demand reduction at peak hours of the day, benefiting grid management. Consumers actively influence their load usage profile patterns according to the change in tariffs (Real-Rime or Time-of-use), but no direct equipment control occurs [24–27].

- Incentive-based DR programmes. In this programme, DR actions are agreed upon in advance between the consumer and the utility or aggregator. In this agreement, specific customer incentives and conditions for load management are defined (e.g., pre-notice time, energy packages to be managed, duration of the event, etc). DR events may result in load interruption or load translation to more convenient time periods for the grid, which may cause undesirable inconveniences for the final users. Some examples are Direct load control (DLC), interruptible/reduced rates, or Demand bidding programmes [28–30].

Over the past years, numerous research studies have been carried out to analyse the actual and future supply and demand energy patterns in the residential sector [31–36]. Research works focus on the characterisation of the energy profile of households (apartments and single-houses), consumers' preferences, and impact over the grid management [37–39] in order to facilitate utilities´ planning activities while benefitting the consumer with a reduction in their energy bills. These studies are mainly concerned with load shape, peak load, and quality of service provided by power companies [40–47], using Energy Management Systems capable of communicating with the Grid Management Operator to facilitate automated demand response actions. However, benefits for individual prosumers are secondary, not reaching their full potential from the individual prosumer perspective.

A revision of the literature reveals several projects in demand-side management [48–55], mainly focused on the benefits that automated load domestic management can offer the grid management system. Table 1 shows significant projects that have served as the basis for the development of demand-side management in the domestic sector around the globe. However, fewer references enhance the individual benefit of the final prosumer, understanding the economic and energy potential benefits from the user perspective.

Enhancing the economic and energy benefits of the consumer by managing their available energy resources (small domestic renewable energy system for self-supply, battery storage system, load, and grid management potential) is the main objective of this methodology. It evaluates two dispatch strategies: economic and self-consumption. The economic dispatch strategy increases the economic savings for the individual consumer (€) based on the balance between the energy packages bought from the retailer and the excess energy sold to the market, while the self-consumption dispatch strategy focuses on getting the most out of the available renewable generation and storage units, aiming for higher rates of self-consumption.

Both dispatch strategies are assessed in the three possible cases that may occur in an hourly-basis operation of the energy system, depending on the energy needs of the individual consumer and the availability of various energy resources. Case 1 represents the scenario when renewable and battery storage systems supply less than needed by the consumer. Case 2 analyses the situation in which the energy demanded by the consumer is similar to the renewable generation supplied by the system, considering the storage units. Finally, Case 3 corresponds to the scenario in which the renewable and storage systems supply more energy than needed by the consumer, therefore generating an energy surplus.

The work presented in this paper is organised into six sections. The Section 1 is a brief introduction to the current situation regarding consumers in residential buildings, while the Section 2 is devoted to a review of the literature related to energy consumption in residential buildings and the potential of demand-side management for domestic customers. Section 3 describes the method used for calculating the two dispatch strategies (economic and self-consumption) in the three cases. Then, Section 4 presents the results of applying the methodology in a case study located in Valencia (Spain), and Section 5 comments on the results of the case study. Lastly, Section 6 presents the conclusions.

Table 1. Previous projects on demand management in the residential sector.

Active DSM Projects	Country	Description
Energy-Smart Pricing Plan (ESPP) (2004) [56]	EEUU	It studies the response of residential consumers to hourly prices imposed by the electricity market, the type of measures taken by consumers, and the definition of the magnitude of the effect. However, no advice is given to the participants, so the savings achieved can only be visualised after the measures have been implemented.
Newmarket Hydro Time-of-Use Pricing Pilot (2008) [57]	Canada	It aims to provide consumers with information on how to manage electricity consumption by shifting the use of household appliances to off-peak periods. In this pilot, all participants are informed about the cost of energy in each period, a comparison between tariffs according to their historical data, a questionnaire on the use of household appliances, and some tips. It is a very comprehensive pilot that incorporates the use of the Electric Vehicle (EV) for charging as an additional device. However, this pilot does not evaluate the self-consumption of the domestic prosumer and their benefits.
GAD Project: Active Demand Management (2010) [58]	Spain	The aim of this project is to research and develop tools to help reduce consumers' electricity bills, and search for devices that inform the user about the price and origin of the energy. Consumers can adjust their consumption profile according to their preferences, thus reducing their electricity bills. Also, it can be integrated into homes, and considers the need to integrate devices to automate the management and information system. However, it does not address the possibility of introducing renewable energies as a source of generation to optimise the consumption curve, avoiding peaks in the electricity system, nor does it introduce the possibility of obtaining tariff incentives thanks to the use of renewable energies.
FLEXCoop (2021) [59]	FLEXCoop Consortium	This European project adapts demand to production but gives flexibility to consumers by allowing energy savings and using the flexibility margin obtained in the electricity market. It focuses on energy communities, so it considers other actors (Producers, Transmission System Operator, Prosumers, Aggregators, etc.) creating an energy market. This enhances the value of creating methodologies and tools to facilitate active demand management, as they would support the creation of energy communities.

3. Methodology

This section presents the methodology that analyses the benefits of the individual small domestic consumer when managing their own energy resources (self-generation, small battery storage system, load management, and grid), independently from nearby consumers or the grid system requirements. This allows small domestic consumers to better comprehend their energy patterns and generation/storage capabilities, so they can reduce their energy bill while increasing their energy awareness and understanding of the actual energy system.

The methodology involves several input parameters that are essential for performing calculations and providing results in accordance with the domestic user (Figure 1). The calculation method quantifies the revenue of small home customers considering their hourly available energy resources using two dispatch strategies: Economic and Self-consumption. In both cases, the small domestic consumer can buy energy packages from the retailer when necessary and sell the excess energy at the energy market price, in accordance with current Spanish legislation.

The first dispatch strategy focuses on enhancing the economic revenue of the customer, considering that the small consumer can use their own resources and may buy and sell energy packages from the retailer at their convenience. The second dispatch strategy enhances the self-consumption capacity of the small consumer, mitigating the exchange of energy with the grid, which acts as a backup when the small consumer cannot make use of its own generation, storage, and load management resources. The calculation for both dispatch strategies is conducted on an hourly basis during a complete year to assess the small consumer benefits.

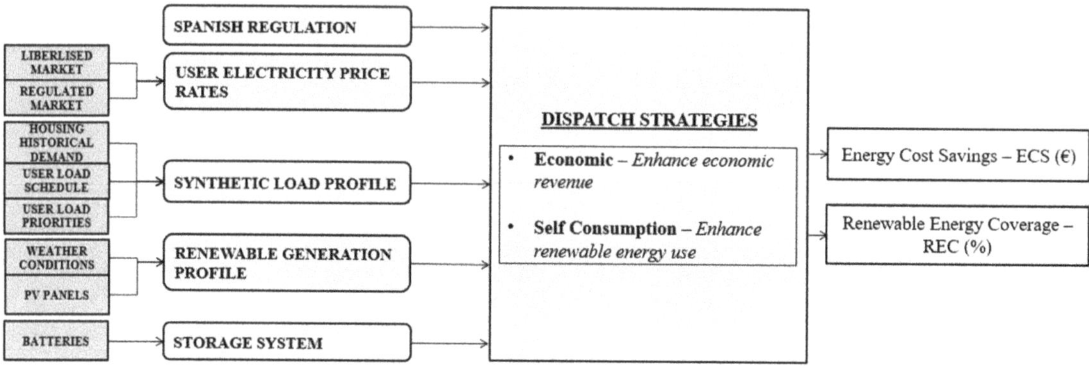

Figure 1. Methodology for energy resource management in small domestic consumers.

This assessment allows grid system operators and other electrical agents, such as aggregators, to compare the benefits obtained by the small domestic customer when managing their own energy resources with the support of the grid, with the benefits offered by utilities for participating in collaborative demand-side management programmes, so competitive and fair rewards for their participation in demand response actions can be offered.

The benefits for the small energy consumer are represented in two main parameters or outputs: Energy Cost Saving (ECS) and Renewable Energy Coverage (REC), calculated for both dispatch strategies to analyse the trade-off between the customer´s final economic revenue and energy independence.

3.1. Input Parameters

Input parameters represent the data and necessary information for executing the calculation method to obtain the desired outputs. Following the rationale in Figure 1, inputs are organized into five modules, which are Spanish regulation, User Electricity Price Rates, Synthetic Load Profile, Renewable Generation Profile, and Storage system. Table 2 indicates the means of data collection for the modules and some sub-modules.

Table 2. Input parameters for the methodology and means of data collection.

Input Parameters	Means of Collection
Spanish regulation	Information gathered from the Spanish national government, Ministry of Ecological Transition, Decree 244/2019 of April 5, which regulates the administrative, technical, and economic conditions of the self-consumption of electrical energy [60]
User Electricity Price Rates	Energy price set by the electricity market (in the case of Spain, the Iberian Energy Market Operator—OMIE) [61]
Housing Historical Demand	Data are collected by the smart metre, owned by the energy distribution company
User Load Schedule	Provided by the user
User Load priorities	Provided by the user
Renewable Generation Profile	Data specifications of the components
Weather conditions	Specifical data of the weather from official webpages
Storage system	Data specifications of the components

- Spanish regulation [60]: Provides information about the policies of the current legislation. At this moment in Spain, the actual self-consumption regulation allows small domestic consumers to adopt a compensation regime, acquiring and delivering energy packages from and to the grid based on the consumer demand needs.
- User Electricity Price Rates: Corresponds to the hourly electricity price paid by the small consumer for consuming electricity at a specific hour of the day. It is obtained from the Spanish System Operator (OMIE) [61], based on the actual energy contract of the consumer (regulated tariff or real-time energy market prices).
- Synthetic Load Profile: Represents the hourly electricity demand of the small domestic consumer. The Synthetic Residential Load Electricity profile is composed of a continuous and a discontinuous load profile. The input parameters are based on historical demand, the consumer's usage preferences associated with their household equipment, the time range of use, and the flexibility of use (whether they are willing to consume the energy package at other times of the day).
- Renewable Generation Profile: This input provides an hourly data vector with 8760-values of the annual electricity generated with the home renewable system. Renewable generation depends on the solar radiation of the location and the characteristics of the photovoltaic solar installation.
    - Weather Conditions: Temperature and solar radiation parameters are used to calculate the power generator with photovoltaic panels.
    - Characteristics of the PV installation: Technical specifications of the whole photovoltaic system (number of panels, inverter, wiring, etc.).
- Storage system: Represents the technical characteristics of the battery storage system and main parameters such as the state of charge, depth of charge, and storage capacity.

3.2. Dispatch Strategies

This section presents the two dispatch strategies defined in the methodology, namely Self-consumption and Economic. The self-consumption dispatch strategy focuses on reducing dependency on the grid, consuming the energy surplus generated by the home renewable energy system in secondary loads initially identified by the domestic consumer. The Economic dispatch strategy enhances the revenue of the consumer, transferring low and medium priority loads to less costly time periods and selling the electrical energy surplus into the power market. Both dispatch strategies are simulated in the three possible cases (demand is greater than self-generation, demand is equal to self-generation, and demand is smaller than self-generation) on an hourly basis during a complete year (Table 3).

Table 3. Cases analysed in the hourly calculation method.

Case	Condition	Description
1	Generation < Demand	Value of Renewable Generation in hour *i* is lower than the value of Demand in hour *i*. Instantly, only part of the demand is supplied; therefore, it is analysed whether or not there is energy available in the storage system to supply the demand completely. • Case 1a—Self-Consumption Dispatch Strategy: Actions are taken to disaggregate demand and transfer it to hours with available generation or battery. • Case 1b—Economic Dispatch Strategy: In this case, the consumer carries out a greater effort, including, in addition to the previous actions, the transfer of unsupplied demand to more economical periods.
2	Generation ≊ Demand	The value of Renewable Generation in hour *i* is equal to the customer´s demand for the hour under study; the demand may be fully satisfied, and there will be no energy surplus.

Table 3. *Cont.*

Case	Condition	Description
3	Generation > Demand	If the value of Renewable Generation in hour *i* is greater than the value for that hour, the consumer's demand can be fully satisfied, and there will be a surplus of energy. • Case 3a—Self-Consumption Dispatch Strategy: The consumer will be able to recharge the batteries, supply secondary consumption to make greater use of the energy generated, and finally, inject surplus energy into the grid. • Case 3b—Economic Dispatch Strategy: In this case, the actions of the previous case are carried out except for supplying secondary consumption. The aim is to inject the surplus completely into the grid.

The different possible cases according to the generation and demand conditions are shown in the following high-level decision-making diagram (Figure 2), which represents the decision path followed by each of the three possible cases. The decision-making path for Case 2 does not imply any action since the demand profile (D0) for that hour matches that of generation; therefore, it is the same for both dispatch strategies. In Case 1 and 3, the decision-making process is different in each dispatch strategy, so it is developed for each dispatch strategy.

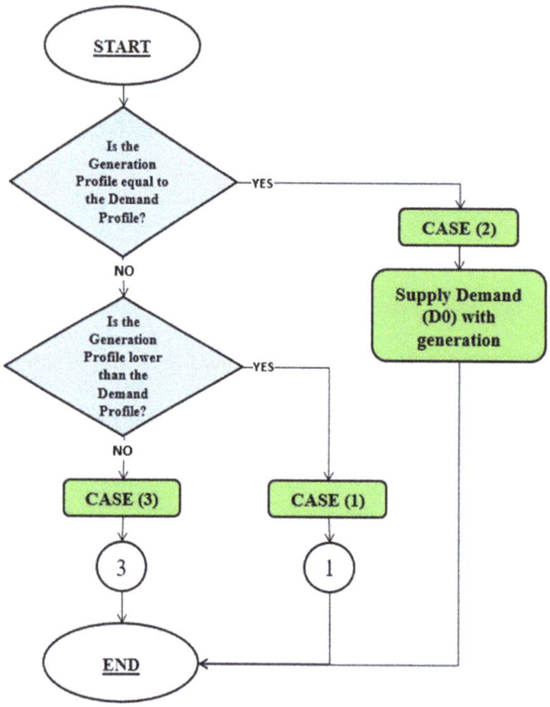

Figure 2. Diagram of the three cases under study.

The decision-making diagrams belonging to the Self-Consumption Dispatch Strategy will be named with the letter "a" (i.e., Case 1a and Case 3a), and those belonging to the Economic Dispatch Strategy with the letter "b" (Case 1b and Case 3b) for ease of understanding.

3.2.1. Self-Consumption Dispatch Strategy

In this strategy, the decision-making process focuses on enhancing self-consumption of the small home renewable system, using an energy storage system to satisfy consumers' peak demand, and supplying energy to secondary loads in case of an energy surplus. It should be noted that the energy resources are variable and depend on the small home renewable equipment (i.e., power of PV panels used, whether batteries are used, etc.), the load priorities set by the domestic consumer, and the number of secondary demand alternatives.

Figure 3 represents the decision-making path for Case 1a in the self-consumption dispatch strategy, for which home renewable generation is less than the consumer demand at a specific hour. As observed in Figure 2 above, if there is a deficit of energy to satisfy the demand, then the energy needed is extracted from the battery storage system, if available, to fulfil the energy requirements of the consumer, according to the diagram in Figure 3. When the energy supplied by the battery storage system is still not enough to satisfy all the demand, it is possible to disaggregate consumption according to the priorities given by the user to make maximum use of the battery. Afterwards, the decision-making process proceeds to transfer low-priority loads to other time-periods according to the user's load schedule preferences. If the energy deficit to satisfy the demand remains after load management, the necessary energy is purchased from the grid.

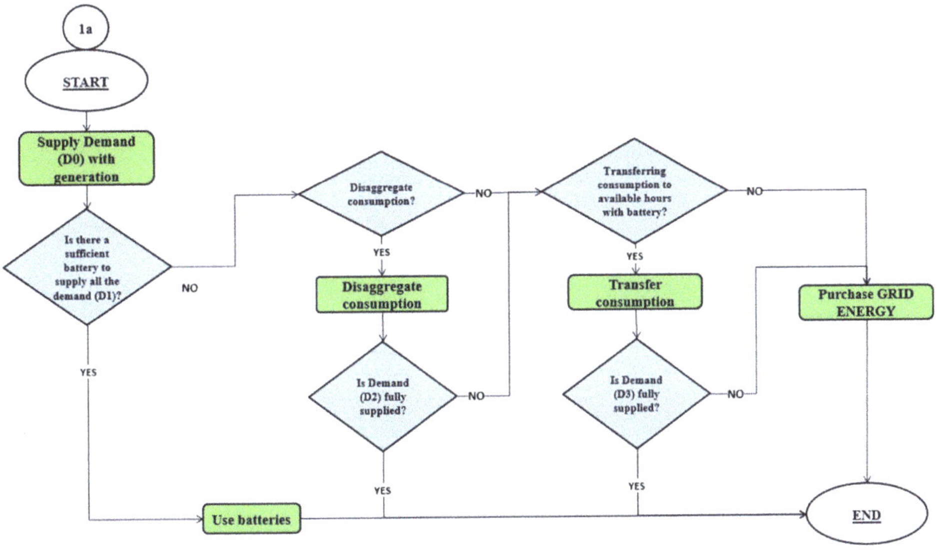

Figure 3. Decision-making diagram for Case 1 in the Self-consumption Dispatch Strategy (a).

Case 3a represents the event in which the energy generation is higher than the energy demand (Figure 4). On this occasion, there is an energy surplus after supplying the energy demand, which is sent to the batteries to be charged. If renewable sources still generate more energy than required by the demand, the secondary consumptions defined by the user are satisfied, and finally, energy is delivered into the grid with economic retribution. Secondary consumption offers greater use of the energy generated, always at the hours when surplus energy is produced. Thus, they allow the user to be more aware of the use and management of their generated energy, naming this dispatch strategy as self-consumption.

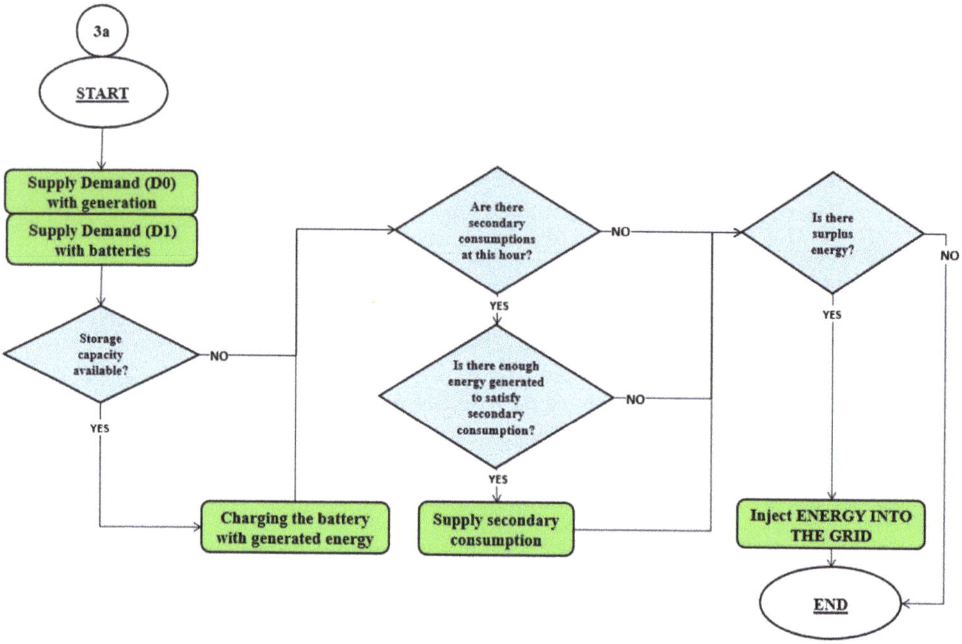

Figure 4. Decision-making diagram for Case 3 in the Self-consumption Dispatch Strategy (a).

3.2.2. Economic Dispatch Strategy

The Economic Strategy also considers the home renewable energy system for self-supply in conjunction with the battery storage system. The main difference from the previous self-consumption strategy is that it prioritises the management of low and medium priority loads during lower-priced timeslots, then, delivers the excess of energy to the grid to obtain economic retribution of the energy sold, instead of satisfying secondary demands of the consumer.

The specific decision-making process for Case 1b, as shown in Figure 5, represents the case in which renewable energy generation is lower than the consumer´s demand. In this event, the demand that is not satisfied by the home renewable energy system is supplied by the battery storage system in the first instance, and if this is not possible, by disaggregating consumption according to the priorities given by the user. In case additional energy is needed to satisfy the demand, loads are shifted to available battery hours (given by the user), and if the demand is fully satisfied, loads are shifted to other time periods with lower prices. In the occasion that further energy is required to satisfy the demand, the remaining energy is purchased from the grid.

Finally, the decision-making diagram for Case 3b of the Economic Dispatch Strategy is provided in Figure 6, corresponding to the case when home renewable generation is higher than the customer's demand for a given hour. In this situation, after satisfying the consumer´s energy demand, the energy surplus is stored in the battery storage system. If the batteries are fully charged, the excess of energy is delivered to the grid with corresponding economic compensation.

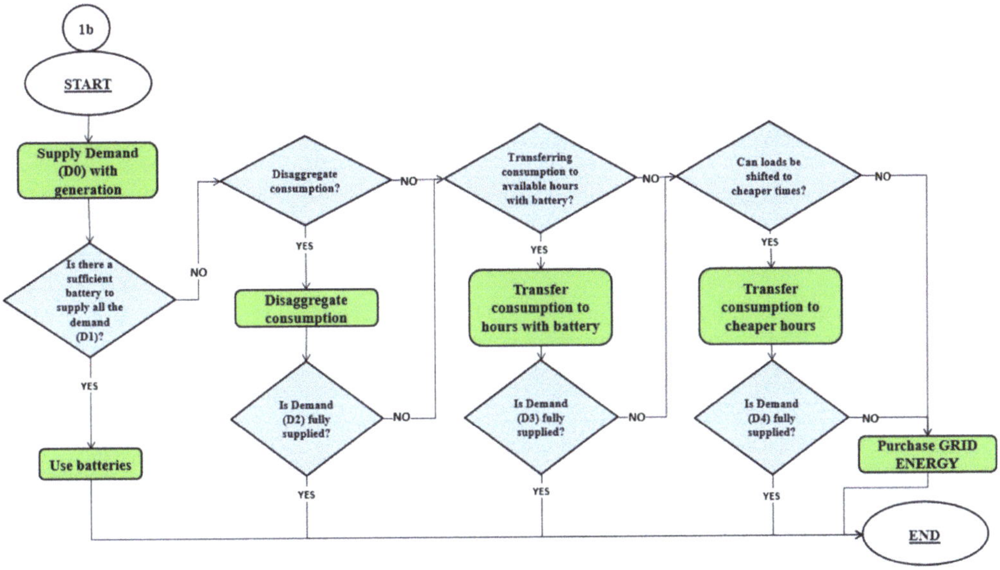

Figure 5. Decision-making diagram for Case 1 in the Economic Dispatch Strategy (b).

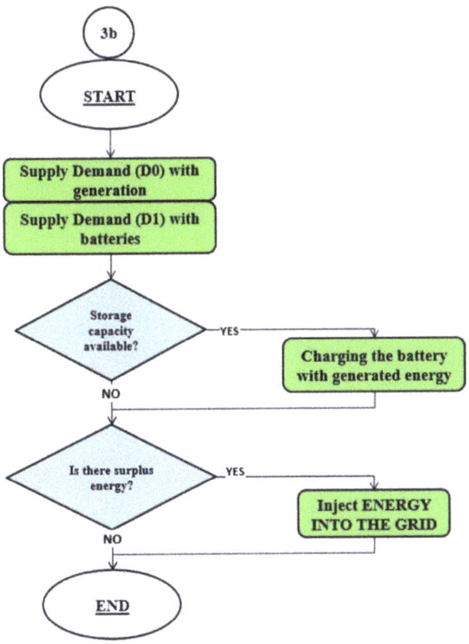

Figure 6. Decision-making diagram for Case 3 in the Economic Dispatch Strategy (b).

3.3. Output Parameters

The output parameters are defined to compare the results in both dispatch strategies (self-consumption and economic) to evaluate the degree of energy independence achieved by the consumer and the consumer´s economic revenue when conducting load management actions to low price timeslots. These would be the main parameters to be provided to the consumers to understand the results of their actions, following the trend of other papers [26,28,29,54,55]. In addition, several complementary parameters have also been defined to provide additional information to the consumer, as can be seen in Table 4.

Table 4. Output parameters for the methodology and units.

Output Parameters	Units
Energy Cost Savings—ECS	€
Renewable Energy Coverage—REC	%
Cost of the total electricity purchased—C_f	€
Revenue of the total electricity sold to the market—$E_{d_i} P_{s_i}$	€
Total electricity purchased from the grid—E_{fc_i}	kWh
Total electricity delivered to the grid—E_{d_i}	kWh

The main two parameters are:

- Energy Cost Savings—ECS (€): Cost savings in the electrical energy purchased from the grid. It compares the cost of purchasing all the customer´s energy needs from the grid with the energy purchased from the grid when the prosumer manages their energy resources, including the home renewable energy system, battery storage system, and load management resource.
- Renewable Energy Coverage—REC (%): It represents the degree of energy independency of the customer from the grid. It represents the energy coverage of the customer's electricity demand with the energy produced by the renewable energy.

Other complementary parameters include:

- Cost of the total electricity purchased C_f, (€)
- Revenue of the total electricity sold to the market $E_{d_i} P_{s_i}$ (€)
- Total electricity purchased from the grid E_{fc_i} (kWh)
- Total electricity delivered to the grid E_{d_i} (kWh)

3.4. Output Calculation Methods

3.4.1. Calculation of Energy Cost Savings (ECS)

Equation (1) defines the calculation of the economic savings parameter (ECS) as the difference between the cost of the energy initially consumed C_0 and the cost of the energy finally consumed C_f after applying the methodology. Equation (1) is divided into Equations (2)–(6) for ease of understanding.

$$ECS\ (€) = C_0 - C_f \quad (1)$$

The calculation of the cost of the energy initially demanded between hour i and the defined period t is made in Equation (2) as the Synthetic Load Profile E_{SLP} for each hour i multiplied by the energy price P_{e_i} set by the market or the trader in hour i. This is the energy demand calculated based on the previous demand historical data without applying the methodology proposed in this paper. The formulation of this parameter is further explained in Equation (12).

$$C_o(€) = \sum_i^t E_{SLP_i} \cdot P_{e_i} \qquad \forall\, i \in [0,\ t] \quad (2)$$

The final costs C_f of the final energy consumed in a period t from hour i after applying the methodology are calculated as the difference between the final energy consumed from the grid E_{fc_i} in each hour i times the energy price P_{e_i} for hour i, minus the surplus energy E_{d_i} that has been injected into the grid in hour i times the price P_{s_i} at which the surplus energy is paid in each hour i. This is presented in Equation (3), the terms of which are found broken down in Equations (4)–(6).

$$C_f\ (€) = \sum_{i}^{t}\left[\left[E_{fc_i}P_{e_i}\right] - \left[E_{d_i}P_{s_i}\right]\right] \qquad \forall\, i \in [0,\ t] \tag{3}$$

Regarding the final energy consumed E_{fc_i} after applying the methodology, it is calculated in Equation (4) as the balance established between Synthetic Load Profile E_{SLP} for each hour i, minus the energy self-consumed E_{sc_i} in that hour. The self-consumed energy term E_{sc_i} is broken down in Equation (5) as the sum of the energy produced from the solar panels E_{pv_i} and that coming from the battery E_{bat_i}.

$$E_{fc_i}(\text{kWh}) = E_{SLP_i} - E_{sc_i} \tag{4}$$

$$E_{sc_i}\ (\text{kWh}) = E_{pv_i} + E_{bat_i} \tag{5}$$

Load transfer conditions:

- They are moved to times set by the consumer when the battery is available.
- In the Economic Dispatch Strategy, the condition that they are transferred to the most economical hour (in the off-peak period) is added.

Finally, the energy delivered to the grid E_{d_i} is calculated in Equation (6) and will be the difference between the energy generated and not consumed by the PV panels $E_{g_{pv_i}}$ in hour i, and the sum of the PV energy consumed E_{pv_i}, the battery energy consumed E_{bat_i}, and, in the case of the Self-Consumption Dispatch Strategy, the energy consumed by the secondary consumptions E_{2con_i}.

$$E_{d_i}\ (\text{kWh}) = E_{g_{pv_i}} - E_{pv_i} - E_{bat_i} - E_{2con_i} \tag{6}$$

3.4.2. Calculation of Renewable Energy Coverage (REC)

The coverage of the renewable energy used to supply demand (REC) is calculated in Equation (7) as the summation in a period t for each hour i of the total self-consumed demand D_{sc_i} over the total demand supplied D_{tot_i} during that period t. In this indicator, energy should be considered from the consumer's point of view, i.e., as the energy that the consumer is demanding to cover the necessary loads in that hour.

$$REC(\%) = \sum_{i}^{t} \frac{D_{sc_i}}{D_{tot_i}} \cdot 100 \qquad \forall\, i \in [0,\ t] \tag{7}$$

The self-consumed demand D_{sc_i} in each hour i will be the sum of the demand supplied by the energy provided by the PV panels E_{pv_i}, by the batteries E_{bat_i}, and by the energy consumed by secondary consumption E_{2con_i}, as represented in Equation (8).

$$D_{sc_i}(\text{kWh}) = E_{pv_i} + E_{bat_i} + E_{2con_i} \tag{8}$$

Initially, Equation (9) defines the terms that compose the total demand D_{tot_i} for an instant I after applying the methodology. The total demand D_{tot_i} for an hour i is calculated as the final demand consumed D_{f_i} and the demand contributed to the grid E_{d_i}. The final demand consumed D_{f_i} is the sum of the energy consumed from the grid D_{fc_i}, the self-consumed demand D_{sc_i}, and the demand transferred D_{trans_i} from other periods to this instant i, as shown in Equation (10). Regarding the transferred appliances D_{trans_i}, according to Equation (11), they will be considered positive when they enter at hour i and negative when they cannot be supplied at hour i and enter at another cheaper hour specified by the user.

$$D_{tot_i}(\text{kWh}) = D_{f_i} + E_{d_i} \tag{9}$$

$$D_{f_i}\ (\text{kWh}) = D_{fc_i} + D_{sc_i} + D_{trans_i} \tag{10}$$

$$D_{trans_i}\ (\text{kWh}) = D_{t-i\ i} - D_{t-o\ i} \tag{11}$$

As mentioned, the secondary consumptions only apply to the self-consumption dispatch strategy. It is used whenever, after having supplied all the demand and transferred loads, there is surplus energy. In this case, those consumptions that the user has considered as secondary can be supplied with the energy surplus for that hour.

3.4.3. Supplementary Terms

Firstly, the method for calculating the Synthetic Load Profile E_{SLPC_i} is defined, which is made up of a continuous E_{SLPC_i} and a discontinuous profile $E_{SLP_{Di}}$.

$$E_{SLP_i}(\text{kWh}) = E_{SLP_{C_i}} + E_{SLP_{Di}} \tag{12}$$

On the one hand, the continuous load profile is a stochastic model that is created from a load profile based on historical demand E_{hd_i} and certain random synthetic perturbation α. The random synthetic perturbation is composed of two terms: the daily perturbation α_d and the instantaneous perturbation α_I. The daily perturbation maintains the same profile but scales up or down according to maximum variability, while the instantaneous perturbation changes the shape of the load profile according to an assigned variability, but without affecting its size. Equation (13) defines the synthetic random perturbation.

$$\alpha = 1 + \alpha_d + \alpha_I \tag{13}$$

In this paper, the model used randomly draws the value of the daily perturbation once a day from a normal distribution with a mean of zero and a standard deviation equal to the daily variability input, and randomly takes the value of the instantaneous perturbation from a normal distribution with a mean of zero and a standard deviation equal to the instantaneous variability input value. Thus, the continuous synthetic load profile is formed from the historical demand E_{hd_i} multiplied by the synthetic perturbation α_i, as shown in Equation (14).

$$E_{SLP_{C_i}}(\text{kWh}) = E_{hd_i}(\text{kWh})\cdot\alpha_i = E_{hd_i}(1 + \alpha_d + \alpha_I) \tag{14}$$

On the other hand, the discontinuous load profile refers to the mobile loads entered per user that depend on the day and the conditions given by the generation profile (Equation (15)). This load profile considers the user's load schedule and load priorities. The user appliances schedule consists of data associated with the time of use of the appliances and the flexibility for their use, providing other time slots in which the appliances can be used. In the user load priorities, the consumer provides data associated with the priorities in the interruption and transferability of loads of household equipment based on three levels of priority: high, medium, and low. High priority means that these loads cannot be transferred to other time slots, while low priority is considered as loads that can be used in some other time slot defined by the user in the User Load Scheduling. Finally, the load transfer term E_{t_i} is calculated as in Equation (11) as the balance between the loads entering E_{t-i_i} for hour i and the loads leaving E_{t-o_i} because they cannot be supplied in that hour.

$$E_{SLP_{D_i}}(\text{kWh}) = \sum E_{jtrans,i} \tag{15}$$

The solar energy produced by the panels E_{gpv_i} for this model is calculated according to Equation (16), considering the irradiation Gh for the area in which the house is located, being proportional to the area S of the panels and the efficiency Ef of the PV modules.

$$E_{gpv_i}(\text{kWh}) = \frac{Gh\left[\frac{\text{Wh}}{\text{m}^2}\right]\cdot S\ (\text{m}^2)\cdot Ef\ (\%)}{1000} \tag{16}$$

Another important term is the State of Charge (SoC) of the battery at each hour i. This term depends on the type of installation, since it does not operate in a similar way if it is a home without self-consumption, with self-consumption without surpluses, or with surpluses. In the case of self-consumption without surplus, the PV modules must not generate energy that cannot be stored by the batteries. However, the equations presented enable the calculation of the state of the battery considering the most complicated case, for a self-consumption installation with surplus compensation. The initial state of the battery E_{ibat} for an hour i will depend on the final state of the battery for the previous hour $E_{fbat_{i-1}}$ and the so-called demand D_o. This demand is the difference between the energy generated by the modules E_{gpv_i} and the historical demand E_{fd_i} for that hour.

$$E_{ibat_i} = E_{fbat_{i-1}} + E_{gpv_i} - E_{SDP_i} = E_{fbat_{i-1}} + D_o \tag{17}$$

The State of Charge (SoC) of the battery is defined in Equations (18) and (19). To determine it, firstly it is checked if the state of charge of the battery for the previous hour plus Demand D_o is higher than the maximum battery E_{bat_max}, which refers to the total capacity of the battery in terms of energy. If the upper limit is exceeded, the battery will adopt the maximum battery value, and the surplus will be used according to the methodology. On the contrary, if the limit is not exceeded, it will proceed according to Equation (19).

$$SoC \begin{cases} E_{bat\_max}, & E_{fbat_{i-1}} + D_o > E_{bat\_max} \\ & E_{fbat_{i-1}} + D_o < E_{bat\_max} \end{cases} \tag{18}$$

The Depth of Discharge (DoD) term appears in Equation (19), which indicates the maximum value of the usable capacity of the battery so that its performance is not impaired. Therefore, the battery will adopt the minimum energy value E_{bat_min} in case the difference between the previous final state of the battery $E_{fbat_{i-1}}$ and the Demand D_o is less than the DoD. Therefore, this will be considered when supplying demand D_1.

$$SoC \begin{cases} E_{fbat_{i-1}} - D_o, & E_{fbat_{i-1}} + D_o > DoD \\ E_{bat\_min}, & E_{fbat_{i-1}} + D_o < DoD \end{cases} \tag{19}$$

Finally, it is important to define the priority dispatch strategy used in the batteries for the rest of the demand (other than D_o). The final state of the battery E_{fbat_i} (kWh) for each hour i depends on the use of different strategies to disaggregate consumption, transfer it to the most economical hour, or use it for secondary consumption. Therefore, the final energy of the battery E_{fbat_i} depends on the initial energy available E_{ibat_i}, and the energy remaining after disaggregating $Bat_{dg_{i,j}}$ and transferring $Bat_{lt_{i,j}}$.

$$E_{fbat_i} = E_{ibat_i} - Bat_{dg_{i,j}} - Bat_{lt_{i,j}} \tag{20}$$

The disaggregate and load transfer terms found in Equation (18) are described below:

- Disaggregate (Bat_{dg_i}): This strategy is common to both Dispatch Strategies (self-consumption and economic). It states that if it is not possible to supply all the demand for an hour i, the battery will supply those devices j in order according to the priority k given by the user, until the maximum available battery is used.

$$Bat_{dg_{i,j}}(i, j, k) \qquad \forall j \in i \cap k$$

- Load transfer (Bat_{lt_i}): This action is also applicable to both Dispatch Strategies and is implemented after the disaggregation one. If devices have not been supplied due to lack of battery power, they can be moved to hours when battery power is available. This is conducted based on the user´s load priorities.

$$Bat_{lt_{i,j}}\left(i, j, k, E_{fbat}\right) \qquad \forall j \in i \cap k \cap E_{fbat}$$

3.5. Case Study

To evaluate the methodology, an application case is carried out in a single-family house with self-consumption and batteries for energy storage located in a municipality of Valencia (Spain). The study analyses the use of different energy resources available in the home to benefit the individual consumer with and without the battery storage system.

The house has PV modules on the roof to generate electricity, as well as a lead-acid battery storage system. The installation was designed with a battery storage system, whose maximum capacity is 150 Ah at a voltage of 12 V. The equipment found in the house is a lead-acid battery with a depth of charge of 40% to extend its useful life. The study analyses the benefits of managing the energy resources in the house for self-consumption and economic revenue of the individual prosumer. According to its location (Valencia, Spain), the annual values of solar energy generation with the annual energy demand of the house are shown in Figure 7.

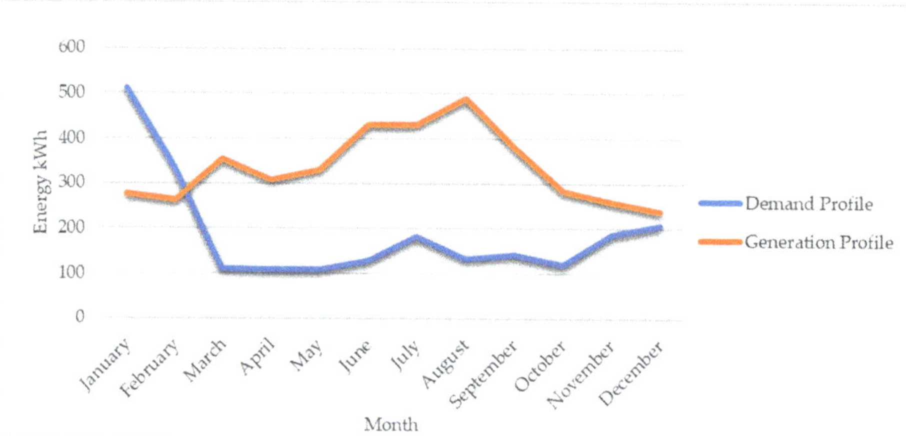

Figure 7. Annual PV Panels generation and household energy demand.

In addition to the information on available energy resources, the consumer establishes their consumption priorities for main home equipment, introducing the hours of use of the appliances and the flexibility range of their use, allowing load management (e.g., shifting consumption from peak to valley).

Figure 8 shows the devices for which the methodology will be applied, the times of use throughout each day, the priorities assigned to each device (high, medium, and low) and the possibility of transferring them to other slots. This table is created for each typical day, and in this case study, summer and winter working days and bank holidays have been established. In addition, secondary devices considered by the user are also included. In this case, they are those intended to activate watering and pool cleaning devices, as they are not strictly necessary, but if they can be used, they will still have a good use and, in addition, will provide greater coverage of the energy generated.

Another input value for the implementation of the methodology is the price of the energy term in addition to the price of the surplus energy that is injected into the grid. Figure 9 provides the values taken in this case study, obtained for the year of study from the Iberian Energy Market Operator.

Devices	Power (kW)	Timetable		Timetable 2		Timetable 3		Time of use (h)	Priority		Interruptibility	Transfer?	Timetable for transferring	
		From	To	From	To	From	To						From	To
Fridge	0.6	0	24					1	1	High	No	No		
Ceramic hob	2.2	14	15	21	22			1	2	High	No	No		
Electric oven	1.8	14	15					1	3	High	No	No		
Washing machine	1	19	20					1	4	Medium	Yes	Yes	16	17
Electric heater	4	7	8	20	21			0.5	5	Medium	Yes	Yes	18	19
Dryer	1.3	21	22					1	6	Medium	No	No		
Iron	1.1	22	23					1	7	Medium	Yes	Yes	7	8
Dishwasher	1.1	15	16					1	8	Medium	Yes	No		
Laptop computer	0.015	20	22					1	9	Low	Yes	Yes	22	23
Television	0.15	15	17	21	24			1	10	Low	Yes	Yes	14	15
Music equipment	0.8	20	21					1	11	Low	Yes	Yes	16	17
Microwave	1.3	8	9	14	15	21	22	1	12	Low	Yes	Yes	22	23

Secondary consumptions									Self-consumption strategy
Devices	Power (kW)	Timetable		Timetable 2		Timetable 3		Time of use (h)	ON/OFF operation
		From	To	From	To	From	To		
Irrigation pump	1	5	7	20	21			1	ON
Purification pump	1	15	17					1	ON
Cleaners	1	1	2					1	ON

Figure 8. Data entered by the user and load priorities.

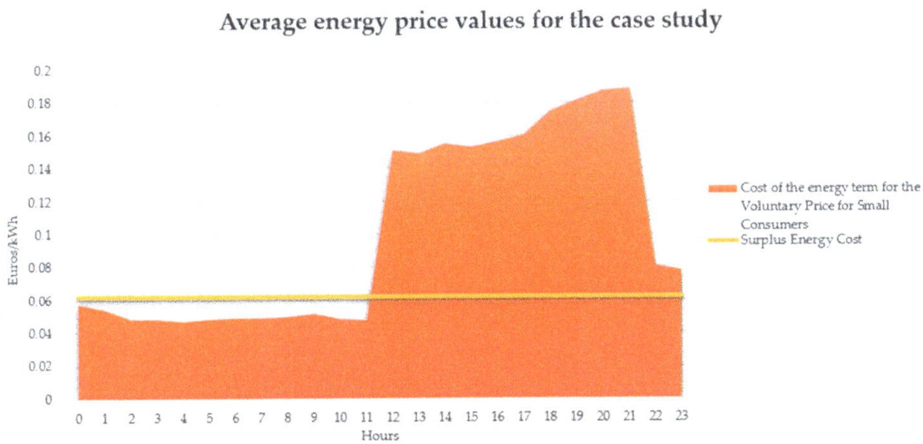

Figure 9. Average energy prices for the case study according to the Energy Market Operator.

Details of the energy resources are summarised in Table 5:

Table 5. Case study: specific input parameters.

Input Parameter	Case Study Data
Type of installation	Self-consumption with surpluses with compensation. Contracted power: P1 = 5.5 kW P2 = 5.5 kW
Tariff group	2.0TD for small consumers
Electrical Market prices	Iberian Energy Market Operator—OMIE
Synthetic Load Profile (SLP)	To define the synthetic load profile, a synthetic perturbation α of up to 1.35 has been considered. The daily α_d and instantaneous α_I perturbation values have been randomly assigned with a maximum value of 20% and 15%, respectively, according to historical residential demand.

Table 5. *Cont.*

Input Parameter	Case Study Data
Generation Profile (PG)	Photovoltaic installation: Peak Power: 1.8 kWp, Efficiency 18%. S = 10 m^2
Weather conditions	Irradiance history for location (39.537–0.632)
Storage Technologies	Maximum capacity: 150 Ah Depth of discharge: 40%
User preferences	Number of available devices, power, weekly hours of use for working days and holidays, usage priorities (high, medium, and low), and flexibility of use for shifting to other times entered by the user.

To illustrate the input data provided and its evolution after applying the methodology, the following figures provide, for a summer working day (Figure 10) and for a winter holiday (Figure 11), the initial and final demand, the battery status, and the energy generation with the solar panels throughout the day.

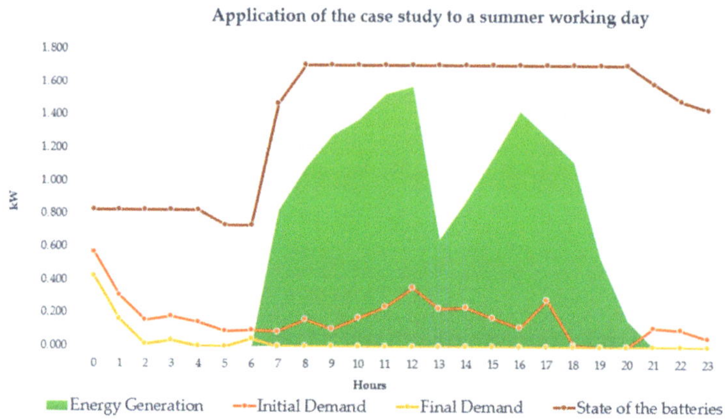

Figure 10. Case study for a summer working day.

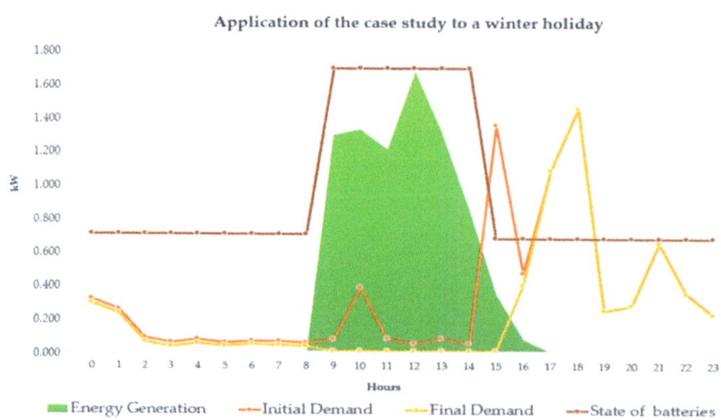

Figure 11. Case study for a winter holiday.

4. Results

The results obtained after applying the methodology in a single house in Valencia (Spain) are shown below. The results are presented with and without batteries for both dispatch strategies. This is intended to provide information on the benefits of applying the two dispatch strategies and the versatility that different energy resources provide for satisfying the demand.

Calculations have been made on an annual basis for each of the outputs defined in the methodology, and monthly graphs have also been provided for the two main outputs: ECS and REC, i.e., economic cost savings and renewable energy coverage of demand. The month selected was July, as it is a summer month with favourable weather conditions for the study of the methodology.

4.1. Case Study without Batteries

This section provides the results of implementing the methodology without the use of battery energy storage. The methodology with the self-consumption and economic dispatch strategies is applied. In this case, it is important to note that since energy cannot be stored in batteries, it is instantaneously used or delivered into the grid.

First, the annual results are shown to illustrate the scope of the methodology. For this purpose, the following table shows the annual values for each of the outputs defined in Section 3.3.

- Energy Cost Savings (ECS) (€): Average economic savings of about 20€ per month are achieved. The economic dispatch strategy presents slightly higher economic savings than self-consumption since it prioritises the transferability of loads to cheaper hours and the sale of energy excess.
- Renewable Energy Coverage (REC) (%): The annual percentage of demand covered by renewable energies is higher than 30% for both criteria, but slightly higher for the self-consumption dispatch strategy since, by introducing secondary consumption, the demand supplied with generation is higher.
- Cost of the total electricity purchased (€): The total energy purchase is about 18€ per month on average, a little higher in the self-consumption strategy because the loads are not shifted to more economical periods.
- Revenue of the total electricity sold to the market (€): This term refers to revenues obtained from the sale of energy. For the economic dispatch strategy, the revenue is equivalent to about 15€ per month. It is slightly lower for the self-consumption strategy. The main difference between the two strategies is that the economic strategy injects all the surplus into the grid, while the self-consumption strategy previously supplies secondary consumption.
- Total electricity purchased from the grid (kWh): This term is provided in terms of energy to be aware of the energy consumed. As can be seen, for both strategies, it is the same value because the demand supplied in terms of energy with respect to the base demand is the same (does not consider secondary consumption).
- Total electricity delivered to the grid (kWh). The energy injected into the grid is superior for the economic strategy since the surplus is not used to supply secondary consumption.

The monthly results obtained in July for the ECS and REC outputs are presented in Figures 12 and 13 below. Monthly results are calculated as the summatory of each hourly electricity value per day of the month represented.

On the one hand, Figure 12 shows that the values of the economic dispatch strategy are higher in terms of energy cost savings (ECS). The use of strategies, such as shifting to more economical periods or grid feeding, is key in this strategy. For this month, a maximum daily saving of up to 1.40€ is obtained.

On the other hand, Figure 13 shows that the demand coverage values for renewables in the self-consumption dispatch strategy are higher or equal to those of the economic dispatch strategy. This is the objective of the self-consumption strategy, to supply a greater part of the demand and use a higher percentage of renewable generation. Specifically, for day 8, demand coverage with renewables is more than 75%.

In relation to day 8, the economic savings are not remarkably high even though the energy used to cover the demand was one of the highest of the month. This is because energy generation was very limited (e.g., cloudy day), so almost all generated energy was used, but no great economic savings were achieved. Another interesting case could be the 29th, where almost 50% of the energy generated is used, and the savings obtained are the highest of the month. Therefore, on this day, it can be said that the energy generated was conveniently managed to cover the demand, as the saving reached a value of 1.40€.

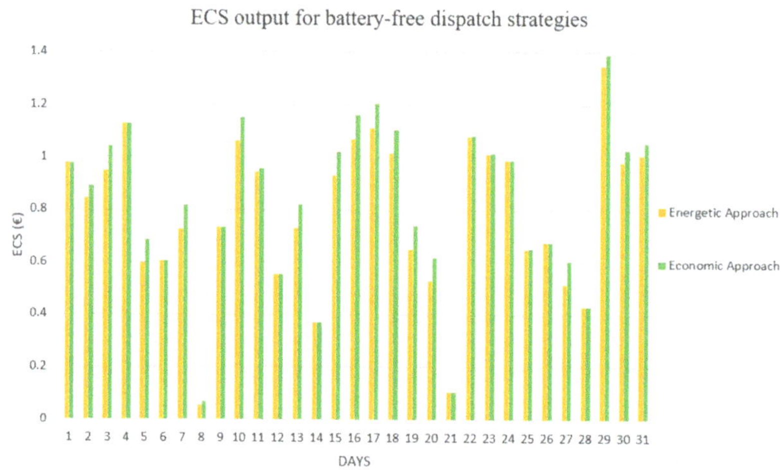

Figure 12. ECS Output (economic savings) for the Self-Consumption and Economic Dispatch Strategies in a summer month for the case without batteries.

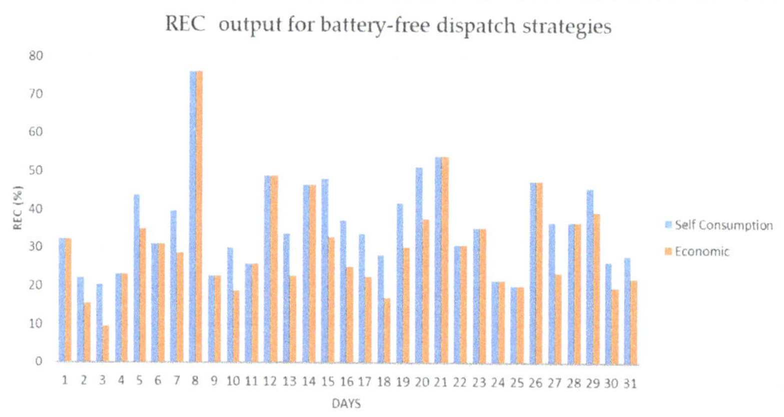

Figure 13. REC Output (use of renewable energy generated) for the Self Consumption and Economic Dispatch Strategies in a summer month for the case without batteries.

In annual terms, the energy generated to supply demand in the case without batteries reaches a value of 36.9% for the energy dispatch strategy and 31.8% for the economic dispatch strategy. Furthermore, this represents savings in energy purchases of 42.65% and 45.02%, respectively.

4.2. Case Study with Batteries

This section provides the results of the case study with the use of batteries for energy storage. As in the case without batteries, annual data are provided for each of the outputs defined in the methodology, as well as monthly graphs of the main outputs. Both cases are compared to analyse the versatility of the battery.

- Energy Cost Savings (ECS) (€): In this case, economic savings are equivalent to about 24€ per month on average, i.e., about 4€ per month more than in the case without a battery. As in the previous case, higher values are obtained in the economic dispatch strategy than in the self-consumption dispatch strategy.
- Renewable Energy Coverage (REC) (%): The coverage of renewable energy over demand is 35% for both criteria, but as in the case of batteries, it is slightly higher in the self-consumption dispatch strategy.
- Cost of the total electricity purchased (€): In the case of batteries, the average monthly energy purchase is about 11€ on average for the self-consumption dispatch strategy and slightly less for the economic one.
- Revenue of the total electricity sold to the market (€): For the economic dispatch strategy, the revenue is equivalent to about 14€ on average per month, and it is slightly lower than the self-consumption strategy.
- Total electricity purchased from the grid (kWh): Demand that must be purchased because it is not supplied by renewable sources. This is identical to the case without batteries.
- Total electricity delivered to the grid (kWh): The energy injected into the grid is superior for the economic strategy, since the surplus is not used to supply secondary consumption.

Finally, Figures 14 and 15 below show the graphs of the ESC and REC outputs for each of the economic dispatch and self-consumption strategies. These graphs have been obtained for a typical summer month, July, to compare the differences between the two strategies with batteries.

Regarding the ESC output, i.e., financial savings, Figure 14 shows cost energy savings on most days of the month. These savings are slightly higher in the economic dispatch strategy, reaching a monthly difference of up to 1.30€. In this case with batteries, the maximum saving obtained is 1.60€ on the 29th, which is 20cts€ more than in the case without batteries.

In the case of the REC output (percentage of renewable energy used), Figure 15 shows higher values for the self-consumption dispatch strategy compared to the economic one for most of the month. However, there are days in which the values are identical because only the measures (load shifting, disaggregate, etc.) common to both dispatch strategies have been used. In this case, energy is managed more equitably throughout the month, obtaining values of up to 63% of demand coverage with the energy generated. These values are higher than in the case of Figure 13 as the use of batteries allows the energy generated to be more flexible. However, for example, on day 8, where in the case without batteries almost all the energy generated was used, in this case, the percentage used is lower. This is because the battery allows deferred demand to be covered, so perhaps energy from the batteries was used to cover the demand, and the rest of the surplus energy could be injected into the grid, thus increasing the savings in the cost of energy for that day.

Therefore, the use of batteries allows flexibility in the use of the energy generated without the need to inject it into the grid, so the REC output is higher, but the correct management of the energy resources also allows the savings in the cost of energy to be higher (by unbundling loads or transferring them to lower-cost periods).

Finally, in annual terms for the case with batteries, the energy generated to supply demand reaches a value of 57.1% for the energy dispatch strategy and 49.7% for the economic dispatch strategy, with the remaining energy fed into the grid. Furthermore, this means savings in energy purchases of 73.40% and 74.40%, respectively, reaching values of over 70%.

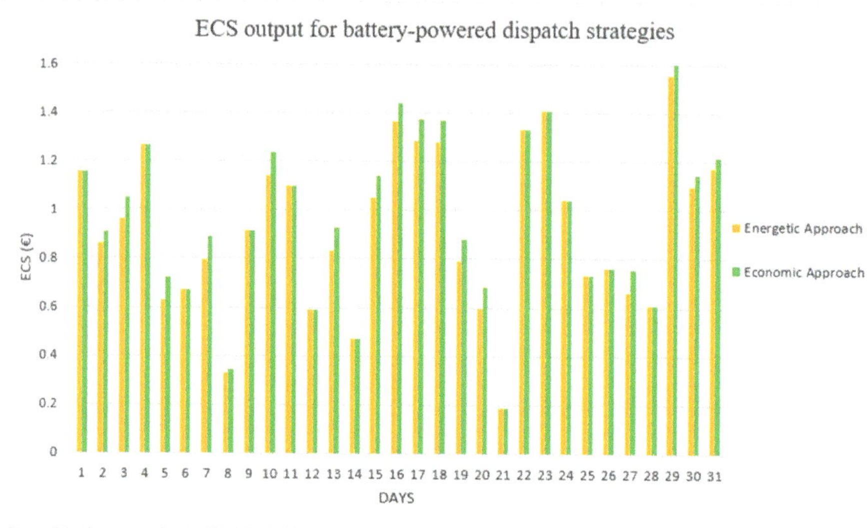

Figure 14. ECS Output (economic savings) for the Self-Consumption and Economic Dispatch Strategies in a summer month for the case with batteries.

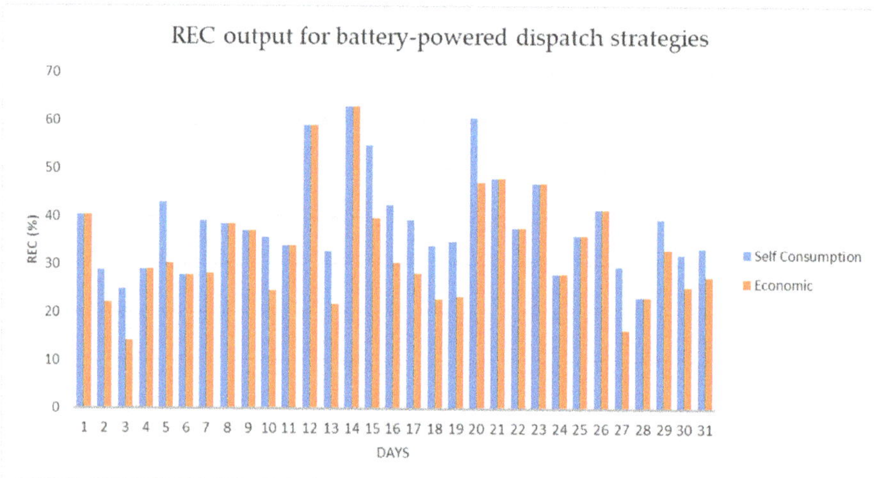

Figure 15. REC Output (use of renewable energy generated) for the Self Consumption and Economic Dispatch Strategies in a summer month for the case with batteries.

5. Discussion

Results of the case study show that the methodology allows for achieving a bill reduction of 70% in comparison to the household´s annual electricity purchased from the grid, by managing all domestic energy resources (generation with photovoltaics, storage system, and demand flexibility). Considering no active participation of the consumer, the electrical bill reduction associated only with self-consumption from the photovoltaic panels reaches a 27% share due to the generation-demand hourly decoupling, while the combination of photovoltaics and battery systems achieves 56% savings in the electrical bill, which includes not only the demand coverage but also selling of the excess of energy into the grid.

Furthermore, the annual results obtained for each of the study cases are analysed: without the use of batteries (Table 6) and with batteries (Table 7) for energy management.

Table 6. Annual output values for battery-free dispatch strategies.

	Economic	Self-Consumption
Energy Cost Savings (ECS) (€)	246.59	233.90
Renewable Energy Coverage (REC) (%)	30.89	36.27
Cost of the total electricity purchased (€)	215.85	218.29
Revenue of the total electricity sold to the market (€)	182.21	169.64
Total electricity purchased from the grid (kWh)	1042.44	1042.44
Total electricity delivered to the grid (kWh)	2603.08	2429.25

Table 7. Annual output values for battery powered dispatch strategies.

	Economic	Self-Consumption
Energy Cost Savings (ECS) (€)	287.40	275.12
Renewable Energy Coverage (REC) (%)	34.15	39.65
Cost of the total electricity purchased (€)	129.41	130.34
Revenue of the total electricity sold to the market (€)	169.65	143.08
Total electricity purchased from the grid (kWh)	467.20	467.20
Total electricity delivered to the grid (kWh)	2421.76	2044.76

Regarding the ECS output or the economic savings obtained, it is noted that the study case with batteries obtains a slightly higher saving compared to the case without batteries, which corresponds to about 4€ per month on average. This saving represents the reduction of energy purchased from the grid, firstly due to greater coverage of the demand with renewable energy (REC output), and secondly because it is possible to manage the loads better and transfer them to other cheaper periods set by the domestic consumer. It can be said that without the use of batteries, the energy generated through the PV panels must be used directly, so the system loses versatility.

The use of batteries allows flexibility in the generation produced by the PV panels to be consumed later when there is no generation. Therefore, it also conditions the REC output, since a higher percentage of the demand can be covered by the energy from the renewable source stored in batteries. In other words, and according to the data provided, 36% of the demand is covered by renewable energies in the case without batteries, and up to almost 40% in the study case with batteries. So, without the use of batteries, all unconsumed energy is delivered into the grid, as there is no other method of dissipating the energy.

Surplus energy in both study cases is also used to supply secondary consumption in the self-consumption dispatch strategy, so that the energy delivered into the grid is less than in the economic dispatch strategy (2.6 kWh in the economic strategy, while in the self-consumption strategy, 2.4 kWh). Thus, according to the economic dispatch strategy, less energy is consumed from the renewable source (REC Output), and the difference is injected into the grid with the consequent economic retribution. This means that the values for the energy delivered to the grid are even more significant in the study case without batteries, and therefore, the revenue from selling energy to the grid is also higher (182.21€ in the economic strategy for the case study without battery versus 169.65€ in the same strategy for the case with batteries).

In the case without batteries the revenue obtained by selling energy to the grid is higher than in the case with batteries, but also, the cost of the energy purchased is higher. In fact, the balance between purchasing and selling energy to the grid in the case without batteries is positive, i.e., the consumer must buy more energy compared to what the consumer sells. However, the balance for the case with batteries is negative, i.e., the consumer can sell more energy than consumes, so this shows that energy management is better with batteries.

On the other hand, the results reported monthly for the two main outputs (ECS and REC) are also the subject of discussion. The monthly values obtained for the ECS output as an indication of daily savings are higher on daily basis in the case study with batteries (Figure 14) than in the case study without batteries (Figure 12). In fact, the maximum daily savings value achieved is 1.6€, while in the case without batteries, it is 1.4€. However, it can be observed that the use of batteries does not affect each of the strategies separately, i.e., the economic and energy dispatch strategy behave similarly, but the values obtained for the case with batteries are proportionally higher. The same applies to the REC output for the case without batteries (Figure 13) and with batteries (Figure 15). Both dispatch strategies (self-consumption and economic) maintain the same behaviour, but the values are higher on average. That is to say, the percentage of demand covered is higher in the case of batteries than in the case of no batteries.

6. Conclusions

In this research work, savings have been obtained in terms of energy consumption from the grid, as well as for the electricity bill. The methodology allows economic savings on electricity bills, greater energy efficiency, and a consequent reduction in pollutant emissions. Considering the active participation of the consumer managing their demand flexibility resources with and without batteries, the methodology allows an average coverage of the demand with renewable energies from 30% (case study without batteries) to almost 40% (case study with batteries) and reaches values of up to 80% in the case with batteries. In addition, it helps to achieve a significant reduction in the annual electricity bill: up to 246,59€ in the case "without batteries" and up to 287,40€ in the case "with batteries", corresponding to a reduction in electricity bills of up to 70% annually. Furthermore, the case study supports that the use of batteries gives versatility to the proposed methodology, increasing the values of savings and renewable coverage obtained for both strategies. Finally, although it is not contemplated within the methodology, managing the use of energy by delivering the surplus into the grid leaves the door open to its subsequent use in what could be a shared self-consumption. This, together with the possibility of integrating the tool into a home using smart grids, creates expectations for the future in which it could be used through an application as a measure for active demand management in the residential sector.

Author Contributions: Conceptualization, V.L.-M., C.A.-M. and E.P.-L.; methodology, V.L.-M., C.A.-M. and E.P.-L.; validation, V.L.-M., C.A.-M., J.Á.S.-J. and E.P.-L.; resources, C.A.-M., J.Á.S.-J. and E.P.-L.; data curation, C.A.-M. and J.Á.S.-J.; writing—original draft preparation, C.A.-M., E.P.-L. and V.L.-M.; writing—review and editing, C.A.-M., V.L.-M. and E.P.-L.; supervision, V.L.-M. and E.P.-L.; project administration, E.P.-L. and V.L.-M.; funding acquisition, E.P.-L. All authors have read and agreed to the published version of the manuscript.

Funding: This research and APC were funded by Generalitat Valenciana, the regional government of Valencian Region, within the ValREM Project (CIAICO/2022/007).

Data Availability Statement: The data presented in this study are available on request from the corresponding author. The data are not publicly available due to privacy.

Conflicts of Interest: The authors declare no conflict of interest. The funders had no role in the design of the study; in the collection, analyses, or interpretation of data; in the writing of the manuscript; or in the decision to publish the results.

Abbreviations

Acronyms	
USA	United States
PURPA	Public Utility Regulatory Policies Act
DSM	Demand Side Management
DR	Demand Response
DLC	Direct Load Control
ESPP	Energy-Smart Pricing Plan
EV	Electric Vehicle
OMIE	Iberian Energy Market Operator
E_{fc_i}	Final energy consumed in each hour i after applying the methodology (kWh)
E_{sc_i}	Self-consumed energy from the renewable source in each hour i after applying the methodology (kWh)
E_{d_i}	Energy delivered from the renewable source to the grid in each hour i after applying the methodology (kWh)
E_{pv_i}	Energy from the PV panels in each hour i after applying the methodology (kWh)
D_{sc_i}	Self-consumed demand in each hour i after applying the methodology (kWh)
D_{tot_i}	Total demand consumed in each hour i after applying the methodology (kWh)
D_{f_i}	Final demand consumed in each hour i after applying the methodology (kWh)
D_{trans_i}	Demand transferred in hour i after application of the methodology (kWh)
E_{SLP_i}	Synthetic profile of the demand initially created to apply the methodology (kWh)
E_{SLPC_i}	Continuous synthetic load profile created from the user's demand history (kWh)
$E_{g_{pv_i}}$	Total energy generated by the PV panels in each hour i (kWh)
E_{ibat_i}	Initial energy of the battery at hour i before applying the methodology (kWh)
E_{fbat_i}	Final energy of the battery at hour i after applying the methodology (kWh)
Variables	
P_{e_i}	Energy price set by the market or the trader in hour i (€/kWh)
P_{s_i}	Surplus energy price set by the market or the trader in hour i (€/kWh)
D_{fc_i}	Demand supplied from the power grid in hour i (kWh)
Symbols	
α	Synthetic perturbation for synthetic load profile modeling.
S	Area covered by the PV Panels (m^2)
α_d	Daily perturbation
α_I	Instantaneous perturbation
t	Hour of the year (complete year includes the 8760 h).
ECS	Energy Cost Savings indicator
REC	Renewable Energy Coverage indicator
PV	Photovoltaic
SLP	Synthetic Load Profile (kWh)
SoC	State of Charge (%)
DoD	Depth of Discharge (%)
Equations	
ECS	Economic costs saved after applying the methodology (€)
REC	Demand covered with renewable energy over total demand supplied (%)
E_{bat_i}	Energy consumed from the battery in hour i (kWh)
E_{2con_i}	Energy consumed by the secondary consumptions in hour i (kWh)
C_0	Cost of the energy initially consumed (€)
C_f	Cost of the energy finally consumed (€)
$D_{t-i\,i}$	Demand transferred from another hour and entered in hour i (kWh)
$D_{t-o\,i}$	Demand leaving hour i and transferred to another hour (kWh)
$E_{j_{trans,i}}$	Historical demand transferred at hour i due to j user-defined devices (kWh)
Gh	Solar irradiation in the corresponding area $\left(\frac{\mathbf{Wh}}{\mathbf{m^2}}\right)$
E_{bat_min}	Minimum battery capacity (kWh)
$Bat_{dg_{i,j}}$	Battery capacity used to supply j equipment transferred at hour i (kWh)
$Bat_{lt_{i,j}}$	Battery capacity used to supply j equipment transferred at hour i (kWh)
Parameters	
D_o	Demand 0 calculated as the difference between the real generation and the historical demand (kWh)
E_{bat_max}	Maximum battery capacity (kWh)
E_{bat_min}	Minimum battery capacity (kWh)
$Bat_{dg_{i,j}}$	Battery capacity used to supply j equipment transferred at hour i (kWh)
$Bat_{lt_{i,j}}$	Battery capacity used to supply j equipment transferred at hour i (kWh)
i	Time variable that includes the 24 h of a day
j	Number of user-defined electrical loads
k	Load transfer priorities according to low, medium, or high.
$\forall$	Universal quantifier meaning "for all".

References

1. International Energy Agency (EA). *World Energy's Outlook*; International Press: Somerville, MA, USA, 2020.
2. EU Building Stock Observatory from European Commission. Available online: https://ec.europa.eu/energy/en/eubuildings (accessed on 9 August 2023).
3. Pérez-Lombard, L.; Ortiz, J.; Pout, C. A review on buildings energy consumption information. *Energy Build.* **2008**, *40*, 394–398. [CrossRef]
4. International Energy Agency (IEA). Final Energy Consumption of Buildings Relative to Other Sectors. Available online: https://www.iea.org/data-and-statistics/charts/final-energy-consumption-of-buildings-relative-to-other-sectors-2022 (accessed on 9 August 2023).
5. EUROSTAT. Change in Electricity Prices for Non-Household Consumers Compared with Previous Year's Same Semester, First Half 2023 (%). Available online: https://ec.europa.eu/eurostat/databrowser/view/nrg_pc_204/default/table?lang=en (accessed on 3 January 2024).
6. GlobalPetrolPrices. Average Cost of Electricity Per Continent. Available online: https://www.globalpetrolprices.com/electricity_prices/#:~:text=The%20prices%20are%20per%20kWh,per%20kWh%20for%20business%20users (accessed on 3 January 2024).
7. Balakrishnan, R.; Geetha, V.; Kumar, M.R.; Leung, M.-F. Reduction in Residential Electricity Bill and Carbon Dioxide Emission through Renewable Energy Integration Using an Adaptive Feed-Forward Neural Network System and MPPT Technique. *Sustainability* **2023**, *15*, 14088. [CrossRef]
8. Motta, L.L.; Ferreira, L.C.B.C.; Cabral, T.W.; Lemes, D.A.M.; Cardoso, G.d.S.; Borchardt, A.; Cardieri, P.; Fraidenraich, G.; de Lima, E.R.; Neto, F.B.; et al. General Overview and Proof of Concept of a Smart Home Energy Management System Architecture. *Electronics* **2023**, *12*, 4453. [CrossRef]
9. Venizelou, V.; Makrides, G.; Efthymiou, V.; Georghiou, G.E. Methodology for deploying cost-optimum price-based demand side management for residential prosumers. *Renew. Energy* **2020**, *153*, 228–240. [CrossRef]
10. Niamir, L.; Ivanova, O.; Filatova, T.; Voinov, A.; Bressers, H. Demand-side solutions for climate mitigation: Bottom-up drivers of household energy behavior change in the Netherlands and Spain. *Energy Res. Soc. Sci.* **2020**, *62*, 101356. [CrossRef]
11. Francisco, A.; Truong, H.; Khosrowpour, A.; Taylor, J.E.; Mohammadi, N. Occupant perceptions of building information model-based energy visualizations in eco-feedback systems. *Appl. Energy* **2018**, *221*, 220–228. [CrossRef]
12. Yoo, S.; Eom, J.; Han, I. Factors driving consumer involvement in energy consumption and energy-efficient purchasing behavior: Evidence from Korean residential buildings. *Sustainability* **2020**, *12*, 5573. [CrossRef]
13. Hargroves, K.; James, B.; Lane, J.; Newman, P. The Role of Distributed Energy Resources and Associated Business Models in the Decentralised Energy Transition: A Review. *Energies* **2023**, *16*, 4231. [CrossRef]
14. Ndebele, T. *Energy Transition and Consumer Support for Green Energy Beyond the 50% Renewable Energy Target: Evidence from Stated Choice Experiments*; Clark University: Worcester, MA, USA, 2023.
15. Ojeda, V.M.M. State of the Art of Electric Energy Management. Master's Thesis, School of Engineering, University of Sevilla, Sevilla, Spain, 2021.
16. Gellings, C.W. The concept of demand-side management for electric utilities. *Proc. IEEE* **1985**, *73*, 1468–1470. [CrossRef]
17. Faruqui, A.; Gellings, C.W. Should demand-side management be a top-down or a bottom-up process. In *Strategic Planning and Marketing for Demand-Side Management: Selected Seminar Papers*; EPREA-4308; Electric Power Research Institute: Palo Alto, CA, USA, 1985.
18. Gellings, C.W.; Chamberlin, J.H. *Demand-Side Management: Concepts and Methods*; U.S. Department of Energy, Office of Scientific and Technical Information: Oak Ridge, TN, USA, 1987. Available online: https://www.osti.gov/biblio/5275778 (accessed on 9 August 2023).
19. Gellings, C.W.; Smith, W.M. Integrating demand-side management into utility planning. *Inst. Electr. Electron. Eng.* **1989**, *77*, 908–918. [CrossRef]
20. Gellings, C.W. Evolving practice of demand-side management. *J. Mod. Power Syst. Clean Energy* **2017**, *5*, 1–9. [CrossRef]
21. Rahman, S. An efficient load model for analyzing demand side management impacts. *IEEE Trans. Power Syst.* **1993**, *8*, 1219–1226. [CrossRef] [PubMed]
22. Qdr, Q. *Benefits of Demand Response in Electricity Markets and Recommendations for Achieving Them*; Technology Reper; US Department of Energy: Washington, DC, USA, 2006; Volume 2006, p. 95.
23. Stephenson, J.; Ford, R.; Nair, N.K.; Watson, N.; Wood, A.; Miller, A. Smart grid research in New Zealand–A review from the GREEN Grid research programme. *Renew. Sustain. Energy Rev.* **2018**, *82*, 1636–1645. [CrossRef]
24. Torriti, J. Price-based demand side management: Assessing the impacts of time-of-use tariffs on residential electricity demand and peak shifting in Northern Italy. *Energy* **2012**, *44*, 576–583. [CrossRef]
25. Rowlands, I.H.; Furst, I.M. The cost impacts of a mandatory move to time-of-use pricing on residential customers: An Ontario (Canada) case-study. *Energy Effic.* **2011**, *4*, 571–585. [CrossRef]
26. Schrammel, J.; Diamond, L.M.; Fröhlich, P.; Mor, G.; Cipriano, J. Influencing residential electricity consumption with tailored messages: Long-term usage patterns and effects on user experience. *Energy Sustain. Soc.* **2023**, *13*, 15. [CrossRef]
27. Andrada-Monrós, C.; Peñalvo-López, E.; León-Martínez, V.; Valencia-Salazar, I. Methodology to optimise electricity demand in the residential sector through efficient load management. *Renew. Energy Power Qual. J.* **2022**, *20*, 719–723. [CrossRef]

28. Panda, S.; Mohanty, S.; Rout, P.K.; Sahu, B.K.; Bajaj, M.; Zawbaa, H.M.; Kamel, S. Residential Demand Side Management model, optimization and future perspective: A review. *Energy Rep.* **2022**, *8*, 3727–3766. [CrossRef]
29. Jasim, A.M.; Jasim, B.H.; Flah, A.; Bolshev, V.; Mihet-Popa, L. A new optimized demand management system for smart grid-based residential buildings adopting renewable and storage energies. *Energy Rep.* **2023**, *9*, 4018–4035. [CrossRef]
30. Mehra, V.; Amatya, R.; Ram, R.J. Estimating the value of demand-side management in low-cost, solar micro-grids. *Energy* **2018**, *163*, 74–87. [CrossRef]
31. Fell, M.J.; Shipworth, D.; Huebner, G.M.; Elwell, C.A. Public acceptability of domestic demand-side response in Great Britain: The role of automation and direct load control. *Energy Res. Soc. Sci.* **2015**, *9*, 72–84. [CrossRef]
32. Cano-Martínez, J.; Peñalvo-López, E.; León-Martínez, V.; Valencia-Salazar, I. Dynamic energy prices for residential users based on Deep Learning prediction models of consumption and renewable generation. *Renew. Energy Power Qual. J.* **2023**, *21*, 76–80. [CrossRef]
33. Hamdani, N.; Muhamad, N.A.; Petra, R.; Ang, S.P. Study on the residential load profiles in Brunei Darussalam based on income and location. *Energy Rep.* **2023**, *9*, 737–747. [CrossRef]
34. Zhou, K.; Yang, S. Understanding household energy consumption behavior: The contribution of energy big data analytics. *Renew. Sustain. Energy Rev.* **2016**, *56*, 810–819. [CrossRef]
35. Andrada-Monrós, C.; Peñalvo-López, E.; Valencia-Salazar, I.; Cárcel-Carrasco, J. Review of the Energy transition assessment methods in touristic municipalities. In Proceedings of the 21th International Conference on Renewable Energies and Power Quality (ICREPQ'23), Madrid, Spain, 24–26 May 2023.
36. Han, X.; Wei, C. Household energy consumption: State of the art, research gaps, and future prospects. *Environ. Dev. Sustain.* **2021**, *23*, 12479–12504. [CrossRef]
37. Cano-Martínez, J.; Peñalvo-López, E.; León-Martínez, V.; Valencia-Salazar, I. Optimisation of an aggregator's customer portfolio based on inter-user compatibility. *Renew. Energy Power Qual. J.* **2022**, *20*, 724–728. [CrossRef]
38. Bhattacharjee, S.; Reichard, G. Socio-economic factors affecting individual household energy consumption: A systematic review. *Energy Sustain.* **2011**, *54686*, 891–901.
39. Borozan, D. Regional-level household energy consumption determinants: The European perspective. *Renew. Sustain. Energy Rev.* **2018**, *90*, 347–355. [CrossRef]
40. Iqbal, M.M.; Waseem, M.; Manan, A.; Liaqat, R.; Muqeet, A.; Wasaya, A. IoT-enabled smart home energy management strategy for DR actions in smart grid paradigm. In Proceedings of the 2021 International Bhurban Conference on Applied Sciences and Technologies (IBCAST), Islamabad, Pakistan, 12–16 January 2021; IEEE: Piscataway, NJ, USA, 2021; pp. 352–357.
41. Palensky, P.; Dietrich, D. Demand side management: Demand response, intelligent energy systems, and smart loads. *IEEE Trans. Ind. Inform.* **2011**, *7*, 381–388. [CrossRef]
42. Meliani, M.; Barkany, A.E.; Abbassi, I.E.; Darcherif, A.M.; Mahmoudi, M. Energy management in the smart grid: State-of-the-art and future trends. *Int. J. Eng. Bus. Manag.* **2021**, *13*, 18479790211032920. [CrossRef]
43. Liu, Z.; Wu, Q.; Huang, S.; Zhao, H. Transactive energy: A review of state of the art and implementation. In Proceedings of the 2017 IEEE Manchester PowerTech, Manchester, UK, 18–22 June 2017; IEEE: Piscataway, NJ, USA, 2017; pp. 1–6.
44. Abrishambaf, O.; Lezama, F.; Faria, P.; Vale, Z. Towards transactive energy systems: An analysis on current trends. *Energy Strategy Rev.* **2019**, *26*, 100418. [CrossRef]
45. Huang, Q.; Amin, W.; Umer, K.; Gooi, H.B.; Eddy, F.Y.S.; Afzal, M.; Shahzadi, M.; Khan, A.A.; Ahmad, S.A. A review of transactive energy systems: Concept and implementation. *Energy Rep.* **2021**, *7*, 7804–7824. [CrossRef]
46. Paterakis, N.G.; Taşcıkaraoğlu, A.; Erdinc, O.; Bakirtzis, A.G.; Catalao, J.P. Assessment of demand-response-driven load pattern elasticity using a combined approach for smart households. *IEEE Trans. Ind. Inform.* **2016**, *12*, 1529–1539. [CrossRef]
47. Ma, Y.; Li, B. Hybridized intelligent home renewable energy management system for smart grids. *Sustainability* **2020**, *12*, 2117. [CrossRef]
48. Sharda, S.; Singh, M.; Sharma, K. Demand side management through load shifting in IoT based HEMS: Overview, challenges and opportunities. *Sustain. Cities Soc.* **2021**, *65*, 102517. [CrossRef]
49. Iqbal, S.; Sarfraz, M.; Ayyub, M.; Tariq, M.; Chakrabortty, R.K.; Ryan, M.J.; Alamri, B. A comprehensive review on residential demand side management strategies in smart grid environment. *Sustainability* **2021**, *13*, 7170. [CrossRef]
50. Jordehi, A.R. Optimisation of demand response in electric power systems, a review. *Renew. Sustain. Energy Rev.* **2019**, *103*, 308–319. [CrossRef]
51. Vega, A.; Amaya, D.; Santamaría, F.; Rivas, E. Active demand-side management strategies focused on the residential sector. *Electr. J.* **2020**, *33*, 106734. [CrossRef]
52. Chakir, A.; Abid, M.; Tabaa, M.; Hachimi, H. Demand-side management strategy in a smart home using electric vehicle and hybrid renewable energy system. *Energy Rep.* **2022**, *8*, 383–393. [CrossRef]
53. Jímenez, R. Active demand management in the residential sector. In *State of the Art of Ongoing Projects*; Universidad Carlos III: Madrid, Spain, 2013.
54. Behzadi, A.; Thorin, E.; Duwig, C.; Sadrizadeh, Y.S. Supply-demand side management of a building energy system driven by solar and biomass in Stockholm: A smart integration with minimal cost and emission. *Energy Convers. Manag.* **2023**, *292*, 117420. [CrossRef]

55. Ebrahimi, J.; Abedini, Y.M. M. A two-stage framework for demand-side management and energy savings of various buildings in multi smart grid using robust optimization algorithms. *J. Build. Eng.* **2022**, *53*, 104486. [CrossRef]
56. Cooperative, C.E.; Summit Blue Consulting. *Evaluation of the Energy-Smart Pricing Plan*; Illinois Department of Commerce and Economic Opportunity: Chicago, IL, USA, 2004.
57. Enright, T.O.N.I.; Faruqui, A. *A Bibliography on Dynamic Pricing of Electricity*; The Battle Group: Cambridge, MA, USA, 2012.
58. CDTI. *Active Demand Management Project Demand*; Cenit GAD "Project": La Coruña, Spain, 2010.
59. FLEXCoop Consortium 2019. Demand Response for Energy Cooperatives. Available online: http://www.flexcoop.eu/about-flexcoop (accessed on 13 July 2021).
60. Ministry of Ecological Transition. Available online: https://www.boe.es/buscar/doc.php?id=BOE-A-2019-5089 (accessed on 14 August 2023).
61. Redeia. Red Electrica. Available online: https://www.ree.es/en/activities/operation-of-the-electricity-systemvoluntary-price-small-consumer-pvpc. (accessed on 21 March 2023).

Disclaimer/Publisher's Note: The statements, opinions and data contained in all publications are solely those of the individual author(s) and contributor(s) and not of MDPI and/or the editor(s). MDPI and/or the editor(s) disclaim responsibility for any injury to people or property resulting from any ideas, methods, instructions or products referred to in the content.

buildings

MDPI

Article

Determinants of Urban Residential Electricity Consumption in Burkina Faso: A Study of Dwelling, Household, Socio-Economic and Appliance-Related Factors

Komlan Hector Seth Tete [1,*], Yrébégnan Moussa Soro [1], Samir Franck Amadou Coulibaly [2], Rory Victor Jones [3] and Sayon dit Sadio Sidibé [1]

1. Laboratoire Energies Renouvelable et Efficacité Energétique, Institut International d'Ingénierie de l'Eau et de l'Environnement (2iE), Rue de la Science, Ouagadougou 01 BP 594, Burkina Faso; moussa.soro@2ie-edu.org (Y.M.S.)
2. Department of Electrical and Computer Engineering, University of Quebec at Trois-Rivieres (UQTR), Trois-Rivieres, QC G8Z 4M3, Canada; samir.coulibaly@uqtr.ca
3. School of the Built Environment, University of Reading, Reading RG6 6UR, UK

* Correspondence: seth.tete@2ie-edu.org; Tel.: +226-57-81-16-93

Abstract: This study analyses the dwelling, household, socio-economic and appliance-related determinants of urban domestic electricity use in Burkina Faso. A survey of 387 households in Ouagadougou was conducted and combined with their electricity use, collected from energy bills. To the authors' knowledge, this is the first large-scale, city-wide household electricity study undertaken in Burkina Faso. Linear regression models were applied to the data to assess the influence of household, dwelling, socio-economic and appliance-related factors on electricity use. Significant and unambiguous predictors of urban domestic electricity use included household income, dwelling tenure type and ownership, and use of lighting fixtures, television sets, fridges, fans, freezers and air conditioners. Dwelling, household and socio-economic factors explained 70.2% and 70.5% of the variance in electricity use, respectively, when combined with appliance ownership and use factors. This study provides an understanding of the driving factors of domestic electricity use and discusses the implications and applications of this research for a range of stakeholders in the electricity sector.

Keywords: residential electricity consumption; household and appliance factors; regression analysis; Burkina Faso; urban households; city-scale survey

1. Introduction

Within the building sector, electricity is mainly used to power lighting, appliances and heating, ventilation and air-conditioning (HVAC) systems. Access to such energy is, therefore, considered an essential social and economic need by this sector [1,2]. In Burkina Faso, the residential sector represents 33% of the overall demand for electricity in the country [3,4], making it a crucial target for demand-side management (DSM) and more efficient use of energy. However, accurate information about households' lifestyles, including socio-economic and dwelling characteristics, appliance ownership and use and factors influencing electricity consumption, which are essential to design effective DSM and energy efficiency programs, remains challenging to find. Contrary to many other countries where residential electricity consumption (REC) has been widely investigated [5–11], there is a great paucity of information on actual REC in the Global South, and Burkina Faso in particular. REC has been investigated in only a few countries in Sub-Saharan Africa, including mainly South Africa [12–14], Ghana [15–18] and Nigeria [19]. In Burkina Faso, the paucity of studies is even more significant. Apart from studies by the National Institute of Statistics (INSD/BF) on the living conditions of the population, which gathers some data on household energy expenditure, and a recently published study by the current authors analysing how households' lifestyles and behaviours relate to the level of electricity use in urban domestic residential buildings [20], almost no studies on REC exist. There is, however, a great interest by the country's government in improving electricity use. In 2016, the National Agency for Renewable Energy and Energy Efficiency (ANEREE) was created to foster the development of renewable energy and energy efficiency in Burkina Faso. In addition, among the grid operators and utility companies, having accurate information on the details on electrical energy end uses, in addition to the factors that influence the dynamics of residential electricity use, is essential for sound management of the country's electricity system and providing good quality services to users [17]. Finally, there is interest from the country and regions' energy research community as accurate information is essential for them, in order to assist the authorities in setting up strategies and policies for DSM or energy efficiency.

This study aims to improve the weak literature on REC in Burkina Faso by providing insights into the determinants of electricity consumption within urban households. For this purpose, this study uses the data collected from a residential electricity consumption survey (RECS) of 387 households in the city of Ouagadougou, which, to the authors' knowledge, is the first large-scale, city-wide household electricity study undertaken in Burkina Faso. As mentioned previously, the first study by the authors [20] uses the results from the study to present an initial overview of the socio-economic background of urban households in Burkina Faso, as well as their level of electricity consumption and related activities and behaviours. This study steps beyond the previous one, by trying to identify the factors that influence such electricity use. More specifically, the present study developed quantitative multivariate models of dwellings, households, and socio-economic factors and appliance ownership and use factors in relation to urban REC in Burkina Faso, to address the following questions: What are the main direct (factors that imply the direct use of electricity: number, daily time of use of appliances and the energy behaviours of users towards appliances and electricity) and indirect (factors that indirectly imply the use of electricity and factors that influence the use of lighting and appliances) determinants of electricity consumption within urban households in Burkina Faso? How can electricity consumption within urban households in Burkina Faso be predicted using such factors? How much variance in electricity consumption do these factors explain? What are the policy implications and applications of this research for stakeholders in the electricity sector?

In the rest of the paper, Section 2 presents the research methodology for conducting this study. Section 3 presents and discusses the results, as well as the policy implications. Finally, Section 4 summarises the study's findings and concludes the paper.

2. Literature Review

Determinants of REC have been widely investigated across the world. The most common approaches used to establish them include statistical, regression and probabilistic techniques, depending on the type of data collected and level of investigation [21]. At the macro level, high-level data are collected, such as demographics, housing statistics and gross domestic product (GDP), and "top-down" approaches are used to derive the causal relationships between these factors and electricity consumption [21,22]. At the micro level, "bottom-up" models are used instead, with data collected at an individual level, in order to determine the causal relationships between the household-level factors and electricity consumption [21,23]. In the current study, attention is paid exclusively to investigations at the micro level using bottom-up models. Therefore, the determinants discussed in this section encompass only household-level driving factors of electricity use, including not only socio-economic and dwelling factors, but also behavioural, technical and appliance-related factors.

2.1. Socio-Economic and Household Factors

The main socio-economic and household driving factors of electricity use investigated in the REC literature include family composition, tenure type, household income, disposable income, household size (number of occupants) and characteristics of the household responsible person (HRP) like age, education level and socio-economic classification [24]. Within such factors, the household size, presence of teenagers, and household and disposable income were found to be unambiguously significantly and positively correlated with electricity use [24].

For example, Sakah et al. [17] made use of a multiple linear regression (MLR) method to analyse questionnaire and monitoring data within 60 urban Ghanian households and found that socio-economic, dwelling, appliance-related and energy behaviour factors explained 56.8% of the variance in electricity consumption. More specifically, household size and income were found, within the socio-economic factors, to be significantly and positively correlated with electricity use, with a one unit rise in these factors leading to an annual increase in electricity use of 580.3 and 2 kWh, respectively. Using logit regression analysis consisting of OLS regression models built upon the Irish Household Budget Survey (HBS) dataset, Leahy and Lyons [25] calculated that an increase in the log of household disposable income by one unit, increased electricity use by 4% per week.

Previous investigations on the influence of family composition on electricity use found that the presence of teenagers was significantly and positively correlated with electricity use [26–28]. For example, Bartusch et al. [28] explored the variation in electricity use within Swedish households using statistical approaches, including independent samples *t*-tests and one-way and univariate independent samples analyses of variance. They found significantly higher annual electricity consumption per m^2 of electrically heated living space for families with teenagers in Sweden.

Tenure type and characteristics of the HRP have also been widely investigated [17,23,25,29–32] and found to have significant effects on electricity use. However, as such factors are investigated using categorical variables, it is not possible to establish a positive or negative correlation with residential electricity use [24]. Furthermore, the age of the HRP has been found to have a significant effect on the electricity consumption in residential buildings [24].

2.2. Dwelling Factors

The main dwelling-related driving factors of electricity use investigated in the REC literature include not only the dwelling type, age and size (floor area or number of rooms), but also the presence in the dwelling of householders and systems like HVAC, DHW and efficient lighting [24]. Dwelling size and age and the presence of air conditioning, electric space heating and water heating systems have been identified as unambiguously, significantly and positively correlated with REC in the literature [24].

For example, Sakah et al. [17] found a positive relationship between the total floor area and electricity use within Ghanaian households, with a recorded increase in annual electricity use by 7.1 kWh for each 1 m^2 increase in floor area. Floor area was also found as significant and positively correlated with REC by Kavousian et al. [32], who used weighted regression models to examine the structural and behavioural determinants of US domestic electricity consumption using a dataset from 1628 households.

Regarding the number of rooms/bedrooms in the dwellings, Bedir et al. [29] used the data collected through a survey of 323 Dutch households and developed three combined regression models, incorporating direct (appliances factors) and indirect (household and dwelling factors) determinants of electricity use and found that the number of rooms, and particularly the number of study/hobby rooms, was positively correlated with electricity consumption.

Dwelling age, which is commonly treated as a categorical variable, also demonstrated an unambiguously positive relationship with REC, meaning that the older the dwelling, the greater the amount of electricity it uses. For example, Leahy and Lyons [25] demonstrated that Irish homes built before 1918 used significantly more electricity per week (6.1%) than those built between 1918 and 1960. Such an increase was explained as resulting from an increase in heat loss associated with less insulation and the use of electric heating and power showers instead of gas. They also found that households built later than 2000 used significantly less electricity than dwellings built in 1918–1960.

2.3. Appliance Factors

The appliance-related driving factors of electricity use investigated in the literature consists of two main parts: ownership and use. The literature clearly demonstrates that the total number of appliances owned, ownership of particular appliances like desktop computers, televisions, electric ovens, fridges, dishwashers, and tumble dryers, and the use of some major appliances like washing machines and tumble dryers, significantly increase residential electricity usage [24].

For example, the number of electric appliances was found to be positively and significantly correlated with electricity use by Danlami [19], who used an OLS regression model to examine electricity use in Nigerian households. Similarly, using MLR analysis of annual electricity consumption by Japanese households, Genjo et al. [9] calculated that a one unit increase in appliance stock significantly increased the annual electricity use by 62 kWh. For Ghanian households, Sakah et al. [17] found that the ownership of a television and refrigerator significantly and positively correlated with electricity use; more specifically, they calculated an increase in the annual electricity use of 226 kWh and 127 kWh for each additional refrigerator and television owned. McLoughlin et al. [31] found that dishwashers, tumble dryers and desktop computers were the appliances that consumed the most electricity in Irish dwellings and were significantly and positively correlated with REC.

However, a part of the literature on REC considers that the number of appliances alone only partially reflects their effect on residential electricity use, which makes it necessary to consider their frequency of use instead [33]. Few pieces of research exist on the influence of appliance use on REC [26,29,31]. For example, the duration of the use of appliances, including information, communication, entertainment, HVAC, washing and laundry appliances were found to explain 37% of the variance in electricity consumption within Dutch households by Bedir et al. [29]. Similarly, Bartiaux and Gram-Hanssen found a significantly positive correlation between the duration and frequency of use of washing appliances and electricity demand in Dutch households [26].

2.4. Other Factors

Various other factors have also been reported in the existing literature, including electricity price, location, ownership of specific systems, activities and energy behaviours of the occupants. For example, Sakah et al. [17] established that awareness and practice of energy conservation can significantly help to decrease annual electricity use by 1277 kWh within urban Ghanian households, while McLoughlin et al. [31] found significantly higher electricity consumption by Irish households who believe they can cut their annual electricity use by high percentages (10–30%). Using a general linear regression model (GLM), Fan et al. [34] identified that the ownership of a swimming pool was found to be one of the significant predictors of Australian residential peak electricity demand. The price of electricity was investigated by O'Doherty et al. [22], who found a 2.1% increase in energy use due to off-peak electricity tariffs.

2.5. Summary of Previous Literature Methods and Results

The existing body of literature suggests that a large range of factors, including socio-economic and household, dwelling, behavioural, appliance-related and other factors, are related to REC. Based on the reviewed literature, statistical analysis and, more specifically, regression analyses, were mostly used by studies dealing with sizeable datasets to provide insights into the determinants of residential electricity use [17]. The major drawback of such studies is the high possible cost of implementation and issues concerning multicollinearity between variables [21].

The current study has used a statistical approach based on a multiple linear regression model to analyse the underlying factors of variations in urban REC in Burkina Faso. The recorded annual electricity consumption of households was used as the dependent variable, whereas household, dwelling, socio-economic and appliance-related factors obtained from a household survey were used as independent variables.

3. Materials and Methods

3.1. Study Area

In Burkina Faso, only 5.9% of the rural population have access to electricity, while at the same time, the urban population, which represents only 26.4% of the total population, have access to electricity services at a rate of 72.5% and use 74% of the total electricity used in the country [35]. The urban areas were, therefore, chosen for our study as they represent a large amount of the total electricity used in the country. Moreover, the city of Ouagadougou (Figure 1) was selected due to its representativeness of the urban areas. Ouagadougou, which is located in the central region of Burkina Faso, is the capital and largest city. With a total population of 2,453,496 inhabitants, the city consists of 12 districts and 55 sectors, and alone represents 45.4% of the country's urban population [36]. The location of Ouagadougou, at longitude 01°31′05″ W and latitude 12°21′58″ N [37], places it in the Sudano-Sahelian climatic zone, conferring the climate with a dry and hot character. Mean monthly temperatures range from 33 °C during the dry and hot seasons (March to June and November), to 25 °C during the dry and cold seasons (December to February) [38]. During the rainy season (July to October), the rainfall reaches up to 200 mm, while the mean relative humidity is estimated as 48.5%. The city was, therefore, chosen because it is a good representant of the urban areas in Burkina Faso. Also, the findings on the city can be used as reference points for predicting the REC in the other urban areas of Burkina Faso.

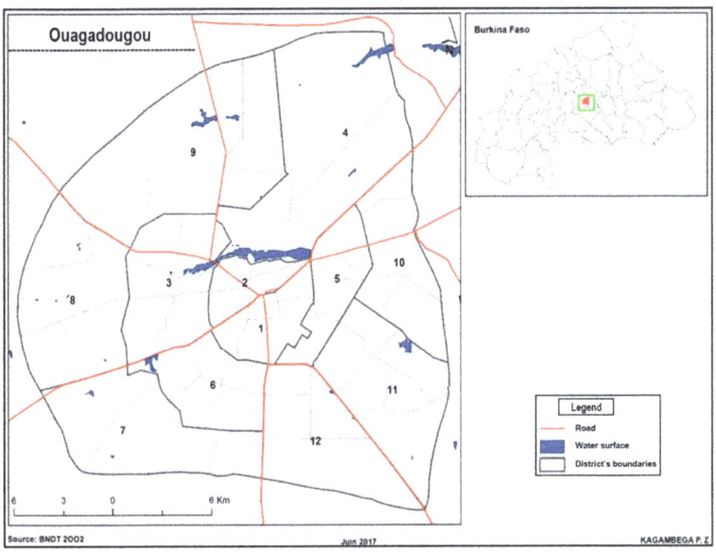

Figure 1. Map of Ouagadougou and its 12 districts (adapted from [39]).

3.2. Design of the Survey and Data Collected

3.2.1. Size of the Studied Sample

In order to make a compromise in the study between accuracy and the available resources, not all the population of the households in Ouagadougou were studied. Instead, sampling was adopted (with the "stratified" random sampling technique used, as the city consists of 12 districts). Indeed, such an approach is known for its numerous advantages, with respect to complete enumeration including reduced cost, greater speed, greater scope and greater accuracy [40]. The sample size was calculated, as suggested by Cochran [40].

Such approach is recognised as not only useful for either research experiments or surveys [41], but it is also flexible when using various types of information (size, proportions or statistics) on the target population [41,42]. Furthermore, it is commonly used in the surveys in the literature, including those on domestic electricity use [19,43,44]. The sample size is given here in its decomposed form by Equations (1) and (2).

$$\mathbf{n}_0 = \mathbf{z}^2\mathbf{pq}/\mathbf{e}^2 \tag{1}$$

$$\mathbf{n} = \mathbf{n_o}/(1 + ((\mathbf{n}_0 - 1)\,/\,\mathbf{N})) \tag{2}$$

In Equation (1), $\mathbf{n}_0$ refers to the non-adjusted sample size, i.e., the sample size for an infinite and unknown population size, **z** represents the z-score corresponding to the chosen confidence interval, **p** refers to a proportion of the population which has a defined attribute, $\mathbf{q} = \mathbf{1} - \mathbf{p}$ and **e** is the margin of error. When the size of the target population, which is **N** here (the number of households in Ouagadougou), is known, a correction factor is introduced by Cochran [40], therefore, Equation (2) is used to correct the sample size. The adjusted/corrected survey sample size (**n**) was, therefore, determined in our study as 384, which is at a 95% confidence interval, with a proportion of 50% and a 5% margin of error (**e**).

3.2.2. Investigation Methods and Results of the Survey

Data was collected during the survey by means of questionnaires sheets, which are known for their suitability regarding data gathering [45]. Moreover, they are often used in studies on energy consumption [46]. Considering the structure of the city of Ouagadougou, 12 districts and 55 sectors, the sample size consisted of a combination of the sample sizes of the districts, with a proportional relationship established between the number of selected households and the total number of households in each district.

Random sampling was used as the household selection method throughout the survey, in order to make sure that each household had an equal chance of being selected. However, household participation also depended on their readiness and willingness to participate (Figure 2). Finally, attention was also paid to the selection of households with different backgrounds and economic levels throughout the sampling process. Table 1 shows the role of the authors in the survey design and implementation. The name of the authors has not been given, instead, a number has been given to them according to their order of appearance on the title page.

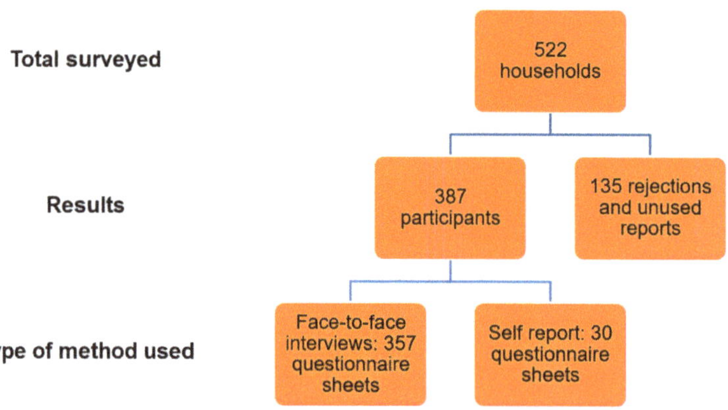

Figure 2. Background of the survey.

Table 1. Authors' contributions to the elaboration and implementation of the survey.

Stage	Activities and Authors Involved		
	Conception	**Validation**	**Obtaining necessary approbations**
Survey Design	Author 1 Author 2 Author 3	Author 2 Author 5 Author 4	Author 1 Author 2
Survey Test	Author 1, Author 3		
Survey Implementation	**Questionnaire distribution and recovery**		**Other complementary data collected**
	Author 1, Author 3, Author 5, Author 2		Author 1

Finally, Table 2 provides the characteristics and number of surveyed households for each district in this study. From such a table, it can be noticed that except for district 9, the number of people per household in each district, from the results in this study, is slightly higher than that from the results of the last census of the country (RGPH, 2019) [36].

Table 2. Ouagadougou's structure and number of participants per district.

N° District	Area (km^2)	Population Size	Number of Households	People per Household	Surveyed Households	People per Surveyed Household
1	18.65	102,016	25,771	4.0	23	6.0
2	12.70	83,436	20,203	4.1	23	4.8
3	60.56	311,406	66,595	4.7	56	5.0
4	82.81	207,647	49,088	4.2	36	6.0
5	21.40	129,984	33,115	3.9	28	5.3
6	29.00	222,854	49,619	4.5	39	6.5
7	32.68	282,837	70,180	4.0	32	5.5
8	66.57	152,880	36,098	4.2	15	4.9
9	88.27	336,483	79,889	4.2	34	3.9
10	27.96	263,969	63,579	4.1	45	4.9
11	48.30	254,928	62,221	4.1	44	5.8
12	44.13	66,826	15,811	4.2	12	6.2
Total	**533.03**	**2,415,266**	**572,169**	**4.2**	**387**	**5.4**

3.2.3. Ethical Considerations in the Data Gathering Process

Ethical considerations related to the data gathering process concerned mainly the participants in the survey. First, the current study is part of a research programme that has been designed and approved according to the research direction of the host institute (2iE Institute) after several screenings by various departments, including the deontology and ethics committee (N°2023/01/DG/SG/DR/HK/fg). In addition, a specific approbation for the implementation of the survey was received by the authors according to the research direction of the host institute. Such an approbation was subject to the following considerations: presentation of an invitation letter to the participants (signed by the Research Director), presentation of an information sheet to the participants (description of the content, scope and objectives of the survey), obtaining permission from the participants, presentation of ID of the host institute to the participants by the researchers before starting the interviews, keeping confidential all data and securely storing the personal data of the participants.

3.2.4. Data Collected

The survey content is shown in Table 3. As can be seen, the survey content was designed around domestic residential electricity use and items characterising household behaviours and lifestyles. Such items were demonstrated to be linked to electricity use in the literature on domestic REC. A household's annual electricity consumption was recorded from the household's electricity bills. The electricity bills were collected for the year 2021 and were used to compute the annual electricity use of the participants. Tables A4–A6 show the characteristics of the appliance stock of the participants, as well as that of their recorded annual electricity use.

Table 3. Items covered by the survey.

Items Investigated	Content or Description
Electricity use	Monthly electricity consumption (amount and cost from bills), electricity sources and electricity account information.
Appliance ownership and use	Appliance stock characteristics (type and number of appliances), appliance purchase condition, typical days appliance pattern of use (weekdays/weekends).
Dwelling characteristics	Tenure type, building's period of construction and type, floor area, number and role of rooms, number of storeys and presence of HVAC/DHW systems.
Household characteristics	Household size, family status, member's age and relation to the household responsible person (HRP), education level, monthly income, employment status of the HRP.
Daily life activities	Household daily life activities involving the presence and use of appliances in dwellings.
Energy behaviour	Appliance labelling awareness and influence on purchase, energy conservation awareness and practices, standby consumption awareness and behaviours.
Others	Preferred appliances for load shifting and curtailment to avoid power outages. Level of satisfaction regarding utility services, etc.

3.3. Data Processing and Analysis

The data processing framework is outlined in Figure 3. Based on the literature review, data were collected on numerous factors (Tables A1–A3) due to their demonstrated effect on REC. An analysis of the correlations (using Pearson's r) helped reduce the number of predictors by retaining only the significantly correlated factors to REC, for the next step in the analysis. In the next step of the analysis, the retained factors were entered into multiple linear regression (MLR) models to find the significant predictors of REC and the amount of variance in the REC explained by such predictors.

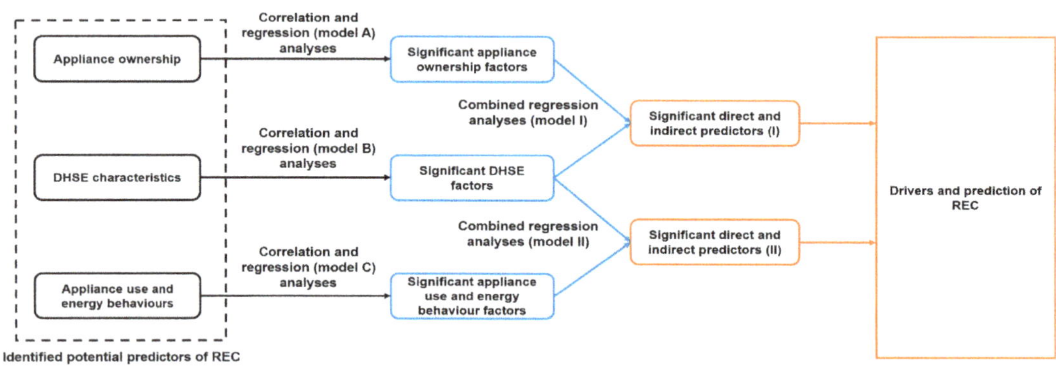

Figure 3. Study framework: data processing methodology.

Simple regression models were used first, using a stepwise method [23] to find the significant predictors of REC for each category of factors, including not only direct effect factors, namely appliance ownership (AO) factors (model A), and appliance use and energy behaviour (AUEB) factors (model C), but also indirect effect factors, namely dwelling, household and socio-economic (DHSE) variables (model B). Following this, the identified significant predictors of the three defined categories (A, B and C) were introduced into two combined regression models. The first combined model (I) used DHSE and AO factors that were found to be significant in the previous simple models (A and B), while the second model (II) used DHSE and AUEB factors (B and C). The two combined models were built separately to provide more insights into the effect of appliance ownership and use, and the amount of variation in the REC that they can explain when investigated separately. This aspect was undertaken as the existing literature lacks comparative studies on these two elements.

The MLR function helps evaluate the linear relationship between a dependent variable (**Y**) and several independent variables or functions of independent variables (**X**) [17]. Linear regression estimates how much Y changes when X changes by one unit. The MLR function is presented by Equation (3):

$$\mathbf{Y} = \beta_0 + \beta_1\mathbf{X}_1 + \beta_2\mathbf{X}_2 + \ldots + \beta_n\mathbf{X}_n + \varepsilon \quad (3)$$

where **Y** is the dependent variable (the log of REC was used in our study as the dependent variable), and $\mathbf{X}_{1\ldots n}$, the independent or explanatory variables, i.e., here DHSE, AO and AUEB variables. Moreover, β_0 stands for the intercept (or constant), i.e., the value of **Y** when the values of all the explanatory variables are 0, while $\beta_{1\ldots n}$ represent the regression coefficients. Here, they reflect the magnitude and direction of change in the REC in a household when the degree of a predictor increases by one unit, while holding all the others constant. A negative coefficient indicates a reduction in the REC, while a positive coefficient indicates an increase in the REC. Additionally, ε is the random error term, which measures how far above or below the true regression line (i.e., the line of means) the actual observation on **Y** lies [17].

The ordinary least squares (OLS) method was used to estimate all the regression coefficients in the model. The proportion of variance in the dependent variable that is predictable from the independent variables is estimated using the coefficient of determination (R^2), which shows how well the values fit the data and is used as a guideline to measure the model's accuracy [47]. The SPSS software (IBM SPSS Statistics 20) was used to perform all the statistical data processing, treatment and analysis.

4. Results

4.1. Preliminary Analyses

Before the models were run, the preliminary assumptions of the linear regression model were checked for each model (A–C, I and II), including the additivity, linearity, normality, homogeneity of variance (homoscedasticity) and independence of the dependent variable. Also, the correlations between the independent variables and the dependent variable (residential electricity consumption: REC) were checked. Finally, other checks such as multicollinearity check between the variables and the influencing observations (outliers) were also run.

4.1.1. Preliminary Assumptions Checking

The additivity, linearity and homoscedasticity were checked together. All of these elements can be assessed using a single graph, as they are all related to errors (or residuals). A scatterplot, which displays the values of these residuals against the corresponding values in the outcome predicted by the regression model [48] in a graph, was used (see the example for the model B in Figure A1. Normality was also checked using graphs [48] (Figure A2), namely the P–P plot (probability–probability plot) was used, which plots the cumulative probability of a variable against the cumulative probability of normal distribution.

Finally, the presence of outliers was checked using both graphs and numbers. The screening of the scatterplot showed the presence of two particular individuals (Figure A1). Further investigations were conducted, including the use of Cook's distance, as well as the converted normal z-scores of the REC for the individuals. For Cook's distance, the maximum distance recorded for all models was 0.182, which is much less than 1, indicating that there was no need to worry about the two individuals [48,49]. The sorting of the converted z-scores of the REC (ignoring whether they are positive or negative) for the individuals indicated that no values were larger than 3.29, only 2% of the values were larger than 1.96 and 0.3% were larger than 2.58. Therefore, no observations violated the outliers' conditions [48] and, consequently, all the observations were included in the following analyses.

4.1.2. Correlation Checking and Variables Retained for the Analyses

Following the assumption checking for the linear regression models, the next step in the data treatment was to find the correlations between the REC and the variables selected from the literature as potential determinants. Tables A7 and A8 present the variables retained for the analyses using the regression models (due to the significant correlations existing between them and REC).

Regarding the AO variables (Table A7), previously identified variables affecting REC, like the ownership of fridges, television sets, desktop computers and large appliances like washing machines, were significantly correlated with REC. Furthermore, the ownership of other appliances, such as fans, air conditioners, laptops and lighting fixtures, were also significantly correlated with REC. Finally, rare factors like the ownership of a printing machine, blender and Wi-Fi routers, were also significant and, therefore, included in the regression analyses.

Regarding the AUEB variables (Table A7), it was evident that appliance use factors were also significant for almost all the appliances where ownership was found to be significant. Furthermore, the energy behaviour factors like energy conservation practices, awareness of standby consumption, number and unplugging of standby devices, appliance purchase and appliance preferences for load shifting during peak demand periods, were significantly correlated with REC. Therefore, they were included in regression model C.

Finally, for the DHSE variables, the common variables in the literature, such as household size, floor area, household income, tenure type, age, education level and socio-economic class of the HRP, were significantly correlated with REC (Table A8). Furthermore, other previously identified variables, such as the type of construction, the number of floors in the dwelling and the individuals' presence in regard to the rooms, were also significantly correlated with REC. Finally, rarely established variables, such as the marital status of the HRP, which has only previously been found in one other Global South study [19], were also significantly correlated with REC. These variables were then entered into the regression model (model B).

Variables not found to be significantly correlated with REC were omitted in the following analyses. For the multicollinearity check, the correlations between the independent variables were investigated, and the values of the coefficients (Pearson's r) reached up to 0.8, suggesting in some cases that there were some issues. In order to go further, the VIF (variance inflation factor) and the tolerance were computed using SPSS to investigate further the relationship between the predictors. The maximum VIF for all models was 4.86, found in model C, which is less than 10, and the minimum tolerance was found to be 0.21, also in model C, and which again is more than 0.1, indicating no concerns related to multicollinearity [48,49].

4.2. Individual Effect of Factors on Residential Electricity Use: Simple Regression Models (A–C)

4.2.1. Individual Effect of Appliance Ownership on Residential Electricity Use: Model A

Table 4 shows the results from the first model on the individual effect of the AO factors on REC. The model (A) was run using all the retained variables (from Table A7) to establish the significant predictors of residential electricity consumption (REC). The significant predictors (at least at the 5% level) of REC included here were the ownership of lighting fixtures, TV sets, fans, fridges, freezers, air conditioners, irons and microwaves. The AO factors explained 65.6% of the variance in the REC.

Table 4. Regression results for the AO determinants (Model A): significant predictors and their effect on REC.

Variables	B	Std. Error	B (Beta)
(Constant)	6.578	0.055	-
Ownership of lighting	0.023	0.005	0.241 ***
Ownership of TV sets	0.037	0.018	0.080 *
Ownership of a fridge	0.222	0.039	0.201 ***
Ownership of a freezer	0.295	0.064	0.165 ***
Ownership of a microwave/oven	0.146	0.068	0.087 *
Ownership of an iron	0.120	0.052	0.083 *
Ownership of air conditioning	0.089	0.030	0.150 **
Ownership of a fan	0.055	0.012	0.229 ***

$R^2 = 0.656$. * $p < 0.05$, ** $p < 0.01$, *** $p < 0.001$.

4.2.2. Individual Effect of Dwelling, Household and Socio-Economic Factors on Residential Electricity Use: Model B

The results from the second model, analysing the determinants of REC concerning the DHSE factors, are presented in Table 5. The significant predictors of REC included here were the floor area, household income, dwelling tenure type, education level of the HRP, presence in the living room during the daytime and the overall presence in the bedroom. The DHSE factors explained 62.1% of the variance in the REC. As many of the DHSE factors were categorical variables, more details on the regression analysis are presented in Table A9.

Table 5. Regression results for the DHSE determinants (Model B): significant predictors and their effect on REC.

Variables	B	Std. Error	B (Beta)
(Constant)	5.047	0.231	-
Total floor area	0.002	0.001	0.202 ***
HRP education level	0.086	0.025	0.120 **
Dwelling tenure type	0.215	0.055	0.159 ***
Household income	0.201	0.019	0.461 ***
Daytime presence in the living room	0.025	0.009	0.098 **
Overall presence in the bedroom	0.034	0.014	0.092 *

$R^2 = 0.621$. * $p < 0.05$, ** $p < 0.01$, *** $p < 0.001$.

4.2.3. Individual Effect of Appliance Use and Energy Behaviours on Residential Electricity Use: Model C

Table 6 shows the results from the AUEB model. The significant predictors of REC included here were the use of lighting fixtures, TVs, fans, fridges, freezers, air conditioners and games consoles. The only energy behaviour factor that was found to be significant was the awareness of standby electricity consumption (SEC). The AUEB model (model C) explained 65.8% of the variance in the REC.

Table 6. Regression results for the AUEB determinants (Model C): significant predictors and their effect on REC.

Variables	B	Std. Error	B (Beta)
Constant	6.636	0.056	-
Use of lighting	0.004	0.001	0.258 ***
Use of TV sets	0.010	0.003	0.135 ***
Use of a fridge	0.014	0.002	0.274 ***
Use of a freezer	0.011	0.003	0.134 ***
Use of a fan	0.011	0.003	0.138 ***
Use of air conditioning	0.048	0.008	0.241 ***

Table 6. *Cont.*

Variables	B	Std. Error	B (Beta)
Use of games consoles	−0.055	0.28	−0.069 *
Awareness of standby electricity consumption	0.091	0.044	0.063 *

$R^2 = 0.658$. * $p < 0.05$, *** $p < 0.001$.

4.3. Combined Effect of Factors on Residential Electricity Use: Combined Regression Models (I and II)

Using the significant predictors identified in the personal effect analyses (Models A to C), two combined models were built to estimate the combined effect of the factors on REC. The combined Model I explored the effect of the DHSE and AO factors on the REC, and the combined Model II explored the effect of the DHSE and AUEB factors on the REC.

4.3.1. Combined Effect of DHSE and AO Factors on REC: Model I

Model I, assessing the effect of the DHSE and AO factors on the REC, explained 70.2% of the variance in the REC (Table 7). The most significant predictors of REC ($p < 0.001$) were household income and ownership of fridges and freezers, followed by the dwelling tenure type, ownership of lighting, fans and air conditioners ($p < 0.01$). The last group of significant predictors included the ownership of irons and presence in the dwelling ($p < 0.05$).

Table 7. Regression results for the effect of DHSE and AO factors (significant predictors) on REC (Model I).

Variables	B	Std. Error	B (Beta)
Constant	5.592	0.176	-
Ownership of a fridge	0.180	0.035	0.164 ***
Ownership of a freezer	0.269	0.057	0.150 ***
Household income	0.128	0.018	0.295 ***
Dwelling tenure type	0.156	0.046	0.115 **
Ownership of lighting	0.012	0.004	0.122 **
Ownership of a fan	0.029	0.011	0.119 **
Ownership of air conditioning	0.068	0.023	0.114 **
Ownership of irons	0.097	0.043	0.068 *
Daytime presence in the living room	0.019	0.008	0.076 *
Overall presence in the bedroom	0.026	0.12	0.071 *

$R^2 = 0.702$.* $p < 0.05$, ** $p < 0.01$, *** $p < 0.001$.

4.3.2. Combined Effect of DHSE and AUEB Factors on REC: Model II

Model II demonstrated that the combined effect of the DHSE and AUEB factors explained 70.5% of the variance in the REC (Table 8). The most significant predictors of REC ($p < 0.001$) included household income, the use of TV sets, fridges and freezers, followed by the use of lighting, air conditioners, the dwelling tenure type and the presence in bedrooms ($p < 0.01$).

Table 8. Regression results for the effect of DHSE and AUEB factors (significant predictors) on REC (Model II).

Variables	B	Std. Error	B (Beta)
Constant	5.565	0.170	-
Household income	0.134	0.018	0.308 ***
Use of TV sets	0.010	0.002	0.130 ***
Use of a fridge	0.011	0.002	0.222 ***
Use of a freezer	0.010	0.003	0.119 ***
Dwelling tenure type	0.150	0.44	0.111 **

Table 8. *Cont.*

Variables	B	Std. Error	B (Beta)
Use of lighting	0.002	0.001	0.144 **
Use of air conditioning	0.020	0.008	0.102 **
Overall presence in the bedroom	0.036	0.11	0.098 **

$R^2 = 0.705$. ** $p < 0.01$, *** $p < 0.001$.

Here again, the combined models are presented with more details about the categorical variables in Tables A10 and A11.

5. Discussion

The statistical analyses were performed in this study to investigate the drivers of REC regarding DHSE characteristics, and the AO and AUEB factors. Three simple regression models were computed, after correlation analysis was performed, seeking to find the significantly correlated variables in regard to REC. The first model (Model A), constructed from the AO factors, explained 65.6% of the variance in the REC, with the most significant ($p < 0.001$) predictors of REC found to be the ownership of lighting, fridges and freezers and fans, followed by the ownership of air conditioners ($p < 0.01$), while the last group ($p < 0.05$) included the ownership of TV sets, microwaves and irons.

The second model, built with the DHSE characteristics, explained 62.1% of the variance in the REC. The most significant predictors ($p < 0.001$) for this model included the floor area, dwelling tenure type and household income, followed by the education level of the HRP and presence in the living room during the daytime ($p < 0.01$) and the overall presence in the bedroom ($p < 0.05$).

The last model, built using the AUEB factors, explained 65.8% of the variance in the REC. The use of lighting, TV sets, fridges, freezers, fans and air conditioners emerged as the most significant predictors ($p < 0.001$) of REC for this model, followed by the awareness of standby electricity consumption (SEC) and the use of games consoles ($p < 0.05$).

Finally, the significant factors from the three models were used to build two combined models (I and II) to investigate, separately, the effect of AO and AUEB factors on the REC when combined with the DHSE factors. On the one hand, the combined model I, which was built using the DHSE and AO factors, explained 70.8% of the variance in the REC, with significant predictors including household income, ownership of fridges and freezers ($p < 0.001$), followed by the dwelling tenure type and ownership of lighting, fans and air conditioners ($p < 0.01$), while the ownership of irons and the presence in rooms were found to be significant at the 5% level. On the other hand, the combined Model II, which was built using the DHES and AUEB factors, explained 70.5% of the variance in the REC. The most significant predictors in this model again included the household income and use of fridges and freezers, and also the use of TV sets ($p < 0.001$), while other significant predictors included the dwelling tenure type, use of lighting, use of air conditioners and the overall presence in the bedroom ($p < 0.01$).

Figure 4 summarises the variance in the REC explained by each model, while the significant factors are discussed below in comparison to the existing literature on determinants of REC and the policy implications and application of the research results.

5.1. *DHSE Determinants of Residential Electricity Use*

Three DHSE factors unanimously (found to be significant in all the models in which it was used as a predictor) emerged as significant predictors of residential electricity consumption (REC), namely household income, dwelling tenure type and overall presence in the bedroom (Tables A9–A11, 5, 7 and 8). The results demonstrate that households living in free (free rent households generally refers to occupiers that do not pay any fees for living in the dwelling, even if they are not the owner of the dwelling. Most of the time they are relatives or friends of the owner) and privately rented dwellings use up to 27% and 15% less electricity than those living in owner-occupied dwellings. These results are similar to those of Hamilton et al. [50], who found that owner-occupied homes use on average 25% more electricity than rented homes. Similarly, Wyatt [51] found that owner-occupied homes use on average 23% more electricity than housing association and council housing homes and 14% more than privately rented homes. Buying land and housing in urban areas in Burkina Faso is expensive, especially for recently built homes (after the 2000s). Therefore, most owner-occupied homes are owned by wealthy people, who can, therefore, also afford to purchase a large range of appliances and use them. Owner-occupied dwellings also tend to be larger than rented dwellings, and therefore consume more electricity, for the use of HVAC systems for example, where installed.

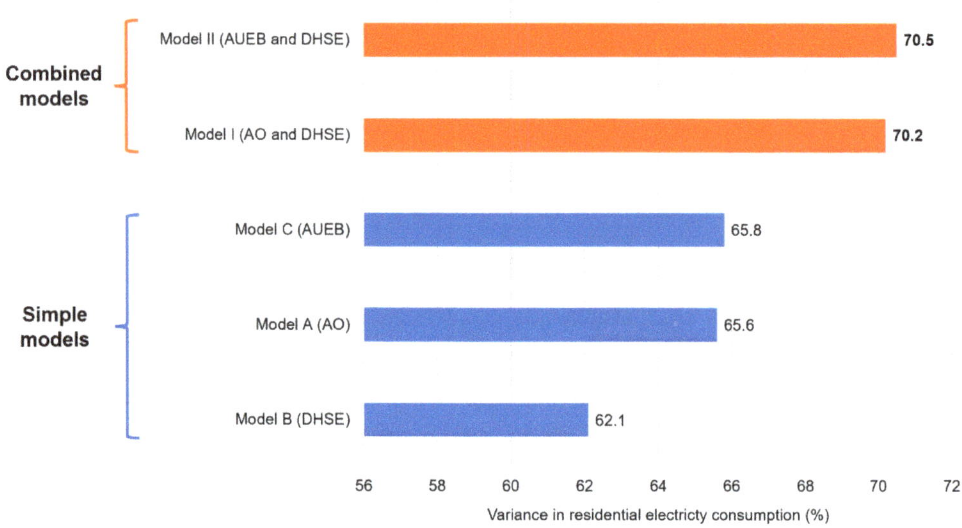

Figure 4. Variance in REC explained by the variables in the studied models.

The results demonstrate that households with a monthly income of more than USD 1426 used 132%, 83% and 48% more electricity than incomes up to USD 168, USD 168–419 and USD 419–671, respectively. These findings are consistent with those of Yohanis et al. [23], Santamouris et al. [52] and Wyatt [51], who found that the highest-income households use 2.5, 1.9 and 1.6 times more electricity than the lowest-income households in Northern Ireland, the UK and Greece, respectively. This finding could be explained as the greater the household economic situation, the more the household can afford to purchase and use appliances.

The overall presence in the bedroom was also identified as a significant predictor of REC. A unit rise (hour) in the presence in the bedroom throughout the day results in an up to 3% increase in the annual REC. This finding is in line with that of Baker and Rylatt [53], who found that the weekly hours of presence in a dwelling positively affected the REC. Indeed, the presence of householders at home for a greater number of hours per week most likely implies the use of additional electricity for activities, such as homework, using a desktop or laptop, or the use of HVAC for comfort.

Some other factors were found to be significant but not unanimously (generally significant in the individual effect models (A–C), but not in the combined effect models (I and II)). Floor area was found to be significant only in Model B (individual effect of the DHSE factors), with a unit rise in floor area (m^2) resulting in an increase in the REC of 0.2%. This complies with the findings by Sakah et al. [17], who found that an increase of 1 m^2 in the floor area increases the REC by 2.1%. Similarly, Zhou and Teng [33] reported an Increase in the REC of 0.6% for a 1% increase in the dwelling size.

Also, the education level of the HRP was also significant in model B. For example, homes with an HRP that completed university studies use 16% and 17% more electricity, respectively, than homes with an HRP that only completed secondary and primary school studies, and 22% more than those that never went to school. Similar results were reported by Zhou and Teng [33] and Leahy and Lyons [25] for Chinese and Irish families.

Finally, presence in the living room was significant only in Models B and I, with a unit rise in the daytime presence in the living room resulting in an increase in the REC of up to 2.5%.

5.2. AO Determinants of Residential Electricity Use

Significant predictors of REC concerning the AO factors include the ownership of common (appliances that demonstrated an ownership rate of at least 70%) appliances, like lighting fixtures, fans and fridges, but also less commonly owned but heavy appliances like irons, freezers and air conditioners (Tables A10, 4 and 7). An additional unit of lighting fixtures, fans and fridges in the appliance stock increases the annual REC by 2%, 6% and 22%, respectively. A positive and significant effect on the REC was also previously reported by Sakah et al. [17] regarding the ownership of such appliances. They found that the REC increases, respectively, by 87 kWh (3%), 226 kWh (7%) and 649 kWh (20%) with a unit rise in the lighting fixtures, fridges and fans. The significant and positive effect of fridges on the REC has also been widely reported in other residential electricity use studies [26,32,33], with for example, Zhou and Teng [33], reporting that households with a fridge use 22% more electricity than those without.

The findings also show that a unit rise in irons, air conditioners and freezers in the appliance stock results in an increase in the REC of 12%, 10% and 29%. Here, freezers and air conditioners were also found to be significant predictors of REC in the literature [31,32]. For example, Sakah et al. [17] reported an increase in the annual REC by 886 kWh (27%) and 1990 kWh (61%), respectively, per unit increase in the ownership of freezers and ACs.

Finally, appliances like television sets and microwaves were significant in the model built using the AO factors only (Model A), but not in the combined effect model using the AO and DHSE factors. A unit rise in the ownership of TV sets was associated with an increase in the REC of 4%, while that of a microwave was associated with an increase in the REC of 15%. Televisions have been widely acknowledged as significant predictors, positively affecting electricity use [17,31,32,53]. To date microwaves have been investigated less, with existing studies reporting no effect on REC [25].

5.3. AUEB Determinants of Residential Electricity Use

Significant predictors of REC regarding the AUEB factors include the use of common appliances like lighting fixtures, TV sets and fridges, but also less common appliances like freezers and air conditioners (Tables A11, 6 and 8). The use of lighting, TV sets and fridges has a positive effect on REC. A unit rise (1 h) in the duration of the use of lighting, TV sets and fridges increases the annual REC by 0.4%, 1% and 1.4%. Even though little research was found in the literature on the effect of appliance use on REC, such findings are similar to those of Bedir et al. [29], who explained 37% and 58%, respectively, of the variance in REC with models built on appliance use and presence factors, and appliance use and DHSE factors. TV use was also reported as having a significant and positive effect on REC within US households by Sanquist [54].

The findings also show a significant and positive effect of air conditioners and freezers on REC, with an increase in the annual REC of 5% and 1%, respectively, associated with a unit rise (1 h) in their daily use. Contrary to such findings, Bedir et al. [29] reported that cooking appliances in which freezers were classified did not affect REC, whilst extra ventilation appliances, including air conditioners, were not found to be correlated with REC in their study.

Other variables, including the use of fans and games consoles and the awareness of standby electricity consumption, were significant in the model built using only the AUEB factors, but not in the combined model using the AUEB and DHSE factors. A unit rise in the use of fans (1 h) is associated with an increase in annual REC by 1.1%. On the other hand, an increase in the use of games consoles (perhaps) unexpectantly resulted in a decrease in the REC by 5%. This finding warrants further investigation. It may be hypothesised that the increased use games consoles results in households avoiding other more electricity-consuming activities.

Also, it was demonstrated that households that are aware of standby electricity consumption use 9% more electricity. This finding may also be unanticipated as understanding that appliances consume electricity in standby mode would presumably lead to households' switching off their appliances rather than using standby. However, this finding compares well to the authors' previous research [20], which established that households headed by an HRP with a higher level of education are more informed of standby consumption, but equally, a higher level of education is associated with a higher economic level and, therefore, higher appliance ownership and use, and increased REC.

Finally, contrary to the opinion of Zhou and Teng [33], and the findings by Bedir et al. [29], who demonstrated that appliance use affects REC more than appliance ownership, the current study found very little difference between the effects of the two categories of factors on REC. Either individually (65.6 and 65.8% of the variance in the REC) or combined with the DHSE factors (70.2 and 70.5% of the variance in the REC), the AO and AUEB factors demonstrated almost the same effect on the REC. This may be explained by the fact that most of the appliances found to be significant predictors of REC had the highest ownership and use in the households' appliance stock. This includes mainly lighting fixtures, fridges, TV sets, fans and ACs.

5.4. Policy Implications and Applications for Research

The current study used data collected from a survey of 387 households in the city of Ouagadougou to analyse the determinants of REC in urban households in Burkina Faso. The analysed determinants included DHSE characteristics, AO and AUEB factors.

Household income and tenure type were identified as unambiguous predictors of REC, affecting it positively. This suggests that the economic level is the main DHSE factor to consider for REC prediction and policy implementation. As a critical objective of countries in the Global South in the coming years, economic development can be expected to lead to an increase in REC. Future policies should, therefore, consider this economic dimension. For example, policy incentives could be used to encourage the purchase of more efficient appliances, or demand-side management (DSM) strategies like load curtailment programmes could also be used.

This study demonstrated similar findings to other studies on Global South countries concerning appliance ownership and use. Indeed, the ownership and use of a range of common domestic appliances were identified as significant predictors of REC, also affecting it positively. Such appliances, including lighting fixtures, TV sets, fans, fridges, freezers and ACs, which should serve as primary targets for energy efficiency/conservation policies and DSM strategies.

Unusually, households in this study that were aware of energy conservation used 9% more electricity. Significant effort is, therefore, needed to ensure energy efficiency information and campaigns are translated into expected demand reduction in future.

The results from this study should be of great interest to a range of stakeholders in the electricity sector, not only Burkina Faso, but also other countries that share common characteristics. Indeed, it should significantly help in estimating current and future residential electricity consumption due to changes in the socio-economic and dwelling characteristics of households, and the patterns of appliance ownership and use. In this sense, the findings should not only interest decisionmakers and policymakers in designing and implementing more tailored energy efficiency/conservation policies in the residential sector, but also (electricity) utility services for future planning concerning the demand on the electrical network in the country, as well as the implementation of DSM strategies, such as demand response programmes. It could also be of key importance to energy modellers and developers of integrated renewable energy systems for buildings as information on electricity use and its influencing factors were revealed. Also, the information could interest householders wishing to review their behaviour or patterns of use in order to reduce their electricity bills.

5.5. Limitations of the Study and Future Research

Aiming to reduce the paucity of information on REC in Burkina Faso and the Global South, this study provides several insights on the influence of household characteristics and behaviours on REC. However, the current findings are limited by some restrictions explained below.

First, the sample size is limited. In the current study, the sample size was designed due to constraints like the costs associated with data collection. Larger sample sizes are needed for future research to validate the results obtained in this study and to extend the research on REC to other areas in Burkina Faso.

Self-reported data provided by households during interviews may have caused bias. Some households may feel observed even after agreeing to participate, leading to differences between their actual characteristics and behaviours and those reported. A solution for future studies is in situ measurements of occupancy and usage patterns of appliances, which will also help conduct research on aspects like occupants' behaviours.

Electricity bills were used for estimating the electricity consumption of households. However, future research could use direct measurements of electricity consumption using sensors. This would provide higher-resolution data, opening up more opportunities for analyses of domestic REC, such as the effects of appliance usage patterns on real-time REC.

Finally, other external factors concerning dwellings, including the effect of the environment, electricity price, policies and dwelling location, which were not part of this study's scope could be investigated.

Although such limitations can be suggested, the current study's findings stand, as the results can serve as a basis for forthcoming studies in Burkina Faso and other countries with similar characteristics. Also, it provides useful insights for energy planners, designers and policymakers as some information, like the prediction of REC with DHSE and appliance factors, is a valuable asset for all electricity sector actors.

6. Conclusions

This study used the data collected from 387 households in the city of Ouagadougou to provide insights into the effects of DHSE, AO and AUEB factors on urban REC in Burkina Faso. Three simple models (A to C) were built first to investigate the individual effects of the DHSE, AO and AUEB factors on REC. With the identified significant predictors of each model, two other models were built to assess the combined effects of the DHSE and AO factors (Model I), and the DHSE and AUEB factors (Model II), on REC. The key findings of the study were:

- The DHSE factors individually explained 62.1% of the variance in the REC and 70.2% and 70.5%, respectively, when combined with the AO and AUEB factors. The unambiguously significant DHSE predictors of REC were household income and dwelling tenure type, with the highest-income households using up to 1.32 times more electricity than the lowest-income households. Owner-occupied households use up to 34% more electricity than rented dwelling households.
- The AO factors individually explained 65.6% of the variance in the REC and 70.2% when combined with the DHSE factors. The unambiguously significant AO predictors of REC were the ownership of lighting fixtures, fans, fridges, irons, freezers and air conditioners. The ownerships of TV sets and microwaves were significant in the model built for the AO factors only.
- Alone, the AUEB factors explained 65.8% of the variance in the REC and 70.5% when combined with the DHSE factors. The unambiguously significant predictors of REC

include the use of lighting fixtures, TV sets, fridges, freezers and air conditioners. The use of fans and games consoles and the awareness of standby electricity consumption were significant factors in the model using the AUEB factors only.

- No differences between the effects of appliance ownership and appliance use on REC were identified, contrary to previous studies. Individually or when combined with the DHSE factors, the AO and AUEB factors explained almost the same share of variance in the REC (up to 70.2% and 70.5%).

Author Contributions: Conceptualization, K.H.S.T., Y.M.S. and S.F.A.C.; methodology, K.H.S.T., Y.M.S. and R.V.J.; software, K.H.S.T.; validation, Y.M.S., R.V.J. and S.d.S.S.; formal analysis, R.V.J. and S.d.S.S.; investigation, K.H.S.T., Y.M.S., S.F.A.C., R.V.J. and S.d.S.S.; resources, Y.M.S. and S.d.S.S.; data curation, K.H.S.T. and S.F.A.C.; writing—original draft preparation, K.H.S.T.; writing—review and editing, Y.M.S., R.V.J. and S.d.S.S.; supervision, Y.M.S., R.V.J. and S.d.S.S.; project administration, Y.M.S.; funding acquisition, Y.M.S. All authors have read and agreed to the published version of the manuscript.

Funding: This work was supported by the World Bank through the Africa Centres of Excellence Project (ACE), especially the Engineering College Project (CoE-2iE) (Grant numbers: IDA 6388-BF/D443-BF).

Data Availability Statement: The raw data supporting the conclusions of this article will be made available by the authors on request.

Conflicts of Interest: The authors declare no conflicts of interest.

Abbreviations

AC	Air conditioner
AO	Appliance ownership
AUEB	Appliance use and energy behaviour
DHW	Domestic hot water
DHSE	Dwelling, household and socio-economic
DSM	Demand-side management
FCFA	Franc of the African Financial Community
HRP	Household responsible person
HVAC	Heating, ventilation and air conditioning
ICE	Information, communication and entertainment
MLR	Multi linear regression
OLS	Ordinary least squares
REC	Residential electricity consumption
RECS	Residential electricity consumption survey
SEC	Standby electricity consumption
Std. Dev	Standard deviation
TV	Television

Appendix A

Tables A1–A3 present the description of the household, dwelling, socio-economic and appliance variables used in this study, while Tables A4–A6 provide the characteristics of the appliance stock, as well as the recorded electricity consumption of the participants.

Table A1. Dwelling, household and socio-economic (DHSE) factors included in the study.

Group of Factors	Variable (Factor/Predictor)	Variable Type	Measurement/Options
	Family composition	Categorical	Single/couple/couple with non-dependent children/family with non-dependent children/couple with dependent children/single-parent family/family with dependent children/other

Table A1. *Cont.*

Group of Factors	Variable (Factor/Predictor)	Variable Type	Measurement/Options
Dwelling, Household and socio-economic (DHSE) characteristics	Household size	Continuous	Number of people in the house
	Household income (monthly): in USD	Categorical	Class G (up to 168)/class F (168–419)/class E (419–671)/class D 671–923/class C (923–1175)/class B (1175–1426)/class A (1426 and more)
	HRP age	Categorical	Up to 28 years old/29–39 years old/40–50 years old/51–61 years old/62 and more years old
	HRP employment status	Categorical	Unemployed/retired/part-time employee/own business/full-time employee
	HRP education level	Categorical	Never attended school/primary school/secondary school/university/college
	HRP socio-economic category	Categorical	Unskilled and agricultural worker/manual-skilled or semi-skilled worker/non-manual worker/business owner/managers and professionals/other
	Gender of HRP	Dummy	Male/female
	Marital status of HRP	Dummy	Married/not married
	Number of active people	Continuous	Number of people working and earning
	Presence of teens	Dummy	Yes/no
	Presence of children (0–12 YO)	Dummy	Yes/no
	Presence of elders/retirees	Dummy	Yes/no
	Presence of housewife	Dummy	Yes/no
	Floor area	Continuous	Floor area in square meters
	Dwelling period of construction	Categorical	Until 1970/1970–1980/1980–1990/1990–2000/after 2000
	Dwelling type of construction	Categorical	Multi-family house/apartment block/semi-detached house/detached house
	Dwelling tenure type	Categorical	Free rented/privately rented/owner occupied
	Dwelling number of storeys	Continuous	Number of storeys in the dwelling
	Dwelling number of rooms	Continuous	Number of rooms in the dwelling
	Dwelling number of bedrooms	Continuous	Number of bedrooms in the dwelling
	Dwelling number of other rooms	Continuous	Number of rooms used for other purposes (other than the living room and bedrooms)
	Presence of a HVAC system	Dummy	Yes/no
	Presence of a DHW system	Dummy	Yes/no
	Overall presence in the living room	Continuous	Duration (hours) of the presence of people in the living room in 24 h
	Daytime presence in the living room	Continuous	Duration (hours) of the presence of people in the living room from 8:00 a.m. to 5:00 p.m.
	Overall presence in the bedrooms	Continuous	Duration (hours) of the presence of people in the bedrooms in 24 h
	Other presence in the bedrooms	Continuous	Duration (hours) of the presence of people in the bedrooms for activities other than resting

Table A2. Appliance ownership (AO) and appliance use, and energy behaviour (AUEB) factors, included in the study.

Group of Factors	Variable (Factor/Predictor)	Variable Type	Measurement/Options
Appliance ownership	Ownership of a Y * type appliance	Continuous	Number of the Y type appliance in the house
	Use of a Y * type appliance	Continuous	Daily duration of use of the Y type appliance (hour)
	Appliance purchase condition	Categorical	New/second hand/mixed

Table A2. *Cont.*

Group of Factors	Variable (Factor/Predictor)	Variable Type	Measurement/Options
Appliance use and energy behaviour	Awareness of appliance labelling	Dummy	Yes/no
	Influence of label on appliance purchase	Dummy	Yes/no
	Practice of energy conservation	Categorical	Yes/no/sometimes or not really
	Energy conservation actions	Categorical	Turn off unused/turn off and unplug unused/usage of efficient light/ usage of solar PV system/none
	Awareness of standby consumption	Dummy	Yes/no
	Ownership of standby appliances	Continuous	Number of standby mode appliances
	Unplugging standby mode appliances	Categorical	Yes/no/sometimes or not really
	Satisfaction related to utility services	Categorical	Not at all/not really/neutral/more or less/yes/yes, very
	Appliance preference for load shifting	Categorical	Fan/fridge or freezer/air conditioner/TV set/others

* Corresponds to the appliances that were found in the households (see Table A2).

Table A3. Activities of the households and corresponding appliances investigated in the study.

Activity	Appliances
Cooling	Humidifier, fan, air conditioner
Cooking/food preserving	Freezer, fridge, blender, kettle, microwave/oven, electric stove
ICE	Television, satellite receiver, sound system, games console, Wi-Fi router, desktop, laptop, radio, DVD/VCD, printing/scanning machine
Lighting	Indoor and outdoor lighting fixtures
Others	Washing machine, iron, others

Table A4. Characteristics of the appliance stock (number) in the households investigated in the study.

Appliance Ownership	Minimum	Mean	Maximum	Std. Dev
Indoor lighting	1	8.70	40	5.96
Outdoor lighting	0	2.51	12	1.85
Fan	0	4.34	24	3.03
Television	0	1.39	8	0.89
Satellite receiver	0	1.12	8	0.80
Fridge	0	0.93	4	0.66
Laptop	0	0.96	8	1.17
Air conditioner	0	0.63	9	1.23
Iron	0	0.30	3	0.51
Radio	0	0.23	3	0.49
Sound system	0	0.21	4	0.49
Blender	0	0.18	2	0.39
Microwave/oven	0	0.19	2	0.43
Freezer	0	0.17	2	0.41
Desktop computer	0	0.17	4	0.49
Kettle	0	0.12	3	0.34
Wi-Fi router	0	0.10	3	0.32
Games console	0	0.10	3	0.36
DVD/VCR player	0	0.08	2	0.31
Printing machine	0	0.04	2	0.20
Humidifier	0	0.03	1	0.17
Washing machine	0	0.03	1	0.16
Electric stove	0	0.04	6	0.34
Others *	0	0.04	3	0.25

* Other devices else than common domestic appliances including working machines (tailoring machines, wood piercer), sports devices, electronic pianos etc.

Table A5. Nominal power of the investigated appliances.

Name of the Appliance	Options	Unitary Power of Use (W)	Unitary Standby Power (W)
Television	Screen size ≤ 32″	52	8
	32″ < Screen size ≤ 40″	93	8
	40″ < Screen size ≤ 65″	125	5
	Screen size > 40″	196	5
Satellite receiver	Type 1	18	16
	Type 2	20	4
	Type 3	6	0.15
Ballast fluorescent lighting fixtures	0.6 m	25	-
	1.2 m	45	-
Compact fluorescent lighting fixtures	-	28	-
LED lighting fixtures	0.6 m	9	-
	1.2 m	18	-
	Bulb	15	-
Other lighting fixtures	-	75	-
Fridge	Capacity ≤ 100 L	64	0.6
	100 L < capacity ≤ 200 L	96	0.6
	100 L < capacity ≤ 300 L	111	1.0
	Capacity > 300 L	121	1.0
Freezer	Capacity ≤ 100 L	68	0.6
	100 L < capacity ≤ 200 L	125	1.0
	100 L < capacity ≤ 300 L	162	1.0
	Capacity > 300 L	305	1.0
Fan	Ceiling	42–70 *	-
	Movable	42	-
Air conditioner	Type 1	656	-
	Type 2	921	-
	Type 3	1302	-
	Type 4	1828	-
	Type 5	2220	-
Desktop computer	-	200	25
Laptop	-	60	2
Radio	-	15	6
DVD/VCR	-	43	4
Sound system	-	57	8
Wi-Fi router	-	11	5
Games console	-	150	7
Printing machine	-	400	10
Iron	-	200	-
Blender	-	125	-
Kettle	-	1500	2
Humidifier	-	150	-
Microwave/oven	-	980	1
Washing Machine	-	800	3
Electric stove	-	1500	-

* Depends on the speed at which the fan is used.

Table A6. Descriptive statistics on the recorded electricity consumption.

Statistics	Recorded Annual Electricity Consumption (kWh)
Mean	2395
Standard deviation	1687
Minimum	181
Maximum	10,188

Appendix B

Tables A7 and A8 present the results of the correlation analysis in terms of the variables retained to be entered into the simple regression models, while Figures A1 and A2 present the results from some of the preliminary checks on the regression models.

Table A7. AO and AUEB variables retained from the correlation analysis.

Variables	Pearson's r with REC	Variables	Pearson's r with REC
Ownership of lighting	0.7 ***	Ownership of freezer	0.4 ***
Ownership of TV set	0.5 ***	Ownership of kettle	0.3 ***
Ownership of desktop computer	0.3 ***	Ownership of blender	0.3 ***
Ownership of laptop	0.4 ***	Ownership of microwave/oven	0.5 ***
Ownership of radio	0.1 **	Ownership of iron	0.3 ***
Ownership of DVD/VCR	0.2 ***	Ownership of washing machine	0.2 ***
Ownership of sound system	0.2 ***	Ownership of electric stove	0.1 **
Ownership of games console	0.3 ***	Ownership of air conditioner	0.6 ***
Ownership of Wi-Fi router	0.3 ***	Ownership of fan	0.7 ***
Ownership of printing machine	0.3 ***	Ownership of other appliances	0.1 **
Ownership of fridge	0.5 ***		
Use of lighting	0.7 ***	Use of iron	0.1 **
Use of TV set	0.5 ***	Use of washing machine	0.2 **
Use of desktop computer	0.2 ***	Use of electric stove	0.2 **
Use of laptop	0.4 ***	Use of air conditioner	0.6 ***
Use of radio	0.2 **	Use of fan	0.5 **
Use of sound system	0.2 ***	Practice of energy conservation	0.2 ***
Use of games console	0.3 ***	Energy conservation actions	−0.1 *
Use of Wi-Fi routers	0.4 ***	Appliance preference for load shifting	0.1 *
Use of printing machine	0.2 ***	Awareness of standby consumption	−0.2 ***
Use of fridge	0.6 ***	Ownership of standby mode appliances	0.6 ***
Use of freezer	0.4 ***	Unplugging of standby mode appliances	0.3 ***
Use of kettle	0.2 ***	Use of microwave/oven	0.2 ***
Use of blender	0.3 ***		

* $p < 0.05$, ** $p < 0.01$, *** $p < 0.001$.

Table A8. DHSE factors retained from the correlation analysis.

Variables	Pearson's r with REC	Variables	Pearson's r with REC
Household size	0.3 ***	Number of floors in the dwelling	0.4 ***
Family composition	0.3 ***	Floor area	0.6 ***
Presence of teens	−0.3 ***	Presence of a hot water system	−0.1 **
Presence of elders/retirees	−0.2 **	Presence of a HVAC system	−0.2 **
HRP education level	0.2 ***	Household income	0.7 **
HRP marital status	−0.2 ***	HRP age	0.3 **
HRP socio-economic category	0.2 **	Daytime presence in the living room	0.3 ***
Number of active people in the house	0.3 ***	Overall presence in the bedrooms	0.4 ***
Dwelling tenure type	0.4 ***	Other presence in the bedrooms	0.3 ***
Dwelling construction type	0.4 ***		

** $p < 0.01$, *** $p < 0.001$.

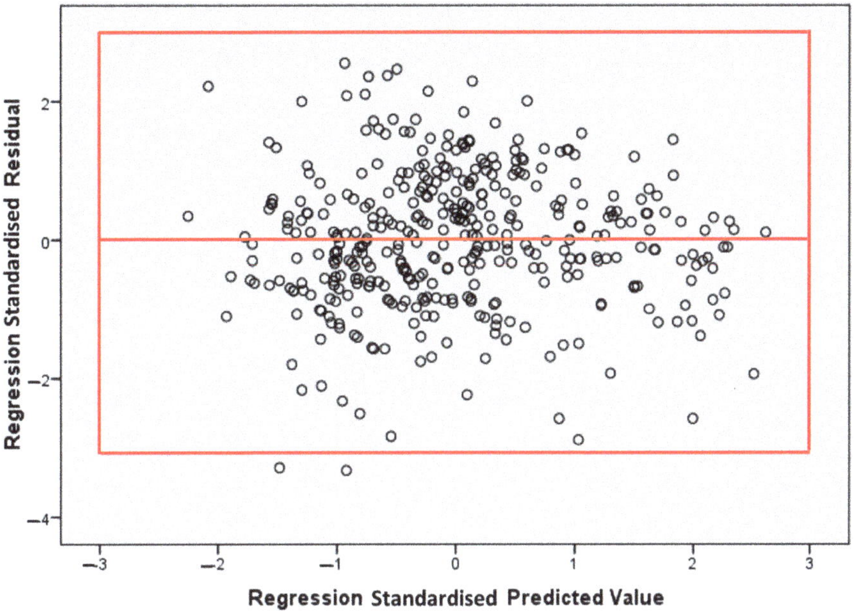

Figure A1. Linearity, additivity and homoscedasticity checking (scatterplot for model B).

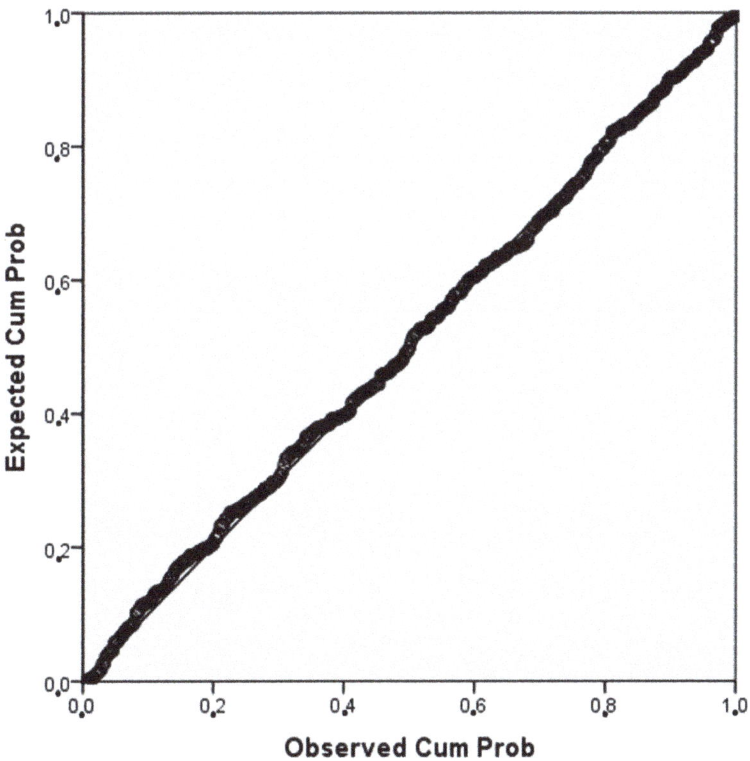

Figure A2. Normal P–P plot for checking the normality assumption (model B).

Appendix C

Tables A9–A11 show the results obtained for Model B on the effect of the DHSE factors on REC and for the combined Models I and II on the effects of the DHSE factors and appliance ownership and use on REC after dummy coding the categorical variables.

Table A9. Significant predictors of REC using DHSE factors (Model B after dummy coding of the categorical variables).

Variables	B	Std. Error
(Intercept)	7.542 ***	0.196
(Edu_Level) Never went to school	−0.215 **	0.077
(Edu_Level) Primary school	−0.176	0.093
(Edu_Level) Secondary school	−0.161 **	0.055
(Tenure type) Free rented	−0.339 *	0.149
(Tenure type) Privately rented	−0.203 **	0.063
(Household_income) Class G	−1.318 ***	0.127
(Household_income) Class F	−0.828 ***	0.106
(Household_income) Class E	−0.479 ***	0.099
(Household_income) Class D	−0.324 **	0.102
(Household_income) Class C	−0.219	0.116
(Household_income) Class B	0.152	0.136
Total floor area	0.002 ***	0.001
Presence in living room during the daytime	0.025 **	0.009
Overall presence in the bedrooms	0.034 *	0.014

R^2 = 0.621. * $p < 0.05$, ** $p < 0.01$, *** $p < 0.001$.

Table A10. Significant predictors of REC using the DHSE and AO factors (combined Model I after dummy coding of the categorical variables).

Variables	B	Std. Error
(Intercept)	7.036 ***	0.178
(Tenure type) Free rented	−0.273 *	0.129
(Tenure type) Privately rented	−0147 **	0.050
(Household_income) Class G	−0.895 ***	0.118
(Household_income) Class F	−0.462 ***	0.100
(Household_income) Class E	−0.172	0.093
(Household_income) Class D	−0.099	0.094
(Household_income) Class C	−0.007	0.105
(Household_income) Class B	−0.060	0.121
Presence in living room during the daytime	0.018 *	0.008
Overall presence in the bedrooms	0.022 *	0.012
Ownership of fan	0.029 **	0.010
Ownership of air conditioner	0.106 ***	0.023
Ownership of fridge	0.122 ***	0.034
Ownership of freezer	0.276 ***	0.054
Ownership of lighting	0.012 **	0.004
Ownership of iron	0.063	0.042

R^2 = 0.702. * $p < 0.05$, ** $p < 0.01$, *** $p < 0.001$.

Table A11. Significant predictors of REC using the DHSE and AUEB factors (combined Model II after dummy coding of the categorical variables).

Variables	B	Std. Error
(Intercept)	6.986 ***	0.178
(Tenure type) Free rented	−0.272 *	0.127

Table A11. *Cont.*

Variables	B	Std. Error
(Tenure type) Privately rented	−0.140 **	0.048
(Household_income) Class G	−0.870 ***	0.119
(Household_income) Class F	−0.478 ***	0.100
(Household_income) Class E	−0.196 *	0.093
(Household_income) Class D	−0.087	0.096
(Household_income) Class C	−0.015	0.106
(Household_income) Class B	−0.010	0.122
Overall presence in the bedrooms	0.033 **	0.011
Use of TV set	0.007 **	0.002
Use of air conditioner	0.031 ***	0.007
Use of fridge	0.009 ***	0.002
Use of freezer	0.010 ***	0.003
Use of lighting	0.002 ***	0.001

$R^2 = 0.705$.* $p < 0.05$, ** $p < 0.01$, *** $p < 0.001$.

References

1. Mutumbi, U.; Thondhlana, G.; Ruwanza, S. Reported behavioural patterns of electricity use among low-income households in Makhanda, South Africa. *Sustainability* **2021**, *13*, 7271. [CrossRef]
2. Wassie, Y.T.; Adaramola, M.S. Socio-economic and environmental impacts of rural electrification with Solar Photovoltaic systems: Evidence from southern Ethiopia. *Energy Sustain. Dev.* **2021**, *60*, 52–66. [CrossRef]
3. UEMOA. Atlas de L'énergie dans L'espace UEMOA. 2020. Available online: https://www.ifdd.francophonie.org/publications/atlas-de-lenergie-dans-lespace-uemoa/ (accessed on 1 October 2021).
4. Wethe, J. *Systèmes Énergétiques: Vulnérabilité—Adaptation—Résilience (VAR): Burkina Faso*; HELIO International: Ho Chi Minh City, Vietnam, 2009; pp. 1–48. Available online: https://www.academia.edu/3567464/Burkina_Faso (accessed on 10 April 2022).
5. Jones, R.V.; Lomas, K.J. Determinants of high electrical energy demand in UK homes: Appliance ownership and use. *Energy Build.* **2016**, *117*, 71–82. [CrossRef]
6. Cabeza, L.F.; Ürge-Vorsatz, D.; Palacios, A.; Ürge, D.; Serrano, S.; Barreneche, C. Trends in penetration and ownership of household appliances. *Renew. Sustain. Energy Rev.* **2018**, *82*, 4044–4059. [CrossRef]
7. Le, V.T.; Pitts, A. A survey on electrical appliance use and energy consumption in Vietnamese households: Case study of Tuy Hoa city. *Energy Build.* **2019**, *197*, 229–241. [CrossRef]
8. Yoshida, A.; Manomivibool, P.; Tasaki, T.; Unroj, P. Qualitative study on electricity consumption of urban and rural households in Chiang Rai, Thailand, with a focus on ownership and use of air conditioners. *Sustainability* **2020**, *12*, 5796. [CrossRef]
9. Genjo, K.; Tanabe, S.; Matsumoto, S.; Hasegawa, K.; Yoshino, H. Relationship between possession of electric appliances and electricity for lighting and others in Japanese households. *Energy Build.* **2005**, *37*, 259–272. [CrossRef]
10. Singh, J.; Mantha, S.S.; Phalle, V.M. Analysis of technical and economic electricity saving potential in the urban Indian households. *Sustain. Cities Soc.* **2018**, *43*, 432–442. [CrossRef]
11. Kim, M.; Jung, S.; Kang, J.W. Artificial neural network-based residential energy consumption prediction models considering residential building information and user features in South Korea. *Sustainability* **2020**, *12*, 109. [CrossRef]
12. Williams, S.P.; Thondhlana, G.; Kua, H.W. Electricity use behaviour in a high-income neighbourhood in Johannesburg, South Africa. *Sustainability* **2020**, *12*, 4571. [CrossRef]
13. Bohlmann, J.A.; Inglesi-Lotz, R. Examining the determinants of electricity demand by South African households per income level. *Energy Policy* **2021**, *148*, 111901. [CrossRef]
14. Ye, Y.; Koch, S.F.; Zhang, J. Determinants of household electricity consumption in South Africa. *Energy Econ.* **2018**, *75*, 120–133. [CrossRef]
15. Diawuo, F.A.; Sakah, M.; Pina, A.; Baptista, P.C.; Silva, C.A. Disaggregation and characterization of residential electricity use: Analysis for Ghana. *Sustain. Cities Soc.* **2019**, *48*, 101586. [CrossRef]
16. Diawuo, F.A.; Sakah, M.; de la Rue du Can, S.; Baptista, P.C.; Silva, C.A. Assessment of multiple-based demand response actions for peak residential electricity reduction in Ghana. *Sustain. Cities Soc.* **2020**, *59*, 102235. [CrossRef]
17. Sakah, M.; de la Rue du Can, S.; Diawuo, F.A.; Sedzro, M.D.; Kuhn, C. A study of appliance ownership and electricity consumption determinants in urban Ghanaian households. *Sustain. Cities Soc.* **2019**, *44*, 559–581. [CrossRef]
18. Akrofi, M.M.; Okitasari, M. Beyond costs: How urban form could limit the uptake of residential solar PV systems in low-income neighborhoods in Ghana. *Energy Sustain. Dev.* **2023**, *74*, 20–33. [CrossRef]
19. Danlami, A.H. Determinants of Household Electricity Consumption in Bauchi State, Nigeria. *Hyperion Econ. J.* **2017**, *5*, 16–28.

20. Tete, K.H.S.; Soro, Y.M.; Sidibé, S.S.; Jones, R. V Urban domestic electricity consumption in relation to households' lifestyles and energy behaviours in Burkina Faso: Findings from a large-scale, city-wide household survey. *Energy Build.* **2023**, *285*, 18. [CrossRef]
21. Mcloughlin, F. Characterising Domestic Electricity Demand for Customer Load Profile Segmentation. Ph.D. Thesis, Technological University Dublin, Dublin, Ireland, 2013. [CrossRef]
22. O'Doherty, J.; Lyons, S.; Tol, R.S.J. Energy-using appliances and energy-saving features: Determinants of ownership in Ireland. *Appl. Energy* **2008**, *85*, 650–662. [CrossRef]
23. Yohanis, Y.G.; Mondol, J.D.; Wright, A.; Norton, B. Real-life energy use in the UK: How occupancy and dwelling characteristics affect domestic electricity use. *Energy Build.* **2008**, *40*, 1053–1059. [CrossRef]
24. Jones, R.V.; Fuertes, A.; Lomas, K.J. The socio-economic, dwelling and appliance related factors affecting electricity consumption in domestic buildings. *Renew. Sustain. Energy Rev.* **2015**, *43*, 901–917. [CrossRef]
25. Leahy, E.; Lyons Sean, S. Energy use and appliance ownership in Ireland. *Energy Policy* **2010**, *38*, 4265–4279. [CrossRef]
26. Bartiaux, F.; Gram-hanssen, K. *Socio-Political Factors Influencing Household Electricity Consumption: A Comparison between Denmark and Belgium*; ECEEE Summer Study; ECEEE: Stockholm, Sweden, 2005; pp. 1313–1325.
27. Gram-Hanssen, K.; Kofod, C.; Petersen, K.N. Different Everyday Lives—Different Patterns of Electricity Use. In Proceedings of the 2004 ACEEE Summer Study on Energy Efficiency in Buildings, Pacific Grove, CA, USA, 22–27 August 2004; pp. 1–13.
28. Bartusch, C.; Odlare, M.; Wallin, F.; Wester, L. Exploring variance in residential electricity consumption: Household features and building properties. *Appl. Energy* **2012**, *92*, 637–643. [CrossRef]
29. Bedir, M.; Hasselaar, E.; Itard, L. Determinants of electricity consumption in Dutch dwellings. *Energy Build.* **2013**, *58*, 194–207. [CrossRef]
30. Esmaeilimoakher, P.; Urmee, T.; Pryor, T.; Baverstock, G. Identifying the determinants of residential electricity consumption for social housing in Perth, Western Australia. *Energy Build.* **2016**, *133*, 403–413. [CrossRef]
31. McLoughlin, F.; Duffy, A.; Conlon, M. Characterising domestic electricity consumption patterns by dwelling and occupant socio-economic variables: An Irish case study. *Energy Build.* **2012**, *48*, 240–248. [CrossRef]
32. Kavousian, A.; Rajagopal, R.; Fischer, M. Determinants of residential electricity consumption: Using smart meter data to examine the effect of climate, building characteristics, appliance stock, and occupants' behavior. *Energy* **2013**, *55*, 184–194. [CrossRef]
33. Zhou, S.; Teng, F. Estimation of urban residential electricity demand in China using household survey data. *Energy Policy* **2013**, *61*, 394–402. [CrossRef]
34. Fan, H.; MacGill, I.F.; Sproul, A.B. Statistical analysis of drivers of residential peak electricity demand. *Energy Build.* **2017**, *141*, 205–217. [CrossRef]
35. MEMC/BF. Tableau de Bord 2020 du Ministère de L'energie, des Mines et des Carrières. 2020. Available online: http://cns.bf/IMG/pdf/tableau_de_bord_2020_memc.pdf (accessed on 12 October 2022).
36. INSD-BF. Cinquième Recensement Général de la Population et de L'habitation du Burkina Faso (5e RGPH): Synthèse des Résultats Définitifs. 2022. Available online: https://www.insd.bf/fr/file-download/download/public/2071 (accessed on 6 February 2023).
37. MEF/BF. Monographie de la Commune Urbaine de Ouagadougou. 2009. Available online: https://ireda.ceped.org/inventaire/ressources/bfa-2006-rec-o2_commune_urbaine_ouagadougou.pdf (accessed on 12 August 2022).
38. Kabore, B.; Kam, S.; Ouedraogo, G.W.P.; Bathiebo, D.J. Etude de l'evolution climatique au Burkina Faso de 1983 à 2012: Cas des villes de Bobo Dioulasso, Ouagadougou et Dori. *Arab. J. Earth Sci.* **2017**, *4*, 50–59.
39. Massieke Soma, A.A.R.; Tapsoba, F.W.; Kabore, D.; Seogo, I.; Tankoano, A.; Dicko, M.H.; Toguyeni, A.; Sawadogo-Lingani, H. Etude sur la capacité de production, du circuit de commercialisation et de la consommation du zoom-koom vendu dans la ville de Ouagadougou au Burkina Faso. *Int. J. Biol. Chem. Sci.* **2018**, *11*, 2294. [CrossRef]
40. Cochran, W.G. Sampling Techniques. 1977. Available online: https://fsapps.nwcg.gov/gtac/CourseDownloads/IP/Cambodia/FlashDrive/Supporting_Documentation/Cochran_1977_Sampling%20Techniques.pdf (accessed on 14 January 2021).
41. Uakarn, C. Sample size estimation using Yamane and Cochran and Krejcie and Morgan and Green formulas and Cohen statistical power analysis by G*power and comparisons. *Apheit Int. J.* **2021**, *10*, 76–88.
42. Adam, A.M. Sample Size Determination in Survey Research. *J. Sci. Res. Rep.* **2020**, *26*, 90–97. [CrossRef]
43. Sukarno, I.; Matsumoto, H.; Susanti, L. Household lifestyle effect on residential electrical energy consumption in Indonesia: On-site measurement methods. *Urban Clim.* **2017**, *20*, 20–32. [CrossRef]
44. Mohammad, A.; Shrestha, P.; Kumar, S. Urban residential energy use in Kandahar, Afghanistan. *Cities* **2013**, *32*, 135–142. [CrossRef]
45. Hong, T.; Yan, D.; D'Oca, S.; Chen, C.-f. Ten questions concerning occupant behavior in buildings: The big picture. *Build. Environ.* **2017**, *114*, 518–530. [CrossRef]
46. Carpino, C.; Mora, D.; De Simone, M. On the use of questionnaire in residential buildings. A review of collected data, methodologies and objectives. *Energy Build.* **2019**, *186*, 297–318. [CrossRef]
47. Aglina, M.K.; Agbejule, A.; Nyamuame, G.Y. Policy framework on energy access and key development indicators: ECOWAS interventions and the case of Ghana. *Energy Policy* **2016**, *97*, 332–342. [CrossRef]
48. Field, A. *Discovering Statistics Using IBM SPSS Statistics*; Sage Publications, Inc.: Thousand Oaks, CA, USA, 2018.
49. Cook, R.D.; Weisberg, S. Residuals and Influence in Regression. 1982, p. 230. Available online: http://conservancy.umn.edu/handle/11299/37076 (accessed on 12 December 2022).

50. Hamilton, I.G.; Steadman, P.J.; Bruhns, H.; Summerfield, A.J.; Lowe, R. Energy efficiency in the British housing stock: Energy demand and the Homes Energy Efficiency Database. *Energy Policy* **2013**, *60*, 462–480. [CrossRef]
51. Wyatt, P. A dwelling-level investigation into the physical and socio-economic drivers of domestic energy consumption in England. *Energy Policy* **2013**, *60*, 540–549. [CrossRef]
52. Santamouris, M.; Kapsis, K.; Korres, D.; Livada, I.; Pavlou, C.; Assimakopoulos, M.N. On the relation between the energy and social characteristics of the residential sector. *Energy Build.* **2007**, *39*, 893–905. [CrossRef]
53. Baker, K.J.; Rylatt, R.M. Improving the prediction of UK domestic energy-demand using annual consumption-data. *Appl. Energy* **2008**, *85*, 475–482. [CrossRef]
54. Sanquist, T.F.; Orr, H.; Shui, B.; Bittner, A.C. Lifestyle factors in U.S. residential electricity consumption. *Energy Policy* **2012**, *42*, 354–364. [CrossRef]

Disclaimer/Publisher's Note: The statements, opinions and data contained in all publications are solely those of the individual author(s) and contributor(s) and not of MDPI and/or the editor(s). MDPI and/or the editor(s) disclaim responsibility for any injury to people or property resulting from any ideas, methods, instructions or products referred to in the content.

buildings

MDPI

Review

Current Developments and Future Directions in Energy-Efficient Buildings from the Perspective of Building Construction Materials and Enclosure Systems

Rahman Azari [1], Ehsan Kamel [2] and Ali M. Memari [3,*]

[1] Department of Architecture, Pennsylvania State University, University Park, PA 16802, USA; razari@psu.edu
[2] Department of Energy Management, New York Institute of Technology, Old Westbury, New York, NY 11568, USA; ehsan.kamel@nyit.edu
[3] Department of Architectural Engineering, Department of Civil and Environmental Engineering, Penn State University, University Park, PA 16802, USA
* Correspondence: amm7@psu.edu

Abstract: The need to design buildings in compliance with the Paris Agreement goal requirements is urgent, and architects and engineers need to consider energy use and operational and embodied carbon requirements in doing so. Building envelopes will be an important element in the next generation of high-performance buildings and there have been significant advancements in recent years to develop building envelopes that help mitigate the building carbon emissions through energy-conserving low-embodied carbon or carbon-sequestering solutions. The key objective of this article is to present an overview of the state-of-the-art in the field of energy-efficient low-carbon buildings with a focus on envelope systems. This article provides a survey of the literature on energy use and carbon emissions of the United States building stock, presents recent advancements in energy-conserving building envelopes, and highlights reuse–reduce–sequester strategies that mitigate the embodied carbon of buildings. As materials are critical in reducing the energy consumption and carbon emissions of buildings, this paper also presents developments on diverse materials and building envelope solutions that have been effective in creating high-performance buildings, from insulation materials to phase-change materials and aerogels. Finally, the characteristics of a selected number of progressive net-zero-energy guidelines such as Passive House Institute (PHI) standards, Passive House Institute US (Phius) standards, the PowerHouse standard, and the BENG standard are discussed. The findings of this work highlight the increased focus on the design, construction, and engineering strategies that aim to mitigate the carbon emissions of buildings based on a holistic whole-life carbon mitigation approach.

Keywords: building envelope; embodied carbon; energy-efficient buildings; building energy modeling tools

1. Introduction

Characterized by extreme weather events and rising global temperature, climate change is one of the pressing challenges of the 21st century. Controlling global warming to below 1.5 °C has therefore become a critical goal for the global community and is emphasized in the Paris Agreement to address climate change. This goal is only achievable through "immediate and deep emissions reduction across all sectors", according to the Intergovernmental Panel on Climate Change [1].

The building sector serves as a significant contributor to carbon emissions and climate change through both operational and embodied energy usage and associated emissions. Operational energy encompasses the day-to-day functions of buildings, such as heating and cooling, while embodied energy pertains to the energy consumed during the construction, maintenance, and demolition phases. In this review paper, we review the pivotal role of building envelopes in reducing energy consumption and carbon emissions and improving buildings' energy efficiency and sufficiency, focusing on innovative engineering solutions aimed at enhancing energy efficiency and sustainability.

Building envelopes play a crucial role in regulating the transfer of heat, air, and moisture between the interior and exterior environments of structures. The effective design and implementation of building envelope technologies offer immense potential for reducing energy consumption and mitigating carbon emissions throughout a building's life cycle. From passive strategies like improved insulation to active measures such as incorporating renewable energy generation systems, a diverse range of engineering solutions exist to enhance the performance of building envelopes. Through an exploration of these innovative technologies, this paper aims to provide insights into how advancements in engineering can drive sustainable practices within the construction industry.

To achieve these objectives, this review paper adopts a comprehensive methodology that involves synthesizing the existing literature on building envelope solutions aimed at reducing energy consumption and embodied carbon emissions. A straightforward search strategy for review papers is adopted, and the relevant literature on building envelope solutions aimed at reducing energy consumption and embodied carbon emissions is identified and collected. Studies focusing on the performance of building envelope solutions across various regions and climate zones, as well as research on novel materials and emerging energy conservation measures (ECMs) in building envelopes are specifically targeted.

For the selection criteria, studies were selected based on their alignment with the topics building envelope solutions, energy efficiency, and embodied carbon reduction. Factors such as climate considerations, cost implications, energy and carbon savings, durability, and constructability of building envelope solutions were considered. The inclusion criteria encompassed research utilizing experimental approaches, physics-based modeling tools, customized methodologies, and data-driven techniques. For data extraction, information regarding the performance of building envelope solutions in terms of energy efficiency and carbon reduction was extracted from selected studies.

The following sections review the envelope-based energy retrofit approaches, use of integrated energy generation systems for energy sufficiency, embodied carbon in buildings, tools for energy and carbon performance of buildings, and net-zero-energy design standards, and this paper is wrapped up by the conclusions.

2. The Current State of Operational and Embodied Energy Use of the United States Building Stock

The global and national economies channel energy into the building sector for two distinct purposes that form the two key sources of building carbon emissions: the operation of buildings (e.g., heating and cooling) and their construction and maintenance over the full life cycle (Figure 1). Referred to as operational and embodied energy or carbon, these two aspects of building energy use and carbon emissions are interrelated and require aligned mitigation. Meanwhile, building envelopes are critical in controlling the flow of mass and energy from and into buildings, while significantly affecting building energy use and operational and embodied carbon emissions. There have been notable advancements over the past decade in the development of building envelope technologies for energy conservation and carbon avoidance as well as corresponding modeling and simulation tools to assess the energy and environmental performance of building envelopes.

It is helpful to first review some of the data on energy use and carbon emissions of buildings. From 100,410 trillion BTUs of total energy consumed in 2022 in the United States, the commercial and residential sectors accounted for 39% [2], using renewable and non-renewable energy (including the energy loss in the electrical system) for their operation. Space heating (42%), water heating (18%), air conditioning (9%), and lighting (3%) consume 72% of the site energy in the residential sector [3], while the energy use in the commercial sector is primarily caused by space heating (32%), ventilation (11%), lighting (10%), and cooling (9%) [4]. Data show that the operational energy-related carbon emissions of buildings (i.e., operational carbon) were responsible for 30% of greenhouse gas emissions in the United States in 2021 [5]. While data, figures, and estimates on the operational energy use of the U.S. building stock are detailed and comprehensive, the embodied energy data are often limited to figures that represent certain construction materials. However, the building sector causes significant amounts of material and energy use and associated carbon emissions. Gursel et al. [6] estimate the carbon emissions associated with the manufacturing of concrete, steel, and wood to represent 2.1% of U.S. emissions. On a global scale, the manufacturing of construction materials is estimated to cause 11% of global emissions [5]. The United Nations Environment Program also reports that the proportion of global energy and process emissions attributable to buildings in the year 2021 was about 37%, comprising 17% residential (direct and indirect), 11% non-residential (direct and indirect), and 9% building construction industry (including 6% for concrete, aluminum, and steel and 3% for brick and glass) (Figure 2) [7].

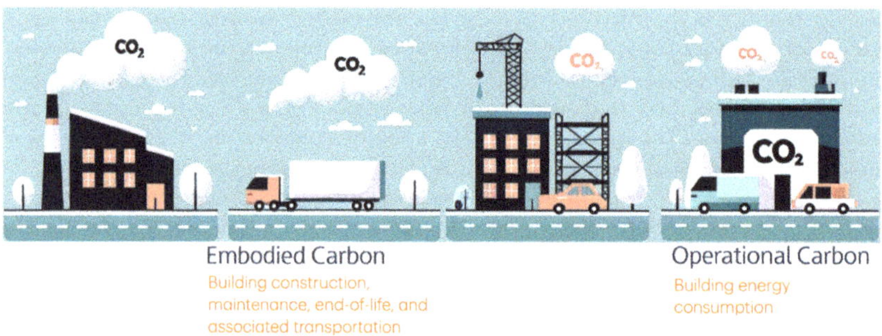

Figure 1. The life cycle of construction and its carbon footprint.

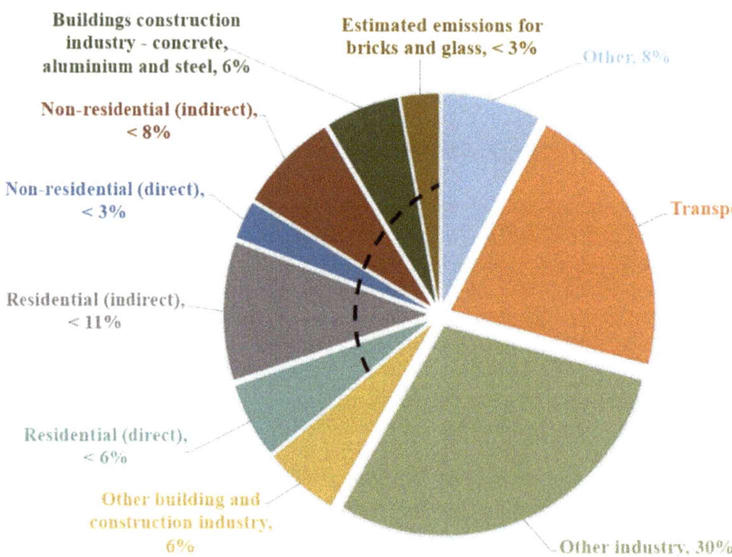

Figure 2. Share of buildings in global energy and process emissions in 2021.

3. Review Methodology

A straightforward search strategy for review papers was adopted, and the relevant literature on building envelope solutions aimed at reducing energy consumption and embodied carbon emissions was identified and collected. Studies focusing on the performance of building envelope solutions across various regions and climate zones, as well as research on novel materials and emerging energy conservation measures (ECMs) in building envelopes were specifically targeted.

For the selection criteria, studies were selected based on their alignment with the topics of building envelope solutions, energy efficiency, and embodied carbon reduction. Factors such as climate considerations, cost implications, energy and carbon savings, durability, and constructability of building envelope solutions are considered in the study. The inclusion criteria encompassed research utilizing experimental approaches, physics-based modeling tools, customized methodologies, and data-driven techniques. For data extraction, information regarding the performance of building envelope solutions in terms of energy efficiency and carbon reduction was extracted from the selected studies.

4. Envelope-Based Approaches to Mitigate Operational Energy Use and Carbon Emissions

The key objective of this article is to present an overview of the state-of-the-art in the field of energy-efficient low-carbon buildings, with a focus on enclosure systems rather than heating, cooling, energy management control systems, or electrical systems. There are a wide range of building envelope-based energy conservation measures (ECMs) that enhance building energy efficiency, from radiative sky cooling to innovative solutions such as phase-change materials (PCMs), vacuum-insulated panels (VIPs), and aerogels. This paper discusses a selection of these solutions, summarized in Table 1 and schematically shown and summarized in Figure 3. These upgrades demonstrate how existing buildings can be modernized to improve energy performance and reduce environmental impact without necessitating complete reconstruction. The illustration emphasizes the feasibility and effectiveness of applying these technologies to retrofit older buildings for sustainability.

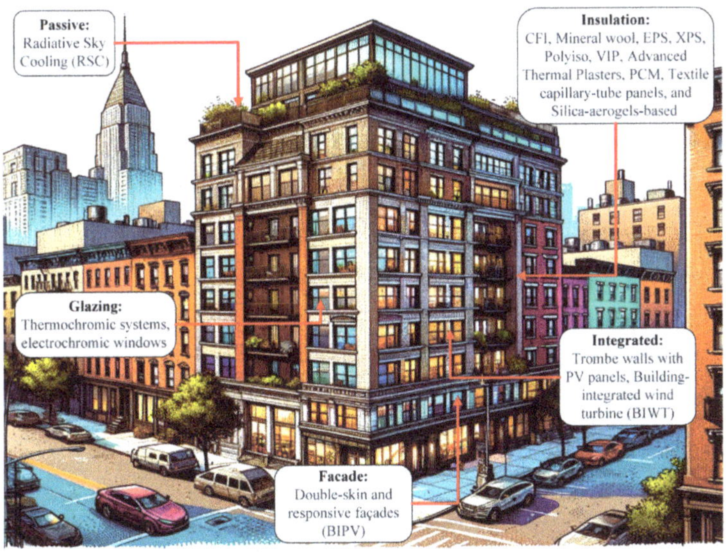

Figure 3. A schematic overview of the reviewed building envelope energy conservation measures.

Radiative sky cooling (RSC) is an emerging ECM in which cooling naturally occurs, allowing objects to release heat into the cooler air by leveraging temperature differences between the surface and the atmosphere. While RSC is not a passive envelope-based ECM in a strict sense, it can be bundled with them to enhance the performance of the building skins. In a discussion of RSC systems for building energy saving and cooling, Wu et al. [8] present a review of previous experimental and simulation-based studies on RSC, highlight its potential for effective energy conservation, and discuss some of the optimization techniques to improve its performance. The authors also point out the challenges associated with implementing RSC, such as the need for sizable radiative cooling panels and the impact of weather conditions on its effectiveness, and conclude that addressing these challenges needs more research [8]. The RSC studies reviewed by Wu et al. [8] are primarily experimental, showing the knowledge gap in simulation-based studies that document the performance of this ECM. This requires new software tools to be developed or the existing software packages to incorporate new features to help simulate the RSC performance.

The energy efficiency achieved with RSC varies by location and season (i.e., climate zone). For example, in Albuquerque, summer savings range from 12% to 21%, and in winter, from 21% to 44%. Similarly, in Atlanta, summer savings are 8% to 17%, while in winter, they are 11% to 25%. Cool roofs' effectiveness depends on building volume and surface area. They save about 4.1–10.2 kWh/m^2 annually or approximately 9% daily. However, RSC's benefits in summer may be offset in winter due to heat loss. Thus, selecting suitable roofs to reduce winter heat loss is crucial. Saving energy relies on location, season, and building type [8].

Chen et al. [9] assessed the thermal efficiency of passive envelope systems integrating thermally insulated glazing materials and radiative sky cooling for cooling purposes. EnergyPlus, a widely utilized whole-building energy simulation tool, was employed to dynamically address mass and energy balances within building spaces, considering variable internal loads and external conditions. The study focused on eight representative cities across diverse ASHRAE climate zones to evaluate the potential benefits of the proposed passive envelope systems, accounting for regional climatic variations and building characteristics. The results showcased climate-dependent performance patterns in response to different passive envelope technologies. A comprehensive analysis of the simulation outcomes identified optimal envelope designs tailored to each climate region, followed by evaluations of economic and environmental performance. The research shows the promising potential of passive envelope strategies in advancing sustainable low-energy building designs [9].

Another emerging ECS approach is building envelope energy retrofit by taking advantage of the thermal properties (e.g., conductivity and specific heat) of certain materials in thin layers of plasters and coatings. Fantucci et al. [10] discussed the rising demand for highly energy-efficient buildings and extensive energy retrofitting of existing structures, aiming to meet the Paris Agreement's 40% emission reduction target by 2030. Various advanced building envelope materials were reviewed, including low-emittance materials, advanced thermal plasters, super insulating materials, phase-change materials, and highly reflective coatings. These materials offer exceptional thermal performance, even in thin layers, making them suitable for scenarios where traditional insulation methods are impractical. However, despite their potential, their adoption remains limited due to high costs and a lack of knowledge and data regarding their thermal properties [10].

Another novel material is aerogel, which has been studied as a potential ECM in buildings since its introduction into the construction industry. Zhong et al. [11] conducted a study on developing a hierarchical cellulose aerogel reinforced with in situ assembled cellulose nanofibers for building cooling. Regarding the tools and techniques related to building envelope retrofit, they used EnergyPlus to simulate the cooling energy saving potential of applying aerogel in building envelopes and also involved theoretical models, Fourier transform infrared spectroscopy (FTIR), X-ray diffraction (XRD) analysis, and mechanical testing to characterize aerogel properties. The results showed that the aerogel exhibited excellent mechanical properties, could achieve significant cooling power during the day, and could achieve a theoretical maximum sub-ambient temperature of 12.0 °C and a cooling power of 58.6 W·m^{-2} during the day [11].

The transparency of aerogel has also made it a suitable ECM in transparent building envelope components such as glazing systems. Abraham et al. [12] discuss the development of a highly transparent aerogel made of silanized cellulose fibers that can be used to improve the energy efficiency of glazing in buildings. The authors used thermal conductivity measurements and numerical energy assessments to evaluate the performance of an aerogel glazing unit. They also used ISO 9050:2003 [13] to determine solar direct transmittance, light transmittance, total solar energy transmittance, ultraviolet transmittance, and other glazing factors. A small Hot-Box apparatus was applied to measure the thermal conductivity of different materials, while a numerical energy assessment was employed to evaluate the energy performance of the aerogel glazing unit, and ISO 9050:2003 was considered to determine the optical properties of the aerogel glazing unit [12].

Ibrahim et al. [14] presented a numerical and experimental investigation into the hygrothermal performance of innovative super-insulating systems that are based on silica aerogels. Environmental inputs and experimental data were collected using sensors, heat flux meters, and a weather station. Numerical models were created using WUFI® Pro 5.3 software, incorporating coupled heat and moisture transfer according to standard EN 15026. Spatial discretization employed the finite volume technique, while the fully implicit scheme handled temporal discretization. The solver iteratively computed temperature and relative humidity profiles until convergence. The results indicated close agreement between the simulated and measured temperature fields, with captured daily dynamics. Statistical parameters, including PRMSE, RMSE, and average absolute difference, were used to assess the simulation–measurement agreement. Overall, the study concluded that these novel super-insulating systems significantly reduced heat flux compared to uninsulated walls, showcasing their potential for efficient thermal insulation [14].

Another category of building envelope ECMs is the integrated systems designed based on the benefits of both passive and active systems, such as Trombe walls and photovoltaic (PV) panels. Duan et al. [15] proposed an innovative Trombe wall system integrating PV panels and aluminum panels layered with photocatalyst and thermal materials to enhance energy harvesting efficiency. This system harnesses solar energy and catalytic oxidation to heat and purify chamber air, thereby enhancing indoor air quality. The results indicate the system's capability to generate heat, ranging from 6.25 kJ/mol to 17.74 kJ/mol, and electricity, from 0.075 kWh to 0.372 kWh, daily from 9:00 to 16:00. Moreover, the system demonstrates a one-way sterilization efficiency against bacterial aerosols from 0.204 to 0.347. Optimal system performance was observed with a spacing of 25 cm. Experimental tools included a hot-wire anemometer, thermocouple, data acquisition instrument, proportional integration differentiation controller, environmental chamber, electric heater, aerosol generator, ultrasonic humidifier, I-V curve tester, and pyranometer [15]. Such systems are typically modeled and simulated using customized tools, and there is potential for existing BEM tools to incorporate more direct simulation options for similar integrated systems for more accurate energy-saving and generation calculations.

Ascione et al. [16] conducted a review of recent research on energy retrofit through double-skin and responsive facades. The study focused on building skins as the primary means of heat gain and loss between indoors and outdoors. The authors reviewed various technologies, including building-integrated photovoltaic (BIPV) systems, double-skin façades (DSFs), and responsive building components. The reviewed studies employed various computer tools, techniques, and software for energy modeling and energy simulation, such as EnergyPlus, TRNSYS®, DesignBuilder®, GenOpt, and MATLAB®. These tools were used to analyze the performance of different façade systems and their impact on energy consumption, thermal comfort, and overall building performance. The review found that the implementation of DSF and responsive façades can lead to significant energy savings. For example, a simple DSF could lead to up to 90 percent reduction in heating loads and 30 percent reduction in cooling loads. However, the authors also identified the obstacles and critical factors, such as lower efficiency of transparent PV compared with traditional PV, the risk of overheating, and greater cost of responsive components [16].

Si et al. [17] introduced an innovative building envelope design featuring variable thermal properties for passive heating systems. Their study aimed to balance solar heat gain and thermal insulation, addressing a common challenge in traditional passive solar setups. Field measurements were undertaken to validate the effectiveness of this transparent building envelope, particularly in cold plateau regions, under a step control operation strategy. Using DesignBuilder software, the researchers conducted numerical simulations to assess the efficiency of the proposed envelope in increasing indoor temperatures. The simulation outcomes mirrored field measurements, confirming that implementing the operation strategy—opening indoor windows at 10:00 a.m. and closing them at 5:00 p.m.—could maximize solar gain and substantially elevate indoor temperatures. The results indicated that rooms utilizing the proposed envelope design exhibited average temperature increases of 2.0 °C (on sunny days) and 1.5 °C (on cloudy days) compared to those employing three other passive solar envelope operation modes. Overall, the building envelope with variable thermal performance holds significant promise for enhancing indoor thermal comfort in cold plateau regions at a minimal expense [17].

Baek et al. [18] investigated an experimental retrofitting approach for seismic and energy improvements in reinforced concrete (RC) and unreinforced masonry (URM) structures. Their retrofit system integrates prefabricated textile capillary-tube panels with existing building envelopes, affixed using mortar or adhesives. Seismic performance was evaluated on scaled masonry-infilled RC frames and URM walls. Thermal efficiency was gauged via in situ experimentation, showing significant effectiveness. The results indicated enhanced seismic performance, delaying masonry infill cracking and diffuse damage. Retrofitted specimens exhibited improved energy dissipation compared to untreated frames, with the TCP-retrofitted wall showcasing notably higher stiffness, load resistance, and displacement capacity. Additionally, the addition of a C-TRM jacket further bolstered stiffness and lateral resistance. This research underscores the practical viability of TCP retrofitting as a cost-efficient solution for integrated seismic and energy retrofitting of existing structures [18].

Cellulose fiber insulation (CFI) consists of paper stocks blended with chemicals to enhance fire and corrosion resistance, among other properties. Typically, its thermal resistance (R-values or RSI) falls within 0.62–0.69 m^2K/W per 25.4 mm (3.5–3.9 ft2·°F·h/BTU per inch) [19]. Mineral wool, on the other hand, is crafted from molten raw materials like stone and slag, turned into fibers, then bonded, and compacted to form insulation batts. It has high fire resistance, high thermal resistance, and low-embodied carbon, making it an attractive insulation choice. Mineral wool's R-value range is generally close to that of cellulose fiber insulation [20].

Apart from loose insulation materials, various types of rigid board insulation are available, including expanded polystyrene (EPS), extruded polystyrene (XPS), and polyisocyanurate (polyiso), each with differing levels of embodied carbon. EPS, commonly used in walls, roofs, and floors, typically offers an R-value ranging from 0.63 to 0.70 m^2K/W per 25.4 mm (3.6–4.0 ft^2·°F·h/BTU per inch). XPS, stronger and denser than EPS, finds application in walls and below-grade slabs, with a semipermeable nature and an R-value of approximately 0.79 to 0.88 m^2K/W per 25.4 mm (4.5–5.0 ft^2·°F·h/BTU per inch). Polyiso, the priciest among these options, has the highest thermal resistance, typically ranging from 1.23 to 1.41 m^2K/W per 25.4 mm (7.0–8.0 ft^2·°F·h/BTU per inch), often featuring reflective foil facing on both sides to enhance resistance to radiant heat transfer. Additionally, advanced systems like vacuum-insulated panels (VIPs) offer highly efficient insulation within minimal thickness, typically with thermal conductivity ranging from 0.004 W/(mK) to 0.008 W/(mK) [21]. Phase-change materials (PCMs) also have immense potential for building envelope applications as they can absorb external heat in the form of latent heat through a phase transition, for example from solid to liquid [22], and release it with a delay to warm up the building.

While insulation materials are key to preventing heat loss from opaque envelop components, glazing systems are the weakest of the building envelope components when it comes to thermal resistance. Because glazing is a major component in the assessment of heating and cooling loads, thermal comfort, and energy demand of a building ([23,24]), proper specification of glazing systems can reduce summer heat gain and winter heat loss [25]. Accordingly, various glazing technologies have been developed and are being introduced to increase such thermal resistance. Of course, because glazing needs to also provide natural lighting and visual transmissibility, such systems are in general evaluated based on the following three main performance attributes: the thermal U-factor, the Solar Heat Gain Coefficient (SHGC), and the Visual Transmittance (Tvis).

U-factor reflects the rate of heat flow (in Btu/(hr * ft^2 * °F)) through conduction, convection, and radiation (NFRC 2005). SHGC quantifies the solar heat transmitted through a material/system, ranging from 0 (no transmission) to 1 (full transmission). Tvis indicates the percentage of visible light passing through glazing, with 1 representing full transmission and 0 indicating no transmission. The selection criteria for glazing systems encompass energy efficiency, sustainability, maintenance, durability, noise control, fire resistance, visual aesthetics, and structural resilience in natural disasters [26,27].

Building envelopes are responsible for approximately 36% of total building energy usage, primarily attributed to heat gain and loss [28]. Accordingly, replacing inefficient single-glazed window units with their newer, more energy-efficient counterparts offers the first step in this direction, which is consistent with the goal of the U.S Department of Energy (DOE) Building Technologies Program (BT) to develop market-viable zero energy buildings (ZEBs) by 2025.

As new emerging areas, dynamic and adaptive curtain wall systems [29] are among the leading technologies, which can be adaptable and adjustable to environmental fluctuations. Adaptive envelopes provide advancements in performance regarding building energy efficiency and economics, through real-time behavior and configuration adjustments [30]. Adaptability is often achieved through a combination of mechanical, electrical, and chemical approaches [29]. These adaptive façades consist of multifunctional highly adaptive systems that can be at material-, component-, and/or system-level implementations [31].

Various technologies employ coatings to decrease glazing emittance, while some incorporate "honeycomb" fill layers for enhanced thermal insulation. Additionally, products utilize polycarbonate or similar materials to boost performance. Thermochromic glazing is recognized as a smart, energy-saving window technology [32]. Advanced window systems such as switchable electrochromic or gas chromic windows are other opportunities to reduce the overall energy loads of buildings [33,34].

5. Use of Locally Renewable Energy Sources for Building Operations to Balance Energy Consumption

With the current trends toward electrification and the desire for greater energy self-sufficiency, there is a growing interest in the adoption of photovoltaic (PV) systems for local energy generation [35]. These systems have the potential to generate a significant portion, if not all, of a household's daily electrical energy demand, either from rooftop installations or as integrated components of the building structure itself [36]. Today, there are a variety of products that differ in configuration and efficiency, but the most common PV systems for houses are roof-mounted PV arrays, generally made up of crystalline silicon cells that are normally installed on special frames attached to roofs.

PV arrays can be categorized into three main types based on the arrangement and assemblies of the silicon cells [35]. The first one is mono-crystalline, which is the most efficient and expensive type of PV cell, with an average efficiency between 18 and 20% [37]. Poly-crystalline panels are less efficient (14–17%) compared to mono-crystalline cells. Then, there are thin films (amorphous), which is a thin layer of silicon deposited on a base material such as metal or glass to create a solar panel. These amorphous solar panels are less efficient (9–10%) than mono- or poly-crystalline panels. However, the advantage of thin-film cells is that they can be applied to different (including curved) surfaces [37].

As a more recent technology, building-integrated photovoltaics (BIPVs) are becoming more popular as they integrate PV arrays with building fabrics [38,39]. When compared to roof-mounted systems, BIPV systems have the added benefit of increasing the potential for renewable energy while also saving money and time during installation by replacing traditional building components. Examples of BIPV systems include shingles, tiles, and glazing [40]. As one example of BIPV application to glazing, photovoltaic (PV) cells are securely laminated between two designated glazing panes, completely integrating PV technology with glass. Its adaptability makes it suitable for many uses where conventional glass would be used, such as facades, skylights, curtain walls, and windows.

The integration of wind turbines with building facades and envelopes represents a creative approach towards sustainable energy generation within urban environments and is predicted by researched to potentially become an integral part of future buildings [41]. Park et al. [42] introduced a novel building-integrated wind turbine (BIWT) setup that leverages the building facade, a traditionally overlooked space in conventional BIWT systems. The setup incorporates a guide vane to capture and amplify wind velocity, complemented by a custom rotor design. Positioned as a potential eco-friendly energy solution for urban settings, the system's projected power generation capacity was assessed in relation to a residential building's energy demands. The analysis suggests that the proposed system could potentially fulfill approximately 6.3% of the building's electricity requirements [42].

Table 1. Summary of reviewed building envelope energy conservation measures and their energy saving potential.

ECM Category	Description	Examples	Study	Energy Savings/Generation
Radiative Sky Cooling (RSC)	Natural cooling leveraging temperature differences	-	Wu et al. [8]	Varies by region/season; e.g., 12–21% in summer and 21–44% in winter in Albuquerque
Passive Envelope Systems	Includes insulated glazing and radiative sky cooling	-	Chen et al. [9]	Dependent on climate zone and building characteristics
Energy Retrofit	Use of high thermal property materials in thin layers	Advanced thermal plasters, PCM	Fantucci et al. [10]	Not specified
Aerogel	Highly insulating material suitable for glazing systems	Hierarchical cellulose aerogel	Zhong et al. [11]	The theoretical maximum sub-ambient temperature of 12.0 °C and cooling power of 58.6 W·m^{-2}
Super-Insulating Systems	Based on silica aerogels	Silica-aerogel-based systems	Ibrahim et al. [14]	Not specified, but reduced heat flux compared to un-insulated reference
Integrated Systems	Combines passive and active systems	Trombe walls with PV panels	Duan et al. [15]	Heat: 6.25–17.74 kJ/mol; electricity: 0.075–0.372 kWh/day
	Integrating wind turbines into the building façade/envelope or the structure	Building-integrated wind turbine (BIWT)	Park et al. [42]	Potentially fulfills about 6.3% of the building's electricity requirements

Table 1. *Cont.*

ECM Category	Description	Examples	Study	Energy Savings/Generation
Responsive Façades	Retrofit via double-skin and responsive façades	BIPV systems, DSF	Ascione et al. [16]	Up to 90% heating load reduction and 30% cooling load reduction
Variable Thermal Performance Envelope	Balances solar gain and insulation in passive solar systems	-	Si et al. [17]	Average temperature increase of 2.0 °C (sunny) and 1.5 °C (cloudy)
Seismic-Plus-Energy Retrofit	Prefabricated panels for seismic and energy performance	Textile capillary-tube panels	Baek et al. [18]	High efficiency noted, specific savings not quantified
Insulation Materials	Prevents heat loss from opaque envelope components	CFI, mineral wool, EPS, XPS, Polyiso, VIP, PCM	Cellulose Insulation Manufacturers Association [19]	Varies by material; e.g., CFI: 4.1–10.2 kWh/m^2 annually
Glazing Systems	Increases thermal resistance of glazing	Thermochromic systems, electrochromic windows	Warwick et al. [32]	Depend on U-factor, SHGC, and Tvis values

6. Embodied Carbon—The Reuse–Reduce–Sequester Approach

As modern buildings now consume less operational energy, thanks to insulation materials and high-performance window systems used in their building skins, the focus pivots toward embodied energy and carbon as the next decarbonization step. Embodied energy is defined as the amount of energy (in the form of fossil fuel and electricity) required to produce the unit weight of a usable material, including the transportation involved [43]. Embodied carbon is the equivalent carbon dioxide of the greenhouse gas emissions released in the partial life cycle (cradle-to-gate) or full life cycle (cradle-to-grave) of a building.

Embodied carbon mitigation is a three-fold strategy involving reuse, reduction, and sequester [44]. The reuse of existing buildings, or parts of them such as substructures and superstructures, can significantly reduce the demand for new materials and, hence, save the embodied carbon. The manufacturing of materials such as concrete releases significant GHG emissions, which is 11% in the case of concrete [44], and the reuse of existing structures can play an important role in reducing the demand for these materials. It is equally vital to reduce embodied carbon by developing low-embodied carbon alternatives to high-impact materials such as concrete. Alternative cementitious materials such as fly ash and slag cement in concrete, calcined clay as binder, and bio-based aggregates are some of the practices used in the industry to reduce the carbon footprint of concrete. Sequestering is another strategy that involves taking advantage of the carbon dioxide stored in the mass of materials such as wood during photosynthesis. Due to their carbon neutral and carbon negative attributes, bio-composite materials are gaining traction in the construction sector. Hemp stands out as a notable material choice, finding application in lumber production [45] and as a key component in hempcrete, a bio-aggregate concrete. Hempcrete consists of hemp hurds, the inner core of the hemp plant, combined with a lime-based binder. This lightweight concrete, composed mainly of lime and hemp shives (fibrous waste), offers a sustainable building solution [46]. Hempcrete (as cast in place or as blocks) is currently mainly used to make walls, although floor slabs, ceiling, and roof insulation can be made as well.

Engineered wood products such as cross-laminated timber (CLT), glue-laminated timber (GLT), and dowel-laminated timber (DLT) have also gained popularity in the construction industry because of their carbon sequestration capabilities and their suitability for diverse residential, educational, commercial, and light industrial applications in the building structures and envelopes. While the predominant use of these products is currently in the construction of mid- to high-rise timber buildings, there is a gradual emergence of examples showing their applications in low-rise single-family houses [47].

7. Tools for Energy Performance Assessment of Building Envelopes

Our review of the building-envelope-based ECM literature demonstrates the application of both the existing energy simulation tools that remain effective and the innovative modeling techniques developed to address the evolving requirements. This work also shows that some ECMs have been studied only experimentally, showing a potential for future numerical or simulation-based studies. The findings are reported below and listed in Table 2 and visualized in Figure 4 based on their use frequency.

Researchers have made strides in developing novel computer tools to enhance the accuracy of analyzing emerging energy conservation measures (ECMs). Yan et al. [45] discussed the complexities and challenges encountered in Building Energy Modeling Programs (BEMPs) while introducing DeST 3.0, a next-generation BEMP. The authors categorized BEMPs into three distinct types: software grounded in a simulation kernel, standalone modules integrated with other software for specific functions, and software utilizing comprehensive room/zone and simplified system models. That paper elaborated on the challenges encountered in building energy-consumption simulations, including capability, calculability, and compatibility issues. DeST 3.0 tackles these challenges by developing advanced simulation modules for building envelopes, occupant behavior, and energy systems used by more than ten national/regional energy-saving standards. They concluded by discussing DeST 3.0's applications in analyzing building energy conservation across extensive projects covering over 200 million m^2, such as Beijing Capital Airport Terminal 3, the National Grand Theater, Beijing Olympic Main Stadium, Beijing Daxing International Airport, and Beijing Winter Olympic Main Stadium [48].

Combining physics-based and data-driven approaches is becoming increasingly common in building energy modeling and simulation, especially in studying the energy savings for various ECMs. Thrampoulidis et al. [49] proposed a machine learning-based surrogate model to yield optimal solutions for building retrofit. The model is designed to reduce the computational cost of retrofit simulations while maintaining accuracy. The authors used DesignBuilder software to simulate the energy performance of a building before and after retrofitting. They then used a machine learning algorithm to train a surrogate model that approximates the energy performance of the retrofitted building. The model was validated using a case study of a Swiss residential building. The results show that the surrogate model can accurately predict the energy performance of the retrofitted building while reducing the computational cost of the simulation. The article highlights the importance of computer modeling and simulation techniques in optimizing building retrofit solutions [49].

Milosevic et al. [50] undertook a study on sustainable retrofit approaches for the "Djuro Djakovic" residential complex in Sarajevo, Bosnia, and Herzegovina. Their main aim was to assess the performance of specific modules within this complex. The research methodologies involved archival and empirical data analysis alongside computational modeling and simulations using BIM software. A comprehensive 3D BIM model of typical residential modules was created based on original drawings and current photographs. The analysis included scrutiny of formal and spatial characteristics, and environmental response assessments were conducted using Revit and Insight 360 simulations. Despite suboptimal environmental adaptation in hillside buildings, the implemented building envelope systems were found to be remarkably advanced for their era [50].

Botticelli et al. [51] conducted a study to investigate how innovative fenestration systems can reduce cooling energy usage in buildings by harnessing passive technologies and understanding the dynamics of building envelopes. The windows in their study featured integrated shading systems and motorized sashes for opening, controllable using a smart building management system. Preliminary findings from the field monitoring and numerical analyses were presented. Field observations revealed that combining solar protection and night ventilation lowered indoor temperatures by 1.4 and 2.2 °C in rooms equipped with new triple and double-glazed units, respectively, compared to older windows. A calibrated numerical model, using monitored data, demonstrated that solar protection and nighttime cooling could achieve up to 76% energy savings. TRNSYS 17, capable of simulating building thermal responses in transient regimes, was utilized for implementing and calibrating the numerical model [51].

Table 2. Summary of computer tools and techniques in the studies on innovative building envelope ECMs.

Study	Main Focus	Tools/Software Used
[8]	Radiative sky cooling for energy conservation	Review of studies, experiments, and simulations
[9]	Passive envelope systems and radiative sky cooling	EnergyPlus for simulations, theoretical models
[10]	Advanced thermal plasters, phase-change materials, low-emittance materials, and high reflective coatings	Focus on materials; no specific software/tools are mentioned
[11]	Cellulose aerogels for building cooling	EnergyPlus, theoretical models, and various characterization tools
[12]	Transparent aerogel glazing	Small Hot-Box apparatus and numerical energy assessment
[14]	Hygrothermal performance of silica-aerogel-based super-insulating systems	WUFI® Pro 5.3 software for numerical modeling
[15]	Trombe wall system with PV panels	Various equipment for experiments and measurements.
[16]	Double-skin and responsive façades	EnergyPlus, TRNSYS®, DesignBuilder®, GenOpt, MATLAB®
[17]	Building envelope with variable thermal performance	DesignBuilder for numerical simulation
[18]	Seismic and energy retrofit for RC and URM structures	Experimental tests and in situ experimentation
[48]	DeST 3.0 for building energy modeling	DeST 3.0 software
[49]	Machine learning-based surrogate model for building retrofit solutions	DesignBuilder for energy simulations and machine learning-based surrogate model
[50]	Sustainable retrofit strategies for brutalist-style residential complex	Revit, Insight 360 simulations
[51]	Impact of innovative fenestration systems on cooling energy consumption using shading systems integrated into the glazing unit and a motorized sash	TRNSYS 17 for numerical modeling and simulation

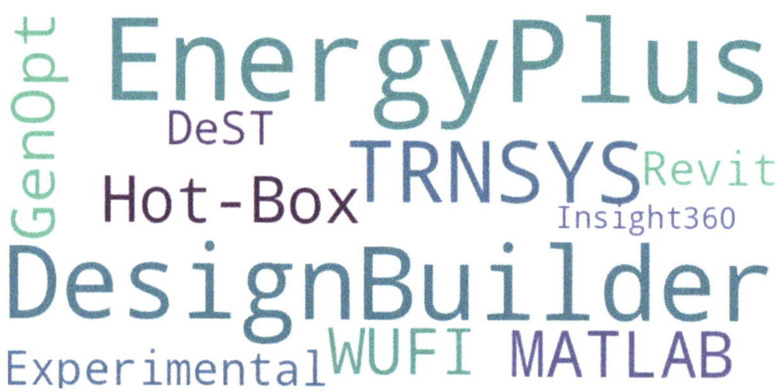

Figure 4. A visual representation of software and tools frequency in energy modeling research for building envelope energy conservation measures.

8. Tools for Embodied Carbon Performance Assessment of Building Skins

To understand and quantify embodied energy and embodied carbon, a life cycle assessment (LCA) is used which helps assess the diverse life cycle environmental impacts of a product or a building, including global warming potential, acidification potential, eutrophication potential, smog formation potential, and other types of impacts. The main life cycle stages that are considered in the life cycle assessment of buildings are defined in compliance with ISO 14040 (ISO 2006) [52], ISO 14044 (ISO 2006) [53], and EN 15978 [54] and include the product stage (A1–A3; including raw material extraction, transportation, and material manufacturing), construction stage (A4–A5), use stage (B1–B5; involving use, maintenance, repair, replacement, and refurbishment), end-of-life stage (C1–C4), and sometimes beyond-life stage (D) [55]. For each stage, the LCA method enables one to capture the types and quantities of consumed materials, primary energy, and the released greenhouse gas emissions; assess the environmental impacts; and identify limitations, opportunities, and aspects to improve upon [56]. A cradle-to-gate LCA concerns the A1–A3 stages of the building life cycle while a cradle-to-grave LCA provides a comprehensive assessment including all life cycle stages. The product stage (A1–A3) is considered to be the life stage that causes the highest embodied carbon due to extensive amounts of consumed energy and emitted carbon involved in the manufacturing of construction materials. The analysis of the use stage, i.e., the stage in which the building is occupied, often involves data availability and data quality challenges mainly because the type and frequencies of the maintenance practices vary significantly from building to building and limited documentation of such practices exist.

There exist a variety of LCA tools for the assessment of embodied carbon and environmental impacts of building envelope systems. Awareness of the strengths and limitations of these tools helps with buildings' LCA to generate results with minimum uncertainty. The state-of-the-art building LCA tools in the market can be classified based on their capability in comprehensive modeling into whole-building LCA tools and non-whole-building tools. The whole-building LCA tools are capable of modeling the entire building including structure, building envelope, and sometimes building systems. Examples of whole-building LCA tools include OpenLCA [57], Sima Pro [58], Gabi [59], EC3 [60], Tally Plugin [61] to Revit, OneClick LCA [62], Athena Impact Estimator for Buildings [63], and other similar tools [64–71], which provide the capability to assess and communicate some or diverse environmental impacts over the complete life cycle of buildings. The non-whole-building LCA tools [71–73] can capture certain building components only. For example, the Structural Carbon Tool [74] is a tool used for the assessment of embodied carbon of structures, while the Kaleidoscope [72] is a tool used for building envelope embodied carbon modeling. The key limitations of non-whole-building LCA tools are their inability to capture all building components, their limited coverage of the full building life cycle, and their limited capturing of non-embodied carbon of environmental impacts. The LCA tools can also be classified based on their capability to couple with energy simulation software. While tools such as ECCOlab [67] provide a fully integrated modeling experience that allows for modeling of both embodied carbon and energy use in one single platform, other tools such as Tally [61] and One Click LCA [62] are available as plugins to build digital modeling tools, which when used along with energy simulation plugins could facilitate both LCA and energy modeling for one digital model. There are also tools such as Athena Impact Estimator for Buildings which, even though they do not present direct coupling with energy simulation, allow for manual inputting of energy use data to assess embodied carbon, operational carbon, and other life cycle environmental impacts. Table 3 provides a comprehensive list of popular LCA tools used to assess the environmental impacts of buildings and their characteristics.

Table 3. The list of popular building LCA tools.

Tool	Building Components	Building Life Stages	Environmental Impacts	Geographical Scope	Tool Type	Reference
General LCA tools: These are comprehensive LCA tools for general construction and non-construction applications. These tools allow the user to define the system boundary and life cycle stages of interest.						
Open LCA	Whole building	A–D	All main impacts	Global	Standalone, free of charge	[57]
Sima Pro	Whole building	A–D	All main impacts	Global	Standalone, commercial	[58]
Gabi	Whole building	A-D	All main impacts	Global	Standalone, commercial	[59]
Construction LCA tools: These are LCA tools tailored for construction industry applications. These tools work with a specific system boundary and the results for all or part of the building life cycle stages are reported.						
EC3	Whole building	A1–A3	Embodied carbon	Global	Standalone, free of charge	[60]
Tally	Whole building	A–D	All main impacts	North America	Revit plugin, commercial with free of charge educational license	[61]
OneClick LCA	Whole building	A–D	All main impacts	Global	Standalone, Revit plugin, commercial with free of charge educational license	[62]
Athena Impact Estimator	Whole building	A–D	All main impacts	North America	Standalone, free of charge	[63]
Beacon	Whole building		Embodied carbon	North America	Revit plugin, free of charge	[64]
eToolLCD	Whole building	A–D		Global	Standalone, commercial, free of charge	[65]
Elodie	Whole building	A–D	All main impacts, except eutrophication	France	Standalone, commercial	[66]
ECCOlab	Whole building	A–D	Embodied carbon, energy	Global	Standalone, commercial with a basic free of charge plan	[67]
eLCA	Whole building	A–D	All main impacts	Germany	Standalone, free of charge	[68]
TOTEM	Whole building	A–D	All main impacts	Belgium	Standalone, free of charge	[69]
LCAbyg	Whole building	A–D	All main impacts	Denmark	Standalone, free of charge	[70]

Table 3. *Cont.*

Tool	Building Components	Building Life Stages	Environmental Impacts	Geographical Scope	Tool Type	Reference
Structural Carbon Tool	Building structure	A–C	Embodied carbon	UK	Standalone, free of charge	[71]
Embodied Carbon Optimizer Tool	Whole building	A1–A3	Embodied carbon, water footprint	North America	Standalone, free of charge	[74]
Kaleidoscope	Envelope	A–D	All main impacts	North America	Standalone, free of charge	[72]
H\B:ERT	Assemblies	A–C	Embodied carbon	Global	Revit plugin, commercial	[73]

The embodied carbon and environmental impacts of construction materials and building products are also reported in detail in the Environmental Product Declaration (EPD) reports. These reports are developed in compliance with EN 15804 + A2 requirements [75] and verified by third parties that use the LCA technique and data and information provided by manufacturers as their primary source of data. EPDs often have a cradle-to-gate scope. EC3 [60] is a searchable tool that compiles with the EPDs of various building products, which could link to some building LCA tools to help customize their LCA results.

9. Net-Zero-Energy Design Standards

There are multiple standards for the design and construction of low-energy and net-zero-energy buildings that affect the way building envelopes are designed, specified, and constructed. This section covers a selected number of such standards for net-zero-energy buildings used in different countries, including those developed by the Passive House Institute (PHI) and Passive House Institute US (Phius), Norway's PowerHouse standard, and the Dutch BENG standard.

Originating in Germany, with roots in the passive buildings of the 20th century in North America, the Passive House Institute (PHI) established stringent requirements for building energy efficiency and thermal comfort by focusing on creating air-tight and super-insulated envelopes with no thermal bridges and with high-efficiency windows as a means of minimizing energy consumption, coupled with heat recovery and optimized solar gain [76]. Buildings with Passive House Certification save more than 75% of the energy used for space heating and cooling when compared with a typical new construction and up to 90% compared to a typical existing home [76]. PHI specifies criteria in three categories of buildings: Passive House, EnerPHit, and PHI Low Energy. The criteria for certified Passive Houses include thresholds in four areas including heating demand/load, cooling demand, airtightness, and renewable primary energy (PER). Additionally, PHI offers the EnerPHit Standard for retrofit projects through which they must meet general airtightness and PER criteria and also choose to meet the requirements of either a component criteria method or an energy demand method. The component criteria method involves climatic-class-based prescriptive requirements for opaque envelope (ground insulation, exterior insulation, interior insulation, and paint color use (for hot and very hot climates)), windows (U-value, SHGC), and window solar loads during cooling periods), and ventilation (minimum heat or ventilation recovery rate) [77]. The PHI Low Energy Building standard involves less stringent requirements and is suitable for buildings that are not able to satisfy the Passive House or EnerPHit requirements.

Developed in the United States, Phius is the voluntary standard that uses the same principles as PHI but involves certification requirements that vary on a climate-specific basis, as compared with the fixed requirements of PHI standards. The Phius standard includes provisions for new buildings (Phius CORE and Phius ZERO) and for retrofit projects (Phius CORE REVIVE and Phius ZERO REVIVE) based on two prescriptive and performance certification paths. In the performance path, thresholds have been established for four criteria, including annual heating demand, annual cooling demand, peak heating demand, and peak cooling demand that vary based on building-related factors (building type, envelope area, floor area, occupant density, and unit density) and project location (i.e., climate zone) [77]. Prescriptive CORE measures prescribe criteria related to airtightness, compactness (i.e., building envelope area), solar protection (SHGC, window area, orientation, and shading), moisture risk limitation, thermal enclosure, mechanical ventilation and systems, lighting, appliances, water heating, and electric vehicle ready [77].

Another noteworthy standard is Norway's Powerhouse Paris Proof voluntary standard, which sets maximum and total CO_2 emissions based on both embodied and operational carbon and certifies buildings that generate more energy than the energy they consume over their life cycle for both their operation and construction [78]. This standard is different from the PHI and Phius standards in that it additionally considers embodied energy and carbon in its certification by promoting recycling, reuse, and low-embodied carbon materials.

The BENG standard that defines "nearly energy-neutral building" requirements is the mandatory building energy code for new residential and non-residential buildings in the Netherlands. The code defines thresholds for three energy indicators, including maximum energy demand per unit of floor area per year, maximum primary fossil fuel use per unit of floor area per year, and minimum share of renewable energy [79]. These requirements vary based on the building type.

10. Conclusions

This work provides a survey of the literature on building envelope developments as related to energy use and embodied carbon. The key purpose is to synthesize existing knowledge and insights on the specific topic of building envelopes, providing readers with a comprehensive overview of the subject matter. Through this approach, this work aims to offer valuable insights and perspectives to researchers, practitioners, and enthusiasts interested in gaining a deeper understanding of building envelope energy retrofit.

This review paper highlights the state-of-the-art and wide range of building envelope solutions that help mitigate energy consumption and embodied carbon emissions of buildings. It shows that the performance of some of these solutions varies with region and climate zone, and architects and engineers employing these solutions must choose them by considering the issues of climate, cost, space requirements, potential energy and carbon saving, constructability, durability, and other factors. This work also shows that the novel materials and building envelopes serving as emerging ECMs have been studied to different extents by experimental approaches, established physics-based approaches (e.g., EnergyPlus, DesignBuilder, and TRNSYS), established customized approaches (e.g., Matlab and GenOpt), numerical or data-driven modeling techniques (e.g., machine learning), or new modeling and simulation tools. Further research is needed to document the performance of some of these ECMs such as RSC using simulation-based methods. With the building energy and embodied carbon regulations becoming more established and stringent in their requirements, the documentation, simulation, modeling, and testing of the building envelopes' hygrothermal and carbon performance are more than ever. The review of the literature also shows the need for future research to develop user-friendly tools to help layman users understand the tradeoffs between cost, thermal behavior and energy performance, embodied carbon, and durability of various building envelope ECMs. Finally, the trend in the field of building construction materials demonstrates the effort to mitigate the embodied carbon by development of low-embodied carbon materials as well as bio-renewable materials, which can provide carbon sequestration capability.

Author Contributions: Conceptualization, R.A., E.K. and A.M.M.; methodology, R.A., E.K. and A.M.M.; software, R.A. and E.K.; validation, R.A. and E.K.; formal analysis, R.A. and E.K.; investigation, R.A., E.K. and A.M.M.; resources, R.A., E.K. and A.M.M.; data curation, R.A., E.K. and A.M.M.; writing—original draft preparation, R.A., E.K. and A.M.M.; writing—review and editing, R.A., E.K. and A.M.M.; visualization, R.A., E.K. and A.M.M.; supervision, R.A., E.K. and A.M.M. All authors have read and agreed to the published version of the manuscript.

Funding: This research received no external funding.

Conflicts of Interest: The authors declare no conflicts of interest.

References

1. IPCC. *Climate Change 2022, Mitigation of Climate Change*; Intergovernmental Panel on Climate Change: Geneva, Switzerland, 2022.
2. EIA. U.S. Energy Facts Explained—Consumption and Production—U.S. Energy Information Administration (EIA). Available online: https://www.eia.gov/energyexplained/us-energy-facts/ (accessed on 27 December 2023).
3. EIA. Residential Energy Consumption Survey (RECS)—Energy Information Administration. Available online: https://www.eia.gov/consumption/residential/ (accessed on 27 December 2023).
4. EIA. Energy Information Administration (EIA)—Commercial Buildings Energy Consumption Survey (CBECS) Data. Available online: https://www.eia.gov/consumption/commercial/data/2018/ (accessed on 27 December 2023).
5. EPA. Sources of Greenhouse Gas Emissions_US EPA. Available online: https://www.epa.gov/ghgemissions/sources-greenhouse-gas-emissions (accessed on 27 December 2023).
6. Gursel, A.P.; Shehabi, A.; Horvath, A. Embodied energy and greenhouse gas emission trends from major construction materials of U.S. office buildings constructed after the mid-1940s. *Build. Environ.* **2023**, *234*, 110196. [CrossRef]
7. United Nations Environment Programme. 2022 Global Status Report for Buildings and Construction: Towards a ZeroEmission, Efficient and Resilient Buildings and Construction Sector. 2022. Available online: https://globalabc.org/sites/default/files/2022-11/FULL%20REPORT_2022%20Buildings-GSR_1.pdf (accessed on 4 February 2024).
8. Wu, Y.; Zhao, H.; Sun, H.; Duan, M.; Lin, B.; Wu, S. A review of the application of radiative sky cooling in buildings: Challenges and optimization. *Energy Convers. Manag.* **2022**, *265*, 115768. [CrossRef]
9. Chen, J.; Gong, Q.; Lu, L. Evaluation of passive envelope systems with radiative sky cooling and thermally insulated glazing materials for cooling. *J. Clean. Prod.* **2023**, *398*, 136607. [CrossRef]
10. Fantucci, S.; Serra, V.; Perino, M. Advanced Materials for Opaque Building Envelopes. 2019. Available online: www.ashrae.org (accessed on 20 January 2024).
11. Zhong, S.; Yuan, S.; Zhang, X.; Zhang, J.; Xu, L.; Xu, T.; Zuo, T.; Cai, Y.; Yi, L. Supplementary Materials for Hierarchical Cellulose Aerogel Reinforced with In situ Assembled Cellulose Nanofibers for Building Cooling. *ACS Appl. Mater. Interfaces* **2023**, *15*, 39807–39817. [CrossRef] [PubMed]
12. Abraham, E.; Cherpak, V.; Senyuk, B.; Hove, J.B.T.; Lee, T.; Liu, Q.; Smalyukh, I.I. Highly transparent silanized cellulose aerogels for boosting energy efficiency of glazing in buildings. *Nat. Energy* **2023**, *8*, 381–396. [CrossRef]
13. *ISO 9050:2003*; Glass in Building—Determination of Light Transmittance, Solar Direct Transmittance, Total Solar Energy Transmittance, Ultraviolet Transmittance and Related Glazing Factors. International Organization for Standardization: Vernier, Geneva, 2003. Available online: https://www.iso.org/standard/35062.html (accessed on 18 June 2024).
14. Ibrahim, M.; Sayegh, H.; Bianco, L.; Wurtz, E. Hygrothermal performance of novel internal and external super-insulating systems: In-situ experimental study and 1D/2D numerical modeling. *Appl. Therm. Eng.* **2019**, *150*, 1306–1327. [CrossRef]
15. Duan, X.; Shen, C.; Liu, D.; Wu, Y. The performance analysis of a photo/thermal catalytic Trombe wall with energy generation. *Renew. Energy* **2023**, *218*, 119361. [CrossRef]
16. Ascione, F.; Bianco, N.; Iovane, T.; Mastellone, M.; Mauro, G.M. The evolution of building energy retrofit via double-skin and responsive façades: A review. *Sol. Energy* **2021**, *224*, 703–717. [CrossRef]
17. Si, P.; Lv, Y.; Rong, X.; Shi, L.; Yan, J.; Wang, X. An innovative building envelope with variable thermal performance for passive heating systems. *Appl. Energy* **2020**, *269*, 115175. [CrossRef]
18. Baek, E.; Pohoryles, D.A.; Kallioras, S.; Bournas, D.A.; Choi, H.; Kim, T. Innovative seismic and energy retrofitting of wall envelopes using prefabricated textile-reinforced concrete panels with an embedded capillary tube system. *Eng. Struct.* **2022**, *265*, 114453. [CrossRef]
19. Cellulose Insulation Manufacturers Association. Standard Practice for Installation of Cellulose Insulation in the United States. Technical Bulletin. February 2023. Available online: https://cellulose.org/wp-content/uploads/2023/02/CIMA-Technical-Bulletin-Standard-Practices-for-Installation-of-Cellulose-Insulation-in-the-United-States.pdf (accessed on 20 January 2024).
20. Papadopoulos, A.M. State of the art in thermal insulation materials and aims for future developments. *Energy Build.* **2005**, *37*, 77–86. [CrossRef]
21. Kalnæs, S.E.; Jelle, B.P. Vacuum insulation panel products: A state-of-the-art review and future research pathways. *Appl. Energy* **2014**, *116*, 355–375. [CrossRef]
22. Casini, M. Smart materials and nanotechnology for energy retrofit of historic buildings. In Proceedings of the International Conference on Advances in Civil, Structural and Construction Engineering, Rome, Italy, 7–8 June 2014; pp. 28–37.
23. Pal, S.; Roy, B.; Neogi, S. Heat transfer modelling on windows and glazing under the exposure of solar radiation. *Energy Build.* **2009**, *41*, 654–661. [CrossRef]

24. Cho, S. Analysis of the performance of vacuum glazing in office buildings in Korea: Simulation and experimental studies. *Sustainability* **2017**, *9*, 936. [CrossRef]
25. Ghoshal, S.; Neogi, S. Advance glazing system—Energy efficiency approach for buildings a review. *Energy Procedia* **2014**, *54*, 352–358. [CrossRef]
26. Jayasinghe, C.; Jayasinghe, C. Embodied energy of alternative building materials and their impact on life cycle cost parameters. In Proceedings of the International Conference on Structural Engineering and Construction Management, Kandy, Sri Lanka, 16–18 December 2011. 16p. [CrossRef]
27. Haynes, R. Embodied Energy Calculations within Life Cycle Analysis of Residential Buildings. 2010. Revised 2013. Available online: https://etoolglobal.com/wp-content/uploads/2012/10/Embodied-Energy-Paper-Richard-Haynes.pdf (accessed on 20 January 2024).
28. Cooperman, A.; Dieckmann, J.; Brodrick, J. Home envelope retrofits. *ASHRAE J.* **2011**, *53*, 82–85.
29. Favoino, F.; Jin, Q.; Overend, M. Towards an ideal adaptive glazed façade for office buildings. *Energy Procedia* **2014**, *62*, 289–298. [CrossRef]
30. Aelenei, D.; Aelenei, L.; Vieira, C.P. Adaptive Façade: Concept, Applications, Research Questions. *Energy Procedia* **2016**, *91*, 269–275. [CrossRef]
31. Barozzi, M.; Lienhard, J.; Zanelli, A.; Monticelli, C. The Sustainability of Adaptive Envelopes: Developments of Kinetic Architecture. *Procedia Eng.* **2016**, *155*, 275–284. [CrossRef]
32. Warwick, M.E.A.; Ridley, I.; Binions, R. The effect of transition gradient in thermochromic glazing systems. *Energy Build.* **2014**, *77*, 80–90. [CrossRef]
33. International Energy Agency. Technology Roadmap Energy Efficient Building Envelopes. Available online: www.iea.org (accessed on 20 January 2024).
34. Raheem, A.A.; Issa, R.R.A.; Olbina, S. Environmental Performance and Economic Analysis of Different Glazing–Sunshade Systems Using Simulation Tools. *J. Comput. Civ. Eng.* **2016**, *30*, C5016001. [CrossRef]
35. PennState. Penn Staters Invited to Review University Self-Study Report. PennState. Available online: https://www.psu.edu/news/administration/story/penn-staters-invited-review-university-self-study-report/ (accessed on 20 January 2024).
36. Guidebook for Solar Photovoltaic Projects in Philadelphia. Available online: https://www.phila.gov/media/20160421161005/solar-photovoltaic-project-guidebook.pdf (accessed on 23 January 2024).
37. Lombardi, V. The Solar Energy System Handbook: A Practical Guide to Solar Power System Design for Homeowners. 2012. Available online: https://readara.com/the-solar-energy-system-handbook-a-practical-guide-to-solar-power-system-design-for-homeowners/9781508433934/philip-crouch/technology-engineering/power-resources-alternative-renewable/book. (accessed on 20 December 2023).
38. Wang, L.; Memari, A.M. The State-of-the-art Review of BIPV Systems for Detached Single-family Homes. In Proceedings of the 7th Residential Building Design and Construction Conference, University Park, PA, USA, 27–28 March 2024; Memari, A.M., Ed.; Penn State University: University Park, PA, USA, 2024; pp. 254–265.
39. Memari, A.M.; Iulo, L.D.; Solnosky, R.L.; Stultz, C.R. Building Integrated Photovoltaic Systems for Single Family Dwellings: Innovation Concepts. *Open J. Civ. Eng.* **2014**, *4*, 102–119. [CrossRef]
40. Tripathy, M.; Sadhu, P.; Panda, S. A critical review on building integrated photovoltaic products and their applications. *Renew. Sustain. Energy Rev.* **2016**, *61*, 451–465. [CrossRef]
41. Bobrova, D. Building-integrated wind turbines in the aspect of architectural shaping. *Procedia Eng.* **2015**, *117*, 404–410. [CrossRef]
42. Park, J.; Jung, H.J.; Lee, S.W.; Park, J. A new building-integrated wind turbine system utilizing the building. *Energies* **2015**, *8*, 11846–11870. [CrossRef]
43. Ashby, M.F. *Materials and the Environment: Eco-Informed Material Choice*; Butterworth-Heinemann: Oxford, UK, 2012.
44. Embodied Carbon Actions—Architecture 2030. Available online: https://www.architecture2030.org/embodied-carbon-actions/ (accessed on 27 December 2023).
45. HEMPWOOD, "HEMPWOOD", HEMPWOOD. Available online: https://hempwood.com/ (accessed on 20 January 2024).
46. Zuabi, W.; Memari, A.M. Review of Hempcrete as a Sustainable Building Material. *Int. J. Archit. Eng. Constr.* **2021**, *10*, 1–17.
47. Jellen, A.C.; Memari, A.M. *Structural Design of a Cross-Laminated Timber (CLT) Single-Family Home*; PHRC: University Park, PA, USA, 2022.
48. Yan, D.; Zhou, X.; An, J.; Kang, X.; Bu, F.; Chen, Y.; Pan, Y.; Gao, Y.; Zhang, Q.; Zhou, H.; et al. DeST 3.0: A new-generation building performance simulation platform. *Build. Simul.* **2022**, *15*, 1849–1868. [CrossRef]
49. Thrampoulidis, E.; Mavromatidis, G.; Lucchi, A.; Orehounig, K. A machine learning-based surrogate model to approximate optimal building retrofit solutions. *Appl. Energy* **2021**, *281*, 116024. [CrossRef]
50. Milosevic, S.; Aksamija, A. *Sustainable Retrofit Strategies for an Existing and Historically Significant Residential Complex: Environmental Response and Facade Performance Analysis*; Check for updates ® 238; Springer: Cham, Switzerland, 2023; Volume 529, pp. 237–252. [CrossRef]
51. Botticelli, M.; Agnoli, S.; Romano, S.; Zinzi, M. *Exploiting Passive Cooling in Office Buildings with Advanced Automated Glazing Systems: Preliminary Analyses from a Field Study*; Springer: Singapore, 2023; pp. 525–534. [CrossRef]
52. *ISO 14040:2006*; Environmental Management—Life Cycle Assessment—Principles and Framework. International Organization for Standardization: Vernier, Geneva, 2006. Available online: https://www.iso.org/standard/37456.html (accessed on 18 June 2024).

53. *ISO 14044:2006*; Environmental Management—Life Cycle Assessment—Requirements and Guidelines. International Organization for Standardization: Vernier, Geneva, 2006. Available online: https://www.iso.org/standard/38498.html (accessed on 18 June 2024).
54. *EN 15978_2011*; Sustainability of Construction Works—Assessment of Environmental Performance of Buildings—Calculation Method. European Standards: Brussel, Belgium, 2011. Available online: https://standards.iteh.ai/catalog/standards/cen/62c22cef-5666-4719-91f9-c21cb6aa0ab3/en-15978-2011 (accessed on 27 December 2023).
55. Bayer, C.; Gamble, M.; Gentry, R.; Joshi, S. *AIA Guide to Building Life Cycle Assessment in Practice Authorship and Acknowledgments A Guide to Life Cycle Assessment of Buildings 2*; The American Institute of Architect: New Yor, NY, USA, 2010.
56. Lehtinen, H.; Saarentaus, A.; Rouhiainen, J.; Pitts, M.; Azapagic, A. A Review of LCA Methods and Tools and Their Suitability for SMEs List of Contents. 2011. Available online: www.biochem-project.eu (accessed on 22 December 2023).
57. OpenLCA. The Open Source Life Cycle and Sustainability Software. 2023. Available online: https://www.openlca.org/open-source/ (accessed on 20 January 2024).
58. SimaPro. Life Cycle Assessment for Informed Changemakers. 2023. Available online: https://simapro.com/ (accessed on 22 December 2023).
59. Sphera. Gabi: Life Cycle Assessment Tool for Experts. 2023. Available online: https://sphera.com/life-cycle-assessment-lca-software/ (accessed on 22 December 2023).
60. EC3. *Embodied Carbon in Construction Calculator (EC3)*; Carbon Leadership Forum: Seattle, WA, USA, 2021; Available online: https://carbonleadershipforum.org/ec3-tool/ (accessed on 20 January 2024).
61. Tally. Tally, Life Cycle Assessment App. 2024. Available online: https://www.ifu.com/lca-software/ (accessed on 22 December 2023).
62. OneClickLCA, "OneClickLCA," OneClickLCA. Available online: https://www.oneclicklca.com/ (accessed on 20 January 2024).
63. Athena. *About Environmental Building Declarations*; Athena Sustainable Materials Institute: Ottawa, ON, Canada, 2018.
64. Beacon. "Beacon", Thornton Tomasetti. 2023. Available online: https://www.thorntontomasetti.com/capability/beacon (accessed on 22 December 2023).
65. ETool LCD. "ETool LCD". 2023. Available online: https://etool.app/ (accessed on 22 December 2023).
66. ELODIE. ELODIE: A Tool for the Environmental Assessment of Building. 2023. Available online: http://www.irbnet.de/daten/iconda/CIB19053.pdf (accessed on 22 December 2023).
67. ECCOLAB. "ECCOLAB". 2023. Available online: https://eccolab.co/overview.html (accessed on 22 December 2023).
68. eLCA. Life Cycle Assessment within BNB1—Online-Tool eLCA and Materials Database OeKOBAU.DAT. BNB. 2023. Available online: http://www.irbnet.de/daten/iconda/CIB_DC28525.pdf (accessed on 22 December 2023).
69. TOTEM. TOTEM: Tool to Optimise the Total Environmental Impact of Materials. 2023. Available online: https://www.totem-building.be/ (accessed on 22 December 2023).
70. LCAbyg. LCAbug. 2023. Available online: https://www.lcabyg.dk/en/ (accessed on 22 December 2023).
71. 2050 Materials. Embodied Carbon Optimizer Tool. 2024. Available online: https://app.2050-materials.com/tools/dot_selection (accessed on 22 December 2023).
72. Payette. Kaleidoscope Embodied Carbon Design Tools. 2023. Available online: https://www.payette.com/kaleidoscope/ (accessed on 22 December 2023).
73. H\B:ERT. H\B:ERT Emissions Reduction Tool. Hawkins\Brown. 2023. Available online: https://www.hawkinsbrown.com/sub-services/hbert-emissions-reduction-tool/ (accessed on 22 December 2023).
74. The Institution of Structural Engineers. The Structural Carbon Tool—Version 2. Available online: https://www.istructe.org/resources/guidance/the-structural-carbon-tool/ (accessed on 20 January 2024).
75. EN 15804+A2; Sustainability of Construction Works—Environmental Product Declarations—Core Rules for the Product Category of Construction Products. 2019. Available online: https://standards.iteh.ai/catalog/standards/cen/c98127b4-8dc2-48a4-9338-3e1366b16669/en-15804-2012a2-2019 (accessed on 20 January 2024).
76. PHI. 25 Years Passive House—Interview with Dr. Wolfgang Feist. Passive House Institute. Available online: https://passivehouse.com/02_informations/01_whatisapassivehouse/01_whatisapassivehouse.htm (accessed on 20 January 2024).
77. Phius 2021 Passive Building Standard Certification Guidebook. Available online: www.Phius.org (accessed on 20 January 2024).
78. PowerHouse. What is a Powerhouse? PowerHouse. Available online: https://www.powerhouse.no/en/what-defines-the-powerhouse-standard/ (accessed on 20 January 2024).
79. Energieprestatie—BENG. Available online: https://www.rvo.nl/onderwerpen/wetten-en-regels-gebouwen/beng (accessed on 27 December 2023).

Disclaimer/Publisher's Note: The statements, opinions and data contained in all publications are solely those of the individual author(s) and contributor(s) and not of MDPI and/or the editor(s). MDPI and/or the editor(s) disclaim responsibility for any injury to people or property resulting from any ideas, methods, instructions or products referred to in the content.

buildings

MDPI

Review

Evaluating the Impact of CO_2 on Calcium SulphoAluminate (CSA) Concrete

Daniel D. Akerele * and Federico Aguayo

Department of Construction Management, University of Washington, Seattle, WA 98195, USA; aguayo12@uw.edu
* Correspondence: dannie70@uw.edu

Abstract: The construction industry is a significant contributor to global CO_2 emissions, primarily due to the extensive use of ordinary portland cement (OPC). In response to the urgent need for sustainable construction materials, calcium sulphoaluminate (CSA) cement has emerged as a promising alternative. CSA cement is renowned for its low carbon footprint, high early-age strength, and superior durability, making it an attractive option for reducing the environmental impact of construction activities. While CSA cement offers benefits in carbon emissions reduction, its susceptibility to carbonation presents challenges. Although the body of literature on CSA cement is rapidly expanding, its adoption rate remains low. This disparity may be attributed to several factors including the level of scientific contribution in terms of research focus and lack of comprehensive standards for various applications. As a result, the present study sets out to track the research trajectory within the CSA cement research landscape through a systematic literature review. The study employed the Prefer Reporting Item for Systematic Review and Meta-Analysis (PRISMA) framework to conduct a literature search on three prominent databases, and a thematic analysis was conducted to identify the knowledge gap for future exploration. The study revealed that while CSA concrete demonstrates superior early-age strength and environmental resistance, its susceptibility to carbonation can compromise structural integrity over time. Key mitigation strategies identified include the incorporation of supplementary cementitious materials (SCMs), use of corrosion inhibitors, and optimization of mix designs. The review also highlights the global distribution of research, with notable contributions from the USA, China, and Europe, emphasizing the collaborative effort in advancing CSA concrete technology. The findings are crucial for enhancing sustainability and durability in the construction sector and advancing CSA binders as a sustainable alternative to traditional cement.

Keywords: concrete; calcium sulphoaluminate cement (CSA); mechanical properties; carbonation (CO_2); durability; sustainability

1. Introduction

The global push for net zero carbon emissions necessitates combined efforts from all sectors. Nañagas et al. (2022) [1] describe CO_2 as a dangerous gas to both humans and the environment. Carbon footprint is one of the most discussed topics in recent times due to climate change challenges such as flooding, heat waves, extreme weather, extinction, and even death [2,3]. Carbon emissions, including all greenhouse gases (GHGs), are major contributors to climate change. Among the sectors contributing to GHG emissions, the construction industry ranks second, accounting for over 40% of global emissions from buildings alone [4].

Concrete production, particularly cement production, is a significant contributor to these emissions. Cement production accounts for over 8% of global CO_2 emissions [5]. For instance, one ton of cement typically produces about 0.8 to 0.9 tons of CO_2 emissions, accounting for approximately 8% of the world's anthropogenic CO_2 emissions [4]. Understanding these emissions and developing strategies to reduce the carbon footprint from concrete production is crucial for contributing to global efforts to mitigate climate change.

Recent studies regarding life cycle assessment of buildings have in fact attributed up to 90% of embodied carbon emission to the A1–A3 (raw material extraction, transportation, and on-site construction) stages of whole building lifecycle assessment in extreme cases, about 25% in conventional buildings, and up to 50% in advanced construction [6]. Understanding and managing these stages are essential for sustainable construction practices.

Concrete is the most widely used building material globally, and its production thus significantly contributes to GHG emissions. The primary source of these emissions is the manufacturing of cement, a key component of concrete [7]. According to Shen et al. (2014) [8], cement production ranks second in global CO_2 emissions, with portland cement production accounting for about 80% of CO_2 emissions in concrete. The CO_2 emissions from portland cement result from the calcination of limestone ($CaCO_3$) and the combustion of fossil fuels, including the energy required in cement production plants. To produce portland cement clinker, limestone is heated with a source of silica in a kiln at temperatures over 1350 °C, making both the process and its byproducts harmful to the environment [9].

Advances in construction have sought means to address this issue, aligning with sustainability development goals by drastically reducing carbon emissions especially at the production point [10]. One approach is to provide alternative binders that produce less carbon than conventional portland cement [11]. Alternative cement materials, such as calcium sulphoaluminate cement (CSA) and other supplementary cementitious materials (such as fly ash, metakaolin, and silica fume), have been proven to produce less carbon [12]. Studies reveal that they reduce carbon emissions at the production point by about 50% compared to portland cement [13–15].

However, one of the challenges with these alternative cement materials is their susceptibility to concrete carbonation compared to portland cement [16]. Although they generally show higher early-age strength and superior properties compared to portland cement [17], carbonation poses significant threats especially to reinforced concrete over its lifespan, including accelerating corrosion of reinforcing rebar [18]. Assessing the impact of carbonation on alternative cement materials like CSA helps better understand the concrete produced with these materials and tackle the problems faced with carbonation. Studies such as those of Guo et al. (2014) [16,17] have highlighted that alternative cement materials like CSA are more prone to carbonation than portland cement, despite their higher early-age strength and superior properties. Additionally, carbonation reactions can pose significant threats to reinforced concrete over its lifespan, including accelerating the corrosion of reinforcing rebar [19]. Korec et al. (2024) [20] conducted an interesting study on predicting carbonation-induced corrosion in concrete with rebars, which included modelling framework for predicting carbonation and understanding the behavior of carbonated concrete; this gives an interesting insight into understanding the behavior of reinforced CSA concrete. Regrettably, the impact of CO_2 on CSA is still not fully understood. These limitations in the existing body of literature serve as a point of departure for the current study. Accordingly, this review sets out to investigate the impact of carbonation on CSA cement through a systematic literature review.

The findings of this study will be valuable to the construction industry and may influence how buildings and transportation infrastructure, including precast structures using CSA concrete, are designed and constructed in the future to provide mitigation against the downside effect of CSA concrete carbonation. Additionally, this study could also inform policy decisions and regulations regarding the use of alternative cement materials, including calcium sulphoaluminate cement, as a means to decarbonize the construction industry. The review could also contribute to the development of new technologies and methods for mitigating the impact of CO_2 on concrete structures, thereby advancing the field of sustainable construction.

2. Systematic Literature Methodology

The current study employed the Prefer Reporting Item for Systematic Review and Meta-Analysis (PRISMA) framework to trace the research output on calcium sulphoaluminate (CSA) cement in cement production domains within a period of about eleven years. Specifically, the study aims to focus on the impact of CO_2 on CSA concrete and exclude articles that delve into discussions of other CSA concepts such as the barriers, applications and adoption as these aspects have been addressed by various authors. As a result, this review followed specific criteria for selecting articles. At first, preference was given to peer-reviewed articles available in open-access databases to enhance the study's replicability and transparency. This approach ensured that other researchers could access and verify the findings, thereby contributing to the overall robustness of the investigation [21].

Additionally, the articles included span from 2012 to the first quarter of 2024. This timeframe was intentionally selected to cover the critical early stages of calcium sulphoaluminate cement development. Furthermore, only manuscripts written in English were chosen to ensure broader accessibility, recognizing English as a widely accepted global language [22]. More importantly, it was a requirement that the words 'CO_2 OR Carbonation', 'Calcium SulphoAluminate', or 'CSA' and 'cement', must be present in the title, abstract, or keyword of the article. However, this condition might exclude some relevant articles. Nevertheless, it ensured that only those related to calcium sulphoaluminate cement were selected, thereby eliminating redundant manuscripts from the selection process.

The literature search was performed across multiple reputable databases, including Google Scholar, Scopus, and Web of Science, covering peer-reviewed journals, conference proceedings, and technical reports. These databases were selected because of their reputation for publishing peer-reviewed materials and the availability of robust studies in the subject area. Additionally, consulting three databases allows for validation and cross-verification, diverse perspectives, and diverse methodologies, and ensures comprehensive data collection. The search terms included "CO_2 impact on CSA concrete", "carbonation of calcium sulphoAluminate cement", "alternative cement materials", and "sustainable construction materials". The search was conducted based on the following criteria: relevance to research questions; focus on CSA cement and its interaction with CO_2; studies published with the selected temporal period; and availability of detailed experimental methodologies and findings.

The data extraction process involved identifying materials used, experimental methods, key findings, novelty of the research, research gaps, and recommendation for future directions. This information was evaluated and synthesized to identify common themes, trends, and literature gaps using a thematic approach [23]. The methodological approach also considered the reliability and validity of the reviewed studies, assessing their study designs, sample sizes, and experimental conditions.

3. Results and Discussion

3.1. Results

The stringent selection criteria of the study led to the identification of 1265 articles. In total, 500 articles were identified from Google Scholar while 375 and 390 articles were identified from Web of Science (WoS) and Scopus, respectively. Figure 1 provides the flowchart illustrating the steps for the article selection and the details of the materials used for qualitative and quantitative synthesis.

Figure 2 illustrates the number of publications per year related to the impact of CO_2 on calcium sulphoaluminate (CSA) concrete from 2012 to the second quarter of 2024. The data indicate fluctuations in research interest over this period, reflecting various factors influencing the research community.

The analysis of the population trend indicates that in the early years (i.e., 2012–2015) there is a moderate number of publications. This period likely represents the early exploration phase of research into CSA cement and its carbonation properties. The initial years show a modest number of publications, indicating the early stages of research into CSA cement. During this period, researchers were primarily focused on understanding the fundamental properties and potential benefits of CSA cement. For instance, studies like those by Mechling et al. (2014) [24] highlighted the basic behaviors of sulphoaluminate cements under various environmental conditions, setting the stage for subsequent research by demonstrating the potential of CSA cement under different environmental conditions However, there is a noticeable increase in the number of publications in the mid period (between 2016 and 2020). This period marks a significant growth phase where CSA cement's potential benefits in reducing CO_2 emissions and enhancing sustainability were increasingly recognized. During this time, studies explored various aspects of CSA, including the chemical interactions between CO_2 and CSA cement, mechanical properties, and durability under different environmental conditions (e.g., [25–27]). In the most recent years (2021–2024), heightened interest and possibly increased funding and policy support for sustainable construction materials were observed. This period coincides with global initiatives to reduce carbon footprints in the construction industry. For example, the work by Ansari et al. (2022) [28] proposed novel approaches to improving the carbonation resistance of CSA cement, showcasing innovative solutions to existing concerns. Sharma et al. (2023) [29] also emphasized the microstructural characteristics and CO_2 uptake of CSA cement, further driving interest in this field.

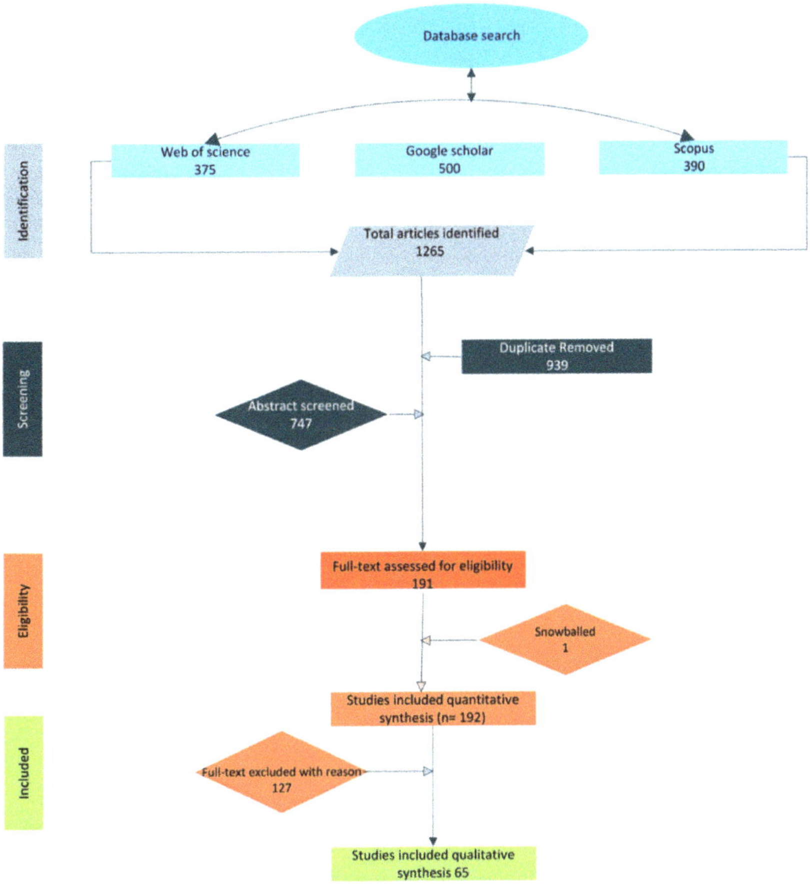

Figure 1. Articles selection process.

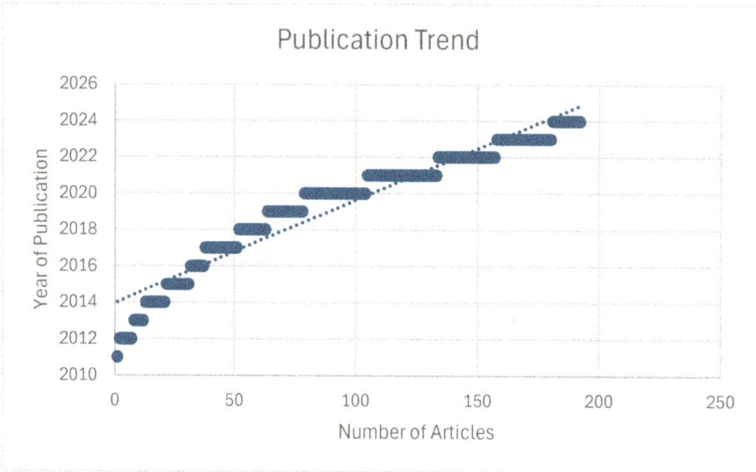

Figure 2. Publication trend.

Figure 3 provides a breakdown of the journal sources for the articles included in the quantitative synthesis. The chart illustrates the top 10 journal sources by the number of publications. The breakdown of journal sources indicates that research on the impact of CO_2 on CSA concrete is published across a variety of journals, with a concentration in a few leading ones. Most of the publications are concentrated in journals such as Construction and Building Materials, Cement & Concrete Composites, and Cement and Concrete Research, accounting for 35% of the total articles selected for eligibility. These journals are highly regarded in the field of construction materials and concrete research, indicating that studies on CSA cement and CO_2 impact are recognized and valued within these specialized publications [30].

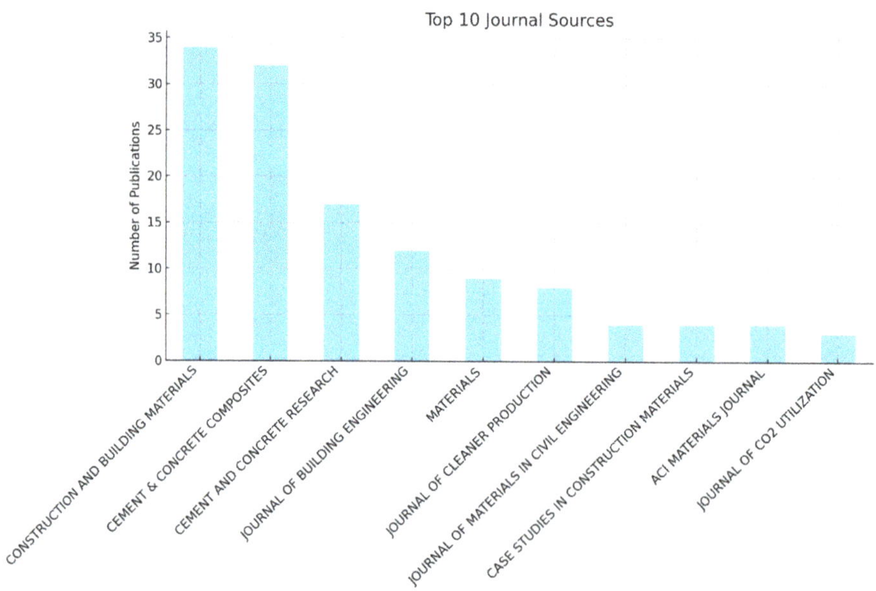

Figure 3. Publication outlet.

The wide range of journals, from Journal of Building Engineering to Journal of Cleaner Production, reflects the interdisciplinary nature of this research area. High-impact journals like Cement and Concrete Composites and Construction and Building Materials lead in publishing major findings, while specialized journals such as Coatings, Sustainability, and Journal of CO_2 utilization provide platforms for focused and interdisciplinary studies. This balanced dissemination ensures comprehensive coverage of both fundamental and applied aspects of CSA cement research, promoting its development as a sustainable construction material.

Figure 4 is used to present a breakdown of journal sources by country. Understanding country distribution of research studies on CSA and its carbonation properties would help in understanding the regions that are contributing the most on the subject, providing an overview of global interests and engagement in this field. Such analysis can help to identify nations with advanced research facilities and significant investment in construction materials. It is also useful to visualize global trends in CSA cement research, such as which countries are championing new methods and technologies. This will also inform researchers and professionals about current state and emerging areas of interest.

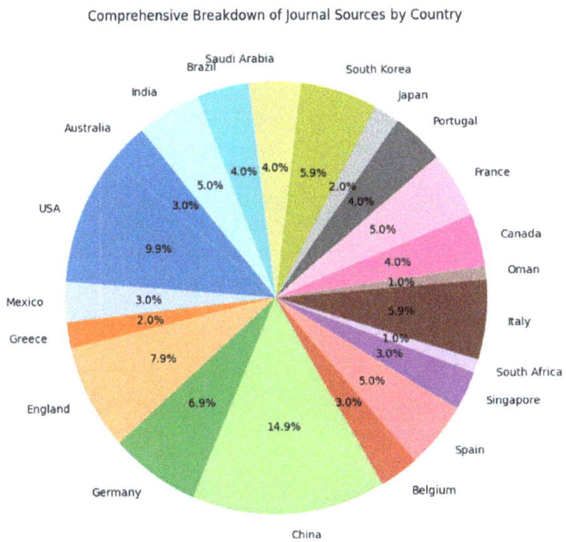

Figure 4. Countries of publications.

Furthermore, policymakers and stakeholders can use the country distribution data to make informed decisions regarding allocation of resources and funding, and to strategically come up with initiatives for sustainable construction. Finally, these data can also inform research collaborations, motivate researchers, and foster technology transfer across industries.

3.2. Discussion

Bibliometric Analysis

Advanced studies have focused on optimizing CSA cement formulations, assessing long-term performance, and scaling up applications for broader industry use. The trend in publications aligns with this study's objectives, which aim to systematically review and synthesize existing research on the impact of CO_2 on CSA concrete. The growing body of literature provides a robust foundation for analyzing trends, identifying gaps, and proposing future research directions. Several key studies highlight the relevance of the observed trend; for example, Bertola et al. (2019) [31] investigated the carbonation resistance of CSA cement and its implications for durability and environmental impact.

Similarly, Monkman (2018) [32] explored the chemical mechanisms of CO_2 interaction with CSA cement, providing insights into optimizing its formulation for enhanced performance, and Vaishnav Kumar Shenbagam et al. (2024) [33] conducted a comprehensive review of sustainable cement materials, emphasizing the importance of CSA cement in reducing the carbon footprint of the construction industry. These studies underscore the importance of understanding the impact of CO_2 on CSA cement and support the increasing research interest reflected in the publication trends.

The top five countries in terms of publications include the USA, China, England, Germany, and Italy, respectively. This reflects the significant contributions from these countries in construction materials and CSA cement research. The research on CSA cement is fairly distributed globally, with contributions from countries across different continents including North America, Europe, Asia, and Australia. This highlights the worldwide interest in sustainable construction materials. The leading countries are often those with strong research institutions and funding for advanced materials research. The USA and China, for example, have numerous universities and research centers dedicated to materials science and engineering. Studies, such as those of Chen et al. (2020) [34], from prominent American institutions, and Wang et al. (2021) [35], from China, showcase the advanced research on the carbonation resistance of CSA cement and its implications for durability. These countries are at the forefront of research and innovation in sustainable construction materials. European countries, including Germany and the UK, have strong research programs in sustainable materials and construction engineering. Publications in leading journals such as Construction and Building Materials and Cement & Concrete Composites reflect the collaborative efforts in advancing sustainable construction practices. Additionally, studies from various countries like Belgium, Spain, Singapore, and Brazil underscore the importance of international collaboration in tackling global sustainability challenges.

The diversity of countries and publication trends involved in the research suggests a high level of international collaboration, and the research on CSA cement and its carbonation properties benefits from insights across different climatic conditions and construction practices. Figure 4 also reveals potential gaps in the research landscape. Countries with emerging construction industries but fewer publications may need more focus on developing sustainable materials like CSA cement.

4. CO_2 Interactions with Calcium SulphoAluminate Cement

Calcium sulphoaluminate (CSA) cement is increasingly recognized for its low carbon footprint and high early-age strength characteristics, making it a sustainable alternative to traditional portland cement. The research on CSA cement has gained momentum globally, focusing on its mechanical properties, durability, and environmental impact, particularly concerning CO_2 emissions and carbonation [17]. Early studies on CSA cement, particularly from China and the USA, have laid the foundation for understanding its unique properties and potential benefits. Zhang et al. (2009) [36] conducted extensive research on the carbonation resistance of CSA cement, demonstrating its superior performance compared to portland cement. Similarly, Pacheco Torgal et al. (2012) [9] highlighted the benefits of CSA cement in reducing CO_2 emissions during production, positioning it as a crucial material in the fight against climate change. The interaction of CSA cement with CO_2, particularly in terms of curing and long-term performance, has been the focus of several studies. This section explores the impact of CO_2 on CSA concrete, covering aspects such as strength, durability, mechanical properties, shrinkage, and environmental benefits.

4.1. *Carbonation of CSA Binders*

Carbonation occurs when CO_2 reacts with calcium hydroxide in concrete and leads to the sequestration of CO_2. This process, called carbonation-induced calcification, forms stable calcium carbonate that locks away CO_2 [37]. Carbonation can increase compressive strength, density, modulus of elasticity, and shrinkage properties of concrete, according to Achal and Mukherjee (2015) [38]. Conversely, Chen, Liu, and Yu (2018) [39] reported that carbonation can potentially reduce compressive strength and increase permeability, making concrete more susceptible to damage from water and chemical attacks.

Carbonation can lead to the breakdown of ettringite in CSA concrete, a mineral that causes expansion and cracking, potentially reducing its durability and strength [40]. Carbonation can also lead to the formation of carbonic acid, which can corrode steel reinforcement, causing a decrease in concrete strength [41]. Carbonation has also been shown to lower the pH of concrete, exposing it to chemical attacks and reducing the lifespan of concrete structures [42]. The effect of carbonation and acidic gases such as SO_2 and NO_2 is to lower concrete pH, leading to steel reinforcement corrosion below a pH of 9.5, with catastrophic corrosion occurring at pH levels below 7 [43].

Carbonation has different effects on alternative cements compared to portland cement. Unlike portland cement (PPC), where porosity decreases due to the formation of calcium carbonate (i.e., calcite) in the pores, carbonation of CSA binders increases concrete porosity by converting ettringite to gypsum, aluminum hydrocalcite, and calcite [25]. This process fills voids in the cement with $Ca(OH)_2$ crystals, enhancing concrete. Zhou and Glasser (2000) [44] studied carbonated synthetic ettringites and found that under certain conditions, hemihydrate can develop and then recrystallize to produce alumina, ferric oxide, and a monosulfate phase. Factors that affect the carbonation of CSA concretes include water–cement (w/c) ratio, pore structure, level of hydration, and relative humidity [45].

In a study on the carbonation of CSA mortars, Hargis et al. (2017) [46] investigated carbonation behavior, the microstructural and chemical factors that direct their rate of carbonation using w/c ratios of 0.65, 0.525, and 0.4. It was found that CSA mortars carbonate more rapidly than PPC mortars and at the same rate as calcium aluminate cement mortars. The anhydrite content of the CSA mortars strongly affected the ye'elimite reaction kinetics, which played an important part in imparting carbonation resistance in CSA mortars. It was noted that the reduction in the w/c ratio of CSA mortars slowed down their carbonation rate. These findings are important in understanding the carbonation behavior of CSA concrete and developing more durable and sustainable concrete mixtures.

Similarly, Fernandez-Carrasco et al. (2012) [47] investigated the carbonation rates of ternary cementitious materials including CSA under different conditions, including temperature, relative humidity, mix ratios, and 4% exposure to CO_2. It was discovered that carbonation rate was influenced by the type of cement used; carbonation rate also increased in CSA compared to PPC. It is difficult to conclude on the behavior of CSA under carbonation because of several factors as afore discussed. Geng et al. (2014) [48] noticed that the carbonation depth of sulphoaluminate cement concrete is lower compared to PPC irrespective of curing time. However, an accelerated carbonation laboratory study using 20% CO_2 and 70% relative humidity showed no significant change in carbonation depths between CSA and portland cement concrete mixes [26].

Studies have shown that the performance of CSA concrete exposed to increased CO_2 concentration is affected by numerous factors. For instance, in the study carried out by Leeman and Moro (2017) [45], there is a relationship between relative humidity, water–cement ratio and carbonation of CSA concrete. The strength of CSA concrete decreases as the CO_2 concentration increases only at a higher relative humidity (above 70%), and no significant difference was observed at 57% relative humidity. Similarly, Chen et al. (2016) also concluded that the porosity of the concrete increases with increased CO_2 concentration [39]. Moreover, T. Mi et al. (2023), upon investigating the corrosion process of reinforced concrete, realized that carbonation can lead to a decrease in the flexural strength and increase in the chloride ion permeability of the concrete [41]. Additionally, the carbonation process can impact the microstructure of concrete, leading to the formation of cracks and the breakdown of ettringite minerals [40]. Research has also shown that the impact of CO_2 concentration on the performance of CSA concrete is influenced by the type of curing used. For instance, Chen, Lui, and Yu (2018) [39] found that air-cured CSA concrete exposed to CO_2 concentration had a lower compressive strength compared to water-cured CSA concrete exposed to the same conditions. The authors attributed this to the fact that the water-cured CSA concrete had a higher degree of hydration and lower porosity.

4.2. Shrinkage Property of CSA Concrete

One of the most remarkable properties of CSA cement is its shrinkage resistance. Shrinkage is the reduction in volume of concrete due to the loss of water or chemical changes in the cement paste [49]. Shrinkage-compensating concrete (SCC) is a type of concrete that compensates for drying shrinkage. Specifically, it is designed to minimize the tendency for cracking caused by drying shrinkage in concrete slabs, pavements, and structures. One of the key materials used in SCC is calcium sulphoaluminate (CSA) cement [50]. CSA concrete, using calcium sulphoaluminate cement as the main binder instead of the conventional portland cement, offers several advantages such as faster setting and hardening, higher early and ultimate strength, lower carbon footprint, and better durability [51]. CSA cement's unique hydration mechanism consumes water molecules during the formation of ettringite crystals, effectively eliminating excess bleed water and preventing drying shrinkage. Unlike ordinary portland cement, CSA cement contains no C_3A (tricalcium aluminate), which maximizes its sulfate resistance [52].

Several factors contribute to the shrinkage resistance of CSA concrete:

i. Water Demand and Chemical Binding: CSA cement requires more water than portland cement for proper hydration, but this water is chemically bound, causing no water surplus, and thus preventing shrinkage cracking and warping.
ii. Chemical and Stable Expansion: During the hardening phase, a chemical and stable expansion occurs that prevents further shrinkage over time.
iii. Low-Shrinkage Phases Formation: CSA cement forms phases such as calcium sulphoaluminate (C_4A_3S), ettringite ($C_6AS_3H_{32}$), and belite (C_2S), which have lower shrinkage potential than phases of portland cement such as alite (C_3S) and free lime (CaO) [53].

Studies have confirmed the shrinkage resistance of CSA concrete under different conditions and applications. Ke et al. (2018) [50] found that CSA concrete showed significantly lower shrinkage than portland cement concrete (PCC) in both sealed and unsealed conditions, as well as under drying and wetting cycles. Similarly, Zhang et al. (2016) [54] found that CSA concrete exhibited much less shrinkage than PCC under various temperatures and relative humidities, with relative humidity being the main influencing factor. Although CSA has proven effective as a shrinkage-compensating concrete, some studies have identified a critical relationship between the carbonation and shrinkage of CSA. Ye et al. (2017) [55] observed shrinkage performance of cement mortars containing various alkali salts at two different drying conditions (nitrogen gas and air); the results show that alkalis drastically increase shrinkage magnitude, but reduce the shrinkage kinetics of these mortars. The findings suggest that alkali enrichment could reduce creep modulus under drying-induced internal stress. Yuya et al. (2021) [56] also identified the relationship between carbonation shrinkage and the calcium/silicate ratio of C-S-H. The result shows that carbonation shrinkage progressively increased at intermediate relative humidities of 43% and 66%. These studies provide an idea of how carbonation can induce shrinkage in CSA concrete. They also highlight various factors such as concrete composition, environmental conditions, and the use of additives influencing this process. However, not many studies have investigated this, and the conclusion is subject to ongoing research.

4.3. Impact on Mechanical Properties

CSA has proven effective in applications requiring early strength and long-term strength [57]. For example, Markosian et al. (2019) [58] reported that CSA used on a prestressed bridge girder allowed the concrete to reach a strength of 4300 pounds per square foot (about 30 MPa) in only 6.5 h, demonstrating CSA's effectiveness for early strength applications. According to Jiaming Wang et al. (2022) [59], the impact of carbonation on the mechanical properties of concrete may vary depending on the type of cement being used. For example, Ashraf et al. (2016) [60] and Zhang et al. (2024) [61] investigated the effect of CO_2 exposure on CSA cement-based mortars and found that exposure to CO_2 caused a reduction in the compressive strength and flexural strength of the mortars. The extent of the reduction depended on the percentage of CSA cement in the mix. SEM and XRD analysis revealed that the CO_2 exposure led to the formation of calcium carbonate, which caused the microstructure of the mortars to become more porous and less dense (see Figure 5). The authors also observed the formation of ettringite, which may have contributed to the reduction in strength.

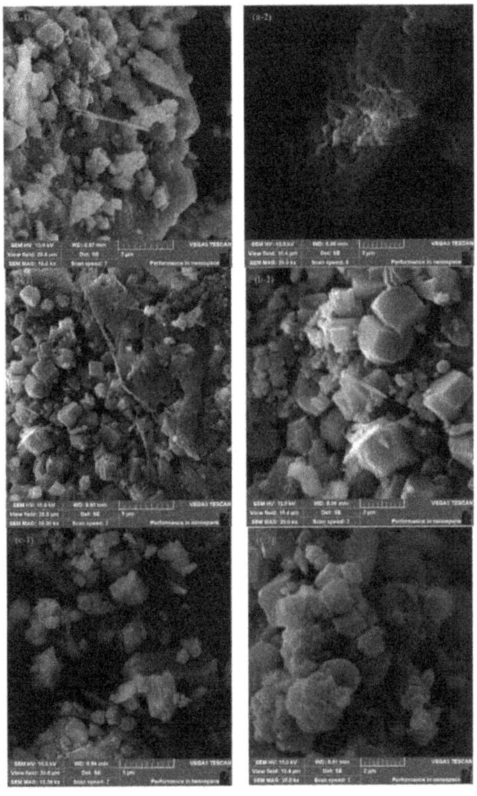

Image (a):
- *Magnification Level (SEM MAG): a-1: 10k, a-2: 20k*
- *Working Distance (WD): a-1: 8.97 mm, a-2: 8.98 mm*
- *View Field: 23.5 pt*
- *SEM Magnification (SEM MAG): a-1: 15.0 k, a-2: 8.34 µm*
- *Detector Type (Det): SE (Secondary Electron Detector)*

Image (b):
- *Magnification Level (SEM MAG): b-1: 10k, b-2: 20k*
- *Working Distance (WD): b-1: 8.95 mm, b-2: 8.36 mm*
- *View Field: 23.5 pt*
- *SEM Magnification (SEM MAG): a-1: 15.0 k, a-2: 8.34 µm*
- *Detector Type (Det): SE (Secondary Electron Detector)*

Image (c):
- *Magnification Level (SEM MAG): c-1: 10k, c-2: 20k*
- *Working Distance (WD): c-1: 8.94 mm, c-2: 8.91 mm*
- *View Field: 23.5 pt*
- *SEM Magnification (SEM MAG): a-1: 15.0 k, a-2: 8.34 µm*
- *Detector Type (Det): SE (Secondary Electron Detector)*

Figure 5. SEM images of carbonated CSA mortars [61].

Similarly, Sevelsted and Skibsted (2015) [62] investigated the effect of CO_2 exposure on the mechanical properties and durability of alkali-activated slag (AAS) mortars. The study found that CO_2 exposure caused a reduction in the compressive strength and increased water absorption of the AAS mortars. The extent of the reduction depended on the type of activator used. The formation of calcium carbonate and other phases in the microstructure of the AAS mortars was also observed, which can be similar to what happens in CSA concrete under CO_2 exposure. The study's findings on potential mitigation measures to improve the resistance of AAS materials to CO_2-induced degradation may also be applicable to CSA concrete.

In another study by Cao et al. (2022) [63] on the carbonation of reactive powder concrete with calcium sulphoaluminate cement, the authors found that the strength of the concrete increased upon carbonation after 7 days. However, it was noticed that the strength of the concrete began to decline after seven days before carbonation. Geng et al. (2014) [48] in their study on the carbonation of CSA cements also noticed something similar: that the compressive strength of CSA-based concrete initially decreased and later increased after prolonged exposure to CO_2, suggesting that carbonation can have complex effects on the mechanical properties of CSA concrete. The formation of a denser microstructure due to carbonation may contribute to the increase in the compressive strength of CSA concrete after longer exposure times to CO_2. However, the accelerated aging conditions used in the study may not fully reflect the long-term effects of CO_2 on CSA concrete. Therefore, further research is needed to better understand the impact of carbonation on the long-term durability of CSA concrete structures.

In a study by Groves et al. (1991) [64], experiments were conducted to investigate the changes in the phase composition, microstructure, and mechanical properties of CSA cement paste after CO_2 exposure. The results indicate that carbonation of CSA cement paste caused the formation of calcium carbonate and ettringite, which led to the degradation of the material's microstructure. The mechanical properties of the CSA cement paste were also affected by carbonation, with a decrease in compressive strength observed after exposure to CO_2. However, a counter argument in a recent study by Wang et al. (2021) [65] suggests that CSA concrete has good carbonation resistance and can even benefit from carbonation in terms of compressive strength. The authors found that the compressive strength of the CSA concrete increased after carbonation due to the formation of calcium carbonate in the pore structure, which helped to fill the pores and improve the mechanical properties. However, the authors also noted that attention needs to be paid to the impact on flexural strength and durability. The flexural strength and durability of the CSA concrete decreased after carbonation, possibly due to the formation of cracks in the interfacial transition zone between the cement paste and the aggregate.

To summarize, the effects of carbonation on the mechanical characteristics of concrete can be complicated and reliant on elements including the composition of concrete, the amount of exposure time, and the ambient conditions. Due to the production of calcium carbonate and other phases in the microstructure, exposure to CO_2 can generally reduce the compressive strength, flexural strength, and durability of concrete. By filling the pores and strengthening the microstructure, carbonation, according to some research, may enhance the compressive strength of specific types of cement, including CSA cement. Further investigation is required to comprehend the long-term impacts of carbonation on the mechanical characteristics of various types of cement and to develop mitigation techniques for carbonation's disadvantageous effects on the resilience of concrete structures.

4.4. Impact on Carbonation Depth

Carbonation depth is the extent to which CO_2 penetrates the concrete, and it can vary depending on factors such as the concrete composition, exposure conditions, and permeability. The depth of carbonation can affect the extent of chemical changes in the concrete and its mechanical properties [66]. Geng et al. (2014) [48] noticed that the carbonation depth of concretes made with CSA cement increases with an increased exposure time to

CO_2, leading to more exposure in the pore structure of the concrete. This indicates that carbonation can be a potential durability concern for CSA concrete structures. Wang et al. (2021) [65] provides key insights into the impact of carbonation on the microstructure and mechanical properties of concrete, specifically in relation to fiber-reinforced concrete (FRC). The study presents experimental results showing that the carbonation depth of FRC is lower than that of conventional concrete, and that the carbonation reduces the porosity and interfacial transition zone thickness of the FRC, resulting in improved compressive strength and modulus of elasticity. The study also highlights the role of fibers in enhancing the resistance of FRC to carbonation-induced deterioration. This study explains that carbonation can impact the properties of CSA concrete, and fiber reinforcement may play a role in mitigating these effects.

Zhang et al. (2009) [36] investigated the effect of w/c ratio and admixtures on the carbonation resistance of sulphoaluminate cement-based high-performance concrete (HPC). The experimental results indicated that decreasing the w/c ratio leads to a significant decrease in carbonation depth and an improvement in carbonation resistance. The addition of slag and fly ash in the ratio 2:1 also enhanced the carbonation resistance of the concrete. In another study, Duan et al. (2013) [67] examined the addition of magnesium aluminate-based layered double hydroxides (LDHs) to sulphoaluminate concrete. This led to a significant decrease in carbonation depth, indicating improved carbonation resistance. It was also found out that the carbonation depth of CSA concretes is less compared to portland cement concrete irrespective of the addition of LDHs. This suggests that CSA concretes may have higher carbonation resistance compared to PC concretes.

Ultimately, carbonation can have a significant impact on how long concrete will last and how it will behave mechanically, including CSA cement-based concrete. Reducing the w/c ratio and adding admixtures can improve the carbonation resistance of concrete. The depth of carbonation can vary depending on several factors, including the concrete composition and exposure conditions. Concrete qualities like compressive strength and modulus of elasticity can be less negatively affected by carbonation by adding fiber reinforcement and LDHs based on magnesium aluminate. In comparison to portland cement concretes, CSA concretes may have stronger carbonation resistance, which suggests a possible advantage of CSA concrete structures in terms of durability over portland cement concrete structures.

4.5. Impact on Chemical Properties of Concrete

Carbonation in concrete, a chemical reaction that leads to the formation of calcium carbonate ($CaCO_3$), significantly impacts the chemical properties of CSA concrete [68]. This process is influenced by the concrete's porosity and moisture content, and it occurs under specific conditions of relative humidity (RH) levels between 40% and 90%. The carbonation process involves three stages: CO_2 reacts with water to produce carbonic acid; dissolved calcium ions react with carbonic acid to produce trioxocarbonate; and finally, $CaCO_3$ precipitates, reducing calcium ions in the solution [69].

Sharma et al. (2023) [29] found that carbonation curing influenced the microstructural characteristics and CO_2 uptake of CSA cement. The low-pH carbonic acid promoted ettringite participation in $CaCO_3$ formation, leading to the formation of different $CaCO_3$ polymorphs, $Al(OH)_3$, and unreacted ye'elimite. The dissolution of hydrated phases moderately affected the strength of the concrete. The study suggests CO_2 can impact the chemical properties of CSA concrete. Taylor et al. (2001) [70] examined the effects of carbonation on CSA cement and found that it can lead to a loss of strength due to the breakdown of ettringite into aluminum hydroxide, gypsum, and calcium carbonate. This reaction is highly sensitive to carbonation, indicating a significant impact on the concrete's chemical properties. Similarly, Zhang et al. (2009) [36] and Cho et al. (2017) [71], using SEM and XRD analysis (see Figure 6), found that ettringite, the main hydration product of sulphoaluminate cement, decomposes after carbonation, leading to a loss of strength in CSA concretes. However, Ansari et al. (2022) [28] proposed a new binder (MCSA) by replacing 20% of CSA with fine cement-sands, which was found to improve the mechanical properties and resistance to carbonation in CSA concretes. These findings suggest the need for further research to develop new materials and techniques to mitigate the negative impact of carbonation on the chemical properties and strength of CSA concrete.

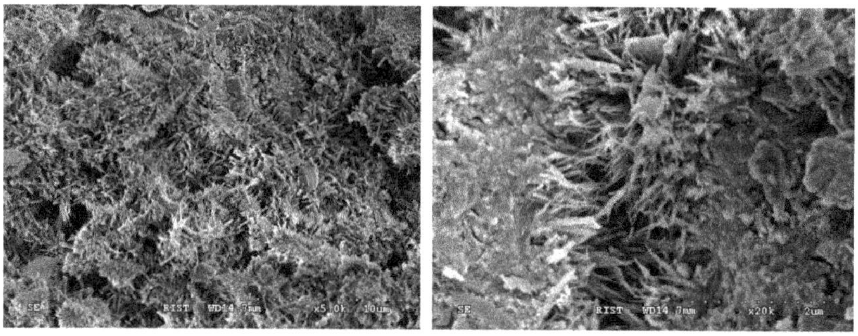

(a) O50B30G20

(b) O50B30AS10G10

Figure 6. SEM images showing ettringite formation [71].

In a recent study by Wang et al. (2021) [65], the impact of carbonation on the chemical properties of CSA cement was investigated. The authors found that carbonation resulted in the formation of calcium carbonate ($CaCO_3$) and the breakdown of ettringite, leading to a reduction in the compressive strength of the cement. Specifically, carbonation caused the decomposition of ettringite into gypsum and calcium carbonate. The authors also found that the carbonation rate of CSA cement was affected by factors such as relative humidity, temperature, and exposure time to CO_2. The study highlights the importance of considering the impact of carbonation on the chemical properties of CSA concrete, particularly in applications where durability and strength are critical factors. In Duan et al. (2013)'s [67] study on how the addition of magnesium aluminate-based layered double hydroxides (LDHs) to CSA cement affects its behavior, it was found out that the LDHs affected the pore structure of the concrete made with calcium aluminate cement, and the exchange and immobilization of carbonate ions is believed to have contributed to the enhanced carbonation resistance.

Gastaldi et al. (2018) [72] explored the effect of carbonation on the microstructure and composition of CSA concrete and compares it to that of ordinary portland cement (OPC) concrete. The study's findings reveal that the carbonation of CSA concrete led to a decrease in pH, a reduction in the amount of calcium hydroxide (CH), and the formation of $CaCO_3$. These changes were like those observed in OPC concrete but occurred at a faster rate in CSA concrete. Also, the study found that the carbonation of CSA concrete decreased the amount of ettringite, a primary hydration product in CSA cement. Moreover, the study found that the carbonation of CSA concrete led to the formation of two distinct layers of $CaCO_3$. The inner layer, formed near the surface of the aggregate particles, was composed of dense and fine-grained $CaCO_3$ crystals. The outer layer, formed in the cement paste, was composed of coarser and more porous $CaCO_3$ crystals. These layers bring about increased porosity in the concrete, which may have contributed to the observed strength decline. Overall, these findings suggest carbonation can significantly impact the chemical properties of CSA concrete, including changes in pH, reduction in CH, and the formation of $CaCO_3$ and ettringite depletion. These chemical and microstructural changes may ultimately lead to a decrease in the strength and durability of CSA concrete.

Quillin (2001) [73] concluded that the chemical properties of CSA concrete are significantly influenced by exposure to CO_2. The author found that the carbonation process in CSA concrete leads to the dissolution of the ettringite phase, resulting in a decrease in pH and an increase in porosity, which leads to a decrease in the compressive strength of the concrete. Furthermore, the author noticed that the extent of carbonation depends on the composition of the concrete mix, with mixes containing higher levels of C_3A being more susceptible to carbonation. According to Fernandez-Carrasco et al. (2012) [47], $Al(OH)_2$ was noticed from the XRD experiment for all the samples tested at early age. However, after 90 days, $Al(OH)_2$ was only noticed in the calcium aluminate samples. The study concluded that carbonation impacts the evolution of the hydrated phases of ternary cements, which tends towards the formation of calcium carbonate, gypsum, and aluminum hydroxide as discussed in prior studies.

In another study, L. Wang et al. (2019) [74] explored the pore solution chemistry of CSA cement and how it impacts steel passivation. The results from the investigation showed that the pore solution of CSA cement had a notably lower pH and increased aluminum and sulfate ion concentrations when compared with ordinary portland cement. It was noticed that the high sulfate concentration leads to the formation of gypsum and ettringite, which can impact the corrosion resistance of steel rebars in concrete. The study also discovered that the addition of silica fume to CSA cement in concrete mix can raise the pH and reduce the sulfate concentration, which can lead to enhanced steel passivation. Carbonation of CSA can also affect the passivation of steel because of the lowered pH, which is also particularly important in understanding the durability of concretes made from CSA cement. It is therefore important to understand the pore solution chemistry of CSA cement. The authors also identified how $Al(OH)_4^-$ can potentially provide protection to steel because of its ability to release hydroxyl during the reactions. The chemical process is presented below:

$$Fe^{2+} + 2H_2O = Fe(OH)_2 + 2H^+$$

$$Al(OH)_4^- \leftrightarrows Al(OH)_3 + OH^- \quad (1)$$

In summary, substantial research has been conducted on the effects of carbonation on the chemical characteristics of concrete, especially in CSA cement. The breakdown of ettringite and the generation of calcium carbonate ($CaCO_3$) because of carbonation might cause the concrete's compressive strength to decrease. Many variables, such as temperature, relative humidity, exposure duration to CO_2, and concrete mix composition, have an impact on the carbonation rate. The chemical and microstructural alterations brought on by carbonation may eventually cause CSA concrete's strength and durability to decline. New binders and additives have been suggested by certain research to improve CSA concrete's resistance to carbonation. Further research is required to develop novel materials and techniques to mitigate the negative impact of carbonation on the chemical properties and strength of CSA concrete.

4.6. Environmental Impact

The impact of CO_2 on CSA concrete is not only limited to its mechanical properties but can also have environmental implications, as the production of CSA concrete can contribute to carbon emissions and climate change [75]. Mechling et al. (2014) [24] highlights the sensitivity of CSA cement mortars to carbonation, which can lead to changes in their mechanical properties and dimensional stability. This information can be used to emphasize the importance of considering the effects of carbonation on the long-term durability and performance of CSA concrete in different environmental conditions. Additionally, the paper highlights the advantages of CSA cement over OPC in terms of energy use and carbon dioxide generation during manufacture, which can be useful in discussing the potential benefits of using CSA concrete as a more sustainable alternative to traditional cementitious materials.

Similarly, Zhang et al. (2009) [36] investigated the impact of CO_2 on the durability of CSA concrete under different environmental conditions, with samples exposed to CO_2 for 7, 14, 28, and 56 days. The results show that the carbonation depth of CSA concrete increases significantly with exposure time, and the carbonation depth is higher in a CO_2-rich environment compared to a normal air environment. The study also suggests that the carbonation of CSA concrete is accelerated in a high humidity and high-temperature environment. Therefore, the study suggests that the durability of CSA concrete should be evaluated by considering the carbonation resistance in different environmental conditions. Vu et al. (2019) [27] also studied the impact of climatic conditions on the carbonation depth of various kinds of concretes under sheltered and unsheltered situations over a five-year period. The results indicate that the general carbonation trend was not much different irrespective of the macroclimate. However, the number of rainy days had a more considerable influence on the progress of carbonation than the total rainfall in the region. This could explain how CSA concrete would perform under similar conditions, since the compositions of the materials tested have similar characteristics.

In another study, Hargis et al. (2017) [46] noticed that portland cement had a greater CO_2 binding capacity than CSA cement, with result disparities of 50 and 19.2. This could suggest that portland cement may be a better option than CSA cement in terms of reducing carbon emissions associated with concrete production. However, this is a subjective statement based on the context of application. Further studies are needed to better understand the carbon footprint of both concrete made from PC and CSA, especially from a cradle-to-grave point of view.

It can be concluded that the carbonation depth of CSA concrete is influenced by climatic conditions. The results show that the general trend of carbonation in CSA concrete is not much different irrespective of the macroclimate, but the number of rainy days has a more considerable influence on the progress of carbonation than the total regional rainfall as presented by [27]. In addition, sheltering the concrete specimens had a significant effect on reducing carbonation. Recent studies, however, suggest that careful consideration should be given to the environmental exposure of CSA concrete structures to minimize the risk of carbonation-induced damage.

4.7. Mitigation Strategies for Carbonation of CSA Concrete

Various mitigation strategies have been proposed to reduce the negative impact of carbonation on CSA concrete. One such strategy is the use of supplementary cementitious materials (SCMs) in the concrete mix. SCMs such as fly ash, slag, and silica fume have been found to reduce the porosity of concrete, leading to reduced carbonation [45]. Also, corrosion inhibitors, such as calcium nitrite, and organic inhibitors have been proposed to reduce the corrosion of steel reinforcement caused by carbonation [41].

Another strategy is the use of coatings or sealers on the surface of the concrete to reduce the ingress of CO_2 and other harmful gases, as described in the work of Chen, Lui, and Yu (2018) [39]. Another study recommended the use of lower w/c ratios in the concrete mix, which can also reduce the porosity of the concrete and limit the extent of carbonation [76]. Based on the findings by Duan et al. (2013) [67], the addition of magnesium aluminate-based layered double hydroxides (LDHs) to CSA concrete improved its carbonation resistance significantly. This was attributed to the modification of the pore structure of the concrete, which reduced the available space for CO_2 to penetrate and react with the cement phases. The LDHs were also found to immobilize carbonate ions, preventing them from reacting with the cement phases. Therefore, incorporating LDHs in CSA concrete could be a promising approach to address the challenges with pore structure and improve the carbonation resistance of CSA concrete. The study by Cho et al. (2017) [71] provides a promising approach by using industrial byproducts like aluminum-rich slag and ground granulated blast furnace slag. These can reduce the carbon footprint by minimizing the need for OPC, which is energy-intensive to produce. The formation of stable hydrates and an improved microstructure also contributes to the durability and longevity of the concrete, further reducing CO_2 emissions over its lifecycle.

In summary, the reviews suggest that carbonation is a complex process that can impact the performance and durability of CSA concrete. While carbonation can have some beneficial effects, such as increased strength and density, it can also lead to a decrease in the long-term strength and durability of the concrete. Consequently, it is important to understand the factors that influence carbonation and to develop effective mitigation strategies to enhance the acceptance of CSA concrete.

Overall, there are many ways that CO_2 has an impact on CSA concrete. It is evident that carbonation can significantly impact the long-term performance of concrete, despite it affecting both the strength and durability of the concrete negatively and positively. According to previous studies, carbonation can cause ettringite minerals to break down, increase porosity, increase chloride ion permeability, and reduce compressive and flexural strength. To increase the adoption of CSA concrete, it is crucial to create efficient mitigating techniques.

Thankfully, studies have shown that there are several ways to lessen the impact of CSA concrete production on the environment and restrict the harmful effects of carbonation. They include the use of corrosion inhibitors, additional cementitious materials, and alternative cementitious materials, as well as carbon capture and storage technologies. Moreover, applying coatings or sealants to the concrete's surface and using less water and cement in the concrete mix can help restrict the amount of carbonation and limit the entry of dangerous gases.

4.8. Summary of Some Studies on CSA Concrete

Some of the studies critically examined for the qualitative analysis are highlighted in following table. The table includes the title of each study, the key findings relating to CSA carbonation and the novelty of each study. This was done to provide transparency and an overview of the studies identified. Additionally, this summary helps to compare novelties and methodologies in different studies to identify trends, gaps, and areas where further research is needed.

Author(s)	Study Title	Findings	Novelty
[28]	A novel approach to improve carbonation resistance of Calcium SulfoAluminate cement by assimilating fine cement-sand mix.	CSA could be more resistant to carbonation with partial replacement of CSA cement with fine particles. Modified CSA samples with fine-cement sand mix showed an over 12% increase in compressive strength upon carbonation as compared to OPC and CSA.	The study proposes an enhanced cement formulation (OMCSA) to reduce carbonation of CSA. It also addresses the main barriers of carbonation in CSA cement by partial replacement with CSA < 2.4 μm powder and shows that mixing OPC or conventional CSA with fine CSA < 2.4 μm cement mortar to various proportions is a viable way to mitigate this issue.

Author(s)	Study Title	Findings	Novelty
[24]	SulfoAluminate cement behaviors in carbon dioxide, warm and moist environments	CSA mortar is highly sensitive to carbonation. Increased water content reduces the mechanical properties and dimensional stability of CSA concrete.	The authors propose a set of mortars at 3 w/c ratios to test the material's performance under different conditions. The study also compares CSA mortars with CEM I portland cement mortars. In addition, the behavior of calcium sulphoaluminate cements in four different environments (ambient air @20 °C and 65% RH, dry air @45 °C and 35% RH, water @45 °C, saturated CO_2 @20 °C) were compared.
[39]	Effects of Environmental Factors on Concrete Carbonation Depth and Compressive Strength	Found out that concrete has three stages of carbonation, each of which corresponds to the presence of different polymorphs.	The authors propose a relationship between concrete carbonation and temperature, CO_2 concentration, and humidity based on XRD, ESEM, and SEM. A novel model to predict carbonation depth and compressive strength in concrete is also proposed.
[60]	Carbonation of cement-based materials: Challenges and opportunities	Noticed that concrete using OPC as the primary binder had a stronger carbonation resistance than concrete with SCMs (such as silica fume, slag, or fly ash) or alternative binder ingredients (i.e., C_2S, AAM). Carbonation has historically been seen as a negative factor for cement-based materials. However, it can also be seen as a chance to develop a CO_2-efficient concrete industry by storing CO_2 in cement-based materials and using carbonate binders.	The study reviews progress in carbonation of cement-based materials by extracting concrete highlights from peer-reviewed journals. In addition to the existing knowledge regarding the carbonation of cement-based materials, areas that require further research are also identified.
[16]	Durability and microstructure of CSA cement-based materials from MSWI fly ash.	Researched the use of fly ash as a raw material for cement. Used MSWI fly ash as a raw material to prepare new type of cement, which was called CSA cement. Successfully used MSWI fly ash as raw material to prepare lime-free sintering cement.	The study explored MSWI fly ash as cementitious material for sustainable construction. It proposes a novel method of MSWI fly ash being transformed into cement that results in durable and effective material that exceeds current global limits for heavy metal leaching. The study also used MSWI fly ash in cement, with hydrated alkali solutions and biomaterials to produce new mortar and building blocks.
[77]	CSA-based portland-free binders to manufacture sustainable concretes for jointless slabs on ground.	The paper explored the possibility of manufacturing eco-friendly shrinkage-compensating mortars by using CSA-based ternary mixtures to replace OPC with SCMs and lime (CH). Replacing OPC with SCM and lime can potentially reduce CO_2 emission and energy consumption by up to 60 percent.	The paper evaluates the impact of water/binder ratio, condition of curing, and tartaric acid dosage on elastic, rheological, and physical properties of CSA concretes, SCM, and anhydride concretes that have shrinkage-compensating properties.

Author(s)	Study Title	Findings	Novelty
[17]	Recent progress and technical challenges in using calcium sulfoAluminate (CSA) cement.	Found out that CSA cement can be used as a replacement for portland cement to reduce CO_2 emissions. The study introduces the hydration mechanism, long-term durability, and new applications of cement made from the CSA minerals.	Profiles CSA cement with respect to its hydration mechanism, mechanical properties, and long-term durability. The study reviewed the literature on the hydration mechanisms of CSA cement and the properties of its hardened cement paste, including strength, durability, and chemical composition. It discussed recent work and potential challenges associated with producing CSA cement at an industrial scale. The potential applications of CSA cement in developing infrastructure in regions where fresh water is scarce was also explored.
[78]	Novel use of calcium sulfoAluminate (CSA) cement for treating problematic soils	The study noticed that using CSA cement in clay-based ground stabilization solutions was more environmentally sustainable than using lime or cement. Identified CSA cement as a sustainable alternative in porous aggregates stabilization. Found out that CSA cement-treated expansive soil exhibited three distinct hydration phases, where the state of hydration was identified through experimentation.	The authors explored a sustainable approach to cement replacement by investigating the effectiveness of CSA cement in ground stabilization. The study proposes CSA cement as an alternative for expansive soil stabilization due to reduced carbon impact. Hydration of CSA cement into expansive soil and its effect on mechanical properties and microstructural change was investigated. The cationic exchange, flocculation, and agglomeration characteristics of CSA cement in expansive soil was also investigated.
[79]	Behavior of blends of CSA and portland cements in high chloride environment	The study achieved high early strength of CSA cement composite in high chloride environments. CSA was found to be effective in showing high mechanical performance even in high chloride conditions. CSA binders could also prove resistant to chloride attack.	Describes a novel approach to cement mix design that uses multi-objective optimization to maximize early strengths and durability in marine environments. The study addressed the durability aspects of CSA cement in concrete. Proposes concrete with CSA replacing some of the OPC in different proportions, suitable for marine environments where seawater interacts with concrete.
[80]	Decomposition of synthesized ettringite by carbonation	The study showed that the carbonation mechanism is different under wet and dry conditions. This can be explained by the excess or lack of water around or in the ettringite. It was discovered that ettringite can capture CO_2 and produce stable solid mineral products.	The authors proposed a theoretical model of carbonation kinetics based on ettringite and its reaction with CO_2 gas. The theoretical model was used to predict carbonation rates using the Jander equation. Proposes a realistic and powerful model for long-term geological sequestration of CO_2.

Author(s)	Study Title	Findings	Novelty
[81]	Cement-based materials eventually reabsorb much of the CO_2 released during creation.	Concrete re-absorbs about 43% of the CO_2 emitted during cement production over its life cycle.	Discovered that concrete reabsorbs a significant amount of the CO_2 emitted during cement production.
[82]	Effects of CSH2 and CH on Strength and Hydration of Calcium SulphoAluminate Cement Prepared from Phosphogypsum	Phosphogypsum increases the strength performance of CSA concrete. The pH of samples increased with the addition of CH. The paper found that adding gypsum dihydrate and calcium hydroxide to cement clinker prepared from phosphogypsum affects the compressive strength and hydration process of the cement.	Investigation of carbonation of CSA with varying w/c ratio. The intervention of optimizing the use of mineral admixtures such as gypsum dihydrate and calcium hydroxide to cement clinker prepared from phosphogypsum.
[63]	Research into Carbon Dioxide Curing's Effects on the Properties of Reactive Powder Concrete with Assembly Unit of SulphoAluminate Cement and Ordinary portland Cement	The findings of the paper are that the addition of sulphoaluminate cement can improve the mechanical strength of RPC at low curing age, but when the curing age is beyond 7 days, the sulphoaluminate cement has a negative effect on the mechanical strength. CO_2 curing increases the mechanical strength and the resistance of RPC to sodium chloride (NaCl) freeze–thaw cycles.	The paper examines the effects of carbon dioxide curing on the properties of reactive powder concrete with assembly unit of sulphoaluminate cement and ordinary portland cement. This study uses a combination of rapid-setting sulphoaluminate cement, OPC, and silica fume as one kind of mineral admixture.
[76]	Durability of calcium sulfoAluminate cement concrete	The paper suggests that CSA concrete has better performance than portland cement concrete in several aspects, including shrinkage and cracking due to restrained shrinkage, freeze–thaw damage, alkali–silica reaction, and sulfate attack.	Reviewed durability of CSA concretes in general.
[45]	Carbonation of concrete: the role of CO_2 concentration, relative humidity, and CO_2 buffer capacity	The carbonation coefficients determined at the different relative humidities (RH) of 57, 70, and 80% clearly indicate that the response of a concrete to carbonation at increased RH is both dependent on cement type and w/c. The carbonation resistance of concrete in sheltered and with restrictions in unsheltered outdoor exposure can be assessed at 4% CO_2 and 57% RH.	The effect of CO_2 concentration and ambient relative humidity (RH) on the accelerated and natural carbonation of eighteen concrete mixtures produced with nine different cement types was studied over days.

Author(s)	Study Title	Findings	Novelty
[74]	Pore Solution Chemistry of Calcium Sulfoaluminate Cement and Its Effects on Steel Passivation.	The pore solution chemistry of CSA can affect the passivation of steel reinforcement, which can have implications for the durability and long-term performance of concrete structures. The study identified a novel ion ($Al(OH)_4^-$) in CSA pore solution, which is not present in OPC. The study found that CSA pore solution chemistry can promote the formation of a protective layer on the steel surface, which can help to prevent corrosion. However, the study also found that certain ions in the pore solution, such as chloride ions, can negatively impact steel passivation and promote corrosion. This highlights the importance of carefully managing the chemistry of the pore solution in CSA concrete to ensure the long-term durability of concrete structures.	The study investigates the effects of CSA cement on the passivation of steel, which has not been extensively studied in previous research. The study also explores the pore solution chemistry of CSA cement, providing insights into the mechanisms by which the cement impacts steel corrosion. Additionally, the study suggests that the use of CSA cement may offer an eco-friendlier alternative to traditional portland cement due to its lower carbon footprint.
[46]	Carbonation of calcium sulfoaluminate mortars	Found that CSA mortars carbonate faster than portland cement mortars and at the same rate as calcium aluminate cement (CAC) mortars. The anhydrite content of the CSA mortars strongly affects the ye'elimite reaction kinetics, which play a vital role in imparting carbonation resistance in CSA mortars. CSA mortars carbonate slower with decreasing water content. Calcium sulfate additions to CSA clinker to produce CSA cement dilutes the clinker content and reduces the amount of CO_2 that the CSA cement can bind. Thermodynamic modeling can be used to predict the carbonation process of CSA mortars.	This study provides a comprehensive investigation of the carbonation process in CSA mortars, including the physical and chemical parameters that govern the carbonation rate and the microstructural and chemical factors that affect the rate of carbonation in CSA cement. Additionally, it also presents a detailed analysis of the part anhydrite content and calcium sulfate additions play in the carbonation resistance of CSA mortars. Finally, it provides thermodynamic modeling to support the experimental results and to better understand the carbonation process in CSA cement. The study gives comprehensive insight to the behavior of CSA cement under carbonation and gives suggestions on how to improve the durability of CSA.

Author(s)	Study Title	Findings	Novelty
[67]	Influence of layered double hydroxides on microstructure and carbonation resistance of sulphoAluminate cement concrete	The addition of magnesium aluminate-based layered double hydroxides (LDH) to CSA concrete improves its carbonation resistance significantly. The LDHs modify the pore structure of the concrete, reducing the amount of available space for CO_2 to penetrate and react with the cement phases. The LDHs can immobilize carbonate ions, preventing them from reacting with the cement phases. The use of fly ash in the concrete mixture further improves the carbonation resistance of the CSA concrete, resulting in a lesser carbonation depth compared to portland cement concrete.	The use of magnesium aluminate-based layered double hydroxides (LDHs) as an additive to improve the carbonation resistance of sulphoaluminate cement (SAC) concrete. Previous studies have focused on the use of various materials to improve the performance of portland cement-based concrete, but this study specifically focuses on SAC concrete, which has different properties and composition. The study also characterizes the effects of LDHs on the pore structure of the concrete using mercury intrusion porosimetry (MIP), which is a technique not commonly used in studies on concrete carbonation resistance. The finding that the addition of LDHs can significantly improve the carbonation resistance of SAC concrete by modifying its pore structure and immobilizing carbonate ions is also novel and contributes to the development of more sustainable and durable concrete materials.
[47]	Carbonation of ternary building cementing materials	The carbonation resistance of ternary building cement materials is higher than that of pure portland cement. The carbonation resistance of the cementing materials increases with the increase in fly ash and limestone powder content. The carbonation depth of the cementing materials decreases as the curing time increases, indicating that the carbonation reaction is a slow process. The C-S-H in the cementing materials are more stable than the CH phase during the carbonation process, which contributes to the carbonation resistance of the materials. The formation of $CaCO_3$ during the carbonation process aids the decrease in porosity and the improvement in the mechanical properties of the cementing materials.	The study investigates the carbonation behavior of ternary cementitious materials, which are composed of different combinations of portland cement, calcium aluminate cement, and calcium sulphoaluminate cement. This study's objective was to provide a comprehensive understanding of the behavior of ternary cements under carbonation, which can potentially lead to the development of more sustainable and durable building materials. Additionally, the use of different techniques such as XRD and thermogravimetric analysis allowed for a detailed characterization of the reaction products and the carbonation process.

Author(s)	Study Title	Findings	Novelty
[65]	Mechanical and Microstructural Characteristics of Calcium Sulfoaluminate Cement Exposed to Early-Age Carbonation Curing	CSA cement pastes' compressive strength increased with briquetting pressure. However, the relationship is not linear; the growth rate of the compressive strength slows down with increasing briquetting pressure. At briquetting pressure less than 20 MPa, the compressive strength of the test block at 3 days does not differ much from the strength at 1 day. When the briquetting pressure exceeds 20 MPa, the gap between the strength at 1 day and 3 days gradually increases. The water/binder ratio also impacts the compressive strength of the cement paste. When the water/binder ratio is 0.05, the compressive strength of the test block after carbonation curing becomes very low. The optimal starting point for carbonation curing is 4 h after the start of hydration. The compressive strengths after carbonation curing are all above 50 MPa, while the strength of the specimens only reaches 37 MPa after 3 days of standard curing. This shows that early-age carbonation curing can greatly improve the early mechanical properties of CSA cement paste. As the carbonation curing starting point progresses, the strength of the specimens after carbonation curing showed a tendency to decrease rather than increase. The occurrence of the carbonization reaction can promote the hydration of the cement paste. As the hydration progresses, the paste becomes denser and the porosity becomes lower, so it is increasingly difficult for CO_2 to diffuse into the paste. When carbonization is carried out at the early stage of hydration, the CO_2 diffuses more easily into the paste, and the hydration and carbonization reactions proceed simultaneously and mutually reinforce each other, thus significantly increasing the compressive strength of the specimen.	The paper discusses the physicochemical reactions that occur during early accelerated carbonization, including phase transition and mass transfer of CO_2, dissolution of solid $Ca(OH)_2$ and mass transfer of dissolved $Ca(OH)_2$, hydration of cement compounds, and carbonation of unhydrated cement compounds and hydration products. The authors note that while there are similar studies on other types of cement and concrete materials, there are relatively few studies on the performance of early carbonation curing of CSA cement. The paper also presents a comprehensive analysis of the hydration carbonization reaction of minerals in the CSA cement paste, indicating that the earlier the carbonation starting point is, the more favorable it is for the reaction. It was also noted that different carbonation starting points will not change the types of carbonization reaction and hydration reaction products but will affect the degree of reaction and macroscopic performance.

5. Analysis of CO_2 Impact on CSA Concrete

This study reviews the impact of CO_2 on concretes made from calcium sulphoaluminate (CSA) cements, highlighting their potential as a sustainable alternative to portland cement in terms of reduced carbon emissions [17,46]. Previous research indicates that CSA concrete exhibits other noteworthy properties, including rapid setting, lower energy consumption, and self-stressing properties [83]. Studies such as [24,27] highlight the high sensitivity of CSA cement mortars to CO_2 environments, showing significant variations in performance based on CO_2 exposure. These works demonstrate that CO_2 curing can improve early strength, but may influence long-term durability. Additionally, the results indicate that CO_2 curing accelerates the hydration process, leading to rapid strength gain. However, this process also introduces complexities in long-term stability that require further investigation. In another study by Honfei Cao et al. (2022) [63], CO_2 curing was found to significantly reduce the shrinkage property of CSA concrete. The study highlights that the carbonation process contributes to the formation of stable phases that limit volume changes. The reduced shrinkage observed together with enhanced durability under CO_2 exposure make CSA concrete a promising material for applications requiring long-term dimensional stability.

Sirtoli et al.'s (2015) [84] study finds CSA cement as an eco-friendly alternative, demonstrating significant mechanical benefits when exposed to CO_2. The cement showed improved compressive strength and reduced environmental impact due to lower CO_2 emissions during production. The mechanical properties under CO_2 curing conditions were notably enhanced, suggesting potential for rapid repair applications in concrete structures. The study by Tan et al. (2020) [76] suggests that CSA concrete performs better compared to traditional cements in CO_2-rich environments, showing resistance to carbonation and improved durability. However, the consensus on long-term performance is not definitive. This underscores the need for comprehensive studies to evaluate the long-term effects of CO_2 exposure on CSA concrete, particularly under varying environmental conditions. In Tan et al.'s (2020) [76] study, the durability of CSA cement was confirmed with improved performance in CO_2 environments compared to ordinary portland cement. The carbonation process contributes positively to the material's resistance against environmental degradation. This also emphasizes the CSA's suitability for use in infrastructures exposed to CO_2, such as pavements and underground structures, where natural carbonation could be leveraged to improve concrete performance.

However, a comprehensive understanding of parameters such as corrosion resistance, expansion mechanisms, and durability concerns is still needed [76]. The reviewed literature suggests that while CSA concrete offers significant advantages, it also presents challenges, particularly related to carbonation and its effects on mechanical properties and long-term durability. Studies show that CSA cement mortars are susceptible to rapid carbonation, which impacts their dimensional stability and mechanical properties [24]. Carbonation curing can increase CO_2 uptake capacity, with a maximum four-fold increase observed at increased w/c ratios [29]. While carbonation can improve mechanical strength and resistance to NaCl freeze–thaw cycles in reactive powder concrete containing CSA cement [63], it may also lead to strength reduction in pure CSA binders due to increased pore volume [85]. The carbonation process in CSA cement involves the formation of calcium carbonate polymorphs, $Al(OH)_3$, and unreacted ye'elimite, which is different from traditional portland cement carbonation [29].

Calcium sulphoaluminate (CSA) cement is complex, and its performance is not fully understood. No specific conclusions have been reached regarding the durability of this binder compared to portland cement. However, due to its composition and inherent qualities, CSA concrete is expected to outperform portland cement concrete in several areas, including shrinkage properties, cracking, freeze–thaw resistance, alkali–silica reaction, and sulfate attack [76]. These findings highlight the complex relationship between CO_2 and CSA cement, emphasizing the need for further research on its long-term durability and performance.

A critical analysis reveals that although some studies report decreased compressive and flexural strength upon carbonation, others suggest improvements in certain conditions. The reliability and validity of these studies vary, with differences in study designs, sample sizes, and experimental conditions impacting their conclusions. Connecting these findings to broader sustainability goals, CSA concrete's reduced carbon footprint and potential for carbon sequestration through carbonation highlight its role in sustainable construction practices. However, effective mitigation strategies for carbonation's detrimental effects are crucial for its broader adoption.

6. Conclusions

The comprehensive review of the literature on the impact of CO_2 on calcium sulphoaluminate (CSA) concrete underscores its significant potential as a sustainable alternative to ordinary portland cement (OPC). CSA concrete offers notable advantages, including lower CO_2 emissions during production, rapid setting times, and enhanced durability under certain conditions. However, the review also highlights critical challenges, particularly related to carbonation.

The study revealed that carbonation's impact on CSA concrete is multifaceted, influencing its mechanical properties, durability, and chemical stability. While CSA concrete shows improved early-age strength and resistance to environmental degradation, its susceptibility to carbonation poses long-term durability concerns, especially concerning steel reinforcement corrosion and dimensional stability. The variability in findings regarding compressive and flexural strength post-carbonation indicates the need for standardized testing protocols and further investigation into long-term performance.

Several mitigation strategies have been identified to enhance CSA concrete's resistance to carbonation, including the incorporation of supplementary cementitious materials (SCMs), use of corrosion inhibitors, use of coatings, and optimization of water–cement ratios. The potential of early-age carbonation curing and the addition of layered double hydroxides (LDHs) to modify pore structure and immobilize carbonate ions present promising avenues for further research.

The geographical distribution of research indicates a global interest in CSA concrete, with significant contributions from the USA, China, and Europe. This international collaboration is crucial for advancing the understanding and application of CSA concrete in various environmental contexts.

For industry professionals and policymakers, the findings emphasize the importance of developing and implementing standards for CSA concrete use, promoting sustainable construction practices, and investing in further research to address the identified gaps. As the construction industry continues to seek low-carbon alternatives, CSA concrete represents a viable and promising option, provided its long-term performance and durability can be ensured through continued innovation and rigorous testing.

In conclusion, while CSA concrete offers substantial environmental benefits and promising performance characteristics, addressing its carbonation susceptibility is critical for its broader adoption. Future research should focus on long-term studies, innovative material formulations, and practical applications to fully realize the potential of CSA concrete in mitigating the environmental impact of construction activities.

Author Contributions: Conceptualization, D.D.A. and F.A.; formal analysis, D.D.A.; investigation, D.D.A.; methodology, D.D.A.; project administration, F.A.; resources, D.D.A.; supervision, F.A.; validation, D.D.A. and F.A.; writing—original draft, D.D.A.; writing—review and editing, D.D.A. and F.A. All authors have read and agreed to the published version of the manuscript.

Funding: This research received no external funding.

Conflicts of Interest: The authors declare that there are no competing interests.

References

1. Nañagas, K.A.; Penfound, S.J.; Kao, L.W. Carbon Monoxide Toxicity. *Emerg. Med. Clin. N. Am.* **2022**, *40*, 283–312. [CrossRef] [PubMed]
2. Rocque, R.J.; Beaudoin, C.; Ndjaboue, R.; Cameron, L.; Poirier-Bergeron, L.; Poulin-Rheault, R.-A.; Fallon, C.; Tricco, A.C.; Witteman, H.O. Health effects of climate change: An overview of systematic reviews. *BMJ Open* **2021**, *11*, e046333. [CrossRef] [PubMed]
3. Di Napoli, C.; McGushin, A.; Romanello, M.; Ayeb-Karlsson, S.; Cai, W.; Chambers, J.; Dasgupta, S.; Escobar, L.E.; Kelman, I.; Kjellstrom, T.; et al. Tracking the impacts of climate change on human health via indicators: Lessons from the Lancet Countdown. *BMC Public Health* **2022**, *22*, 663. [CrossRef] [PubMed]
4. How Can We Reduce the Construction Industry's Carbon Footprint? World Economic Forum. Available online: https://www.weforum.org/agenda/2021/07/construction-industry-doesn-t-know-where-it-stands-when-it-comes-to-carbon-emissions/ (accessed on 24 May 2023).
5. Watts, J. Concrete: The Most Destructive Material on Earth. *Guardian*. 2019. Available online: https://www.theguardian.com/cities/2019/feb/25/concrete-the-most-destructive-material-on-earth (accessed on 16 January 2023).
6. Greer, F.; Raftery, P.; Brager, G.; Horvath, A. A perspective on tools for assessing the building sector's greenhouse gas emissions and beyond. *Environ. Res. Infrastruct. Sustain.* **2023**, *3*, 043001. [CrossRef]
7. Sousa, V.; Bogas, J.A.; Real, S.; Meireles, I. Industrial production of recycled cement: Energy consumption and carbon dioxide emission estimation. *Environ. Sci. Pollut. Res.* **2023**, *30*, 8778–8789. [CrossRef] [PubMed]
8. Shen, L.; Gao, T.; Zhao, J.; Wang, L.; Wang, L.; Liu, L.; Chen, F.; Xue, J. Factory-level measurements on CO_2 emission factors of cement production in China. *Renew. Sustain. Energy Rev.* **2014**, *34*, 337–349. [CrossRef]
9. Torgal, F.P.; Miraldo, S.; Labrincha, J.A.; De Brito, J. An overview on concrete carbonation in the context of eco-efficient construction: Evaluation, use of SCMs and/or RAC. *Constr. Build. Mater.* **2012**, *36*, 141–150. [CrossRef]
10. Concrete needs to lose its colossal carbon footprint. *Int. J. Sci.* **2021**, *597*, 593–594. [CrossRef]
11. Bishnoi, S. Carbon Emissions and Their Mitigation in the Cement Sector. In *Carbon Utilization: Applications for the Energy Industry*; Goel, M., Sudhakar, M., Eds.; Springer: Singapore, 2017; pp. 257–268. [CrossRef]
12. Nie, S.; Zhou, J.; Yang, F.; Lan, M.; Li, J.; Zhang, Z.; Chen, Z.; Xu, M.; Li, H.; Sanjayan, J.G. Analysis of theoretical carbon dioxide emissions from cement production: Methodology and application. *J. Clean. Prod.* **2022**, *334*, 130270. [CrossRef]
13. Glasser, F.P.; Zhang, L. High-performance cement matrices based on calcium sulfoaluminate–belite compositions. *Cem. Concr. Res.* **2001**, *31*, 1881–1886. [CrossRef]
14. von Greve-Dierfeld, S.; Lothenbach, B.; Vollpracht, A.; Wu, B.; Huet, B.; Andrade, C.; Medina, C.; Thiel, C.; Gruyaert, E.; Vanoutrive, H.; et al. Understanding the carbonation of concrete with supplementary cementitious materials: A critical review by RILEM TC 281-CCC. *Mater. Struct.* **2020**, *53*, 136. [CrossRef]
15. Tekin, İ.; Dirikolu, İ.; Gökçe, H.S. A regional supplementary cementitious material for the cement industry: Pistachio shell ash. *J. Clean. Prod.* **2021**, *285*, 124810. [CrossRef]
16. Guo, X.; Shi, H.; Hu, W.; Wu, K. Durability and microstructure of CSA cement-based materials from MSWI fly ash. *Cem. Concr. Compos.* **2014**, *46*, 26–31. [CrossRef]
17. Tao, Y.; Rahul, A.V.; Mohan, M.K.; De Schutter, G.; Van Tittelboom, K. Recent progress and technical challenges in using calcium sulfoaluminate (CSA) cement. *Cem. Concr. Compos.* **2023**, *137*, 104908. [CrossRef]
18. Al Fuhaid, A.F.; Niaz, A. Carbonation and Corrosion Problems in Reinforced Concrete Structures. *Buildings* **2022**, *12*, 586. [CrossRef]
19. Durisety, H.; Palcham, K.; Babu, K.P. The Concrete Incorporated with Zeolite for Reducing Atmospheric Carbon Dioxide. *Int. J. Recent Technol. Eng.* **2020**, *8*, 2117–2121. [CrossRef]
20. Korec, E.; Mingazzi, L.; Freddi, F.; Martínez-Pañeda, E. Predicting the impact of water transport on carbonation-induced corrosion in variably saturated reinforced concrete. *Mater. Struct.* **2024**, *57*, 91. [CrossRef]
21. Sarkis-Onofre, R.; Catalá-López, F.; Aromataris, E.; Lockwood, C. How to properly use the PRISMA Statement. *Syst. Rev.* **2021**, *10*, 117. [CrossRef] [PubMed]
22. Park, H.Y.; Suh, C.H.; Woo, S.; Kim, P.H.; Kim, K.W. Quality Reporting of Systematic Review and Meta-Analysis According to PRISMA 2020 Guidelines: Results from Recently Published Papers in the *Korean Journal of Radiology*. *Korean J. Radiol.* **2022**, *23*, 355. [CrossRef]
23. McCombes, S. How to Write a Literature Review | Guide, Examples, & Templates. Scribbr. Available online: https://www.scribbr.com/dissertation/literature-review/ (accessed on 8 June 2023).
24. Mechling, J.-M.; Lecomte, A.; Roux, A.; Le Rolland, B. Sulfoaluminate cement behaviours in carbon dioxide, warm and moist environments. *Adv. Cem. Res.* **2014**, *26*, 52–61. [CrossRef]
25. Moffatt, E.G.; Thomas, M.D.A. Effect of Carbonation on the Durability and Mechanical Performance of Ettringite-Based Binders. *ACI Mater. J.* **2019**, *116*, 95–102. [CrossRef]
26. Alapati, P.; Kurtis, K. Carbonation in Alternative Cementitious Materials: Implications on Durability and Mechanical Properties. 2019. Available online: https://www.semanticscholar.org/paper/Carbonation-in-Alternative-Cementitious-Materials:-Alapati-Kurtis/03575798bd8236d3229714af4bf106d68c7b79eb (accessed on 20 February 2023).

27. Vu, Q.H.; Pham, G.; Chonier, A.; Brouard, E.; Rathnarajan, S.; Pillai, R.; Gettu, R.; Santhanam, M.; Aguayo, F.; Folliard, K.J.; et al. Impact of different climates on the resistance of concrete to natural carbonation. *Constr. Build. Mater.* **2019**, *216*, 450–467. [CrossRef]
28. Ansari, W.S.; Chang, J.; Rehman, Z.U.; Nawaz, U.; Junaid, M.F. A novel approach to improve carbonation resistance of Calcium Sulfoaluminate cement by assimilating fine cement-sand mix. *Constr. Build. Mater.* **2022**, *317*, 125598. [CrossRef]
29. Sharma, R.; Kim, H.; Lee, N.K.; Park, J.-J.; Jang, J.G. Microstructural characteristics and CO_2 uptake of calcium sulfoaluminate cement by carbonation curing at different water-to-cement ratios. *Cem. Concr. Res.* **2023**, *163*, 107012. [CrossRef]
30. Cagno, E.; Neri, A.; Marta, N.; Bassani, C.A.; Lampertico, T. Applied Sciences | Free Full-Text | The Role of Digital Technologies in Operationalizing the Circular Economy Transition: A Systematic Literature Review. *J. Appl. Sci.* **2021**, *11*, 3328. [CrossRef]
31. Bertola, F.; Gastaldi, D.; Irico, S.; Paul, G.; Canonico, F. Influence of the amount of calcium sulfate on physical/mineralogical p roperties and carbonation resistance of CSA-based cements. *Cem. Concr. Res.* **2022**, *151*, 106634. [CrossRef]
32. Monkman, S.; Grandfield, K.; Langelier, B. On the Mechanism of Using Carbon Dioxide as a Beneficial Concrete Admixture. In *SP-329: Superplasticizers and Other Chemical Admixtures in Concrete Proceedings Twelfth International Conference, Beijing, China;* American Concrete Institute: Farmington Hills, MI, USA, 2018. [CrossRef]
33. Shenbagam, V.K.; Shaji, P.; Eswita, Y.; Cepuritis, R.; Chaunsali, P. Carbonation of calcium sulfoaluminate belite binder: Mechanism and its implication on properties. *J. Sustain. Cem.-Based Mater.* **2024**, *13*, 938–950. [CrossRef]
34. Chen, T.; Gao, X.; Gao, X. Use of carbonation curing to improve mechanical strength and durability of pervious concrete. *ACS Sustain. Chem. Eng.* **2020**, *8*, 3872–3884. [CrossRef]
35. Wang, L.; Ma, H.; Li, Z.; Ma, G.; Guan, J. Cementitious composites blending with high belite sulfoaluminate and medium-heat Portland cements for largescale 3D printing. *Addit. Manuf.* **2021**, *46*, 102189. [CrossRef]
36. Zhang, D.; Xu, D.; Cheng, X.; Chen, W. Carbonation resistance of sulphoaluminate cement-based high performance concrete. *J. Wuhan Univ. Technol.-Mat. Sci. Edit.* **2009**, *24*, 663–666. [CrossRef]
37. Kamal, N.L.M.; Itam, Z.; Sivaganese, Y.; Beddu, S. Carbon dioxide sequestration in concrete and its effects on concrete compressive strength. *Mater. Today Proc.* **2020**, *31*, A18–A21. [CrossRef]
38. Achal, V.; Mukherjee, A. A review of microbial precipitation for sustainable construction. *Constr. Build. Mater.* **2015**, *93*, 1224–1235. [CrossRef]
39. Chen, Y.; Liu, P.; Yu, Z. Effects of Environmental Factors on Concrete Carbonation Depth and Compressive Strength. *Materials* **2018**, *11*, 2167. [CrossRef]
40. Sirtoli, D.; Wyrzykowski, M.; Riva, P.; Tortelli, S.; Marchi, M.; Lura, P. Shrinkage and creep of high-performance concrete based on calcium sulfoaluminate cement. *Cem. Concr. Compos.* **2019**, *98*, 61–73. [CrossRef]
41. Mi, T.; Li, Y.; Liu, W.; Dong, Z.; Gong, Q.; Min, C.; Xing, F.; Wang, Y.; Chu, S. The effect of carbonation on chloride redistribution and corrosion of steel reinforcement. *Constr. Build. Mater.* **2023**, *363*, 129641. [CrossRef]
42. Lin, X. Effect of Early Age Carbonation on Strength and pH of Concrete. Master's Thesis, McGill University, Montréal, QC, Canada, 2007.
43. Ahmad, S. Reinforcement corrosion in concrete structures, its monitoring and service life prediction—-A review. *Cem. Concr. Compos.* **2003**, *25*, 459–471. [CrossRef]
44. Zhou, Q.; Glasser, F.P. Kinetics and mechanism of the carbonation of ettringite. *Adv. Cem. Res.* **2000**, *12*, 131–136. [CrossRef]
45. Leemann, A.; Moro, F. Carbonation of concrete: The role of CO_2 concentration, relative humidity and CO_2 buffer capacity. *Mater. Struct.* **2017**, *50*, 30. [CrossRef]
46. Hargis, C.W.; Lothenbach, B.; Müller, C.J.; Winnefeld, F. Carbonation of calcium sulfoaluminate mortars. *Cem. Concr. Compos.* **2017**, *80*, 123–134. [CrossRef]
47. Fernández-Carrasco, L.; Torréns-Martín, D.; Martínez-Ramírez, S. Carbonation of ternary building cementing materials. *Cem. Concr. Compos.* **2012**, *34*, 1180–1186. [CrossRef]
48. Geng, H.; Duan, P.; Chen, W.; Shui, Z. Carbonation of sulphoaluminate cement with layered double hydroxides. *J. Wuhan Univ. Technol.-Mat. Sci. Edit.* **2014**, *29*, 97–101. [CrossRef]
49. Lai, M.H.; Binhowimal, S.A.M.; Griffith, A.M.; Hanzic, L.; Chen, Z.; Wang, Q.; Ho, J.C.M. Shrinkage, cementitious paste volume, and wet packing density of concrete | Semantic Scholar. *Struct. Concr.* **2022**, *23*, 488–504. [CrossRef]
50. Ke, G.; Zhang, J.; Liu, Y. Shrinkage characteristics of calcium sulphoaluminate cement concrete. *Constr. Build. Mater.* **2022**, *337*, 127627. [CrossRef]
51. Concrete Counter Top Institute. CSA Cements in Concrete Countertops: Rapid Strength with a Low Carbon Footprint. Concrete Countertop Institute. Available online: https://concretecountertopinstitute.com/free-training/csa-cements-in-concrete-countertops-rapid-strength-with-a-low-carbon-footprint/ (accessed on 6 November 2023).
52. Tam, V.W.Y.; Butera, A.; Le, K.N. An investigation of the shrinkage, concrete shrinkage reversibility and permeability of CO_2-treated concrete. *Constr. Build. Mater.* **2023**, *365*, 130120. [CrossRef]
53. Chaunsali, P.; Mondal, P. Influence of Calcium Sulfoaluminate (CSA) Cement Content on Expansion and Hydration Behavior of Various Ordinary Portland Cement-CSA Blends. *J. Am. Ceram. Soc.* **2015**, *98*, 2617–2624. [CrossRef]
54. Zhang, D.; Shao, Y. Effect of early carbonation curing on chloride penetration and weathering carbonation in concrete. *Constr. Build. Mater.* **2016**, *123*, 516–526. [CrossRef]

55. Ye, H.; Radlińska, A.; Neves, J. Drying and carbonation shrinkage of cement paste containing alkalis. *Mater. Struct.* **2017**, *50*, 132. [CrossRef]
56. Suda, Y.; Tomiyama, J.; Saito, T.; Saeki, T. Phase Assemblage, Microstructure and Shrinkage of Cement Paste during Carbonation at Different Relative Humidities. *J. Adv. Concr. Technol.* **2021**, *19*, 687–699. [CrossRef]
57. Lootens, D.; Bentz, D.P. On the relation of setting and early-age strength development to porosity and hydration in cement-based materials. *Cem. Concr. Compos.* **2016**, *68*, 9–14. [CrossRef] [PubMed]
58. Markosian, N.; Thomas, R.; Maguire, M.; Sorensen, A. Calcium Sulfoaluminate Cement Concrete for Prestressed Bridge Girders: Prestressing Losses, Bond, and Strength Behavior. Utah State University, MPC-560. Available online: https://digitalcommons.usu.edu/etd/7474/ (accessed on 23 April 2024).
59. Wang, J.; Lord, T.; Wang, Y.; Black, L.; Li, Q. Effects of carbonation on mechanical properties of two types of concrete under extreme loadings of high temperature and impact. *Proc. Inst. Civ. Eng.* **2022**, *175*, 44–56. [CrossRef]
60. Ashraf, W. Carbonation of cement-based materials: Challenges and opportunities. *Constr. Build. Mater.* **2016**, *120*, 558–570. [CrossRef]
61. Zhang, Y.; Yang, H.; Zhang, Q.; Zhang, C.; Wu, K.; Shen, P. Microstructural Evolution of Calcium Sulfoaluminate Cement during the Wet-Carbonation Process. *Buildings* **2024**, *14*, 343. [CrossRef]
62. Sevelsted, T.F.; Skibsted, J. Carbonation of C–S–H and C–A–S–H samples studied by 13C, 27Al and 29Si MAS NMR spectroscopy. *Cem. Concr. Res.* **2015**, *71*, 56–65. [CrossRef]
63. Cao, H.; Liang, Z.; Peng, X.; Cai, X.; Wang, K.; Wang, H.; Lyu, Z. Research into Carbon Dioxide Curing's Effects on the Properties of Reactive Powder Concrete with Assembly Unit of Sulphoaluminate Cement and Ordinary Portland Cement. *Coatings* **2022**, *12*, 209. [CrossRef]
64. Groves, G.W.; Brough, A.; Richardson, I.G.; Dobson, C.M. Progressive Changes in the Structure of Hardened C3S Cement Pastes due to Carbonation. *J. Am. Ceram. Soc.* **1991**, *74*, 2891–2896. [CrossRef]
65. Wang, W.; Wei, X.; Cai, X.; Deng, H.; Li, B. Mechanical and Microstructural Characteristics of Calcium Sulfoaluminate Cement Exposed to Early-Age Carbonation Curing. *Materials* **2021**, *14*, 3515. [CrossRef]
66. Carbonation—An Overview | ScienceDirect Topics. Available online: https://www.sciencedirect.com/topics/engineering/carbonation (accessed on 8 June 2023).
67. Duan, P.; Chen, W.; Ma, J.; Shui, Z. Influence of layered double hydroxides on microstructure and carbonation resistance of sulphoaluminate cement concrete. *Constr. Build. Mater.* **2013**, *48*, 601–609. [CrossRef]
68. Aguayo, F.M.; Drimalas, T.; Folliard, K.J. Natural Carbonation of Concrete. *Spec. Publ.* **2015**, *305*, 2.1–2.12. [CrossRef]
69. Tang, J.; Wu, J.; Zou, Z.; Yue, A.; Mueller, A. Influence of axial loading and carbonation age on the carbonation resistance of recycled aggregate concrete. *Constr. Build. Mater.* **2018**, *173*, 707–717. [CrossRef]
70. Taylor, H.F.W.; Famy, C.; Scrivener, K.L. Delayed ettringite formation. *Cem. Concr. Res.* **2001**, *31*, 683–693. [CrossRef]
71. Cho, B.S.; Lee, H.H.; Choi, Y.C. Effects of aluminate rich slag on compressive strength, drying shrinkage and microstructure of blast furnace slag cement. *Constr. Build. Mater.* **2017**, *140*, 293–300. [CrossRef]
72. Gastaldi, D.; Bertola, F.; Canonico, F.; Buzzi, L.; Mutke, S.; Irico, S.; Paul, G.; Marchese, L.; Boccaleri, E. A chemical/mineralogical investigation of the behavior of sulfoaluminate binders submitted to accelerated carbonation. *Cem. Concr. Res.* **2018**, *109*, 30–41. [CrossRef]
73. Quillin, K. Performance of belite–sulfoaluminate cements. *Cem. Concr. Res.* **2001**, *31*, 1341–1349. [CrossRef]
74. Wang, L.; Zhan, S.; Tang, X.; Xu, Q.; Qian, K. Pore Solution Chemistry of Calcium Sulfoaluminate Cement and Its Effects on Steel Passivation. *Appl. Sci.* **2019**, *9*, 1092. [CrossRef]
75. Jain, J.; Seth, A.; Decristofaro, N. Environmental impact and durability of carbonated calcium silicate concrete. *Proc. Inst. Civil. Eng. Constr. Mater.* **2019**, *172*, 179–191. [CrossRef]
76. Tan, B.; Okoronkwo, M.U.; Kumar, A.; Ma, H. Durability of calcium sulfoaluminate cement concrete. *J. Zhejiang Univ. Sci. A* **2020**, *21*, 118–128. [CrossRef]
77. Coppola, L.; Coffetti, D.; Crotti, E.; Pastore, T. CSA-based Portland-free binders to manufacture sustainable concretes for jointless slabs on ground. *Constr. Build. Mater.* **2018**, *187*, 691–698. [CrossRef]
78. Pooni, J.; Robert, D.; Giustozzi, F.; Setunge, S.; Xie, Y.M.; Xia, J. Novel use of calcium sulfoaluminate (CSA) cement for treating problematic soils. *Constr. Build. Mater.* **2020**, *260*, 120433. [CrossRef]
79. Bertola, F.; Gastaldi, D.; Irico, S.; Paul, G.; Canonico, F. Behavior of blends of CSA and Portland cements in high chloride environment. *Constr. Build. Mater.* **2020**, *262*, 120852. [CrossRef]
80. Nishikawa, T.; Suzuki, K.; Ito, S.; Sato, K.; Takebe, T. Decomposition of synthesized ettringite by carbonation. *Cem. Concr. Res.* **1992**, *22*, 6–14. [CrossRef]
81. Concrete Jungle Functions as Carbon Sink, UCI and Other Researchers Find. UCI News. Available online: https://news.uci.edu/2016/11/21/concrete-jungle-functions-as-carbon-sink-uci-and-other-researchers-find/ (accessed on 26 February 2023).
82. Zhang, J.; Chang, J.; Zhang, P.; Wang, T. Effects of C$H2 and CH on Strength and Hydration of Calcium Sulphoaluminate Cement Prepared from Phosphogypsum. *Buildings* **2022**, *12*, 1692. [CrossRef]
83. Zhang, S.; Wang, Q.; Zhou, W.; Lu, Y.; Liu, X.; Chang, X. A fast-setting and eco-friendly superhydrophobic high belite sulphoaluminate cement mortar. *J. Mater. Res. Technol.* **2023**, *23*, 2690–2702. [CrossRef]

84. Sirtoli, D.; Tortelli, S.; Riva, P.; Marchi, M.; Cucitore, R.; Nangah, M. Mechanical and Environmental Performances of Sulpho-Based Rapid Hardening Concrete. *SP-305 Durab. Sustain. Concr. Struct.* **2015**, *305*, 47.1–47.8.
85. Shenbagam, V.K.; Chaunsali, P. Influence of calcium hydroxide and calcium sulfate on early-age proper ties of non-expansive calcium sulfoaluminate belite cement. *Cem. Concr. Compos.* **2022**, *128*, 104444. [CrossRef]

Disclaimer/Publisher's Note: The statements, opinions and data contained in all publications are solely those of the individual author(s) and contributor(s) and not of MDPI and/or the editor(s). MDPI and/or the editor(s) disclaim responsibility for any injury to people or property resulting from any ideas, methods, instructions or products referred to in the content.

buildings MDPI

Review

Artificial Intelligence in Net-Zero Carbon Emissions for Sustainable Building Projects: A Systematic Literature and Science Mapping Review

Yanxue Li [1], Maxwell Fordjour Antwi-Afari [2], Shahnawaz Anwer [3,*], Imran Mehmood [3], Waleed Umer [4], Saeed Reza Mohandes [1], Ibrahim Yahaya Wuni [5], Mohammed Abdul-Rahman [3] and Heng Li [3]

1 Department of Mechanical, Aerospace and Civil Engineering, Faculty of Science and Engineering, University of Manchester, Manchester M13 9PL, UK; 17822901 31@163.com (Y.L.); saeedreza.mohandes@manchester.ac.uk (S.R.M.)
2 Department of Civil Engineering, College of Engineering and Physical Sciences, Aston University, Birmingham B4 7ET, UK; m.antwiafari@aston.ac.uk
3 Department of Building and Real Estate, The Hong Kong Polytechnic University, Hung Hom, Kowloon, Hong Kong SAR, China; imran.mehmood@connect.polyu.hk (I.M.); 18042561r@connect.polyu.hk (M.A.-R.); heng.li@polyu.edu.hk (H.L.)
4 Department of Architecture and Built Environment, Northumbria University, Newcastle upon Tyne NE1 8ST, UK; waleed.umer@northumbria.ac.uk
5 Department of Architectural Engineering and Construction Management, King Fahd University of Petroleum and Minerals, Dhahran 31261, Saudi Arabia; ibrahim.wuni@kfupm.edu.sa
* Correspondence: shah-nawaz.anwer@polyu.edu.hk

Abstract: Artificial intelligence (AI) has emerged as an effective solution to alleviate excessive carbon emissions in sustainable building projects. Although there are numerous applications of AI, there is no state-of-the-art review of how AI applications can reduce net-zero carbon emissions (NZCEs) for sustainable building projects. Therefore, this review study aims to conduct a systematic literature and science mapping review of AI applications in NZCEs for sustainable building projects, thereby expediting the realization of NZCEs in building projects. A mixed-method approach (i.e., systematic literature review and science mapping) consisting of four comprehensive stages was used to retrieve relevant published articles from the Scopus database. A total of 154 published articles were retrieved and used to conduct science mapping analyses and qualitative discussions, including mainstream research topics, gaps, and future research directions. Six mainstream research topics were identified and discussed. These include (1) life cycle assessment and carbon footprint, (2) practical applications of AI technology, (3) multi-objective optimization, (4) energy management and energy efficiency, (5) carbon emissions from buildings, and (6) decision support systems and sustainability. In addition, this review suggests six research gaps and develops a framework depicting future research directions. The findings contribute to advancing AI applications in reducing carbon emissions in sustainable building projects and can help researchers and practitioners to realize its economic and environmental benefits.

Keywords: artificial intelligence; net-zero carbon emissions; science mapping approach; sustainable buildings; systematic literature review

1. Introduction

Due to global warming (e.g., climate change), humans and other living organisms face severe crises; thus, proactive interventions are critically needed to mitigate these environmental issues [1] Many industries (e.g., manufacturing, automobile, and construction) have driven the expansion of the global economy and significantly impacted both the natural and built environments [2–4]. The built environment is a major source of most greenhouse gases worldwide [5–7]. Increases in global industrialization and development have led to the release of greenhouse gases, which may lead to high temperatures and environmental degradation [8–10]. Therefore, there is an urgent need to achieve net-zero carbon emissions (NZCEs) within industries and across the world.

NZCEs are greenhouse gases produced by human activities and can be reabsorbed by the atmosphere [11–13]. Achieving NZCEs would help to protect the ecological environment and reduce damage to the natural environment. To achieve sustainable development and industrialization, attaining NZCEs by 2050 is a global goal [14], which requires efforts from everyone, including government policies, individuals, and corporate bodies. Although it may seem challenging, continuous improvement, planned goals, long-term efforts, and persistence are of utmost priority. For instance, some countries have already set policies to achieve NZCEs [15–19] Italy's long-term strategy shows that deep decarbonization is achieved with the support of government policies and increased investment in technology, research, and development [20]. According to a comprehensive assessment model, Brazil and the United States achieved NZCEs earlier than the global average [16,17].

Sustainable construction processes help to promote economic justice while protecting the natural environment [21,22]. In other words, they help minimize damage to the natural environment during project development [23]. Sustainable construction, which is an advanced construction method, has an adverse impact on the harmonious development of humans and nature. Despite its global recognition and usefulness, it is still regarded as a complex process or method owing to financial constraints, non-economic construction projects, or sustainable resource management [16,24]. Three strategies and methods for achieving sustainable building projects have been reported: resource conservation, cost efficiency, and design for human adaptation [25]. Based on these strategies, the implementation of advanced digital technologies is crucial. However, the construction industry has a slow adoption rate compared with other industries in digital transformation [26].

Artificial intelligence (AI) refers to the theory and development of computer systems that can perform tasks that often require human intelligence, such as visual perception, speech recognition, decision-making, and language translation [27]. Within the architecture, engineering, and construction (AEC) sector, there are many AI applications, such as genetic programming, neural networks, fuzzy logic, heuristic search, and computer software [28–30]. Previous studies have explored how AI can improve practical projects in the AEC sector [31–33]. Building information modeling (BIM) can help manage a centralized database for the entire life cycle of a building and combining it with AI can also be used to estimate costs, identify product issues, and increase construction quality [34]. Darko et al. [35] reported on state-of-the-art research on AI in the AEC industry. They found that genetic algorithms, neural networks, fuzzy logic, fuzzy sets, and machine learning were the predominant AI techniques employed in the field of AEC. Similarly, Pan and Zhang [36] conducted an in-depth review of AI applications in construction engineering and management. They found that a variety of AI methods have greatly contributed to the transformation of the construction industry. In construction engineering and management, AI has been able to automate and expedite learning, reasoning, and perception processes using extensive data. This has significant promise for addressing various engineering projects based on their unique attributes. Moreover, unlike conventional computational approaches and expert judgments, highly promising AI excels in handling intricate and ever-changing situations amidst significant ambiguity and vast amounts of data. Consequently, they are more inclined to provide precise and persuasive outcomes for tactical decision-making. These studies have suggested that various AI-related techniques can ultimately accomplish three primary functions (model and pattern detection, prediction, and optimization) that are helpful to the AEC industry in terms of automation, risk reduction, digitalization, enhanced efficiency, and computer vision [35–37].

Artificial intelligence has revolutionized the management of carbon emissions in sustainable building projects. AI techniques, such as machine learning, neural networks, and genetic algorithms, are increasingly being employed to optimize energy efficiency, predict carbon footprints, and enhance decision-making processes in construction projects. The integration of AI in these processes not only accelerates the path toward achieving NZCEs but also significantly reduces the environmental impact of construction activities. This study highlights how AI-driven technologies can automate complex processes, providing more accurate and data-driven insights that were previously unattainable through traditional methods [32,35]. Thus, the application of AI represents a critical advancement in carbon emissions management, offering the potential to transform the construction industry into a more sustainable and environmentally conscious sector.

Previous review studies have focused on NZCEs and the application of AI in other sectors [34,38–40]. Using decision-making and prediction algorithms, Ahmed et al. [41] presented a state-of-the-art review application of artificial neural networks (ANNs) for promoting sustainable construction projects within three sustainable development aspects: environmental, economic, and social. Additionally, Chen et al. [42] conducted a comprehensive literature review to analyze AI-based solutions for climate change. They discovered that the life cycle assessment is widely employed in the construction industry to evaluate environmental effects, with carbon emissions as the primary assessment indication. Owing to the continual advancement of computer technology and big data technology, random forests and neural networks have been utilized in recent years to quantify and estimate the carbon emissions of the construction industry [43]. The progression of green construction to attain net-zero carbon emissions aligns with worldwide sustainability goals. Manzoor et al. [34] conducted a systematic review based on a case study to develop techniques for incorporating BIM into sustainable building projects. The findings indicated that workshops, lectures, and conference events are employed to augment public consciousness, while improved information on the expenses and advantages of sustainable materials and reinforced sustainable development were identified as the most crucial tactics to enhance sustainable progress in construction projects. Moreover, Adunadepo and Sunday [44] conducted a review to explore novel methods to facilitate the implementation of AI to attain the sustainable development of intelligent buildings. Their recommendation involves utilizing integrated project delivery and creative green architecture to effectively execute these concepts and to promote the sustainable growth of artificial intelligence in buildings. Their suggestions could aid in mitigating adverse effects such as the degradation of the built environment and the exacerbation of global warming resulting from the mounting pressures of fast-expanding worldwide populations. Despite the relevance of previous review studies, they have limited applications of AI in reducing carbon emissions in sustainable building projects. Owing to the challenges of global warming, it is crucial to identify mainstream research topics on how to apply AI (especially machine learning) in combating NZCEs for sustainable building projects. In recent times, versatile applications of machine learning have extended far beyond energy efficiency, impacting various scientific and engineering challenges. Recent studies in the construction industry have highlighted its role in predicting energy consumption and optimizing building performance. For example, research such as "Machine learning as a surrogate to building performance simulation: Predicting energy consumption under different operational settings" emphasizes the ability of ML models to accurately forecast energy needs, thereby enabling better resource management [45]. Additionally, the study "Machine Learning Method Based on Symbiotic Organism Search Algorithm for Thermal Load Prediction in Buildings" explores advanced algorithms to predict thermal loads, ensuring optimal energy use and comfort in buildings [46,47]. These examples demonstrate the multifaceted potential of ML in revolutionizing the construction industry, making it an indispensable tool in modern engineering problem-solving.

In addition, none of the previous review studies adopted a mixed-method approach consisting of a systematic literature review (i.e., qualitative method) and a science mapping approach (i.e., quantitative method) on the application of AI in NZCEs for sustainable building projects. This review research paper distinguishes itself from the existing academic literature by adopting a mixed-method approach that integrates both a systematic literature review and science mapping analysis, which is not commonly found in prior studies. While previous research has extensively examined artificial intelligence (AI) applications in various sectors, including construction and environmental management, this study specifically focuses on the role of AI in achieving net-zero carbon emissions (NZCEs) in sustainable building projects. By identifying mainstream research topics, research gaps, and future research directions, this study provides a comprehensive framework that has largely been absent from the current literature. The novelty lies in its dual methodological approach and targeted exploration of AI's potential to mitigate carbon emissions in the construction industry, making it a valuable resource for both academics and practitioners in this evolving field [39,48].

In summary, this study aims to conduct a systematic literature and science mapping review of published articles related to the application of AI in NZCEs for sustainable building projects and to discuss mainstream research topics, research gaps, and future research directions. To achieve this aim, the specific research objectives of this review are as follows:

1. Analyze the annual publication trends of published articles and select peer-reviewed journals on AI in NZCEs for sustainable building projects.
2. Apply a science mapping approach to analyze influential keywords and document analyses of AI in NZCEs for sustainable building projects.
3. Identify and discuss mainstream research topics related to AI in NZCEs for sustainable building projects.
4. Develop a framework for depicting research gaps and future research directions on AI in NZCEs for sustainable building projects.

This study is significant to both researchers and professionals in the sense that analyzing the annual publication trends and selecting peer-reviewed journals will provide insights into the field's evolution and identify key journals, aiding in strategic publishing and funding acquisition for researchers [49]. A science mapping approach to influential keywords and document analyses also helps researchers and professionals understand critical terminologies and research interconnections, thus enabling strategic research planning [50]. On the other hand, identifying mainstream research topics helps narrow the focus to impactful areas, aiding benchmarking and curriculum development [51]. Developing a framework for depicting research gaps and future research directions highlights under-researched areas and emerging trends, fostering innovation and informing policies and strategies [52]. Consequently, this review acts as a comprehensive resource, guiding research, professional practices, and collaboration at the intersection of AI and sustainable building projects.

The remainder of this paper is organized as follows. The rationale for using a systematic literature review and science mapping approach and the methodological steps are discussed in Section 2. Section 3 presents the results of this review based on annual publication trends, the selection of relevant peer-reviewed journals, the co-occurrence analysis of keywords, and document analysis. Moreover, Section 4 discusses the mainstream research themes, gaps, and future research directions in the research domain. Lastly, conclusions, contributions, limitations, and further studies are presented in Section 5.

2. Research Methods

This review study aims to conduct a systematic literature and science mapping review of AI applications in NZCEs for sustainable building projects, thereby discussing mainstream research topics and identifying research gaps and future research directions. As such, this study adopted a mixed-method approach, consisting of a systematic literature review (i.e., a qualitative approach based on PRISMA guidelines) and a science mapping review (i.e., a quantitative approach based on VOSviewer, version 1.6.17). A systematic literature review can collect a significant number of articles from various perspectives and provide literature references on a particular domain [53]. The goal of a systematic literature review is to minimize the diverse results of published articles through a thorough literature search [54]. By providing a comprehensive process for identifying literature samples, precise bibliographic information and novel contributions to specific research domains can be obtained [55,56]. On the other hand, science mapping reviews are characterized by their ability to manage vast volumes of measurement data while serving as descriptive and diagnostic tools [57]. Adopting a mixed-method approach can combine the benefits of qualitative and quantitative research methods by focusing on the relevance and value of the study object, data analysis, and objective connections [58]. Specifically, the proposed mixed-method approach allows for the use of simultaneous datasets and thorough analyses using relevant tools [59]. Figure 1 provides an overview of the research methods used in this study. As shown in Figure 1, the review was conducted in four main stages: search strategy, selection criteria, science mapping analysis, and qualitative discussion.

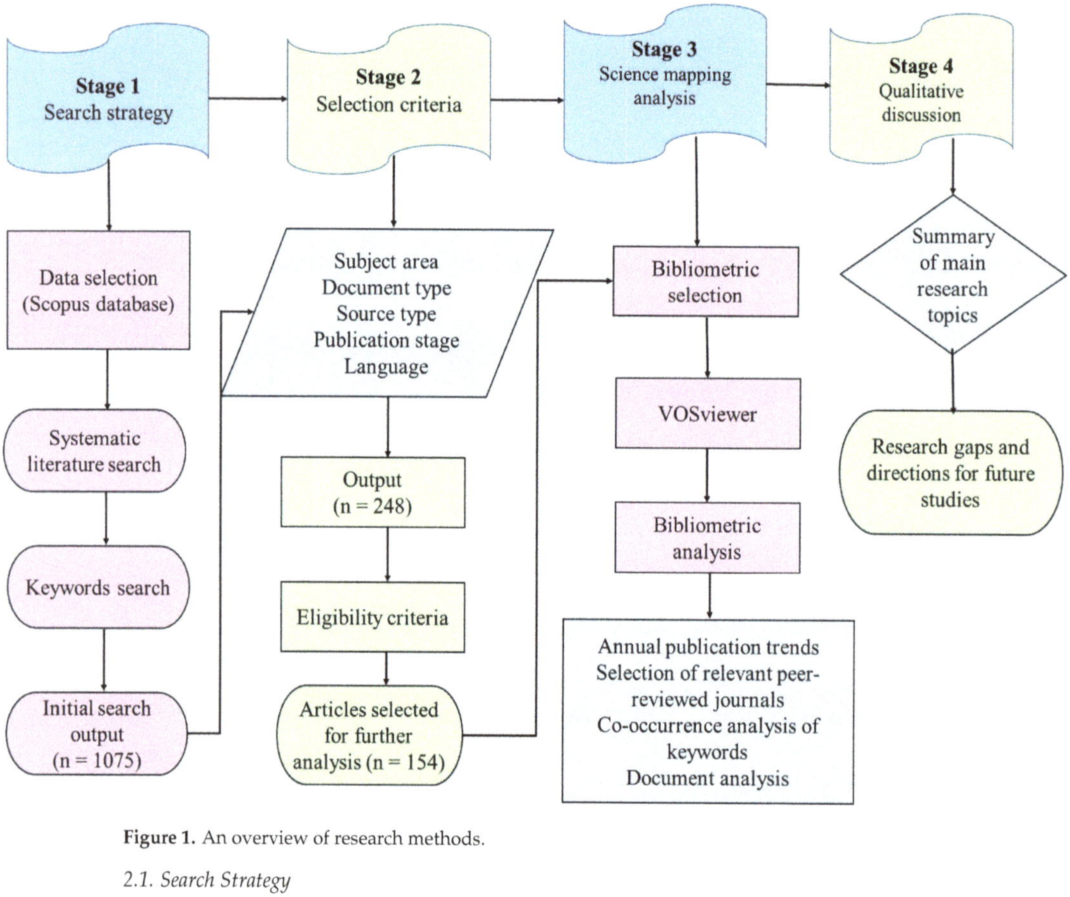

Figure 1. An overview of research methods.

2.1. Search Strategy

Many online bibliographic databases are used as storage platforms for literature samples (e.g., published articles, conferences, and books). Examples of these databases include the Web of Science (WoS), Scopus, Google Scholar and Science Direct, PubMed, and CINAHL. Although these databases overlap, the Scopus database provides wider coverage of publications and improves the efficiency of the indexing process [60]. Therefore, the Scopus database was selected to retrieve literature samples in this study. First, a systematic literature search was performed under the "title/abstract/keywords" section in the Scopus database. Based on the purpose of this review study, the main search terms used are "AI", "net-zero carbon emissions", and "sustainable building projects". The full search string was (TITLE-ABS-KEY (artificial OR intelligence OR AI) AND TITLE-ABS-KEY (net-zero OR carbon OR emissions) AND TITLE-ABS-KEY (sustainable OR building OR project) DOCTYPE (all) AND ACCTYPE (all). A total of 1075 bibliographic records, including articles, books, and reviews, were identified. It is worth noting that the search was conducted on 25 July 2023.

2.2. Selection Criteria

The second step of the proposed mixed-method approach is the selection of relevant bibliographic records using the following criteria: First, the subject category was limited to "engineering", and 624 other subject areas (e.g., business, economics, etc.) were excluded. Second, bibliographic records related to reviews, conference papers, and book chapters were excluded. In other words, only articles were included in this review. Consequently, 179 bibliographic records were excluded. In the third criterion, the source of publication type was limited to journals. Five articles from trade journals and book series were excluded from the analysis. According to Smith and Johnson [61], scientific journals focus on rigorous, peer-reviewed academic research, whereas trade journals provide practical, industry-specific information for professionals. Next, the fourth criterion was to limit the publication stage to "final", meaning online published journal articles. Four articles were excluded after meeting the fourth criterion. Finally, the fifth criterion was limiting published journal articles to only those written in English. Thus, 15 articles written in other languages, such as Chinese and Spanish, were excluded. After applying the selection criteria, 248 articles were obtained. The 248 articles were then comprehensively screened by reading each article's title, abstract, and full text to identify relevant articles related to the studied domain. For example, some articles were removed during the screening process because they did not focus on emission reductions in the construction industry. Moreover, some articles were excluded because they did not focus on the role of AI in reducing carbon emissions. Two independent reviewers (YL and MFAA) screened the titles, abstracts, and full texts of 248 articles based on the selection criteria. Any disagreements between them were resolved via meeting and consensus. As of 25 July 2023, 94 articles were excluded, and 154 were used for further analysis. All the included articles were exported and stored using a CSV file. Next, the two reviewers extracted relevant data from the 154 included articles. These include publication year, author keywords, authors' full names, affiliations, article titles, and source titles.

2.3. Science Mapping Analysis

In the third step of the proposed mixed-method approach, the science mapping approach was used to analyze 154 articles related to AI applications in NZCEs for sustainable building projects. Bibliometrics plays an important role in evaluating and analyzing scientific production [62] and is specifically divided into two analytical methods: performance analysis and science mapping analysis [63]. The former mainly uses literature data and documentary indicators [64], whereas the latter focuses on the correlation between various field parts while considering the time factor [65,66]. Therefore, this study used science mapping analysis to analyze influential AI articles in NZCEs for sustainable building projects. Many tools have been used for science mapping analysis, including Bibexcel (version 1.0.3), CiteSpace (version 6.2.R4), CoPalRed (version 1.7.5), Science of Science (Sci 2) Tool, Gephi, and VOSviewer (version 1.6.17). Choosing a suitable science-mapping tool can help achieve relevant study results. For instance, CiteSpace can analyze trends in scientific literature and provide visualization functions. Gephi is a tool for gaining insight into networks [14]. VOSviewer can build and visualize any co-occurrence data using a mapping technique that is divided into different views, such as network visualization, overlay visualization, and density visualization. In addition, these tools are freely available and easy to use. Therefore, this review adopted the VOSviewer tool for science mapping analysis for the reasons listed. In this study, four science-mapping analyses were performed using VOSviewer. These include (1) annual publication trends, (2) selection of relevant peer-reviewed journals, (3) co-occurrence analysis of keywords, and (4) document analysis.

2.4. *Qualitative Discussion*

The final step after the science mapping analysis was a qualitative discussion. In this step, an in-depth systematic review was carried out on the included articles to identify the mainstream research topics, gaps, and future research directions using methods recommended by [67]. Moreover, the theoretical and practical contributions of this study are summarized.

3. Results

3.1. *Annual Publication Trend*

Figure 2 shows the number of annual publications of the 154 included articles. As shown in Figure 2, the number of published articles within the study domain was very small from 2006 to 2018. From 2006 to 2018 (years inclusive), only 32 articles were published, with less than five articles in other years, except 2010. Between 2019 and July 2023, the number of published articles increased significantly, reaching its highest value of 28 articles in 2021. This significant growth may be linked to the recent research interest in AI and carbon emissions among scholars from different disciplines. It is worth mentioning that there was a decrease in the number of publications by 2022. However, 26 articles were published in 2023 (as of 25 July 2023). These findings indicate that AI plays a critical role in reducing carbon emissions from sustainable buildings; thus, more relevant research articles are expected to be published before the end of 2023.

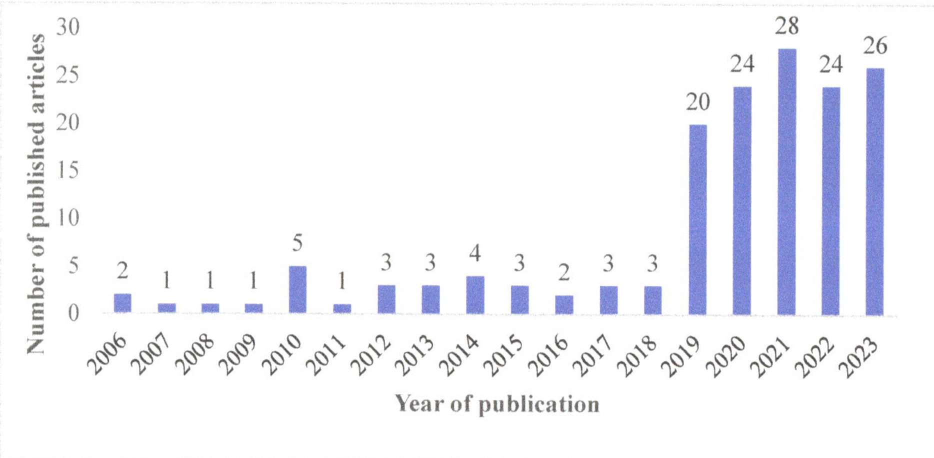

Figure 2. Number of publications from 2006 to July 2023.

3.2. *Selection of Relevant Peer-Reviewed Journals*

Table 1 lists the selection of relevant peer-reviewed journals during the study period. The peer-reviewed journals listed in Table 1 are those with two or more published articles focusing on AI in NZCEs for sustainable building projects during the study period. As shown in Table 1, only 15 peer-reviewed journals were selected. The top three peer-reviewed journals, according to their number of published articles, were the *Journal of Cleaner Production*, *Applied Energy*, and *Energy and Buildings*, representing 35.06% of the total published articles on AI in NZCEs for sustainable building projects. The *Journal of Cleaner Production* had the highest number of published articles (i.e., 24), indicating its impact on AI in NZCEs for sustainable buildings. The were two relevant articles published in *Computers and Industrial Engineering*, the *International Journal of Low-Carbon Technologies*, and the *Journal of Building Engineering*. Other peer-reviewed journals that published a single article constituted 29.87% of the included articles.

Table 1. Selection of relevant peer-reviewed journals from 2006 to 2023 (26 July 2023).

Journal Name	Number of Relevant Articles	% Total Publications
Journal of Cleaner Production	24	15.58
Applied Energy	17	11.04
Energy and Buildings	13	8.44
Energy	10	6.49
Sustainability (Switzerland)	9	5.84
Building and Environment	5	3.25
Buildings	5	3.25
Energies	5	3.25
Sustainable Cities and Society	5	3.25
Construction and Building Materials	3	1.95
Engineering Applications of Artificial Intelligence	3	1.95
Sensors	3	1.95
Computers and Industrial Engineering	2	1.30
International Journal of Low-Carbon Technologies	2	1.30
Journal of Building Engineering	2	1.30
Others	46	29.87
Total	**154**	

3.3. Co-Occurrence Analysis of Keywords

This review conducted a co-occurrence analysis of keywords to construct a network diagram and map the scientific knowledge domain of AI in NZCEs for sustainable building projects. By using "author keywords" as the unit of analysis, "full counting" as the counting method, and choosing the minimum number of occurrences of a keyword to be 3, 33 keywords met the threshold out of 655.

The second round of text mining of the selected keywords was conducted to eliminate general keywords. For example, "ann" and "ghg emissions" were excluded. Some other keywords with the same semantic meaning such as "artificial neural networks" and "artificial neural network", were combined. Finally, 23 items, six clusters, 51 links, and 57 link strengths were generated in the network of co-occurrence of keywords, as shown in Figure 3.

As presented in Figure 3, the font sizes of "artificial neural network", "carbon emission", "artificial intelligence", and "machine learning" are larger than other keywords, indicating that these keywords occur more frequently in the studied domain. The connecting lines in the figure indicate the interrelationships between two or more keywords. For example, "machine learning" and "building energy performance" are closely related. Through the presentation of different colors, it was found that the keywords in this review study could be categorized into six main clusters of keywords, representing mainstream research topics in the field of AI in NZCEs for sustainable building projects.

Table 2 summarizes the list of selected keywords and relevant network data, such as occurrences, average publication year, links, average citations, average normalized citation, and total link strength. The selected keywords in Table 2 are ranked according to their total link strengths. It was found that "artificial intelligence", "machine learning", "artificial neural network", and "energy efficiency" were the most frequently used keywords, indicating that they have been widely researched in the field of AI in NZCEs for sustainable building projects. The total link strength indicates the total strength linked to a specific keyword [68]. The results revealed that keywords such as "machine learning", "artificial intelligence", "life cycle assessment", and "sustainability" are the four keywords with the highest total link strength. The findings show that keywords with the highest occurrences do not necessarily have the highest total link strength. Keywords such as "building energy performance", "energy consumption", and "decision support system" had the highest average normalized citations, and "multi-objective optimization", "artificial intelligence", and "renewable energy" were the most recently used keywords according to the average publication year. These studies have focused on using ML to rate the energy performance of buildings. For instance, ML algorithms were used to analyze building energy data and perform prediction functions to reduce energy consumption [69–72].

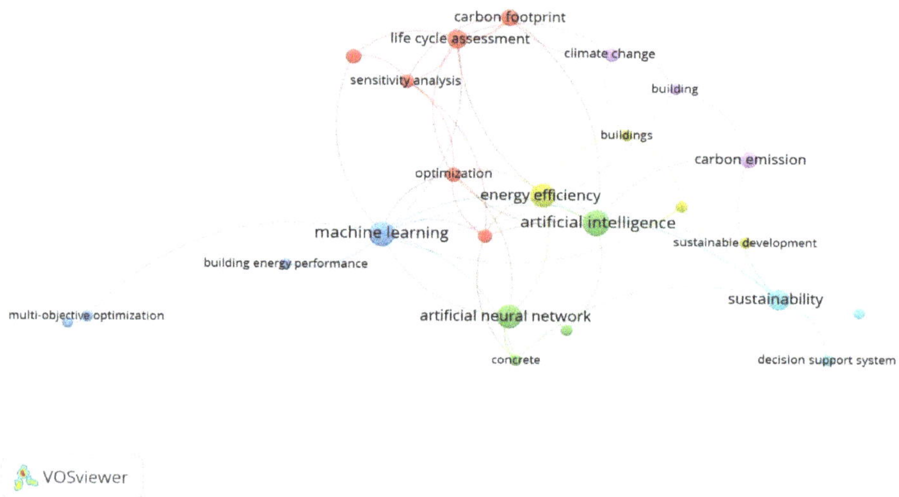

Figure 3. A network of co-occurrence of keywords related to AI in NZCEs for sustainable building projects.

Table 2. List of selected keywords and relevant network data.

Keywords	Occurrences	Average Publication Year	Links	Average Citations	Average Normalized Citations	Total Link Strength
Machine learning	14	2021	12	29.79	1.08	14
Artificial intelligence	15	2022	9	16.40	0.67	10
Life cycle assessment	8	2019	7	28.25	0.79	9
Sustainability	9	2020	8	32.78	1.31	9
Optimization	5	2020	7	19.00	1.65	7
Carbon footprint	6	2019	4	28.00	0.78	6
Energy consumption	4	2019	6	55.00	1.96	6
Artificial neural network	12	2021	5	28.25	1.47	6
Sensitivity analysis	4	2021	5	29.25	1.46	5
Concrete	3	2021	4	33.67	1.78	5
Energy efficiency	12	2020	5	19.83	0.69	5
Renewable energy	3	2023	4	0.67	0.54	4

Table 2. *Cont.*

Keywords	Occurrences	Average Publication Year	Links	Average Citations	Average Normalized Citations	Total Link Strength
Carbon emission	6	2021	3	14.83	1.45	4
Climate change	5	2016	4	36.40	0.88	4
Embodied carbon	5	2021	3	16.60	0.70	3
Buildings	3	2018	3	25.67	0.71	3
Building energy performance	3	2020	1	58.00	2.19	2
Sustainable development	3	2018	2	8.33	0.94	2
Energy conservation	3	2020	2	14.33	0.86	2
Multi-objective optimization	3	2023	2	0.67	0.54	2
Decision support system	3	2012	1	118.00	1.80	1
Thermal energy storage	3	2017	1	11.00	1.18	1
Compressive strength	3	2020	1	19.33	0.78	1

The observations in Figure 3 and Table 2 may lead to the following keyword clusters, which represent the dominant mainstream research topics related to AI in NZCEs for sustainable building projects:

1. Building eco-friendly, efficient, and energy-efficient structures can significantly reduce the problems associated with excessive carbon emissions. It has been shown that quantifying and analyzing the carbon footprint of public buildings over their life cycle can reduce negative environmental impacts [73]. Tushar et al. [74] applied sensitivity analysis to reduce the carbon footprint, thus improving energy efficiency. Developing implicit databases is also a good way to reduce carbon emissions and can be combined with machine and deep learning algorithms to combat climate change and resource scarcity [75]. It has also been reported that embodied carbon can be used throughout the life cycle of a building to improve the safety and environmental impact of a building project [76–79]. Additionally, the heating and cooling aspects of buildings consume more energy; therefore, the development of intelligent control systems is necessary. To reduce emissions, scalability should be the focus [69].
2. The use of AI to minimize carbon emissions in construction projects is the second cluster of research. AI can be used to create smart energy networks and reduce energy costs [80]. By applying AI techniques, building energy and carbon footprints can be used to predict energy consumption and CO_2 emissions [81–83]. Deep learning and ML are branches of AI techniques that are widely used as data analytics techniques for reducing NZCEs for sustainable building projects. For example, ANN has been used to quantify environmental costs in residential buildings and optimize commercial building design [84,85]. To achieve this goal, Palladino [86] studied the use of ANN in specific energy strategies in the Umbria Region. It has been reported that the application of ML can reduce the power consumption of buildings and help optimize building performance in the design and development of smart buildings [87,88].
3. A multi-objective optimization technique is proposed to reduce residential construction carbon emissions, accomplishing the dual goals of economic development and environmental conservation, and conforming to the sustainable development principle [89]. Multi-objective optimization combined with AI technology, can contribute to the development of sustainable buildings in terms of building material selection, retrofitting energy systems, and decision-making in building construction [90]. For example, the combination of an ANN with a multi-objective genetic algorithm can optimize the design of residential buildings [91,92]. Clustering techniques are integrated

with multi-objective optimization to identify urban structures based on their energy performance. This strategy can be replicated in other cities to increase energy efficiency and execute carbon-cutting initiatives [70]. Multiple goals can help sustainable buildings achieve NZCEs.

4. Improving energy consumption efficiency and strengthening building energy management are critical for mitigating the greenhouse effect and global warming trend [93]. Reduced carbon emissions, green buildings, and sustainable development have emerged as major concerns worldwide [2,94]. On the one hand, renewable energy-driven building systems based on solar and wind resources can reduce environmental effects and costs [95,96]. Building carbon emissions must be minimized to achieve energy sustainability [97]. However, focusing on building carbon emissions throughout their life cycle, including the design, transportation, construction, and operation stages, and quantifying them as environmental and carbon costs, can contribute to the long-term development of the construction industry [98]. In summary, reducing energy consumption can contribute to economic benefits and achieve sustainable development [77,99].
5. In the face of serious problems posed by climate change, efficient ways to minimize carbon emissions in the construction sector are receiving considerable attention. China is attempting to assess the feasibility of NZCEs, provide a path to reduce emissions, adjust and optimize the industrial structure, and achieve the policy goals of green development and carbon neutrality [1,100]. The prediction of carbon emission intensity in different countries can help policymakers devise environmental policies to address the adverse environmental effects of climate change [101,102]. Enhancing building management systems and promoting smart buildings will also help reduce the energy footprint and continuously optimize building performance [88]. Carbon capture and storage technologies currently play an essential role in lowering carbon dioxide emissions; however, they face problems such as high costs and regulatory issues, and related technologies still need to be developed [103].
6. Consider a structural design scheme for upgrading a building based on the decision support system (DSS). Carbon capture and storage technologies have been demonstrated in previous studies [104]. On the other hand, environmental considerations can be evaluated to assess building sustainability. As a result, the entire decision-making process can be optimized [105]. Simultaneously, DSS, combined with the predictive capabilities of ML to investigate the proper concrete mix proportions, can aid in assessing the impact of a building over its full life cycle, both in terms of environmental and financial expenses [72,106].

3.4. Document Analysis

Document analysis describes the most captivating research topics in a particular field, allowing researchers and practitioners to understand the specific findings and references cited in published articles. VOSviewer was used as a scientific mapping tool to generate a document analysis of the included articles retrieved from the Scopus database. Using "citation" as the type of analysis and "documents" as the unit of analysis in VOSviewer, and by setting the minimum number of citations of a document to 40, 24 documents met the thresholds out of the total 154 documents [68]. Table 3 summarizes the highly cited published articles related to AI in NZCEs for sustainable building projects. Although 24 documents met the threshold, the articles listed in Table 3 only included those with a normalized citation greater than one [107]. As such, 22 articles were listed and arranged based on normalized citations, whereas other articles (for example, Jiang et al. [108]; and Roaf et al. [109]) were manually excluded. As shown in Table 3, the top three articles with the highest total citations during the study period were Juan et al. [110] (238 citations), Acheampong and Boateng [101] (125 citations), and Çay et al. [111] (115 citations). A study by Fraga-lamas et al. [112], which focused on green IoT and edge AI toward a smart circular economy, received the highest normalized citations during the study period.

Most of the influential articles listed in Table 3 focus on the impacts of AI techniques on sustainable development [101,111–114]. Specifically, AI techniques, such as ML algorithms, were most applied in this studied research field [90,93,101,113,114]. Articles with fewer normalized citations during the study period included those by Petrovic et al. [115] (1.08) and Gobakis et al. [116] (1.00) were excluded.

Table 3. Summary of highly cited published articles related to AI in NZCEs for sustainable building projects.

Article	Title	Total Citations	Normalized Citations
[112]	Green IoT and edge AI as key technological enablers for a sustainable digital transition toward a smart circular economy: An industry 5.0 use case	65	3.60
[110]	A hybrid decision support system for sustainable office building renovation and energy performance improvement	238	2.88
[101]	Modeling carbon emission intensity: Application of artificial neural network	125	2.82
[74]	An integrated approach of BIM-enabled LCA and energy simulation: The optimized solution toward sustainable development	48	2.66
[114]	Comparison of linear regression and artificial neural networks models to predict heating and cooling energy demand, energy consumption, and CO_2 emissions	100	2.63
[117]	Modeling heating and cooling energy demands for building stock using a hybrid approach	47	2.61
[90]	Machine learning modeling for predicting non-domestic buildings energy performance: A model to support deep energy retrofit decision-making	73	2.54
[118]	Designing sustainable concrete mixture by developing a new machine learning technique	66	2.29
[111]	Prediction of engine performance and exhaust emissions for gasoline and methanol using artificial neural network	115	2.25
[80]	Developing novel 5th generation district energy networks	63	2.19
[119]	Design and implementation of cloud analytics-assisted smart power meters considering advanced artificial intelligence as edge analytics in demand-side management for smart homes	97	2.19
[120]	Life cycle greenhouse gas emissions and energy use of polylactic acid, bio-derived polyethylene, and fossil-derived polyethylene	58	2.02
[113]	A machine-learning-based approach to predict residential annual space heating and cooling loads considering occupant behavior	55	1.91
[70]	Grading buildings on energy performance using city benchmarking data	84	1.9
[121]	Analyzing the influence factors of the carbon emissions from China's building and construction industry from 2000 to 2015	81	1.83
[122]	The hourly life cycle carbon footprint of electricity generation in Belgium, bringing a temporal resolution in life cycle assessment	84	1.76
[93]	Comparative study of machine learning-based multi-objective prediction framework for multiple building energy loads	47	1.63
[123]	Development of electrochromic evacuated advanced glazing	49	1.38
[124]	Low-carbon cold chain logistics using ribonucleic acid-ant colony optimization algorithm	61	1.38
[125]	Data-driven strategic planning of building energy retrofitting: The case of Stockholm	50	1.13

Table 3. *Cont.*

Article	Title	Total Citations	Normalized Citations
[115]	Life cycle assessment of a wooden single-family house in Sweden	48	1.08
[116]	Development of a model for urban heat island prediction using neural network techniques	69	1.00

4. Discussion

Following the science mapping analysis of the included articles, this section mainly focuses on discussing mainstream research topics, identifying research gaps, and summarizing future research directions in the field of AI in NZCEs for sustainable building projects.

4.1. Mainstream Research Topics on AI in NZCEs for Sustainable Building Projects

As shown in Figure 3, the clusters of keywords have many connections with each other, and the keywords from different clusters are also closely connected. Based on the keyword analysis, the mainstream research topics related to AI in NZCEs for sustainable building projects can be summarized as follows:

4.1.1. Life Cycle Assessment and Carbon Footprint

Owing to the complexity of building projects and the variety of construction materials, the environmental impacts of different phases need to be considered. The life cycle assessment (LCA) method can be used to investigate the total carbon footprint of a building and is a widely known environmental impact assessment method [75,126]. The LCA method can be combined with BIM to reduce energy consumption [74]. Yan et al. [127] examined the carbon footprint of building energy systems and the factors affecting carbon emissions using sensitivity analysis. The development of novel building materials with sustainable and energy-efficient qualities can reduce the carbon footprints of buildings [128,129]. Simultaneously, energy efficiency standards have been established, which contribute to improving the quality of life of the population [41,48,73]. Combining AI techniques with energy prediction models can improve the cost of the environmental impact of buildings at the design stage [84]. Intelligent control systems are being developed to address the issue of excessive energy use in the provision of heating and cooling in buildings [69]. Specifically, there are differences in the total carbon footprints of heating- and cooling-dominated buildings in terms of their economic, environmental, and operational aspects [130]. As a result, the selection of heat-carrying fluids should be based on actual building characteristics to avoid environmental impact and ensure system stability. When combined with a building energy simulation tool (EnergyPlus^{TM}, Department of Energy, Springfield, IL, USA), AI technology can predict the annual cooling energy consumption of a building in a shorter period, thereby enhancing the prediction efficiency [131]. In addition to the operational energy, buildings consume implicit energy. Thus, it is important to calculate the implicit carbon footprints of buildings. LCA methods can be used to provide data support for energy consumption throughout the life cycle of a building, to reduce carbon emissions during construction [132]. All the aforementioned studies can assist project practitioners in making decisions to reduce the total carbon footprint, thus improving the sustainability of the building's life cycle.

4.1.2. Practical Applications of AI Techniques in Sustainable Buildings

AI techniques are increasingly used in the construction sector, particularly in the design and operation phases of buildings, providing practical solutions in the construction industry [133]. AI techniques have been previously adopted in traditional industrial sectors, but their benefits have been widely recognized in recent years. Zhang and Zhu [134] examine the effects of industrial robots on carbon emissions in the three largest economies—China, Japan, and the United States. They employed the classical linear regression model, ordinary least squares (OLS), to analyze the relationship between robot installations and robot density, which is measured as ownership per thousand manufacturing individuals. They concluded that the growth of industrial intelligence not only enhanced the efficiency of the energy system but also enhanced environmental quality and contributed to the achievement of ecologically sustainable development. In the Stockholm case, a data-driven approach to building retrofitting was empirically tested to achieve deep decarbonization, energy efficiency, and sustainable development [125]. AI can analyze data from many perspectives to optimize energy-intensive building systems [135]. When applied to outdoor lighting, AI techniques can improve energy efficiency [136]. The implications of living density on energy and CO2 emissions can also be investigated using a robust deep learning technique of the long short-term memory (LSTM) model, as well as how to improve living quality [82]. In addition, AI plays an important predictive role in helping sustainable buildings save energy and reduce their carbon emissions. Combined with carbon capture and storage (CCS) technology, it can better fulfill its predictive role in contributing to CO2 storage and combating climate change [103]. In contrast to LCA, ANN focuses on developing predictive models through data analysis, dealing with complicated nonlinear interactions, and is appropriate for predicting building-specific carbon emissions. ANN in an environmental impact cost assessment model can be used to estimate a building's energy consumption and calculate the configuration that will result in NZCEs in a shorter period [85,100]. It is beneficial in helping construction practitioners estimate early costs and achieve sustainability goals using an ANN model [84]. Additionally, ML can forecast certain elements and the total carbon footprint of a structure during the early design phase, thereby assisting designers in making sustainable design decisions [133].

Beyond the broad application of machine-learning algorithms, the research corpus highlights several specific AI techniques that are instrumental in advancing sustainable building projects. One such technique involves the use of neural networks, which are employed for the predictive modeling of energy consumption and optimization of heating, ventilation, and air conditioning (HVAC) systems. These models learn from historical data to forecast energy needs under varying conditions, thereby enabling more efficient resource management [137]. Another AI technique gaining traction is genetic algorithms, which are utilized to optimize building designs to minimize energy consumption and maximize natural lighting and ventilation [138]. Furthermore, fuzzy logic systems have been applied to manage uncertainties in building performance simulations, providing more robust decision-making tools for sustainable construction practices [139]. These AI techniques are integral for achieving more efficient, cost-effective, and environmentally friendly building solutions.

4.1.3. Multi-Objective Optimization

Multi-objective optimization is a powerful decision-making tool in the field of sustainable construction. It is efficient and time-saving compared to single-objective forecasting and can reduce carbon emissions during building operations while maintaining comfort [93,95]. In recent years, the combination of multi-objective optimization and ML to evaluate construction cases has seen a gradual increase in interest in the construction industry. For example, it can provide energy optimization for non-residential buildings [90]. Innovative structural designs of buildings through environmental impact assessment to improve the sustainability of buildings [75]. The effect of changes in the insulation installation position on carbon emissions as the temperature changes, to reduce energy consumption [87].

4.1.4. Energy Management and Energy Efficiency

Energy efficiency has a significant effect on carbon emissions. Managers focusing on energy management and improving energy efficiency can contribute to the sustainable development of the built environment. Most importantly, AI contributes positively to ecological and social sustainability through energy efficiency [1,99]. By comparing data from nine countries, including China and the UK, it was found that energy consumption significantly impacts the environment [140]. They suggested that using renewable energy sources, smart energy modeling using AI, and raising public awareness of environmental protection could foster the implementation of sustainable development policies [141]. A holistic assessment is required to achieve the goal of net-zero energy consumption in buildings. For instance, appropriate emission reduction programs need to be formulated in the light of the actual situation, thereby aiding in effectively reducing the total carbon emissions [74,121]. Using AI techniques in energy modeling can be a smart way to investigate a building's carbon footprint and explain and manage the elements that influence energy consumption [142]. According to the above findings, improving energy efficiency is the foundation and key to the development of sustainable buildings. The building sector can achieve environmental and social sustainability through the implementation of innovative technologies and strategies using AI.

4.1.5. Carbon Emissions from Buildings

In recent years, industrialization has severely influenced the environment while encountering the economy, and the increase in carbon dioxide emissions has contributed to global warming [134]. As a major contributor to global carbon emissions, the construction industry is a key sector in meeting emissions reduction targets; therefore, optimizing building performance is vital to improving climate conditions [88]. Many countries have committed to carbon-neutral policies to address the excessive carbon dioxide emissions [1]. Carbon emissions from countries such as the United States and the United Kingdom decreased significantly in 2017. Asian countries such as China and India account for a larger share of global emissions [101]. The energy optimization of German building types is aimed at improving energy efficiency and reducing carbon emissions [135]. The UK has developed AI-driven green building practices to mitigate urban heat [143]. AI-enabled building control systems have also been developed in the USA to reduce energy consumption and carbon emissions [69].

4.1.6. Decision Support System (DSS) and Sustainability

The DSS can be used to assess carbon emissions, energy consumption, design decisions, and material selection for buildings, thereby improving the energy efficiency of buildings. DSS has been reported to play a role in design decisions for curtain walls in high-rise buildings, helping designers make environmentally friendly decisions [144]. Moreover, DSS has been applied to the design of retrofitted residential complexes in European cities [145]. Compared to traditional structural design solutions, sustainable structural design considers the impact of the building on the environment, and combining it with DSS can help to increase the flexibility of design decisions [105]. As a common material in construction, high-compressive strength concrete can lead to increased carbon emissions, so it is crucial to balance the need between compressive strength and other properties. Low-carbon concrete is a good option for optimizing the structural design of buildings. It can also reduce carbon emissions from construction and promote sustainable buildings [146].

4.2. Research Gaps of AI in NZCEs for Sustainable Buildings

4.2.1. Existing Problems of the Life Cycle Assessment Method

Studies based on LCA methods consider the contribution of sustainable materials and energy prediction modeling to the reduction of the carbon footprint [74,131,132]. Sensitivity analysis can be used as a decision-making tool for predicting energy consumption in future buildings and exploring the factors affecting carbon emissions [74,127]. However, only a few studies have been conducted on various climatic conditions and buildings of different sizes. For instance, distributed energy systems are limited in their use and are not suitable for analyzing the carbon footprint of cities or countries [127]. When evaluating a client's energy requirements, future research should evaluate the feasibility and cost considerations to optimize energy savings considering a limited global carbon budget [147].

4.2.2. Opportunities and Challenges Faced by AI Techniques in Sustainable Buildings

Previous studies applying AI techniques in the building sector have focused on predicting carbon emissions from buildings [85,103]. However, the approach of combining ANN with genetic algorithms can be extended to a wider range of building types [85]. Compared to traditional regression techniques, ANN can use multiple parameters in decision analyses with high reliability [114]. However, there is a lack of case or empirical studies on whether the application of AI techniques could promote economic growth while optimizing the environment [134]. The application of AI techniques also faces the challenges of international regulatory issues, military conflicts, and insufficient legal frameworks, which require international cooperation to innovate technologies and address opportunities and challenges [103].

4.2.3. Scope of Application of Multi-Objective Modeling

When using multi-objective modeling, because the costs of occupancy and carbon emissions are fixed for a particular building, its parameters remain constant and do not apply to other building types [89,127]. In the future, there is still a need to refine the ML framework and add more sophisticated algorithms to improve decision-making in building design and prediction during the construction phase [90]. The findings of future research can assist policymakers in balancing competing goals and developing more effective carbon reduction plans.

4.2.4. Improvements in Energy Management and Efficiency

In general, improvements in energy management and efficiency can effectively reduce carbon emissions from buildings. Research on energy efficiency to improve carbon emissions has been conducted in other sectors [148,149]. For instance, innovative production and consumption patterns have been developed in the chemical industry to reduce energy consumption and achieve sustainable development [148]. AI techniques have also been demonstrated to improve energy efficiency in the logistics industry [149]. However, in the building sector, few studies have been conducted on how to enhance the correlation of energy emissions among different project life cycle phases to ensure optimized energy efficiency. Moreover, the estimation of the development density and degree of urbanization should be studied in the future. These two important indicators for assessing carbon emissions are complex, and new models must be created for their application in sustainable buildings [121].

4.2.5. Raise Awareness of Reducing Carbon Emissions

The negative environmental impact of buildings is of great concern to professionals and stakeholders in the construction industry. Many countries and industries have focused on low-carbon developments, with government departments actively involved and providing policy support. However, the design and implementation of low-carbon buildings should also consider economic factors, occupant comfort, and material availability [150]. These factors should be considered in future studies to create awareness among other researchers, practitioners, policymakers, and the general public.

4.2.6. Sustainable Development of Buildings

Similar to research on multi-objective optimization, research on DSS and sustainability has focused on design decisions, energy optimization, and sustainable material selection [125,151]. Future research on eco-concrete should consider more environmental parameters, such as methane and sulfur dioxide, and focus on the environmental friendliness and sustainability of building materials [118]. The use of bio-composites can also effectively improve the thermal performance of buildings and reduce energy consumption and carbon emissions [152].

4.3. Research Trends of AI in NZCEs for Sustainable Building Projects

After discussing the mainstream research topics and research gaps in the studied research domain, this section highlights recommendations for future research direction in the field of AI in NZCEs for sustainable building projects. Figure 4 illustrates the proposed framework for future research directions on AI in NZCEs for sustainable building projects. Numerous recommendations for future research include the following:

1. Various factors, such as energy savings, emissions reduction, and the feasibility of financial costs, should be considered when adopting LCA methods.
2. Improving the legal framework and international regulatory regime for the application of AI techniques to reduce carbon emissions.
3. Balancing carbon emission reduction with other sustainability objectives in response to changes in building parameters.
4. Empirical research on energy optimization strategies for different building scenarios.
5. Construction industries and practitioners should actively implement carbon-neutral policies.
6. Countries can share their experiences and work together to promote the development of sustainable buildings.
7. Using DSS to provide data analyses and forecasts should incorporate more environmental parameters to enable decision-makers to make sustainable development decisions.
8. Increased attention to decision-making processes and the implementation of program design to reduce carbon emissions.

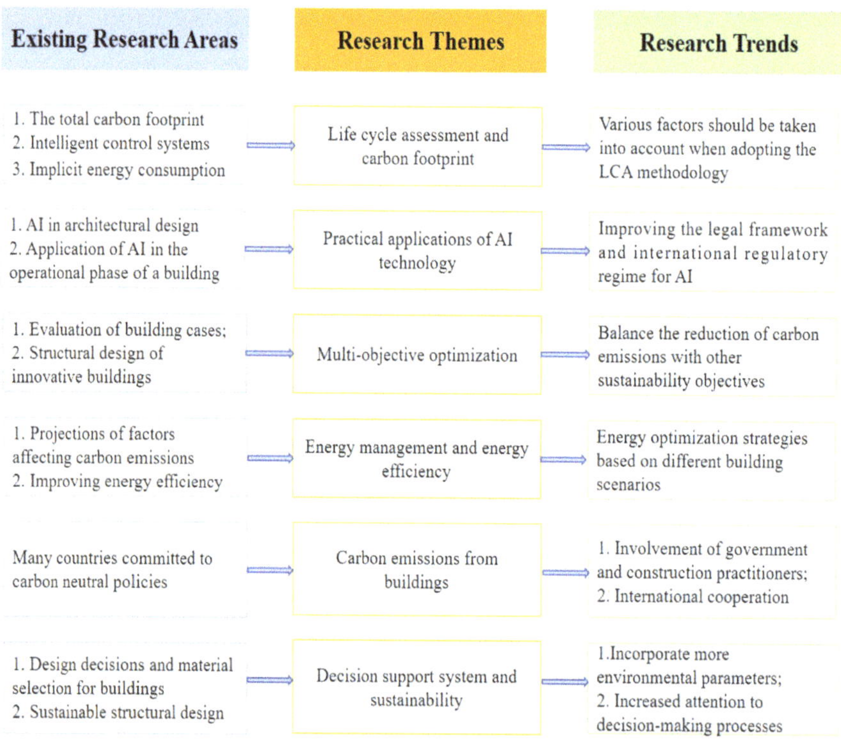

Figure 4. Proposed framework for the future research direction of AI in NZCEs for sustainable building projects.

5. Conclusions

In this review study, we conduct a systematic literature and science mapping review of AI applications in NZCEs for sustainable building projects. We employed a mixed-method approach (i.e., systematic literature review and science mapping) consisting of four comprehensive stages: search strategy, selection criteria, science mapping analysis, and qualitative discussion. The Scopus database was used to retrieve relevant articles, and 154 articles were included in further analysis. The results revealed a significant increase in relevant published articles in the studied research domain since 2019. Moreover, the most influential journals that published several articles during the study period were the *Journal of Cleaner Production, Applied Energy, and Energy and Buildings*. According to the keyword co-occurrence analysis, "building energy performance", "energy consumption", and "decision support system" had the highest average normalized citations, whereas "multi-objective optimization", "artificial intelligence", and "renewable energy" were the most recently used keywords. In addition, countries such as China, Australia, Malaysia, and the United Kingdom had the highest total link strength, suggesting strong connections with other countries in the studied research domain.

The qualitative discussion focuses on mainstream research topics and gaps and summarizes future research directions in AI in NZCEs for sustainable building projects. Six mainstream research topics were identified and discussed. Next, this review study discussed six main research gaps: (1) existing problems of the life cycle assessment method, (2) opportunities and challenges faced by AI techniques in sustainable buildings, (3) scope of application of multi-objective modeling, (4) improvements in energy management and efficiency, (5) raising awareness of reducing carbon emissions, and (6) sustainable development of buildings. Finally, this review study proposed a research framework for future research directions that could help researchers and practitioners advance related research areas in this domain.

5.1. Study Implications and Contributions

This study has several theoretical and practical implications. First, it emphasizes the reduction in carbon emissions through advanced technological innovations. The application of AI can effectively minimize the negative environmental impact of the construction industry, transition the construction industry toward low-carbon practices, and achieve sustainable development. Second, this review provides other scholars with new research methods and models to gain insights into the factors affecting carbon emissions. For example, it suggests potential areas that require attention to achieve NZCEs. Third, the findings of this review would assist local governments and project practitioners (e.g., project managers, workers, etc.) in raising awareness and actively participating in reducing carbon emissions. One potential initiative is to introduce stringent building codes that mandate the use of energy-efficient materials and technologies. These codes can be coupled with incentives such as tax breaks or grants for developers who exceed the minimum sustainability requirements [153]. Additionally, local governments could establish green certification programs that recognize and reward buildings to achieve high standards for energy efficiency and carbon footprint reduction. Another key policy could involve providing subsidies or low-interest loans to support the adoption of AI-based tools in building design and management, facilitating the transition toward more sustainable construction practices [119]. By enacting these policies, local governments can drive the adoption of the best practices identified in the literature and contribute significantly to reducing the environmental impact of the built environment.

The differences in building types in different regions are also considered, which provides managers in the construction industry with effective management solutions to improve existing energy management systems and focus on environmental sustainability indicators. One prominent management solution involves the use of AI-powered building energy modeling, which allows precise predictions of energy consumption under various operational conditions. This technology enables building managers to optimize energy use, reduce waste, and reduce overall carbon emissions [154]. Another solution involves the integration of smart sensors and IoT (Internet of Things) devices, which provide real-time data on building performance, enabling proactive adjustments to improve efficiency [155]. When implemented effectively, these technologies offer powerful tools for achieving and maintaining high energy efficiency in buildings. Achieving NZCEs is a focus of attention for construction projects and companies and practical action for the entire construction industry. As such, it would help the construction industry and other related industries (e.g., manufacturing, automobile) adopt smart and environmentally friendly approaches to improve project efficiency and achieve economic and environmental benefits.

Traditionally, life cycle assessment (LCA) has focused mainly on operational emissions; however, embodied carbon emissions in LCA are increasingly recognized as essential for a more comprehensive assessment of a building's environmental impact. Embodied carbon refers to greenhouse gas emissions associated with the extraction, production, transportation, and installation of building materials and their disposal at the end of a building's life. By integrating both embodied and operational emissions into LCA, stakeholders can better understand the total carbon footprint of a building across its entire life cycle. This holistic approach is crucial, especially in modern construction, where energy-intensive materials and processes significantly contribute to a building's overall carbon emissions [156,157]. Such an integrated assessment enables more informed decision-making and supports the development of strategies aimed at reducing both types of emissions, ultimately contributing to more sustainable building practices [158]. In addition, this research also calls for countries worldwide to learn from each other's successful experiences and work together to address the challenges of climate change.

5.2. Limitations and Future Research Directions

Although the application of AI in sustainable building projects offers substantial benefits, it also presents several challenges that need to be addressed. One significant challenge is the requirement for high-quality training data. AI models, particularly those based on machine learning, rely heavily on large datasets for learning and making accurate predictions. However, acquiring such datasets can be difficult, especially in the construction industry, where data collection is often inconsistent, and data quality can vary significantly [159]. Another challenge is the "black box" nature of some AI algorithms, particularly deep learning models. These models can produce highly accurate predictions; however, their internal workings are often opaque, making it difficult to understand how specific decisions are made. This lack of transparency can be problematic in the construction industry, where accountability and trust are critical [160]. Addressing these challenges requires ongoing research and development of methods to enhance the interpretability of AI models and improve data quality in the construction sector.

Despite the contributions of this study, it still has some limitations. This study only used one database (Scopus), and the search was limited to journal articles written in English only. It is recommended to utilize other databases in order to obtain more complete and comprehensive results. Moreover, the source publication type was limited to journals. Integrating multiple databases and document types in review studies presents significant challenges, particularly regarding data harmonization. Differences in data formats, metadata standards, and terminologies across databases can create inconsistencies that can complicate data analysis and interpretation. Harmonizing these data requires extensive preprocessing, which can be time-consuming and prone to errors if not meticulously performed [161]. Additionally, the variation in document types, ranging from structured databases to unstructured text, further complicates the integration process, as different approaches and tools are often needed to extract and standardize relevant information [162]. Addressing these challenges is critical for ensuring the accuracy and reliability of research findings in review studies that rely on diverse data sources. However, it is beneficial to incorporate multiple data sources if the risks mentioned here can be minimized to enhance the authenticity and usability of these findings; future research should consider more databases, document types, multiple languages, and source publication types. Future research could also use the PRISMA diagram format to summarize data from the extracted articles.

Author Contributions: Y.L.: Conceptualization, Formal analysis, Data curation, Methodology, Writing—original draft. M.F.A.-A.: Conceptualization, Resources, Supervision, Writing—review & editing. S.A.: Data curation, Resources, Project administration, Writing—review & editing. I.M.: Methodology, Resources, Writing—review & editing. W.U.: Resources, Visualization, Writing—review & editing. S.R.M.: Methodology, Visualization, Writing—review & editing. I.Y.W.: Methodology, Resources, Writing—review & editing. M.A.-R.: Data curation, Methodology, Resources, Writing—review & editing. H.L.: Methodology, Resources, Writing—review & editing. All authors have read and agreed to the published version of the manuscript.

Funding: This research was funded by the Postdoc Matching Fund Scheme [PolyU, UGC] grant number P0044276 and P0050845.

Data Availability Statement: The data supporting this study's findings are available from the corresponding author upon reasonable request.

Acknowledgments: The authors are grateful to the Department of Mechanical, Aerospace and Civil Engineering, University of Manchester, UK, for supporting this research. The authors are also grateful to the Editor and reviewers for their comments to improve the quality of this paper. This paper forms part of the first author's research project where similar research backgrounds but different scopes and/or analyses may have been published.

Conflicts of Interest: The authors declare no conflicts of interest.

References

1. Fan, R.; Zhang, X.; Bizimana, A.; Zhou, T.; Liu, J.; Meng, X. Achieving China's carbon neutrality: Predicting driving factors of CO_2 emission by artificial neural network. *J. Clean. Prod.* **2022**, *362*, 132331. [CrossRef]
2. Chen, L.; Chen, Z.; Zhang, Y.; Liu, Y.; Osman, A.I.; Farghali, M.; Hua, J.; Al-Fatesh, A.; Ihara, I.; Rooney, D.W.; et al. Artificial intelligence-based solutions for climate change: A review. *Environ. Chem. Lett.* **2023**, *21*, 2525–2557. [CrossRef]
3. Rymarczyk, J. Technologies, opportunities and challenges of the industrial revolution 4.0: Theoretical considerations. *Entrep. Bus. Econ. Rev.* **2020**, *8*, 185–198. [CrossRef]
4. Wang, C.; Zhan, J.; Xin, Z. Comparative analysis of urban ecological management models incorporating low-carbon transformation. *Technol. Forecast. Soc. Change* **2020**, *159*, 120190. [CrossRef]
5. Bottaccioli, L.; Aliberti, A.; Ugliotti, F.; Patti, E.; Osello, A.; Macii, E.; Acquaviva, A. Building Energy Modelling and monitoring by integration of IOT devices and building information models. In Proceedings of the 2017 IEEE 41st Annual Computer Software and Applications Conference (COMPSAC) [Preprint], Turin, Italy, 4–8 July 2017. [CrossRef]
6. Lützkendorf, T.; Frischknecht, R. (Net-) zero-emission buildings: A typology of terms and definitions. *Build. Cities* **2020**, *1*, 662–675. [CrossRef]
7. Wu, W.; Skye, H.M. Residential net-zero energy buildings: Review and perspective. *Renew. Sustain. Energy Rev.* **2021**, *142*, 110859. [CrossRef]
8. Sarkodie, S.A.; Owusu, P.A.; Leirvik, T. Global effect of urban sprawl, industrialization, trade and economic development on carbon dioxide emissions. *Environ. Res. Lett.* **2020**, *15*, 034049. [CrossRef]
9. Shakoor, A.; Ashraf, F.; Shakoor, S.; Mustafa, A.; Rehman, A.; Altaf, M.M. Biogeochemical transformation of greenhouse gas emissions from terrestrial to atmospheric environment and potential feedback to climate forcing. *Environ. Sci. Pollut. Res.* **2020**, *27*, 38513–38536. [CrossRef]
10. Yoro, K.O.; Daramola, M.O. CO_2 emission sources, Greenhouse Gases, and the global warming effect. In *Advances in Carbon Capture*; Woodhead Publishing: Sawston, UK, 2020; pp. 3–28. [CrossRef]
11. Muhammad Ashraf, W.; Moeen Uddin, G.; Afroze Ahmad, H.; Ahmad Jamil, M.; Tariq, R.; Wakil Shahzad, M.; Dua, V. Artificial intelligence enabled efficient power generation and emissions reduction underpinning net-zero goal from the coal-based power plants. *Energy Convers. Manag.* **2022**, *268*, 116025. [CrossRef]
12. Shirinbakhsh, M.; Harvey, L.D.D. Net-zero energy buildings: The influence of definition on greenhouse gas emissions. *Energy Build.* **2021**, *247*, 111118. [CrossRef]
13. Supriya; Chaudhury, R.; Sharma, U.; Thapliyal, P.C.; Singh, L.P. Low-CO_2 emission strategies to achieve net zero target in cement sector. *J. Clean. Prod. [Online]* **2023**, *417*, 137466. [CrossRef]
14. Ohene, E.; Chan, A.P.C.; Darko, A. Review of global research advances towards net-zero emissions buildings. *Energy Build.* **2022**, *266*, 112142. [CrossRef]

15. AlKheder, S.; Almusalam, A. Forecasting of carbon dioxide emissions from power plants in Kuwait using United States Environmental Protection Agency, Intergovernmental Panel on Climate Change, and Machine Learning Methods. *Renew. Energy* **2022**, *191*, 819–827. [CrossRef]
16. Fearnside, P.M. Challenges for sustainable development in Brazilian Amazonia. *Sustain. Dev.* **2018**, *26*, 141–149. [CrossRef]
17. Van Soest, H.L.; den Elzen, M.G.J.; van Vuuren, D.P. Net-zero emission targets for major emitting countries consistent with the Paris Agreement. *Nature Communications* **2021**, *12*, 2140. [CrossRef]
18. Wang, H.; Yu, X. Carbon dioxide emission typology and policy implications: Evidence from machine learning. *China Econ. Rev.* **2023**, *78*, 101941. [CrossRef]
19. Wang, P.; Zhong, Y.; Yao, Z. Modeling and estimation of CO_2 emissions in China based on Artificial Intelligence. *Comput. Intell. Neurosci.* **2022**, *2022*, 6822467. [CrossRef]
20. Gaeta, M.; Nsangwe Businge, C.; Gelmini, A. Achieving net zero emissions in Italy by 2050: Challenges and opportunities. *Energies* **2021**, *15*, 46. [CrossRef]
21. Kabisch, N.; Frantzeskaki, N.; Pauleit, S.; Naumann, S.; Davis, M.; Artmann, M.; Haase, D.; Knapp, S.; Korn, H.; Stadler, J.; et al. Nature-based solutions to climate change mitigation and adaptation in urban areas: Perspectives on indicators, knowledge gaps, barriers, and opportunities for action. *Ecol. Soc.* **2016**, *21*. [CrossRef]
22. McCauley, D.; Ramasar, V.; Heffron, R.J.; Sovacool, B.K.; Mebratu, D.; Mundaca, L. Energy justice in the transition to low carbon energy systems: Exploring key themes in interdisciplinary research. *Appl. Energy* **2019**, *233*, 916–921. [CrossRef]
23. Oh, T.H.; Hasanuzzaman, M.; Selvaraj, J.; Teo, S.C.; Chua, S.C. Energy policy and alternative energy in Malaysia: Issues and challenges for sustainable growth—An update. *Renew. Sustain. Energy Rev.* **2018**, *81*, 3021–3031. [CrossRef]
24. Karakhan, A.A.; Gambatese, J.; Simmons, D.R.; Nnaji, C. How to improve workforce development and sustainability in construction. In Proceedings of the Construction Research Congress 2020, Tempe, AZ, USA, 8–10 March 2020. [CrossRef]
25. Akadiri, P.O.; Chinyio, E.A.; Olomolaiye, P.O. Design of a sustainable building: A conceptual framework for implementing sustainability in the building sector. *Buildings* **2012**, *2*, 126–152. [CrossRef]
26. Nawari, N.O.; Ravindran, S. Blockchain and Building Information Modeling (BIM): Review and Applications in Post-Disaster Recovery. *Buildings* **2019**, *9*, 149. [CrossRef]
27. Knowles, E. (Ed.) *The Oxford Dictionary of Phrase and Fable*; OUP Oxford: Oxford, UK, 2006.
28. Alwetaishi, M.; Shamseldin, A. The use of Artificial Intelligence (AI) and big-data to improve energy consumption in existing buildings. *IOP Conf. Ser. Mater. Sci. Eng.* **2021**, *1148*, 012001. [CrossRef]
29. An, Y.; Li, H.; Su, T.; Wang, Y. Determining uncertainties in AI applications in AEC sector and their corresponding mitigation strategies. *Autom. Constr.* **2021**, *131*, 103883. [CrossRef]
30. Khaleel, M.; Ahmed, A.A.; Alsharif, A. Artificial Intelligence in Engineering. *Brill. Res. Artif. Intell.* **2023**, *3*, 32–42. [CrossRef]
31. Momade, M.H.; Durdyev, S.; Estrella, D.; Ismail, S. Systematic review of application of artificial intelligence tools in architectural, engineering and construction. *Front. Eng. Built Environ.* **2021**, *1*, 203–216. [CrossRef]
32. Pan, Y.; Zhang, L. Roles of artificial intelligence in construction engineering and management: A critical review and future trends. *Autom. Constr.* **2021**, *122*, 103517. [CrossRef]
33. Zhang, F.; Chan, A.P.; Darko, A.; Chen, Z.; Li, D. Integrated applications of building information modeling and artificial intelligence techniques in the AEC/FM industry. *Autom. Constr.* **2022**, *139*, 104289. [CrossRef]
34. Manzoor, B.; Othman, I.; Durdyev, S.; Ismail, S.; Wahab, M.H. Influence of artificial intelligence in civil engineering toward Sustainable Development—A systematic literature review. *Appl. Syst. Innov.* **2021**, *4*, 52. [CrossRef]
35. Darko, A.; Chan AP, C.; Ameyaw, E.E.; Owusu-Manu, D.-G.; Edwards, D.J. Artificial intelligence in the AEC industry: State-of-the-art review and future trends. *Autom. Constr.* **2020**, *110*, 103010. [CrossRef]
36. Pan, Y.; Zhang, L. Applications of artificial intelligence in construction engineering and management: A review. *Autom. Constr.* **2021**, *126*, 103671. [CrossRef]
37. Pan, Y.; Zhang, L. Integrating bim and AI for Smart Construction Management: Current status and future directions. *Arch. Comput. Methods Eng.* **2022**, *30*, 1081–1110. [CrossRef]
38. Ahmed, A.; Ge, T.; Peng, J.; Yan, W.; Tee, B.T.; You, S. Assessment of the renewable energy generation towards net-zero energy buildings: A Review. *Energy Build.* **2022**, *256*, 111755. [CrossRef]
39. Chen, L.; Huang, L.; Hua, J.; Chen, Z.; Wei, L.; Osman, A.I.; Fawzy, S.; Rooney, D.W.; Dong, L.; Yap, P.-S. Green construction for low-carbon cities: A Review. *Environ. Chem. Lett.* **2023**, *21*, 1627–1657. [CrossRef]
40. Delanoë, P.; Tchuente, D.; Colin, G. Method and evaluations of the effective gain of artificial intelligence models for reducing CO_2 emissions. *J. Environ. Manag.* **2023**, *331*, 117261. [CrossRef]
41. Ahmed, M.; AlQadhi, S.; Mallick, J.; Kahla, N.B.; Le, H.A.; Singh, C.K.; Hang, H.T. Artificial Neural Networks for sustainable development of the construction industry. *Sustainability* **2022**, *14*, 14738. [CrossRef]
42. Chen, L.; Chen, Z.; Zhang, Y.; Liu, Y.; Osman, A.I.; Farghali, M.; Hua, J.; Ai-Fatesh, A.; Ihara, I.; Rooney, D.W.; et al. Artificial Intelligence-based solutions for climate change: A Review. *Environ. Chem. Lett.* **2023**, *21*, 2525–2557. [CrossRef]
43. Brown, T.; Smith, J. Advanced computational methods for carbon emission estimation in construction. *J. Sustain. Constr.* **2021**, *14*, 123–135.
44. Adunadepo, A.M.D.; Oladiran, S. Artificial intelligence for sustainable development of intelligent buildings. In Proceedings of the 9th CIDB Postgraduate Conference, Cape Town, South Africa, 1–4 February 2016; pp. 1–4.

45. Ali, A.; Jayaraman, R.; Mayyas, A.; Alaifan, B.; Azar, E. Machine learning as a surrogate to building performance simulation: Predicting energy consumption under different operational settings. *Energy Build.* **2023**, *286*, 112940. [CrossRef]
46. Sohani, A.; Sayyaadi, H.; Miremadi, S.R.; Samiezadeh, S.; Doranehgard, M.H. Thermo-Electro-Environmental Analysis of a photovoltaic solar panel using machine learning and real-time data for smart and Sustainable Energy Generation. *J. Clean. Prod.* **2022**, *353*, 131611. [CrossRef]
47. Nejati, F.; Zoy, W.O.; Tahoori, N.; Abdunabi Xalikovich, P.; Sharifian, M.A.; Nehdi, M.L. Machine learning method based on symbiotic organism search algorithm for thermal load prediction in buildings. *Buildings* **2023**, *13*, 727. [CrossRef]
48. Ahmed, S.; Jones, T.; Brown, K. Application of artificial neural networks in promoting sustainable construction projects. *J. Constr. Eng. Manag.* **2022**, *148*, 05022001. [CrossRef]
49. Smith, J. Annual publication trends in AI and NZCEs. *J. Sustain. Build. Res.* **2020**, *15*, 234–245.
50. Johnson, L. Science mapping and influential keywords in AI for sustainable projects. *AI Sustain.* **2019**, *12*, 98–112.
51. Williams, R.; Brown, T. Mainstream research topics in AI and sustainable building. *Sustain. Technol. Rev.* **2018**, *10*, 321–335.
52. Davis, P.; Miller, S.; Thompson, H. Framework for research gaps in AI and NZCEs. *J. Environ. Technol.* **2021**, *18*, 45–59.
53. Linnenluecke, M.K.; Marrone, M.; Singh, A.K. Conducting systematic literature reviews and bibliometric analyses. *Aust. J. Manag.* **2020**, *45*, 175–194. [CrossRef]
54. Tranfield, D.; Denyer, D.; Smart, P. Towards a methodology for developing evidence-informed management knowledge by means of systematic review. *Br. J. Manag.* **2003**, *14*, 207–222. [CrossRef]
55. Hosseini, M.R.; Martek, I.; Zavadskas, E.K.; Aibinu, A.A.; Arashpour, M.; Chileshe, N. Critical evaluation of off-site construction research: A Scientometric analysis. *Autom. Constr.* **2018**, *87*, 235–247. [CrossRef]
56. Keele, S. Guidelines for Performing Systematic Literature Reviews in Software Engineering. 2007. Available online: https://legacyfileshare.elsevier.com/promis_misc/525444systematicreviewsguide.pdf (accessed on 21 August 2024).
57. Tijssen, R.J.W.; Van Raan, A.F.J. Mapping changes in science and Technology. *Eval. Rev.* **1994**, *18*, 98–115. [CrossRef]
58. Ashley, P.; Boyd, B.W.E. Quantitative and qualitative approaches to research in Environmental Management. *Australas. J. Environ. Manag.* **2006**, *13*, 70–78. [CrossRef]
59. Clark VL, P.; Creswell, J.W.; Green DO, N.; Shope, R.J. Mixing quantitative and qualitative approaches. *Handb. Emergent Methods* **2008**, *363*, 363–387.
60. Meho, L.I.; Rogers, Y. Citation counting, citation ranking, and *h*-index of Human-Computer Interaction Researchers: A comparison of scopus and web of science. *J. Am. Soc. Inf. Sci. Technol.* **2008**, *59*, 1711–1726. [CrossRef]
61. Smith, J.; Johnson, L. Understanding the Differences Between Scientific and Trade Journals. *J. Res. Methods* **2022**, *29*, 456–467.
62. Moral-Munoz, J.A.; Cobo, M.J.; Chiclana, F.; Collop, A.; Herrera-Viedma, E. Andrew Collop, and Enrique Herrera-Viedma. Analyzing highly cited papers in Intelligent Transportation Systems. *IEEE Trans. Intell. Transp. Syst.* **2016**, *17*, 993–1001. [CrossRef]
63. Cobo, M.J.; López-Herrera, A.G.; Herrera-Viedma, E.; Herrera, F. Science mapping software tools: Review, analysis, and cooperative study among tools. *J. Am. Soc. Inf. Sci. Technol.* **2011**, *62*, 1382–1402. [CrossRef]
64. Alonso, S.; Cabrerizo, F.J.; Herrera-Viedma, E.; Herrera, F. Hg-index: A new index to characterize the scientific output of researchers based on the H- and G-indices. *Scientometrics* **2010**, *82*, 391–400. [CrossRef]
65. Morris, S.A.; Van der Veer Martens, B. Mapping research specialties. *Annu. Rev. Inf. Sci. Technol.* **2009**, *42*, 213–295. [CrossRef]
66. Noyons, E.C.M.; Moed, H.F.; Luwel, M. Combining mapping and citation analysis for evaluative bibliometric purposes: A bibliometric study. *J. Am. Soc. Inf. Sci.* **1999**, *50*, 115–131. [CrossRef]
67. Pollock, A.; Berge, E. How to do a systematic review. *Int. J. Stroke* **2018**, *13*, 138–156.
68. Van Eck, N.J.; Waltman, L. Software survey: VOSviewer, a computer program for bibliometric mapping. *Scientometrics* **2010**, *84*, 523–538. [CrossRef]
69. Deng, Z.; Wang, X.; Jiang, Z.; Zhou, N.; Ge, H.; Dong, B. Evaluation of deploying data-driven predictive controls in buildings on a large scale for greenhouse gas emission reduction. *Energy* **2023**, *270*, 126934. [CrossRef]
70. Papadopoulos, S.; Kontokosta, C.E. Grading buildings on energy performance using City Benchmarking Data. *Appl. Energy* **2019**, *233–234*, 244–253. [CrossRef]
71. Tam, V.W.; Butera, A.; Le, K.N.; Silva LC, D.; Evangelista, A.C. A prediction model for compressive strength of CO_2 concrete using regression analysis and artificial neural networks. *Constr. Build. Mater.* **2022**, *324*, 126689. [CrossRef]
72. Wenninger, S.; Kaymakci, C.; Wiethe, C. Explainable long-term building energy consumption prediction using QLattice. *Appl. Energy* **2022**, *308*, 118300. [CrossRef]
73. Kairies-Alvarado, D.; Muñoz-Sanguinetti, C.; Martínez-Rocamora, A. Contribution of energy efficiency standards to life-cycle carbon footprint reduction in public buildings in Chile. *Energy Build.* **2021**, *236*, 110797. [CrossRef]
74. Tushar, Q.; Bhuiyan, M.A.; Zhang, G.; Maqsood, T. An integrated approach of BIM-enabled LCA and energy simulation: The optimized solution towards sustainable development. *J. Clean. Prod.* **2021**, *289*, 125622. [CrossRef]
75. D'Amico, B.; Myers, R.; Sykes, J.; Voss, E.; Cousins-Jenvey, B.; Fawcett, W.; Richardson, S.; Kermani, A.; Pomponi, F. Machine learning for sustainable structures: A call for data. *Structures* **2019**, *19*, 1–4. [CrossRef]
76. Kanyilmaz, A.; Tichell, P.R.; Loiacono, D. A genetic algorithm tool for conceptual structural design with cost and embodied carbon optimization. *Eng. Appl. Artif. Intell.* **2022**, *112*, 104711. [CrossRef]
77. Sharif, S.A.; Hammad, A. Simulation-based multi-objective optimization of institutional building renovation considering energy consumption, life-cycle cost and life-cycle assessment. *J. Build. Eng.* **2019**, *21*, 429–445. [CrossRef]

78. Sun, C.; Wang, H.; Liu, C.; Zhao, Y. Real Time Energy Efficiency Operational Indicator: Simulation Research from the perspective of life cycle assessment. *Proc. Inst. Mech. Eng. Part M J. Eng. Marit. Environ.* **2020**, *235*, 763–772. [CrossRef]
79. Zhang, X.; Wang, F. Hybrid input-output analysis for life-cycle energy consumption and carbon emissions of China's building sector. *Build. Environ.* **2016**, *104*, 188–197. [CrossRef]
80. Revesz, A.; Jones, P.; Dunham, C.; Davies, G.; Marques, C.; Matabuena, R.; Scott, J.; Maidment, G. Developing novel 5th Generation District Energy Networks. *Energy* **2020**, *201*, 117389. [CrossRef]
81. Arsiwala, A.; Elghaish, F.; Zoher, M. Digital Twin with machine learning for predictive monitoring of CO_2 equivalent from existing buildings. *Energy Build.* **2023**, *284*, 112851. [CrossRef]
82. Chen, C.; Chai, K.K.; Lau, E. AI-assisted approach for building energy and Carbon Footprint Modeling. *Energy AI* **2021**, *5*, 100091. [CrossRef]
83. Chen, Y.Y.; Lin, Y.H.; Kung, C.C.; Chung, M.H.; Yen, I.H. Design and implementation of cloud analytics-assisted Smart Power Meters considering advanced artificial intelligence as edge analytics in demand-side management for Smart Homes. *Sensors* **2019**, *19*, 2047. [CrossRef]
84. Hamida, A.; Alsudairi, A.; Alshaibani, K.; Alshamrani, O. Environmental impacts cost assessment model of residential building using an artificial neural network. *Eng. Constr. Archit. Manag.* **2021**, *28*, 3190–3215. [CrossRef]
85. McKinstray, R.; Lim, J.B.; Tanyimboh, T.T.; Phan, D.T.; Sha, W.; Brownlee, A.E. Topographical eijingmon of single-storey non-domestic steel framed buildings using photovoltaic panels for net-zero carbon impact. *Build. Environ.* **2015**, *86*, 120–131. [CrossRef]
86. Palladino, D. Greening Umbria's future: Investigation of the retrofit measures' potential to achieve energy goals by 2030 in the Umbria region. *Buildings* **2023**, *13*, 1039. [CrossRef]
87. Farouk, N.; Babiker, S.G. A comprehensive study on thermal reinforcement of Saudi Arabia buildings considering CO_2 emissions and Capital Cost Using Machine Learning. *Eng. Anal. Bound. Elem.* **2023**, *148*, 351–365. [CrossRef]
88. Genkin, M.; McArthur, J.J. B-smart: A reference architecture for artificially intelligent autonomic smart buildings. *Eng. Appl. Artif. Intell.* **2023**, *121*, 106063. [CrossRef]
89. Xue, Q.; Wang, Z.; Chen, Q. Multi-objective optimization of building design for life cycle cost and CO_2 Emissions: A case study of a low-energy residential building in a severe cold climate. *Build. Simul.* **2021**, *15*, 83–98. [CrossRef]
90. Seyedzadeh, S.; Pour Rahimian, F.; Oliver, S.; Rodriguez, S.; Glesk, I. Machine learning modelling for predicting non-domestic Buildings Energy Performance: A model to support deep energy retrofit decision-making. *Appl. Energy* **2020**, *279*, 115908. [CrossRef]
91. Saryazdi SM, E.; Etemad, A.; Shafaat, A.; Bahman, A.M. Data-driven performance analysis of a residential building applying Artificial Neural Network (ANN) and multi-objective Genetic Algorithm (GA). *Build. Environ.* **2022**, *225*, 109633. [CrossRef]
92. Tang, Y.X.; Lee, Y.H.; Amran, M.; Fediuk, R.; Vatin, N.; Kueh, A.B.; Lee, Y.Y. Artificial neural network-forecasted compression strength of alkaline-activated slag concretes. *Sustainability* **2022**, *14*, 5214. [CrossRef]
93. Luo, X.; Oyedele, L.O.; Ajayi, A.O.; Akinade, O.O. Comparative study of machine learning-based multi-objective prediction framework for multiple building energy loads. *Sustain. Cities Soc.* **2020**, *61*, 102283. [CrossRef]
94. Adel, T.K.; Pirooznezhad, L.; Ravanshadnia, M.; Tajaddini, A. Global policies on Green Building Construction from 1990 to 2019. *J. Green Build.* **2021**, *16*, 227–245. [CrossRef]
95. Behzadi, A.; Alirahmi, S.M.; Yu, H.; Sadrizadeh, S. An efficient renewable hybridization based on hydrogen storage for peak demand reduction: A rule-based energy control and optimization using machine learning techniques. *J. Energy Storage* **2023**, *57*, 106168. [CrossRef]
96. Behzadi, A.; Gram, A.; Thorin, E.; Sadrizadeh, S. A hybrid machine learning-assisted optimization and rule-based energy monitoring of a green concept based on low-temperature heating and high-temperature cooling system. *J. Clean. Prod.* **2023**, *384*, 135535. [CrossRef]
97. Moraliyage, H.; Dahanayake, S.; De Silva, D.; Mills, N.; Rathnayaka, P.; Nguyen, S.; Alahakoon, D.; Jennings, A. A robust artificial intelligence approach with explainability for measurement and verification of energy efficient infrastructure for net zero carbon emissions. *Sensors* **2022**, *22*, 9503. [CrossRef] [PubMed]
98. Luo, W.; Zhang, Y.; Gao, Y.; Liu, Y.; Shi, C.; Wang, Y. Life cycle carbon cost of buildings under carbon trading and carbon tax system in China. *Sustain. Cities Soc.* **2021**, *66*, 102509. [CrossRef]
99. Zhang, X.; Liu, P.; Zhu, H. The impact of industrial intelligence on energy intensity: Evidence from China. *Sustainability* **2022**, *14*, 7219. [CrossRef]
100. Chen, S.; Liu, Y.; Guo, Z.; Luo, H.; Zhou, Y.; Qiu, Y.; Zhou, B.; Zang, T. Deep reinforcement learning based research on low-carbon scheduling with distribution network schedulable resources. *IET Gener. Transm. Distrib.* **2023**, *17*, 2289–2300. [CrossRef]
101. Acheampong, A.O.; Boateng, E.B. Modelling carbon emission intensity: Application of artificial neural network. *J. Clean. Prod.* **2019**, *225*, 833–856. [CrossRef]
102. Liu, J.; Liu, L.; Qian, Y.; Song, S. The effect of artificial intelligence on carbon intensity: Evidence from China's Industrial Sector. *Socio-Econ. Plan. Sci.* **2022**, *83*, 101002. [CrossRef]
103. Qerimi, Q.; Sergi, B.S. The case for global regulation of carbon capture and storage and artificial intelligence for climate change. *Int. J. Greenh. Gas Control* **2022**, *120*, 103757. [CrossRef]

104. Sun, L.; Chen, W. The improved CHINACCS decision support system: A case study for Beijing–Tianjin–Hebei region of China. *Appl. Energy* **2013**, *112*, 793–799. [CrossRef]
105. Hou, S.; Li, H.; Rezgui, Y. Ontology-based approach for structural design considering low embodied energy and carbon. *Energy Build.* **2015**, *102*, 75–90. [CrossRef]
106. Shobeiri, V.; Bennett, B.; Xie, T.; Visintin, P. A generic framework for augmented concrete mix design: Optimisation of geopolymer concrete considering environmental, financial and mechanical properties. *J. Clean. Prod.* **2022**, *369*, 133382. [CrossRef]
107. Leydesdorff, L.; Bornmann, L. How to normalize counts of citations? An evaluation of six methods. *Scientometrics* **2011**, *87*, 545–562.
108. Jiang, J.A.; Su, Y.L.; Shieh, J.C.; Kuo, K.C.; Lin, T.S.; Lin, T.T.; Wei, F.; Chou, J.J.; Wang, J.C. On application of a new hybrid maximum power point tracking (MPPT) based photovoltaic system to the Closed Plant Factory. *Appl. Energy* **2014**, *124*, 309–324. [CrossRef]
109. Roaf, S.; Nicol, F.; Humphreys, M.; Tuohy, P.; Boerstra, A. Twentieth Century standards for Thermal comfort: Promoting high energy buildings. *Archit. Sci. Rev.* **2010**, *53*, 65–77. [CrossRef]
110. Juan, Y.-K.; Gao, P.; Wang, J. A hybrid decision support system for Sustainable Office Building Renovation and Energy Performance Improvement. *Energy Build.* **2010**, *42*, 290–297. [CrossRef]
111. Çay, Y.; Korkmaz, I.; Çiçek, A.; Kara, F. Prediction of engine performance and exhaust emissions for gasoline and methanol using artificial neural network. *Energy* **2013**, *50*, 177–186. [CrossRef]
112. Fraga-Lamas, P.; Lopes, S.I.; Fernández-Caramés, T.M. Green IOT and Edge Ai as key technological enablers for a sustainable digital transition towards a smart circular economy: An industry 5.0 use case. *Sensors* **2021**, *21*, 5745. [CrossRef]
113. Li, X.; Yao, R. A machine-learning-based approach to predict residential annual space heating and cooling loads considering occupant behaviour. *Energy* **2020**, *212*, 118676. [CrossRef]
114. Pino-Mejías, R.; Pérez-Fargallo, A.; Rubio-Bellido, C.; Pulido-Arcas, J.A. Comparison of linear regression and artificial neural networks models to predict heating and cooling energy demand, energy consumption and CO_2 emissions. *Energy* **2017**, *118*, 24–36. [CrossRef]
115. Petrovic, B.; Myhren, J.A.; Zhang, X.; Wallhagen, M.; Eriksson, O. Life cycle assessment of a wooden single-family house in Sweden. *Appl. Energy* **2019**, *251*, 113253. [CrossRef]
116. Gobakis, K.; Kolokotsa, D.; Synnefa, A.; Saliari, M.; Giannopoulou, K.; Santamouris, M. Development of a model for urban heat island prediction using neural network techniques. *Sustain. Cities Soc.* **2011**, *1*, 104–115. [CrossRef]
117. Li, X.; Yao, R. Modelling heating and cooling energy demand for building stock using a hybrid approach. *Energy Build.* **2021**, *235*, 110740. [CrossRef]
118. Naseri, H.; Jahanbakhsh, H.; Hosseini, P.; Moghadas Nejad, F. Designing sustainable concrete mixture by developing a new machine learning technique. *J. Clean. Prod.* **2020**, *258*, 120578. [CrossRef]
119. Chen, M.; Xu, B. The role of government in the development of AI technologies: Policies and initiatives for sustainable building. *Sustainability* **2019**, *11*, 2197. [CrossRef]
120. Benavides, P.T.; Lee, U.; Zarè-Mehrjerdi, O. Life cycle greenhouse gas emissions and energy use of polylactic acid, bio-derived polyethylene, and fossil-derived polyethylene. *J. Clean. Prod.* **2020**, *277*, 124010. [CrossRef]
121. Wu, P.; Song, Y.; Zhu, J.; Chang, R. Analyzing the influence factors of the carbon emissions from China's building and Construction Industry from 2000 to 2015. *J. Clean. Prod.* **2019**, *221*, 552–566. [CrossRef]
122. Messagie, M.; Mertens, J.; Oliveira, L.; Rangaraju, S.; Sanfelix, J.; Coosemans, T.; Rangaraju, S.; Mierlo, J.V.; Macharis, C. The hourly life cycle carbon footprint of electricity generation in Belgium, bringing a temporal resolution in life cycle assessment. *Appl. Energy* **2014**, *134*, 469–476. [CrossRef]
123. Papaefthimiou, S.; Leftheriotis, G.; Yianoulis, P.; Hyde, T.; Eames, P.C.; Fang, Y.; Pennarun, P.Y.; Jannasch, P. Development of electrochromic evacuated advanced glazing. *Energy Build.* **2006**, *38*, 1455–1467. [CrossRef]
124. Zhang, L.Y.; Tseng, M.L.; Wang, C.H.; Xiao, C.; Fei, T. Low-carbon cold chain logistics using ribonucleic acid-ant colony optimization algorithm. *J. Clean. Prod.* **2019**, *233*, 169–180. [CrossRef]
125. Pasichnyi, O.; Levihn, F.; Shahrokni, H.; Wallin, J.; Kordas, O. Data-driven strategic planning of building energy retrofitting: The case of Stockholm. *J. Clean. Prod.* **2019**, *233*, 546–560. [CrossRef]
126. Pittau, F.; Giacomel, D.; Iannaccone, G.; Malighetti, L. Environmental consequences of refurbishment versus demolition and reconstruction: A comparative life cycle assessment of an Italian case study. *J. Green Build.* **2020**, *15*, 155–172. [CrossRef]
127. Yan, Y.; Zhang, H.; Meng, J.; Long, Y.; Zhou, X.; Li, Z.; Wang, Y.; Liang, Y. Carbon footprint in building distributed energy system: An optimization-based feasibility analysis for potential emission reduction. *J. Clean. Prod.* **2019**, *239*, 117990. [CrossRef]
128. Azevedo, I.; Bataille, C.; Bistline, J.; Clarke, L.; Davis, S. Net-zero emissions energy systems: What we know and do not know. *Energy Clim. Change* **2021**, *2*, 100049. [CrossRef]
129. Meddah, M.; Benkari, N.; Al-Saadi, S.; Al Maktoumi, Y. Sarooj mortar: From a traditional building material to an engineered Pozzolan -mechanical and thermal properties study. *J. Build. Eng.* **2020**, *32*, 101754. [CrossRef]
130. Bartolini, N.; Casasso, A.; Bianco, C.; Sethi, R. Environmental and economic impact of the antifreeze agents in geothermal heat exchangers. *Energies* **2020**, *13*, 5653. [CrossRef]
131. Mui, K.W.; Wong, L.T.; Satheesan, M.K.; Balachandran, A. A hybrid simulation model to predict the cooling energy consumption for residential housing in Hong Kong. *Energies* **2021**, *14*, 4850. [CrossRef]

132. Opher, T.; Duhamel, M.; Posen, I.D.; Panesar, D.K.; Brugmann, R.; Roy, A.; Zizzo, R.; Sequeira, L.; Anvari, A.; MacLean, H.L. Life cycle GHG assessment of a building restoration: Case study of a heritage industrial building in Toronto, Canada. *J. Clean. Prod.* **2021**, *279*, 123819. [CrossRef]
133. Płoszaj-Mazurek, M.; Ryńska, E.; Grochulska-Salak, M. Methods to optimize carbon footprint of buildings in regenerative architectural design with the use of machine learning, Convolutional Neural Network, and parametric design. *Energies* **2020**, *13*, 5289. [CrossRef]
134. Zhang, X.; Zhu, H. The impact of industrial intelligence on carbon emissions: Evidence from the three largest economies. *Sustainability* **2023**, *15*, 6316. [CrossRef]
135. Gökçe, H.U.; Gökçe, K.U. Intelligent Energy Optimization System Development and validation for German building types. *Int. J. Low-Carbon Technol.* **2021**, *16*, 1299–1316. [CrossRef]
136. Atis, S.; Ekren, N. Development of an outdoor lighting control system using expert system. *Energy Build.* **2016**, *130*, 773–786. [CrossRef]
137. Fan, C.; Xia, X. A multi-objective optimization model for energy-efficiency building envelope retrofits considering cost, energy, and thermal comfort. *Energy Build.* **2017**, *142*, 431–441. [CrossRef]
138. Caldas, L.G.; Norford, L.K. Genetic algorithms for optimization of building envelopes and the design and control of HVAC systems. *J. Sol. Energy Eng.* **2003**, *125*, 343–351. [CrossRef]
139. Zhang, Y.; Wang, S.; Xia, X.; Wang, D. Fuzzy logic-based decision-making model for smart building energy management. *Energy Build.* **2018**, *158*, 1672–1682. [CrossRef]
140. Nejat, P.; Jomehzadeh, F.; Taheri, M.M.; Gohari, M.; Majid MZ, A. A global review of energy consumption, CO_2 emissions and policy in the residential sector (with an overview of the top ten CO_2 emitting countries). *Renew. Sustain. Energy Rev.* **2015**, *43*, 843–862. [CrossRef]
141. Bildirici, M.; Ersin, Ö.Ö. Nexus between industry 4.0 and environmental sustainability: A Fourier Panel Bootstrap Cointegration and causality analysis. *J. Clean. Prod.* **2023**, *386*, 135786. [CrossRef]
142. Herbinger, F.; Vandenhof, C.; Kummert, M. Building Energy Model Calibration using a surrogate neural network. *Energy Build.* **2023**, *289*, 113057. [CrossRef]
143. He, B.-J. Green building: A comprehensive solution to urban heat. *Energy Build.* **2022**, *271*, 112306. [CrossRef]
144. Kassem, M.; Dawood, N.; Mitchell, D. A decision support system for the selection of curtain wall systems at the design development stage. *Constr. Manag. Econ.* **2012**, *30*, 1039–1053. [CrossRef]
145. Torabi, M.; Hashemi, S.; Saybani, M.R.; Shamshirband, S.; Mosavi, A. A Hybrid clustering and classification technique for forecasting short-term energy consumption. *Environ. Prog. Sustain. Energy* **2019**, *38*, 66–76. [CrossRef]
146. Faridmehr, I.; Nehdi, M.L.; Huseien, G.F.; Baghban, M.H.; Sam, A.R.; Algaifi, H.A. Experimental and informational modeling study of sustainable self-compacting geopolymer concrete. *Sustainability* **2021**, *13*, 7444. [CrossRef]
147. Rasmussen, F.N.; Birkved, M.; Birgisdóttir, H. Low- carbon design strategies for new residential buildings—Lessons from architectural practice. *Archit. Eng. Des. Manag.* **2020**, *16*, 374–390. [CrossRef]
148. Fiorentino, G.; Zucaro, A.; Ulgiati, S. Towards an energy efficient chemistry. Switching from fossil to bio-based products in a life cycle perspective. *Energy* **2019**, *170*, 720–729. [CrossRef]
149. Su, Y.; Fan, Q.-M. The Green Vehicle Routing problem from a Smart Logistics Perspective. *IEEE Access* **2020**, *8*, 839–846. [CrossRef]
150. Hossain, Y.; Marsik, T. Conducting life cycle assessments (LCAs) to determine carbon payback: A case study of a highly energy-efficient house in rural Alaska. *Energies* **2019**, *12*, 1732. [CrossRef]
151. Dell'Anna, F.; Bottero, M.; Becchio, C.; Corgnati, S.P.; Mondini, G. Designing a decision support system to evaluate the environmental and extra-economic performances of a nearly zero-energy building. *Smart Sustain. Built Environ.* **2020**, *9*, 413–442. [CrossRef]
152. Ahmad, M.R.; Chen, B.; Dai, J.; Kazmi, S.M.S.; Munir, M.J. Evolutionary Artificial Intelligence Approach for performance prediction of bio-composites. *Constr. Build. Mater.* **2021**, *290*, 123254. [CrossRef]
153. Zhang, X.; Platten, A.; Shen, L. Green property development practice in China: Costs and barriers. *Build. Environ.* **2011**, *46*, 2153–2160. [CrossRef]
154. Mocanu, E.; Nguyen, P.H.; Gibescu, M.; Kling, W.L. Deep learning for estimating building energy consumption. *Sustain. Energy Grids Netw.* **2018**, *6*, 91–99. [CrossRef]
155. Li, W.; Liu, X.; Chen, J. The role of IoT and smart sensors in improving building energy efficiency: A review. *Sustain. Cities Soc.* **2019**, *45*, 543–552. [CrossRef]
156. Dixit, M.K.; Fernández-Solís, J.L.; Lavy, S.; Culp, C.H. Need for an embodied energy measurement protocol for buildings: A review paper. *Renew. Sustain. Energy Rev.* **2012**, *16*, 3730–3743. [CrossRef]
157. Hammond, G.P.; Jones, C.I. Embodied energy and carbon in construction materials. *Proc. Inst. Civ. Eng. Energy* **2008**, *161*, 87–98. [CrossRef]
158. Cabeza, L.F.; Rincón, L.; Vilariño, V.; Pérez, G.; Castell, A. Life cycle assessment (LCA) and life cycle energy analysis (LCEA) of buildings and the building sector: A review. *Renew. Sustain. Energy Rev.* **2014**, *29*, 394–416. [CrossRef]
159. Yang, X.; Xu, T.; Zhao, Y. The challenges and opportunities of artificial intelligence for sustainable development in the construction industry. *Sustainability* **2020**, *12*, 6053. [CrossRef]
160. Rai, A. Explainable AI: From black box to glass box. *J. Acad. Mark. Sci.* **2020**, *48*, 137–141. [CrossRef]

161. Moinat, M.; Papez, V.; Denaxas, S. Data Integration and Harmonisation. In *Clinical Applications of Artificial Intelligence in Real-World Data*; Springer: Cham, Switzerland, 2023; pp. 51–67. [CrossRef]
162. Chen, M.; Mao, S.; Liu, Y. Big data: A survey. *Mob. Netw. Appl.* **2018**, *23*, 171–209. [CrossRef]

Disclaimer/Publisher's Note: The statements, opinions and data contained in all publications are solely those of the individual author(s) and contributor(s) and not of MDPI and/or the editor(s). MDPI and/or the editor(s) disclaim responsibility for any injury to people or property resulting from any ideas, methods, instructions or products referred to in the content.

buildings

MDPI

Article

Urban Management for Building-Sector Decarbonization: Focusing on the Role of Low-Carbon Policies

Jianxin Tang [1], Pengpeng Yang [2], Kai Tang [3,*] and Sibo Wang [4,*]

[1] Shanghai Construction Management Vocational College, Shanghai 201702, China; jiantjx02@163.com
[2] Architecture and Urban Planning Design & Research Institute of HUST Co., Ltd., Wuhan 430070, China; pengpeng_young@163.com
[3] Dibenqi Architectural Design Firm (Shanghai) Co., Ltd., Shanghai 201101, China
[4] School of Architecture and Urban Planning, Huazhong University of Science and Technology, Wuhan 430070, China
* Correspondence: kai.tang@globaldbc.com (K.T.); 1992010206@hust.edu.cn (S.W.)

Abstract: The building sector is a major source of anthropogenic carbon emissions worldwide. While existing studies have extensively explored the socioeconomic and technological impacts on carbon emissions generated from building operations, few have assessed the effectiveness of low-carbon policies in curbing the increasing trend of building sector carbon emissions. This study examines the impacts of low-carbon policy intensity on building sector carbon emissions using a two-way fixed effects model on a 6-year panel (2015–2020) dataset for 286 cities in China. Our findings indicate that, on average, the aggregated intensity of low-carbon policies fails to pose any significant impacts on carbon emissions from building operations. This is partly due to the variations in different types of policy. Specifically, a 10% increase in the intensity of energy conservation policy results in a 0.05% decrease in carbon emissions, whereas capacity utilization policies are associated with an increase in building-operation carbon emissions. Moreover, these policy–emission relationships vary across building types and end-use sources. In particular, energy conservation policies are negatively associated with emissions from cooking and heating, but positively related to emissions generated from appliances and cooling. In comparison, capacity utilization policies tend to encourage additional emissions from most sources. This study highlights the partial effectiveness of energy conservation policies in curbing building sector carbon emissions and underscores the need for additional efforts in tackling the rebound effects to realize building sector decarbonization.

Keywords: building-operation carbon emissions; low-carbon policy; urban management; heterogenous effects; two-way fixed effects model

1. Introduction

Anthropogenic carbon emissions are known as a leading cause of global climate change, severely challenging the goal of achieving sustainable development [1–4]. The building operations sector is becoming a main source of anthropogenic carbon emissions worldwide. For example, it is estimated that around 36% of the total carbon emissions globally come from the building sector, and this number is slightly higher in developed countries than in developing countries [5]. Given this background, a series of policies have been employed to curb the increasing trend of carbon emissions generated from the building operations sector. For example, in Europe, the Emission Trading Scheme (ETS) has been designed to mitigate carbon emissions from the building sector [6]. Similar ETS projects have also been adopted in several pilots in China, together with some other carbon emission regulations for the building sector [7]. Despite the wide application of these policies and instruments, rigorous evaluations of their effectiveness in mitigating carbon emissions are still sparse, hindering sustainable urban and environmental management.

Among the existing literature, mixed evidence on the impacts of policy measures on building sector emissions has been documented. On the one hand, a group of studies have shown that policy measures can effectively mitigate carbon emissions generated from the building operations sector [8,9]. On the other hand, another group of studies suggests that the adoption of policy measures or other instruments does not have significant emission-mitigation impacts and may even bring about emission-intensifying effects [10,11]. Such disparities may be explained by three reasons. First, there is a dualistic nature to certain types of policy measures affecting carbon emissions. For example, while improved energy efficiency may lead to energy savings and, subsequently, carbon emission reductions [12,13], it is also likely to induce additional energy use via the rebound effect, which have been widely documented in the Europe and elsewhere in the world [14,15]. Second, different sources of energy consumption may not respond in the same way to certain policy measures. For example, China's ETS is only applicable to carbon emissions generated from public and commercial buildings [16]. Therefore, it is less likely that the ETS could curb the increasing trend of carbon emissions from residential buildings at the current stage. Third, since the estimation results reflect the average treatment effects of policy measures, they can hardly capture the spatial heterogeneities underlying policy impacts on carbon emissions. For these reasons, a comprehensive understanding of how various types of policy measures affect carbon emissions generated from different sources of energy use for different cities during building operations holds significant value in paving the way toward building-sector decarbonization, motivating this study.

In this study, we ask four research questions: (1) how large are the impacts of low-carbon policy intensity on building sector carbon emissions?; (2) how do such impacts vary across cities?; (3) how do different types of policies affect building sector carbon emissions?; and (4) how do different types of policies affect carbon emissions from various end-use sources? For the analysis, we examine the impacts of the intensity of different emission-related policies on the carbon emissions generated from various sources of building operations, using a two-way fixed effects model on a 6-year panel dataset (2015–2020) for 286 Chinese cities.

We focus on China for two reasons. First, China, at present, is the largest carbon emitter in the world and has made an emission mitigation commitment to the world with its ambitious carbon emission mitigation plans [17,18]. Considering the carbon emissions and climate change are regional or even global issues, the findings drawn from this research are expected to also hold international implications. Second, as one of the largest developing countries in the world, China is experiencing rapid urbanization and soaring demand for buildings, particularly in the large cities with substantial migration inflows [19,20]. In this sense, our study would carry long-lasting policy and practical implications for the China's growing building sector.

The academic merit of this study is twofold. First, this study enriches the literature on building sector carbon emission control by quantifying the emission impacts of low-carbon policies. Previous studies have focused mainly on socioeconomic and technological impacts on building sector carbon emissions while neglecting the role of policy measures [21,22]. Second, the use of up-to-date policy intensity and building operation carbon emissions data for the analysis allows us to explore more details underlying the policy impacts on building sector carbon emissions. In particular, the policy intensity measures extracted from textual data used in this study can accurately capture variations in policy stringency, and the emissions data provide information on building types and end-use sources.

2. Materials and Methods

2.1. Empirical Model

A two-way fixed effects model is adopted to examine the impacts of the intensity of low-carbon policies on building sector carbon emissions. The two-way fixed effects model has the advantage of addressing the estimation biases brought by unobserved time-invariant confounding variables (e.g., topographic features, cultural factors) and time-varying confounders that are applied to all samples (e.g., implementation of national-level standards and regulations), using city-and time-specific fixed effects terms. Specifically, a baseline model is specified as follows:

$$CE_{it} = \beta_0 + \beta_1 PI_{it} + \mathbf{X}_{it}\gamma + \sigma_i + \mu_t + \varepsilon_{it} \quad (1)$$

where CE_{it} is the carbon emissions generated from the building-operations sector (in logarithm form) in city i at time t; PI_{it} is the policy intensity for low-carbon policies implemented at national, provincial, and prefecture levels (in logarithm form) in city i at time t; $\mathbf{X}_{it}$ is a vector of control variables (in logarithm form), which are key drivers of energy consumption as suggested in previous empirical studies [23–25], including population, GDP per capita, GDP share of tertiary industry, temperature, wind speed, and precipitation, in city i at time t; β_0, β_1, γ are the coefficient or vector of coefficients to be estimated; σ_i and μ_t are the city- and year-fixed effects terms; and ε_{it} is the error term.

Incorporating climatic variables into the regression analysis investigating the policy impacts on building sector carbon emissions is essential for several reasons. First, climatic factors, such as temperature, humidity, and precipitation, significantly influence energy demand patterns in buildings, particularly for heating and cooling requirements. By accounting for these variables, one can better isolate the effects of specific policies on carbon emissions, ensuring that the analysis accurately reflects the interplay between environmental conditions and energy consumption behaviors. Furthermore, the inclusion of climatic variables enhances the robustness of the model, allowing for a more comprehensive understanding of how external factors may exacerbate or mitigate the effectiveness of low-carbon policies.

Considering that various policies may have different impacts, in terms of the signs and magnitude of the effect on building sector carbon emissions, we further disaggregate the indicator of energy-related policy intensity to four specific policy measures in Equation (2):

$$CE_{it} = \beta_0 + \mathbf{PI}_{it}\boldsymbol{\tau} + \mathbf{X}_{it}\gamma + \sigma_i + \mu_t + \varepsilon_{it} \quad (2)$$

where $\mathbf{PI}_{it}$ is a vector of intensity measures for carbon reduction policies, energy conservation policies, capacity utilization policies, and technology policies in city i at time t; $\boldsymbol{\tau}$ is a vector of coefficients to be estimated; and all other notations are the same as defined in Equation (1).

To examine the heterogeneous effects underlying the policy–emission relationship, we estimate the regression model specified in Equation (3):

$$CE_{it} = \beta_0 + \beta_1 PI_{it} + \beta_2 PI_{it} \cdot D_i + \mathbf{X}_{it}\gamma + \sigma_i + \mu_t + \varepsilon_{it} \quad (3)$$

where D_i is a binary dummy indicating whether city i is among the first half of the samples with a larger population size, a greater GDP per capita, or a higher temperature; β_2 is the coefficient capturing the heterogeneous effects, and all other notations are the same as defined in Equation (1).

2.2. Data

A panel dataset for 286 Chinese cities has been developed for the period 2015–2020 from various public sources. A full list of variables, their definitions, and summary statistics are presented in Table 1, and the spatial distribution of sample cities is displayed in Figure 1a (since over 90% of the population resides in eastern China, socioeconomic data from statistical yearbooks is often missing for western Chinese cities). Specifically, the data on building-related carbon emissions from 2015 to 2020 are estimated and provided by Yu et al. (2024) [26]. This dataset contains the total carbon emissions generated from building operations and disaggregated carbon emissions for different building types (e.g., public and commercial buildings, urban residential buildings, and rural residential buildings) and end-use sources (e.g., appliances, lighting, cooling, cooking, heating) available on a yearly basis at the city level. As shown in Figure 1b, carbon emissions from building operations are generally higher in northern Chinese cities and first- and second-tier cities than others. This may be explained by their greater energy consumption associated with winter heating and higher demand for socioeconomic activities.

Policy intensity data for low-carbon policies from 2007 to 2022 at the city level are calculated and offered by Dong et al. (2024) [27]. Specifically, natural language processing (NLP) techniques are applied to policy documents to develop policy intensity measures. The dataset contains aggregated policy intensity measures for low-carbon policies and disaggregated policy intensity measures for carbon reduction policies, energy conservation policies, capacity utilization policies, and technology policies. As shown in Figure 1c, the aggregated policy intensity for low-carbon policies tends to be higher in the eastern coastal region.

Finally, we collect socioeconomic and climate data from the China Urban Statistical Yearbooks and the National Centers for Environmental Information (NCEI) climate data online, respectively. Specifically, the data on population, GDP per capita, GDP share of tertiary industry, temperature, wind speed, and precipitation are used for the analysis.

Table 1. List of variables, variable definitions, and detailed statistics.

Variable	Definition	Obs	Mean	Std. Dev
Dependent Variables				
TCE	Total carbon emissions from building operation (10^3 tons)	1716	6713.47	9123.20
PCCE	Carbon emissions from public and commercial building operations (10^3 tons)	1716	2855.58	4640.28
URCE	Carbon emissions from urban residential building operation (10^3 tons)	1716	2630.85	4018.84
RRCE	Carbon emissions from rural residential building operation (10^3 tons)	1716	1227.05	1182.79
ACE	Carbon emissions from appliances (10^3 tons)	1716	1482.89	2417.97
LCE	Carbon emissions from lighting (10^3 tons)	1716	751.65	897.63
CCE	Carbon emissions from cooling (10^3 tons)	1716	571.96	1036.11
CWCE	Carbon emissions from cooking and water heating (10^3 tons)	1716	1203.48	1433.17
DHCE	Carbon emissions from distributed heating (10^3 tons)	1716	1202.77	1627.34
CHCE	Carbon emissions from central heating (10^3 tons)	1716	1501.12	3518.24
Policy Intensity Measures				
PIA	Total intensity for all low-carbon policies	1716	52.91	42.85
PICR	Intensity for carbon reduction policies	1716	19.81	16.59
PIEC	Intensity for energy conservation policies	1716	15.03	14.07
PICU	Intensity for capacity utilization policies	1716	8.13	8.78
PITEC	Intensity for technology policies	1716	9.95	12.32
Covariates				
POP	Population within the city proper (million persons)	1716	453.02	325.89
GDPPC	Gross domestic product per capita (RMB)	1716	58,580.44	33,940.44
T_SHR	GDP share of tertiary industry (%)	1716	45.65	8.90
TEMP	Annual average temperature (°C)	1716	15.87	5.34
WS	Annual average wind speed (m/s)	1716	4.92	0.98
PREC	Annual total precipitation (mm)	1716	1156.13	589.90

Note: Obs refers to number of observations; std. dev refers to standard deviations.

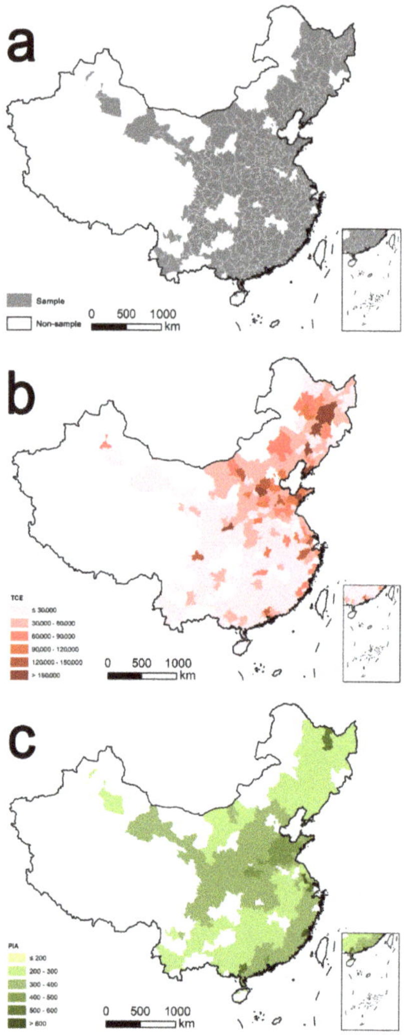

Figure 1. Spatial distribution of sample cities, building sector carbon emissions, and policy intensity for low-carbon policies. (**a**) Spatial distribution of sample cities; (**b**) building sector carbon emissions of sample cities; and (**c**) policy intensity for low-carbon policies of sample cities.

3. Results

3.1. Baseline Results

This section attempts to answer the first research question. Specifically, the baseline model results show that the aggregated policy intensity for low-carbon policies fails to pose any significant impacts on carbon emissions generated from building operations (Table 2). As shown in column 1, the coefficient of interest is negative but not significant at the 10% or higher level. This may be understood in several ways. First, there are different types of low-carbon policies that may potentially have conflicting impacts on building-operation carbon emissions. Second, residents in different cities may respond in different ways to low-carbon policies. Third, even within the same city, people's behavioral responses to low-carbon policies may vary across building types and end-use sources. Finally, there may be data limitation and policy design issues. For example, the document-based policy intensity data may not accurately capture the policy design features.

The results presented in column 2 confirm the first hypothesis. Among the four different types of policy measures, energy conservation policies tend to be negatively associated with building-operation carbon emissions, reflecting the effectiveness of this type of policy in promoting carbon neutrality in China's context. Specifically, a 1% increase in the intensity of energy conservation policies leads to a −0.005% decrease in carbon emissions from building operations. In contrast, capacity utilization policies are positively associated with carbon emissions generated from building operations sector, with a coefficient of 0.010. This may be explained by the rebound effect, which refers to an increasing demand for energy associated with improved energy efficiency [28]. The remaining two types of policies, carbon reduction policies and technology policies, do not exhibit any significant impacts on building operation-related carbon emissions. Furthermore, to address potential multicollinearity problems, we calculate the variance inflation factor (VIF) value for each independent variable. As shown in the last column of Table 2, all estimated VIF values are substantially below the threshold of 10 [29,30], suggesting that the independent variables included in the regression model are not likely to be highly correlated with each other and cause estimation biases.

As for control variables, in both columns, the coefficients for population, GDP per capita, and wind speed are positive and statistically significant at the 5% or higher level, and the coefficient for precipitation is negative and statistically significant at the 1% level, following our expectations.

Table 2. Model results: baseline regression.

	LHS Variable: TCE		
	(1)	**(2)**	**VIF**
Policy Intensity Measures			
PIA	−0.004 (0.005)		
PICR		0.001 (0.004)	1.86
PIEC		−0.005 * (0.003)	1.86
PICU		0.010 *** (0.003)	1.16
PITEC		0.004 (0.004)	1.89
Covariates			
POP	0.668 *** (0.151)	0.657 *** (0.148)	1.21
GDPPC	0.230 *** (0.044)	0.236 *** (0.043)	1.25
T_SHR	0.090 (0.062)	0.092 (0.062)	1.27
TEMP	0.038 (0.029)	0.040 (0.029)	1.83
WS	0.166 ** (0.068)	0.170 ** (0.068)	1.39
PREC	−0.106 *** (0.018)	−0.107 *** (0.018)	1.60
City fixed effects	YES	YES	
Year fixed effects	YES	YES	
Constant	0.418 (1.407)	0.404 (1.389)	
R^2	0.990	0.990	
Observations	1716	1716	

Note: * 10% statistical significance; ** 5% statistical significance; *** 1% statistical significance; standard errors (in parentheses) are clustered at the city level.

3.2. Heterogeneous Effects

In this section, we investigate the spatial heterogeneities in the carbon emission effects of low-carbon policy intensity, answering the second research question. As demonstrated in Figure 2a,b, the estimated carbon emission impacts of policy intensity vary across space. Specifically, most cities in the northeast provinces (Heilongjiang, Jilin, and Liaoning) show positive coefficients, while most cities in the Guangxi, Henan, Jiangsu, Jiangxi, and Shandong provinces have negative coefficients (Figure 2c). This may be explained by the higher energy demand for winter heating and slow economic growth in the northeastern cities.

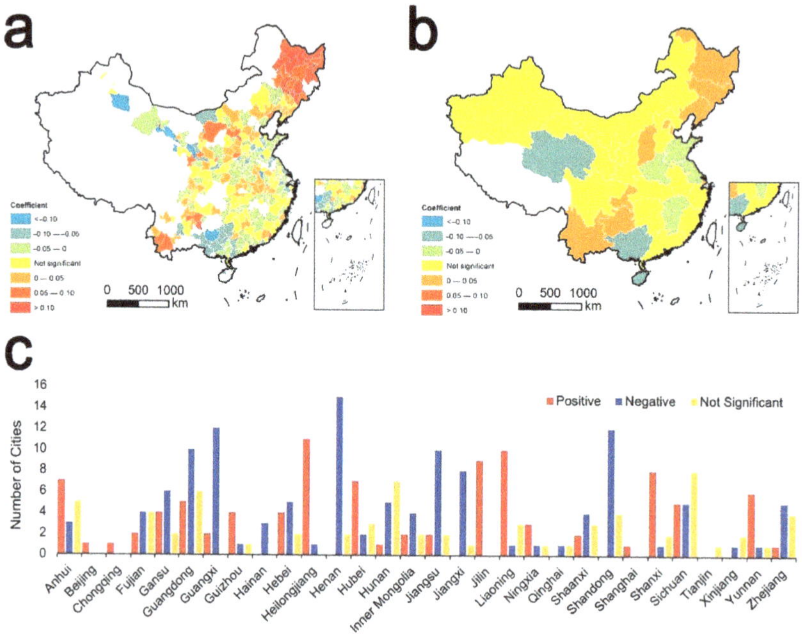

Figure 2. City- and province-specific estimates of policy intensity impacts on building sector carbon emissions. (**a**) City-specific estimates of the PIA impact on TCE; (**b**) province-specific estimates of the PIA impact on TCE; and (**c**) number of cities with positive, negative, and not significant coefficients within each province.

Then, we further investigate whether the spatial heterogeneity in the carbon emission impacts of policy intensity can be explained by socioeconomic and climate factors (Table 3). Among the three models displayed in columns 1 to 3, the coefficients for the interaction terms, PIA × H_GDPPC and H_TEMP, are negative and significant at the 1% level, suggesting that low-carbon policies have greater emission-mitigation impacts in wealthier and warmer cities, with estimated elasticities of −0.016 and −0.022, respectively. At one level, residents in wealthier and larger cities usually have a higher awareness of environmental problems [31,32]. At another level, energy demand for winter heating, a main source of building-operation carbon emissions, is inelastic to policy measures in cities with cold winters [33,34]. In addition, the coefficient for the interaction term, PIA × H_POP, is not significant at the 10% level, suggesting that the policy impacts on carbon emissions are indifferent across cities with various population sizes.

Table 3. Model results: heterogeneous effect.

	LHS Variable: TCE		
	(1)	**(2)**	**(3)**
Policy Intensity Measures			
PIA	−0.001 (0.006)	0.001 (0.006)	0.010 (0.007)
Interaction terms			
PIA × H_POP	−0.008 (0.006)		
PIA × H_GDPPC		−0.016 *** (0.006)	
PIA × H_TEMP			−0.022 *** (0.006)
Covariates			
POP	0.659 *** (0.149)	0.647 *** (0.147)	0.623 *** (0.142)
GDPPC	0.226 *** (0.044)	0.240 *** (0.044)	0.197 *** (0.044)
T_SHR	0.087 (0.062)	0.109 (0.062)	0.063 (0.061)
TEMP	0.039 (0.029)	0.038 (0.029)	0.043 (0.029)
WS	0.166 ** (0.068)	0.174 ** (0.069)	0.142 ** (0.069)
PREC	−0.106 *** (0.018)	−0.098 *** (0.018)	−0.103 *** (0.018)
City fixed effects	YES	YES	YES
Year fixed effects	YES	YES	YES
Constant	0.545 (1.403)	0.343 (1.376)	1.238 (1.358)
R^2	0.990	0.990	0.990
Observations	1716	1716	1716

Note: ** 5% statistical significance; *** 1% statistical significance; standard errors (in parentheses) are clustered at the city level; H_POP: binary dummy variable equals one if the city has a population size greater than the sample average; H_GDPPC: binary dummy variable equals one if the city has a GDP per capita greater than the sample average; H_TEMP: binary dummy variable equals one if the city has a higher temperature than the sample average.

3.3. Results for Different Building Types

To answer the third research question, we examine the emission effects of low-carbon policy intensity for different building types. Overall, the policy impacts on building-operation carbon emissions vary across building types (Table 4). Specifically, carbon emissions generated from public and commercial buildings are positively affected by the intensity of carbon reduction and capacity utilization policies, with a 1% increase in carbon reduction and capacity utilization policy intensity contributing to a 0.01% and a 0.009% increase in emissions from public and commercial buildings, respectively. In comparison, those generated from rural residential buildings are positively affected by the intensity of capacity utilization and technology policies, with estimated elasticity of 0.014 and 0.015, respectively. However, carbon emissions generated from urban residential buildings are not significantly impacted by low-carbon policies. These results suggest that China's low-carbon policies, in general, cannot mitigate carbon emissions from building operations and even result in a rebound effect, leading to increased carbon emissions.

Table 4. Model results: building type.

	LHS Variable: PCCE	LHS Variable: URCE	LHS Variable: RRCE
	(1)	(2)	(3)
Policy Intensity Measures			
PICR	0.010 * (0.006)	−0.000 (0.006)	−0.003 (0.005)
PIEC	−0.005 (0.004)	−0.005 (0.004)	−0.005 (0.004)
PICU	0.009 ** (0.004)	0.007 (0.005)	0.014 *** (0.005)
PITEC	0.000 (0.005)	0.004 (0.005)	0.015 *** (0.005)
Covariates			
POP	0.829 *** (0.200)	0.360 *** (0.130)	0.448 *** (0.137)
GDPPC	0.433 *** (0.070)	0.187 *** (0.050)	0.267 (0.038)
T_SHR	0.130 (0.100)	0.260 *** (0.070)	−0.178 ** (0.070)
TEMP	0.004 (0.042)	0.050 (0.043)	0.048 * (0.027)
WS	0.233 *** (0.086)	0.187 ** (0.082)	0.176 ** (0.088)
PREC	−0.114 *** (0.018)	−0.086 *** (0.024)	−0.095 *** (0.021)
City fixed effects	YES	YES	YES
Year fixed effects	YES	YES	YES
Constant	−4.184 ** (1.900)	1.459 (1.338)	3.677 *** (1.334)
R^2	0.983	0.987	0.985
Observations	1716	1716	1716

Note: * 10% statistical significance; ** 5% statistical significance; *** 1% statistical significance; standard errors (in parentheses) are clustered at the city level.

3.4. Results for Different End-Uses

Finally, we answer the last research question by showing how the intensity of low-carbon policies affects carbon emissions generated from different end-use sources during building operations (Table 5). As shown in columns 1 and 2, energy conservation and capacity utilization policies serve as emission abators and intensifiers, respectively, for cooling, water heating, and distributed heating. This coincides with the results presented in Table 2, reflecting that energy conservation policies are a more favorable choice than capacity utilization in mitigating carbon emissions generated from cooling and heating during building operations. Additionally, the intensity of low-carbon technology policies is negatively associated with carbon emissions generated from central heating. This is probably because central heating is subsidized by the central government and will be influenced more by technological changes rather than residents' behavioral changes.

Furthermore, as shown in Table 6, energy conservation policies are found to be positively associated with carbon emissions generated from the use of appliances and cooking, and capacity utilization policies are positively associated with the use of appliances, lighting, and cooking. These results reflect that low-carbon policies tend to produce rebound effects relating to carbon emissions generated from end-use sources that are central to residents' daily lives.

Table 5. Model results: end-use sources (cooling and heating).

	LHS Variable: CWCE	LHS Variable: DHCE	LHS Variable: CHCE
	(1)	(2)	(3)
Policy Intensity Measures			
PICR	0.003 (0.006)	0.007 (0.006)	−0.009 (0.019)
PIEC	−0.015 *** (0.005)	−0.012 ** (0.006)	0.011 (0.014)
PICU	0.020 *** (0.006)	0.015 ** (0.007)	0.035 (0.025)
PITEC	0.006 (0.006)	0.001 (0.005)	−0.026 * (0.016)
Covariates			
POP (log)	0.703 *** (0.179)	0.968 *** (0.240)	0.694 (0.559)
GDPPC (log)	0.279 *** (0.067)	0.409 *** (0.094)	−0.155 (0.145)
T_SHR (log)	0.021 (0.105)	−0.087 (0.115)	−0.048 (0.220)
TEMP (log)	0.073 * (0.041)	−0.120 *** (0.043)	−0.225 *** (0.086)
WS (log)	0.042 (0.103)	0.188 (0.120)	−0.345 (0.264)
PREC (log)	−0.055 * (0.028)	−0.184 *** (0.034)	0.021 (0.083)
City fixed effects	YES	YES	YES
Year fixed effects	YES	YES	YES
Constant	−2.039 (1.838)	−5.153 ** (2.397)	0.710 (4.675)
R^2	0.973	0.996	0.982
Observations	1716	1716	1716

Note: * 10% statistical significance; ** 5% statistical significance; *** 1% statistical significance; standard errors (in parentheses) are clustered at the city level.

Table 6. Model results: end-use sources (others).

	LHS Variable: ACE	LHS Variable: LCE	LHS Variable: CCE
	(1)	(2)	(3)
Policy Intensity Measures			
PICR	−0.002 (0.008)	−0.001 (0.003)	−0.002 (0.005)
PIEC	0.016 ** (0.006)	0.000 (0.002)	0.009 ** (0.004)
PICU	0.013 ** (0.006)	0.006 *** (0.002)	0.010 *** (0.003)
PITEC	−0.000 (0.006)	0.003 (0.002)	0.001 (0.004)
Covariates			
POP (log)	−0.053 (0.147)	0.467 *** (0.096)	0.256 *** (0.084)
GDPPC (log)	0.273 *** (0.068)	0.124 *** (0.027)	0.134 *** (0.046)
T_SHR (log)	0.234 (0.150)	−0.063 (0.046)	0.056 (0.109)
TEMP (log)	0.014 (0.051)	0.029 (0.018)	0.013 (0.029)
WS (log)	0.239 * (0.130)	0.155 *** (0.044)	0.118 ** (0.058)
PREC (log)	−0.040 (0.031)	−0.037 *** (0.012)	−0.049 ** (0.021)
City fixed effects	YES	YES	YES
Year fixed effects	YES	YES	YES
Constant	3.061 * (1.815)	1.242 (0.999)	1.944 (1.243)
R^2	0.979	0.996	0.993
Observations	1716	1716	1716

Note: * 10% statistical significance; ** 5% statistical significance; *** 1% statistical significance; standard errors (in parentheses) are clustered at the city level.

4. Discussion

There are several important policy and urban management implications that can be drawn from the findings of this study. First, by comparing the effectiveness of different low-carbon policies in curbing building-related carbon emissions, we find that implementing energy conservation policies tends to mitigate carbon emissions generated from building operations, while capacity utilization policies are associated with rebound effects that drive up carbon emissions (Figure 3). Specifically, a 1% increase in energy conservation policy intensity would result in a 0.005%, a 0.015%, and a 0.012% decrease in total building-operation carbon emissions, carbon emissions from cooking and water heating, and carbon emissions from distributed heating, respectively, despite the fact that it could slightly increase carbon emissions from appliances and cooling. In contrast, a 1% increase in capacity utilization policy intensity tends to contribute to a 0.006% to 0.020% increase in carbon emissions from different building operations-related sources. This highlights the need to further promote energy conservation policies to curb the increasing trend of building sector carbon emissions and calls for additional efforts from policymakers to further consider rebound effects in the building sector when designing and refining capacity utilization policies. Addressing the rebound effect necessitates a multi-faceted approach that incorporates demand-side management strategies, such as incentivizing behavioral changes among occupants, which can mitigate the unintended consequences of increased capacity utilization. Policymakers should also explore the implementation of limits on energy consumption per capita in certain high-demand areas, ensuring that energy efficiency gains are not offset by increased usage.

In addition, it is also reflected in Figure 3 that the policy impacts on building-related carbon emissions can vary across building types and end-use sources. For example, the intensity of technology policies is positively associated with carbon emissions generated from rural residential buildings, but is negatively associated with those from central heating. This underscores the necessity for policy refinements targeting specific emission sources. Policymakers should invest in the research and development of targeted interventions that consider the unique characteristics of various building types, ensuring that policies are not only effective but also equitable across diverse populations. Furthermore, developing a framework for collaborative governance that involves stakeholders from local communities can enhance the effectiveness of these targeted interventions, ensuring they align with the needs and capacities of different populations.

Furthermore, the results of spatial heterogeneous effects show that the estimated carbon emission impacts vary largely across space. This reveals that context-specific low-carbon policies should be designed to adapt to local conditions. For example, low-carbon policies, which focus on regulating energy demand, fail to deliver any carbon emission mitigation in northeast Chinese cities, partly due to the inelastic heating demand during winter seasons. In this context, providing alternative clean energy for winter heating may be a potential solution that comes from the supply side. Furthermore, integrating building energy efficiency measures with broader urban planning strategies can enhance resilience against climate impacts, aligning with China's national commitments to reduce carbon emissions as outlined in its 14th Five-Year Plan and its long-term goals for carbon neutrality by 2060.

Finally, the findings of this study also have broader implications for China's national and international commitments to climate change mitigation. By adopting and refining effective low-carbon policies, China can not only meet its internal targets but also strengthen its role in global climate agreements, such as the Paris Agreement. This presents an opportunity for China to showcase its leadership in sustainable urban development and carbon management, potentially influencing other nations to adopt similar strategies.

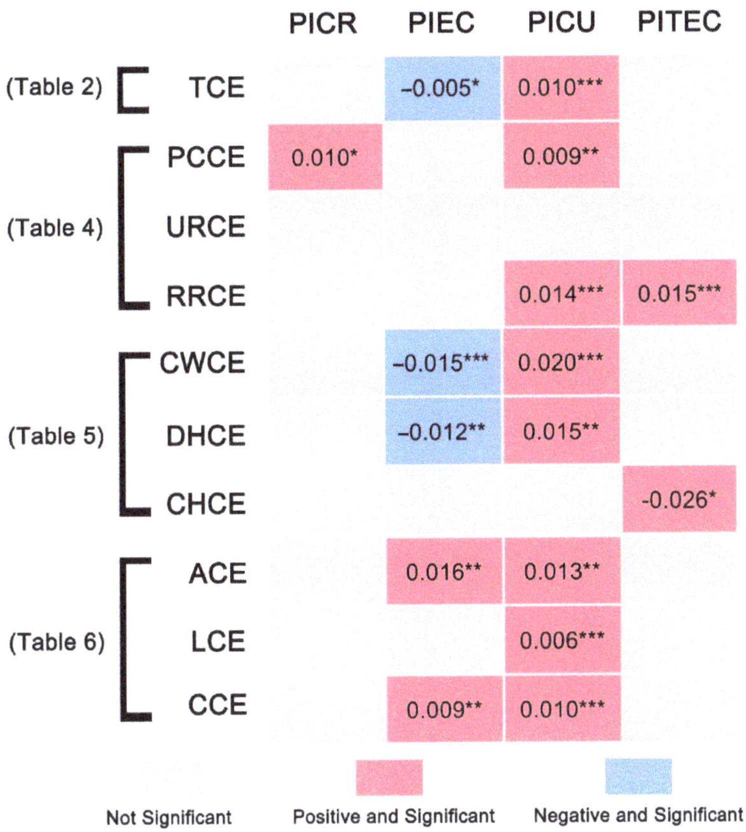

Figure 3. Summary of results. * 10% statistical significance; ** 5% statistical significance; *** 1% statistical significance; PICR: intensity for carbon reduction policies; PIEC: intensity for energy conservation policies; PICE: intensity for capacity utilization policies; PITEC: intensity for technology policies; TCE: total carbon emissions from building operation; PCCE: carbon emissions from public and commercial building operations; URCE: carbon emissions from urban residential building operation; RRCE: carbon emissions from rural residential building operation; ACE: carbon emissions from appliances; LCE: carbon emissions from lighting; CCE: carbon emissions from cooling; CWCE: carbon emissions from cooking and water heating; DHCE: carbon emissions from distributed heating; CHCE: carbon emissions from central heating.

There are also two limitations of this study that are worth noting. First, the findings drawn from this study are only one piece of the puzzle for carbon emission control, since low-carbon policies could also influence carbon emissions generated from other sectors, such as transportation [35,36], construction [37,38], and manufacturing [39,40]. In addition, carbon policies could also bring about broader environmental impacts, such as air quality co-benefits and ecological diversity enhancement, which subsequently deliver positive socioeconomic and health outcomes [41–47], but have not been examined in the present study. Second, some empirical studies suggest that when environmental indicators are used as the dependent variable, there may be potential spatial interdependency problems [48,49].

Looking ahead, future research could explore the integration of low-carbon policies with renewable energy sources and smart grid technologies, assessing their combined effects on carbon emissions across different sectors. Moreover, longitudinal studies that track the long-term impacts of specific policies on urban carbon emissions will provide valuable insights into their effectiveness over time. It would also be beneficial to investigate the social dimensions of policy implementation, such as public perception and acceptance, to ensure that low-carbon strategies are embraced by local communities. This multifaceted approach will enhance understanding of carbon emission dynamics and contribute to the development of more robust, adaptable policies.

5. Conclusions

This study has explored the impact of low-carbon policy intensity on building sector carbon emissions in 286 Chinese cities over a six-year period (2015–2020). Using a two-way fixed effects model, we have enhanced the current understanding of how different types of low-carbon policies affect carbon emissions from building operations.

Our findings show that, on average, the aggregated intensity of low-carbon policies does not significantly impact carbon emissions from building operations. This suggests that simply increasing the overall intensity of low-carbon policies may not be sufficient to achieve meaningful reductions in building-sector emissions.

The effectiveness of low-carbon policies varies significantly depending on the type of policy implemented. Specifically, we found that a 1% increase in the intensity of energy conservation policies resulted in a slight decrease in carbon emissions. This highlights the potential of targeted energy conservation measures to contribute to decarbonization efforts. Conversely, capacity utilization policies are associated with an increase in building-operation carbon emissions. This counterintuitive result suggests that such policies may bring rebound effects and require further adjustments and refinements.

This study also reveals that the relationship between policy intensity and carbon emissions varies across different building types and end-use sources. These variations underscore the importance of tailoring low-carbon policies to the specific characteristics and needs of different building categories.

Overall, this study provides a comprehensive assessment of the impacts of low-carbon policy intensity on building sector carbon emissions in Chinese cities. It offers valuable guidance for urban managers and policymakers striving to promote building sector decarbonization by shedding light on the variations in policy effectiveness. The insights gained from this research can contribute to the development of more effective and sustainable urban management practices, supporting global climate change mitigation.

Author Contributions: Conceptualization, J.T., S.W., P.Y. and K.T.; methodology, J.T. and S.W.; software, J.T. and P.Y.; validation, K.T., J.T. and S.W.; formal analysis, J.T. and S.W.; investigation, J.T. and S.W.; resources, S.W.; data curation, J.T.; writing—original draft preparation, J.T. and P.Y.; writing—review and editing, K.T. and S.W.; visualization, J.T. and S.W.; supervision, K.T. and S.W. All authors have read and agreed to the published version of the manuscript.

Funding: This research received no external funding.

Data Availability Statement: The dataset utilized in this study will be made available upon reasonable request.

Conflicts of Interest: Author Pengpeng Yang was employed by the company Architecture and Urban Planning Design & Research Institute of HUST Co., Ltd. Author Kai Tang was employed by the company Dibenqi Architectural Design Firm (Shanghai) Co., Ltd. The remaining authors declare that the research was conducted in the absence of any commercial or financial relationships that could be construed as a potential conflict of interest.

References

1. Solomon, S.; Plattner, G.K.; Knutti, R.; Friedlingstein, P. Irreversible climate change due to carbon dioxide emissions. *Proc. Natl. Acad. Sci. USA* **2009**, *106*, 1704–1709. [CrossRef] [PubMed]
2. Hansen, J.; Kharecha, P.; Sato, M.; Masson-Delmotte, V.; Ackerman, F.; Beerling, D.J.; Hearty, P.J.; Hoegh-Guldberg, O.; Hsu, S.L.; Parmesan, C.; et al. Assessing "dangerous climate change": Required reduction of carbon emissions to protect young people, future generations and nature. *PLoS ONE* **2013**, *8*, e81648. [CrossRef] [PubMed]

3. Silva, N.; Fuinhas, J.A.; Koengkan, M.; Kazemzadeh, E. What are the causal conditions that lead to high or low environmental performance? A worldwide assessment. *Environ. Impact Assess. Rev.* **2024**, *104*, 107342. [CrossRef]
4. Koengkan, M.; Fuinhas, J.A.; Radulescu, M.; Kazemzadeh, E.; Alavijeh, N.K.; Santiago, R.; Teixeira, M. Assessing the role of financial incentives in promoting eco-friendly houses in the Lisbon Metropolitan Area—Portugal. *Energies* **2023**, *16*, 1839. [CrossRef]
5. Chen, C.; Bi, L. Study on spatio-temporal changes and driving factors of carbon emissions at the building operation stage-A case study of China. *Build. Environ.* **2022**, *219*, 109147. [CrossRef]
6. Braungardt, S.; Bei der Wieden, M.; Kranzl, L. EU emissions trading in the buildings sector–an ex-ante assessment. *Clim. Policy* **2024**, 1–15. [CrossRef]
7. Shen, L.; Song, X.; Wu, Y.; Liao, S.; Zhang, X. Interpretive Structural Modeling based factor analysis on the implementation of Emission Trading System in the Chinese building sector. *J. Clean. Prod.* **2016**, *127*, 214–227. [CrossRef]
8. Tan, X.; Lai, H.; Gu, B.; Zeng, Y.; Li, H. Carbon emission and abatement potential outlook in China's building sector through 2050. *Energy Policy* **2018**, *118*, 429–439. [CrossRef]
9. Ahmed Ali, K.; Ahmad, M.I.; Yusup, Y. Issues, impacts, and mitigations of carbon dioxide emissions in the building sector. *Sustainability* **2020**, *12*, 7427. [CrossRef]
10. Lin, B.; Liu, H. A study on the energy rebound effect of China's residential building energy efficiency. *Energy Build.* **2015**, *86*, 608–618. [CrossRef]
11. Bye, B.; Fæhn, T.; Rosnes, O. Residential energy efficiency policies: Costs, emissions and rebound effects. *Energy* **2018**, *143*, 191–201. [CrossRef]
12. Li, J.; Colombier, M. Managing carbon emissions in China through building energy efficiency. *J. Environ. Manag.* **2009**, *90*, 2436–2447. [CrossRef] [PubMed]
13. de Melo, C.A.; de Martino Jannuzzi, G. Cost-effectiveness of CO_2 emissions reduction through energy efficiency in Brazilian building sector. *Energy Effic.* **2015**, *8*, 815–826. [CrossRef]
14. Greening, L.A.; Greene, D.L.; Difiglio, C. Energy efficiency and consumption—The rebound effect—A survey. *Energy Policy* **2000**, *28*, 389–401. [CrossRef]
15. Bourrelle, J.S. Zero energy buildings and the rebound effect: A solution to the paradox of energy efficiency? *Energy Build.* **2014**, *84*, 633–640. [CrossRef]
16. Song, X.; Pan, C.; Yuan, H.; Wang, Z. Does emission trading system (ETS) deserve further promotion in the building sector? Evidence from China. *Renew. Energy* **2024**, *237*, 121713. [CrossRef]
17. Mallapaty, S. How China could be carbon neutral by mid-century. *Nature* **2020**, *586*, 482–483. [CrossRef]
18. Davidson, M.; Karplus, V.J.; Zhang, D.; Zhang, X. Policies and institutions to support carbon neutrality in China by 2060. *Econ. Energy Environ. Policy.* **2021**, *10*. [CrossRef]
19. Hong, L.; Zhou, N.; Feng, W.; Khanna, N.; Fridley, D.; Zhao, Y.; Sandholt, K. Building stock dynamics and its impacts on materials and energy demand in China. *Energy Policy* **2016**, *94*, 47–55. [CrossRef]
20. Garriga, C.; Hedlund, A.; Tang, Y.; Wang, P. Rural-urban migration, structural transformation, and housing markets in China. *Am. Econ. J. Macroecon.* **2023**, *15*, 413–440. [CrossRef]
21. Erdogan, S. Dynamic nexus between technological innovation and building sector carbon emissions in the BRICS countries. *J. Environ. Manag.* **2021**, *293*, 112780. [CrossRef] [PubMed]
22. Xu, F.; Li, X.; Yang, Z.; Zhu, C. Spatiotemporal characteristics and driving factor analysis of embodied CO_2 emissions in China's building sector. *Energy Policy* **2024**, *188*, 114085. [CrossRef]
23. Shahbaz, M.; Loganathan, N.; Muzaffar, A.T.; Ahmed, K.; Jabran, M.A. How urbanization affects CO_2 emissions in Malaysia? The application of STIRPAT model. *Renew. Sustain. Energy Rev.* **2016**, *57*, 83–93. [CrossRef]
24. Auffhammer, M.; Mansur, E.T. Measuring climatic impacts on energy consumption: A review of the empirical literature. *Energy Econ.* **2014**, *46*, 522–530. [CrossRef]
25. Vázquez, D.; Guimerà, R.; Sales-Pardo, M.; Guillén-Gosálbez, G. Automatic modeling of socioeconomic drivers of energy consumption and pollution using Bayesian symbolic regression. *Sustain. Prod. Consum.* **2022**, *30*, 596–607. [CrossRef]
26. Yu, Y.; You, K.; Cai, W.; Feng, W.; Li, R.; Liu, Q.; Chen, L.; Liu, Y. City-level building operation and end-use carbon emissions dataset from China for 2015–2020. *Sci. Data* **2024**, *11*, 138. [CrossRef]
27. Dong, X.; Wang, C.; Zhang, F.; Zhang, H.; Xia, C. China's low-carbon policy intensity dataset from national-to prefecture-level over 2007–2022. *Sci. Data* **2024**, *11*, 213. [CrossRef]
28. Berkhout, P.H.; Muskens, J.C.; Velthuijsen, J.W. Defining the rebound effect. *Energy Policy* **2000**, *28*, 425–432. [CrossRef]
29. O'brien, R.M. A caution regarding rules of thumb for variance inflation factors. *Qual. Quant.* **2007**, *41*, 673–690. [CrossRef]
30. Ou, Y.; Chen, K.; Ma, L.; He, B.J.; Bao, Z. Coordinating public and government responses to air pollution exposure: A multi-source data fusion approach. *J. Environ. Manag.* **2024**, *370*, 123024. [CrossRef]
31. Zheng, S.; Cao, J.; Kahn, M.E.; Sun, C. Real estate valuation and cross-boundary air pollution externalities: Evidence from Chinese cities. *J. Real Estate Financ. Econ.* **2014**, *48*, 398–414. [CrossRef]
32. Ou, Y.; Zheng, J.; Liang, Y.; Bao, Z. When green transportation backfires: High-speed rail's impact on transport-sector carbon emissions from 315 Chinese cities. *Sustain. Cities Soc.* **2024**, *114*, 105770. [CrossRef]

33. Almond, D.; Chen, Y.; Greenstone, M.; Li, H. Winter heating or clean air? Unintended impacts of China's Huai river policy. *Am. Econ. Rev.* **2009**, *99*, 184–190. [CrossRef]
34. Fan, M.; He, G.; Zhou, M. The winter choke: Coal-fired heating, air pollution, and mortality in China. *J. Health Econ.* **2020**, *71*, 102316. [CrossRef] [PubMed]
35. Karlsson, I.; Rootzén, J.; Johnsson, F. Reaching net-zero carbon emissions in construction supply chains–Analysis of a Swedish road construction project. *Renew. Sustain. Energy Rev.* **2020**, *120*, 109651. [CrossRef]
36. Ou, Y.; Song, W.; Nam, K.M. Metro-line expansions and local air quality in Shenzhen: Focusing on network effects. *Transp. Res. D Transp. Environ.* **2024**, *126*, 103991. [CrossRef]
37. Weigert, M.; Melnyk, O.; Winkler, L.; Raab, J. Carbon emissions of construction processes on urban construction sites. *Sustainability* **2022**, *14*, 12947. [CrossRef]
38. Sun, S.; Chen, Y.; Wang, A.; Liu, X. An evaluation model of carbon emission reduction effect of prefabricated buildings based on cloud model from the perspective of construction supply chain. *Buildings* **2022**, *12*, 1534. [CrossRef]
39. Gutowski, T.G.; Allwood, J.M.; Herrmann, C.; Sahni, S. A global assessment of manufacturing: Economic development, energy use, carbon emissions, and the potential for energy efficiency and materials recycling. *Annu. Rev. Environ. Resour.* **2013**, *38*, 81–106. [CrossRef]
40. Akbarnezhad, A.; Xiao, J. Estimation and minimization of embodied carbon of buildings: A review. *Buildings* **2017**, *7*, 5. [CrossRef]
41. Liu, D.; Lu, Y.; Yang, L. Exploring non-linear effects of environmental factors on the volume of pedestrians of different ages using street view images and computer vision technology. *Travel Behav. Soc.* **2024**, *36*, 100814. [CrossRef]
42. Ou, Y.; Li, X.; Nam, K.M. Rail transit disruptions, traffic generations, and adaptations: Quasi-experimental evidence from Hong Kong. *Transp. Res. D Transp. Environ.* **2024**, *135*, 104381. [CrossRef]
43. Koengkan, M.; Kazemzadeh, E.; Fuinhas, J.A.; Tash, M.N.S. Heterogeneous impact of eco-innovation on premature deaths resulting from indoor and outdoor air pollution: Empirical evidence from EU29 countries. *Environ. Sci. Pollut. Res.* **2023**, *30*, 2298–2314. [CrossRef] [PubMed]
44. Liu, D.; Jiang, Y.; Wang, R.; Lu, Y. Establishing a citywide street tree inventory with street view images and computer vision techniques. *Comput. Environ. Urban Syst.* **2023**, *100*, 101924. [CrossRef]
45. Liu, D.; Wang, R.; Grekousis, G.; Liu, Y.; Lu, Y. Detecting older pedestrians and aging-friendly walkability using computer vision technology and street view imagery. *Comput. Environ. Urban Syst.* **2023**, *105*, 102027. [CrossRef]
46. Wei, D.; Lu, Y.; Zhou, Y.; Ho, H.C.; Jiang, B. The effect of peri-urban parks on life expectancy and socioeconomic inequalities: A 16-year longitudinal study in Hong Kong. *Landsc. Urban Plan.* **2024**, *252*, 105192. [CrossRef]
47. Liu, M.; Wei, D.; Chen, H. Consistency of the relationship between air pollution and the urban form: Evidence from the COVID-19 natural experiment. *Sustain. Cities Soc.* **2022**, *83*, 103972. [CrossRef]
48. Nam, K.M.; Ou, Y.; Kim, E.; Zheng, S. Air pollution and housing values in Korea: A hedonic analysis with long-range transboundary pollution as an instrument. *Environ. Resour. Econ.* **2022**, *82*, 383–407. [CrossRef]
49. Ou, Y.; Kim, E.; Liu, X.; Nam, K.M. Delineating functional regions from road networks: The case of South Korea. Environ. *Plan. B Urban Anal. City Sci.* **2023**, *50*, 1677–1694. [CrossRef]

Disclaimer/Publisher's Note: The statements, opinions and data contained in all publications are solely those of the individual author(s) and contributor(s) and not of MDPI and/or the editor(s). MDPI and/or the editor(s) disclaim responsibility for any injury to people or property resulting from any ideas, methods, instructions or products referred to in the content.

buildings

MDPI

Systematic Review

Artificial Intelligence Approaches to Energy Management in HVAC Systems: A Systematic Review

Seyed Abolfazl Aghili [1], Amin Haji Mohammad Rezaei [1], Mohammadsoroush Tafazzoli [2], Mostafa Khanzadi [1,*] and Morteza Rahbar [3]

[1] School of Civil Engineering, Iran University of Science & Technology, Tehran 16846-13114, Iran; sa_aghili@civileng.iust.ac.ir (S.A.A.); haji_amin@civileng.iust.ac.ir (A.H.M.R.)
[2] Department of Civil Engineering and Construction, Georgia Southern University, 201 COBA Dr, Bldg. 232, Statesboro, GA 30460, USA; mtafazzoli@georgiasouthern.edu
[3] School of Architecture and Environmental Design, Iran University of Science & Technology, Tehran 16846-13114, Iran; rahbarm@iust.ac.ir
* Correspondence: khanzadi@iust.ac.ir

Abstract: Heating, Ventilation, and Air Conditioning (HVAC) systems contribute a considerable share of total global energy consumption and carbon dioxide emissions, putting them at the heart of the issues of decarbonization and removing barriers to achieving net-zero emissions and sustainable development goals. Nevertheless, the effective implementation of artificial intelligence (AI)-based methods to optimize energy efficiency while ensuring occupant comfort in multifarious settings remains to be fully realized. This paper provides a systematic review of state-of-the-art practices (2018 and later) using AI algorithms like machine learning (ML), deep learning (DL), and other computation-based techniques that have been deployed to boost HVAC system performance. The review highlights that AI-driven control strategies can reduce energy consumption by up to 40% by dynamically adapting to environmental conditions and occupancy levels. Compared to other work that focuses on single aspects of HVAC management, this work deals with the methods of control and maintenance in a comprehensive manner. Rather than focusing on abstract applications of machine learning models, this study underlines their applicability in HVAC systems, bridging the science–practice gap. This study highlights the prospective role AI could play, on the one hand, by enhancing HVAC systems' incorporation, energy consumption, and building technologies, while, on the other hand, also addressing the potential uses AI can have in practical applications in the future, bridging gaps and addressing challenges.

Keywords: artificial intelligence (AI); energy management; heating, ventilation, and air conditioning (HVAC) system

1. Introduction

With city growth, a rise in the number of factories and people, and improved living standards, energy use in buildings has sharply increased. Currently, buildings account for more than 40% of the world's annual energy consumption and are responsible for 40% of total carbon dioxide emissions [1,2]. Due to global climate change, we may experience more extreme weather in the future. Therefore, when designing buildings, it is crucial to consider both climate change and future energy demands [3]. Nearly 40% of total energy is consumed by HVAC systems, including air conditioners, heat pumps, ventilation, and refrigeration equipment, making them significant contributors to energy use in commercial buildings [4,5]. Consequently, reducing energy costs and the carbon footprint requires a more efficient use of energy in buildings. Managing a building's energy consumption is complex due to the influence of various factors, such as the building's location, weather conditions, desired indoor temperature and humidity, the number of people present and their time spent indoors, the building's thermal insulation, and its purpose. Since most people spend the majority of their time indoors [6], ensuring indoor comfort is essential for quality of life and productivity.

For HVAC systems to effectively transfer heat and mass, they have to contain specific elements such as air supply ducts, heating/cooling coils, boilers, and chillers. Their operations will depend on conditions such as air temperature, flow velocity, and humidity. These features, among others, are said to be influenced by other external factors that include weather conditions, sunlight, and outdoor temperature. The American Society of Heating, Refrigerating and Air-Conditioning Engineers (ASHRAE) has extensively studied HVAC systems and their applications [6]. HVAC systems consume nearly half of a building's energy, so they must operate effectively to create a comfortable indoor environment. Enhancing the energy efficiency of HVAC systems has, therefore, become vital to improving the overall energy performance of buildings. Many researchers have worked on reducing the energy consumption of HVAC systems while maintaining occupant comfort.

The integration of Building Energy Management Systems (BEMSs) enables autonomous control of building activities, including HVAC [7], lighting, and equipment [8]. Studies suggest that, by employing intelligent technologies and analytics, office buildings can achieve 18% energy savings, while retail buildings can reach 12% [9]. BEMSs are designed to ensure the efficient operation of building systems, equipment, and services, thereby providing a more comfortable environment for occupants while reducing energy use and operational costs. BEMSs have greatly enhanced automated monitoring and control of HVAC systems, reducing the need for manual interventions and improving building efficiency, making them a key development in smart building technologies.

Different computer and software technologies bring more and more industries to automation; AI is playing a great role in it nowadays [1]. With the growing demand for optimization of energy usage, waste reduction, and maintaining comfort–security–health standards, the demand for smart buildings has appeared [2]. With the growing interest in building monitoring and control technologies like AI and the Internet of Things (IoT), the market for "smart buildings" is expected to expand rapidly. Both academia and industry are adopting AI in the building sector to improve design, construction, operations, and maintenance processes [8,9]. These could be AI algorithms merged with sensors that track and control indoor environmental conditions in real time. Such systems automatically analyze data, support predictions about performance into the future, and provide information to make informed decisions satisfactorily [3]. The McKinsey Global Institute [10] conducted a comparative study of AI adoptions across industries, further emphasizing the need for the building and construction sector to accelerate digital transformation and the adoption of AI. Investigating AI techniques is crucial for improving building energy efficiency and addressing related challenges, including the slow pace of AI adoption in this domain. There is a need for comprehensive analysis and evaluation of AI's current applications and its potential in future building solutions. Given rising global energy demands, HVAC systems have become focal points in urban energy management strategies. This paper examines recent advancements in AI technologies designed to enhance HVAC efficiency in line with stricter global standards and policy reforms. Advanced control methods with AI and HVAC optimization through maintenance management have been studied in recent times, and machine learning algorithms were found to be highly effective in optimizing energy efficiency [11,12]. Further, it has been demonstrated through studies that integrating smart control algorithms with real-time analytics is very effective in optimizing HVAC, reducing energy consumption while keeping the interior comfortably cooled [13].

In this paper, we present a clear research hypothesis, demonstrate the novelty of our approach, and identify how our findings contribute to the field of AI-enhanced HVAC systems. Our hypothesis suggests that integrating advanced AI techniques, such as machine learning (ML) and deep learning (DL), into HVAC systems can improve energy savings while maintaining occupant comfort. This hypothesis shapes our research objectives and drives our comprehensive investigation into the application of these technologies across various building types.

Unlike previous studies that often focus on a single aspect of HVAC energy management, this research provides an integrated review of recent advancements in AI for HVAC systems. Our approach examines AI applications across multiple HVAC components, focusing on the dual goals of enhancing energy efficiency and operational effectiveness.

This paper primarily contributes to the academic and practical understanding of AI's potential to transform energy management within buildings. Unlike prior studies that focus on isolated aspects of HVAC management, this work provides a comprehensive overview of both control and maintenance techniques. Rather than focusing on the theoretical aspects of machine learning models, this study emphasizes their practical applications in HVAC systems, bridging the gap between research and implementation. As such, this work is a valuable resource for researchers and practitioners looking to implement AI solutions in pursuit of broader sustainability and energy efficiency goals. This study is a general overview of AI in HVAC optimization, with more emphasis laid on summing up methodologies, trends, and challenges than in listing specific case studies. Practical implementations are mentioned, however, so that readers can refer to real-world applications when needed.

This research offers a comprehensive and systematic literature review (SLR) on the integration of AI techniques in HVAC energy management. The novelty of this study lies in its rigorous synthesis of current research across diverse databases, highlighting AI-driven approaches for optimizing HVAC systems' energy efficiency while ensuring occupant comfort. Beyond merely reviewing existing literature, this study categorizes and assesses various AI algorithms and their applications within this domain, providing deeper insights into both control and maintenance strategies. Additionally, this research innovatively identifies significant research gaps and outlines potential future research directions, emphasizing the role of AI in enhancing both the operational efficiency and predictive maintenance of HVAC systems. Our systematic approach helps map the landscape of AI applications in HVAC energy management, making it a valuable resource for researchers and practitioners working toward improved building energy efficiencies. The significance of this research is further discussed in Section 5.

2. Review Methodology

2.1. Research Questions

An SLR process begins with defining the research questions (RQs). In this paper, we identify the following RQs:

RQ1. *What methods are available that use AI techniques to improve the energy performance of HVAC systems?*

RQ2. *What AI algorithms are used most widely in each method of improving HVAC energy efficiency?*

RQ3. *What are the reported outcomes and effectiveness of AI techniques in improving the energy efficiency of HVAC systems?*

2.2. Literature Search

In computer science and engineering, systematic literature reviews are commonly conducted using string-based searches in various digital libraries. Utilizing multiple databases to encompass a wide range of evidence during the database searching is strongly advised. As seen in Figure 1, this research searches the literature for pertinent contributions linked to the RQs outlined in Section 2.1 utilizing a database search based on three of the most used digital libraries for performing SLRs. Multiple databases, namely Science Direct, Google Scholar, and Scopus are discussed for conducting a systematic literature review (SLR). Since Science Direct has the highest level of coverage for an SLR, allowing it to facilitate reproducible and efficient searches through queries, filters, and hands-on searching, it is considered the main search engine. Unlike Scopus and Web of Science (WoS), Science Direct provides full-text access to several journals and books. Considering that there exists, in parallel, literature coming from other publishers, Google Scholar was utilized as a supplementary search tool, thereby retrieving roughly the same number of citations as WoS and Scopus. Scopus still had wider coverage of engineering records, which made it more appropriate to carry out evidence synthesis on engineering applications. Hence, Scopus was chosen as an additional specialized search database in the present study. Additionally, a forward and backward citation search was performed to ensure that all relevant studies, including those indexed in other databases, such as IEEE Xplore and Web of Science, were included. This approach ensured a comprehensive review of the available literature.

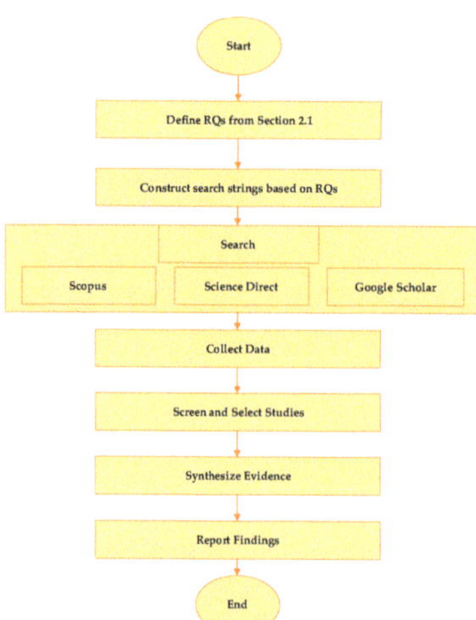

Figure 1. Search script flow diagram.

To conduct corresponding searches on the selected digital libraries described above, "Machine AND Learning" OR "Deep AND Learning" OR "AI" AND "Buildings AND Energy AND HVAC" search strings are defined based on the RQs introduced in Section 2.1.

2.3. Inclusion/Exclusion Criteria

The SLR undertaken in this study relies predominantly on peer-reviewed publications from reputable journals, books, and international conference proceedings, considered reliable sources of scholarly literature. Additionally, relevant articles from the grey literature may also be incorporated. To emphasize the contemporary advancements and trends in HVAC energy management utilizing AI methods and optimization, only articles published within the six-year timeframe from 2018 and later are included throughout this paper.

2.4. Literature Search Results

A comprehensive search of academic literature was carried out utilizing Figure 2 to locate relevant articles from the three specified databases mentioned in Section 2.2, namely Science Direct, Scopus, and Google Scholar. The search yielded a total of 374 articles. Articles that concentrated on applying AI approaches to enhance HVAC energy management were included in the analysis. Only peer-reviewed journal articles, conference papers, and books published by respectable publishers were regarded as high-quality sources to guarantee the research's validity and legitimacy. The inclusion of such sources is illustrated in Table 1.

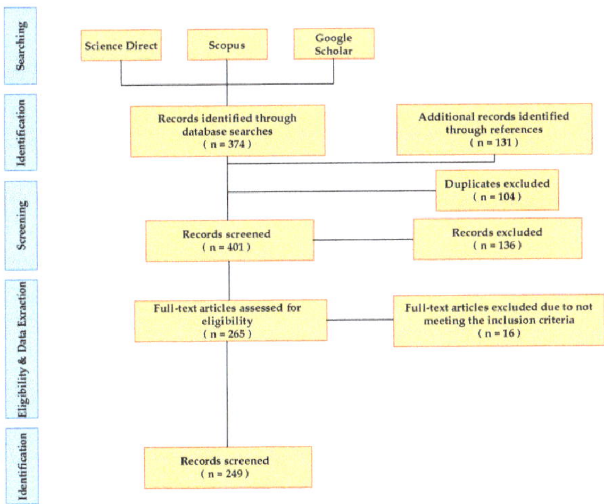

Figure 2. Literature search flow diagram.

Table 1. Top cited publications.

Title	Contribution	Importance	Novelty	Institution
Gradient boosting machine for modeling the energy consumption of commercial buildings [14]	Significant for modeling energy consumption	High impact on commercial building design	Introduces gradient boosting, an advanced machine learning technique, for energy modeling. This method offers a sophisticated approach compared to traditional models, improving prediction accuracy.	Lawrence Berkeley National Laboratory [USA]

Table 1. *Cont.*

Title	Contribution	Importance	Novelty	Institution
Using machine learning techniques for occupancy-prediction-based cooling control in office buildings [15]	Key for energy-efficient cooling control	Critical for reducing office energy usage	Applies machine learning techniques to predict occupancy and optimize cooling control, providing a modern solution for building management and improving energy efficiency.	ETH Zürich [Switzerland]
Whole building energy model for HVAC optimal control: A practical framework based on deep reinforcement learning [16]	Practical framework for HVAC control	High impact on HVAC optimization	Utilizes deep reinforcement learning for HVAC control, a cutting-edge method that offers a significant innovation in optimizing energy usage and system performance.	Carnegie Mellon University [USA]
Deep Reinforcement Learning for Smart Home Energy Management [17]	Advances smart home energy management	Important for residential energy efficiency	Applies deep reinforcement learning to optimize energy management in smart homes, introducing a novel approach to improving energy efficiency and control.	National Natural Science Foundation [China]
Predictive modeling for US commercial building energy use: A comparison of existing statistical and machine learning algorithms using CBECS microdata [18]	Comparison of various models	Significant for improving predictive accuracy	Compares statistical and machine learning models for energy prediction, providing insights into the effectiveness and advancements of newer algorithms in the field.	Northeastern University [USA]
Generative adversarial network for fault detection diagnosis of chillers [19]	Fault detection using advanced techniques	Essential for fault diagnosis in chillers	Innovatively applies generative adversarial networks (GANs) for diagnosing faults in chillers, a novel approach that enhances fault detection by leveraging GANs' unique capabilities.	National University of Singapore [Singapore]
Multi-Agent Deep Reinforcement Learning for HVAC Control in Commercial Buildings [20]	Multi-agent approach to HVAC control	Important for complex building systems	Introduces a multi-agent deep reinforcement learning approach for controlling HVAC systems, offering a sophisticated method for managing complex systems with multiple agents.	Nanjing University [China]

Table 1. *Cont.*

Title	Contribution	Importance	Novelty	Institution
Deep reinforcement learning to optimize indoor temperature control and heating energy consumption in buildings [21]	Optimization of temperature control	Crucial for energy savings in buildings	Uses deep reinforcement learning to optimize indoor temperature and heating energy consumption, presenting a novel and advanced method for improving building energy efficiency.	Politecnico di Torino [Italy]
Intelligent multi-zone residential HVAC control strategy based on deep reinforcement learning [22]	Multi-zone HVAC control strategy	High relevance for residential HVAC systems	Applies deep reinforcement learning to control multiple zones in residential HVAC systems, offering a new approach for enhanced and efficient multi-zone control.	University of Tennessee [USA]
A review of studies applying machine learning models to predict occupancy and window-opening behaviors in smart buildings [23]	Comprehensive review of machine learning	Important for understanding occupancy models	Provides a comprehensive review of machine learning applications for predicting occupancy and window-opening behaviors, summarizing significant advancements and methodologies.	Tianjin University [China]
Gnu-RL: A precocial reinforcement learning solution for building HVAC control using a differentiable MPC policy [24]	New reinforcement learning solution	Significant for optimizing HVAC control	Presents a novel reinforcement learning solution with a differentiable MPC policy for HVAC control, offering an innovative approach to managing building systems.	Carnegie Mellon University [USA]
Intelligent building control systems for thermal comfort and energy-efficiency: A systematic review of artificial intelligence-assisted techniques [10]	Systematic review of AI in building control	Valuable for understanding AI applications	Reviews various AI-assisted techniques for building control, offering valuable insights into innovative methods and their applications in achieving thermal comfort and energy efficiency.	Smart Systems Laboratory [Morocco]
Deep Comfort: Energy-Efficient Thermal Comfort Control in Buildings Via Reinforcement Learning [25]	Energy-efficient comfort control	Important for residential thermal comfort	Utilizes reinforcement learning to achieve energy-efficient thermal comfort control in buildings, presenting a novel approach to balancing comfort and energy use.	Nanjing University of Science and Technology [China]

Table 1. *Cont.*

Title	Contribution	Importance	Novelty	Institution
Towards optimal control of air handling units using deep reinforcement learning and recurrent neural network [26]	Optimal control of air handling units	High impact on HVAC systems	Introduces a novel method combining deep reinforcement learning and recurrent neural networks for optimizing air handling units, providing an advanced approach to building control.	New York University [USA]
OCTOPUS: Deep reinforcement learning for holistic smart building control [27]	Holistic approach to building control	Significant for smart building management	Employs deep reinforcement learning for comprehensive smart building control, offering an innovative approach to managing various aspects of building systems holistically.	University of California [USA]
Model input selection for building heating load prediction: A case study for an office building in Tianjin [28]	Model input selection for heating prediction	Important for accurate heating load forecasting	Proposes a new methodology for selecting model inputs in heating load prediction, providing a novel approach to improving prediction accuracy for building heating systems.	Tianjin University [China]
Comparison of machine learning models for occupancy prediction in residential buildings using connected thermostat data [29]	Comparison of ML models for occupancy	Important for improving occupancy predictions	Compares various machine learning models for predicting occupancy using thermostat data, offering valuable insights into the effectiveness of different algorithms.	University of Toronto [Canada]
Building HVAC scheduling using reinforcement learning via neural network based model approximation [30]	Scheduling HVAC systems using reinforcement learning	Crucial for efficient HVAC scheduling	Applies reinforcement learning with neural network model approximation for HVAC scheduling, presenting a novel approach to optimizing building energy management.	University of Southern California [USA]
Occupancy-based HVAC control systems in buildings: A state-of-the-art review [31]	Comprehensive review of occupancy-based control	Important for understanding current technologies	Reviews state-of-the-art occupancy-based HVAC control systems, summarizing recent advancements and providing a comprehensive overview of current technologies.	Concordia University [Canada]

Table 1. *Cont.*

Title	Contribution	Importance	Novelty	Institution
Deep-learning-based fault detection and diagnosis of air-handling units [32]	Fault detection in HVAC systems	Important for HVAC maintenance and efficiency	Utilizes deep learning techniques for fault detection and diagnosis in air-handling units, offering a novel approach to improving maintenance and system reliability.	National Taipei University of Technology [China]
A novel deep reinforcement learning based methodology for short-term HVAC system energy consumption prediction [33]	Short-term prediction of HVAC energy consumption	Useful for short-term energy management	Introduces a novel deep reinforcement learning methodology for predicting short-term HVAC energy consumption, providing a sophisticated approach to energy forecasting.	Zhengzhou University [China]
Practical implementation and evaluation of deep reinforcement learning control for a radiant heating system [34]	Practical implementation of DRL for heating	Important for radiant heating control	Evaluates the practical implementation of deep reinforcement learning for controlling radiant heating systems, offering a novel and practical approach to system optimization.	
Learning-based CO_2 concentration prediction: Application to indoor air quality control using demand-controlled ventilation [35]	CO_2 prediction for air quality control	Important for indoor air quality management	Uses learning-based methods for predicting CO_2 concentrations to enhance indoor air quality control, presenting a novel approach to demand-controlled ventilation.	Indiana University-Purdue University Indianapolis [USA]
Can HVAC really learn from users? A simulation-based study on the effectiveness of voting for comfort and energy use optimization [36]	Simulation study on user interaction with HVAC	Relevant for user-centric HVAC optimization	Simulates user voting for HVAC control and energy optimization, offering a novel approach to understanding user preferences and their impact on system performance.	Universidade de Lisboa [Portugal]
Application of deep Q-networks for model-free optimal control balancing between different HVAC systems [37]	Model-free HVAC control with deep Q-networks	Significant for balancing multiple HVAC systems	Applies deep Q-networks for model-free control balancing different HVAC systems, introducing a novel method for optimizing complex HVAC controls.	College of Engineering, Seoul National University [Republic of Korea]

Table 1. *Cont.*

Title	Contribution	Importance	Novelty	Institution
Improving Energy Consumption of a Commercial Building with IoT and Machine Learning [38]	IoT and ML for improving energy consumption	Important for integrating IoT in buildings	Combines IoT and machine learning to improve energy consumption in commercial buildings, providing a novel approach to energy management through advanced technologies.	COMSATS Institute of Information Technology [Pakistan]
A deep reinforcement learning approach to using whole building energy model for HVAC optimal control [39]	Deep reinforcement learning for HVAC control	High impact on building energy management	Applies deep reinforcement learning to a whole building energy model for HVAC control, offering an advanced and novel approach to optimizing building systems.	

In contrast, non-peer-reviewed articles were omitted from consideration. Following the removal of duplicates, a total of 401 articles remained. These articles then underwent a screening process, in which we explicitly evaluated their abstracts to determine their relevance to the research questions outlined in Section 2.1. Subsequently, the full text of 267 relevant articles was meticulously analyzed to assess their quality and eligibility. A thorough review led to the exclusion of 16 studies that failed to address the research questions adequately. Our selection criteria also accounted for excluding articles that focused solely on AI methods in energy forecasting, indoor comfort, occupancy detection, control strategies, and fault detection without addressing the enhancement of HVAC energy management. Furthermore, articles were excluded if they demonstrated poor writing quality, needed more methodological clarity, inadequately explained their approaches, or presented insignificant results.

3. Scientometric Analysis

3.1. Publication Source

Journal articles are widely regarded as the most indispensable form of scholarly publication in terms of sources of dissemination. Based on estimations, a total of 56 scientific journals presently exist. Consequently, Figure 3 exclusively showcases the six journals that have published the highest volume of scientific articles. Among these, Building and Environment, Energies, and Energy and Buildings are the top three journals, having produced 16, 16, and 13 articles, respectively. These journals can be considered valuable references in the domain of HVAC systems and the efficacy of building energy systems. Furthermore, the fact that these journals are edited by Elsevier and MDPI enhances their credibility and authority within their specialized fields.

3.2. Most Cited Publications

Furthermore, research was conducted to assess the exact quantity of citations each paper obtained. A selection of over twenty citations from a total of carefully chosen articles that included both targeted research on energy saving and broad HVAC literature were included in the analysis. Examining each publication's contribution, importance, and novelty in Table 1 sheds light on new developments in the area. It explains why some studies have attracted a lot of interest and citations. A significant finding of this analysis is the close relationship between a research paper's citation count and uniqueness. Higher citation counts are typically seen in papers that present new uses of sophisticated machine learning methods, such as generative adversarial networks (GANs), deep reinforcement learning, and gradient boosting. Research that pushes the frontiers of what is feasible in the subject, offers new methodology, or dramatically improves current approaches is highly valued by the academic community based on this pattern.

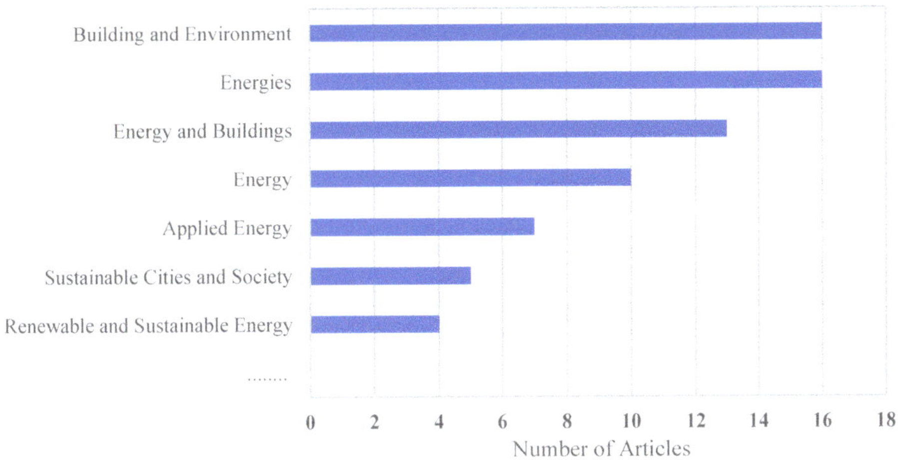

Figure 3. Publication sources.

It also emphasizes how crucial real-world applications are to maximizing the influence of research. Many of the most referenced publications deal with practical issues and provide answers that may be used immediately to build system management. Papers on occupancy-based control techniques, HVAC system defect detection, and predictive modeling for energy consumption, for instance, are essential because they address the pressing concerns of engineers and building managers. In addition to adding to the body of knowledge, these studies offer resources and techniques that may be applied to save expenses, increase energy efficiency, and improve the dependability of building systems. Additionally, the data demonstrate that more attention is often given to research that connects theoretical developments with real-world applications. Research and development efforts in energy management are frequently guided by studies that effectively integrate state-of-the-art AI and machine learning approaches. The studies that present novel frameworks for HVAC control—like those that use deep reinforcement learning—clarify this. They present fresh approaches to optimizing energy usage while preserving comfort and efficiency.

In conclusion, Table 1 shows how the uniqueness of the approaches presented and their applicability to real-world situations are directly related to the influence and significance of research in the field of energy management. Papers that present novel, creative solutions to practical issues are typically mentioned more frequently, which is a testament to their importance in advancing the discipline and benefiting society in concrete ways. This review emphasizes the importance of research that advances theoretical understanding while providing workable solutions for building energy system management.

3.3. Keywords Co-Occurrence

In order to investigate keywords with multiple interpretations, a thorough examination was conducted using varying search combinations, such as "Machine AND Learning", "Deep AND Learning", and "AI", along with "Buildings AND Energy AND HVAC". Subsequently, a network analysis was implemented to establish the interconnectedness among 27 frequently appearing keywords out of the total pool of 832 identified keywords in VOSviewer 1.6.18. The resulting visual representation in Figure 4 depicts the frequency of keyword occurrences through node size, while the thickness of the links illustrates the degree of co-occurrence between the keywords. In the constructed network, comprised of three main entities, namely "Machine Learning", "HVAC", and "Energy Efficiency", the keyword co-occurrence network reveals notable associations between keywords like "Thermal Comfort", "Artificial Intelligence", "Deep Learning", and "Smart Buildings" with the keyword "Machine Learning". This correlation signifies that enhancing efficiency can be achieved by implementing machine learning techniques to assess its influence on the domains represented by these associated keywords. This analysis extends beyond theoretical advancements, offering practical implications for enhanced building codes and energy policies. Stakeholders, including policymakers and technology developers, are poised to benefit from these insights through improved regulatory frameworks and operational efficiencies.

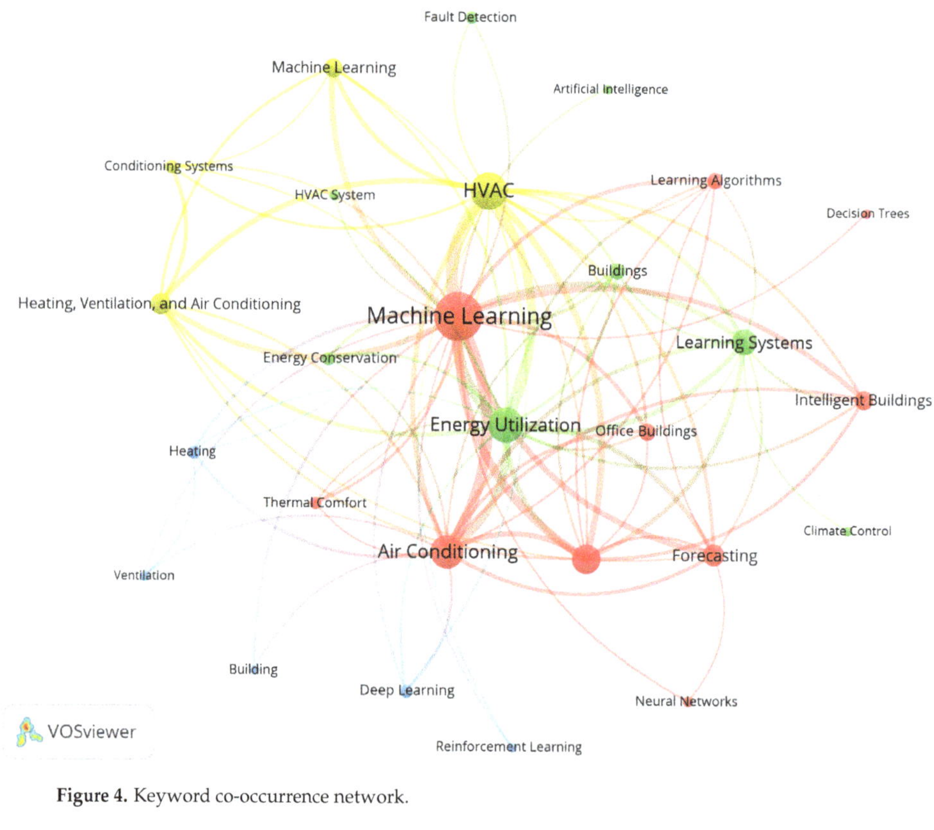

Figure 4. Keyword co-occurrence network.

3.4. Co-Authorship

The network visualization in Figure 5 displays co-authorship connections among countries, with various colors representing different connections. A major hub of this network is the USA, which indicates its extensive collaborative connections with numerous other nations. Lines between each country's nodes show co-authorship links in academic or research publications.

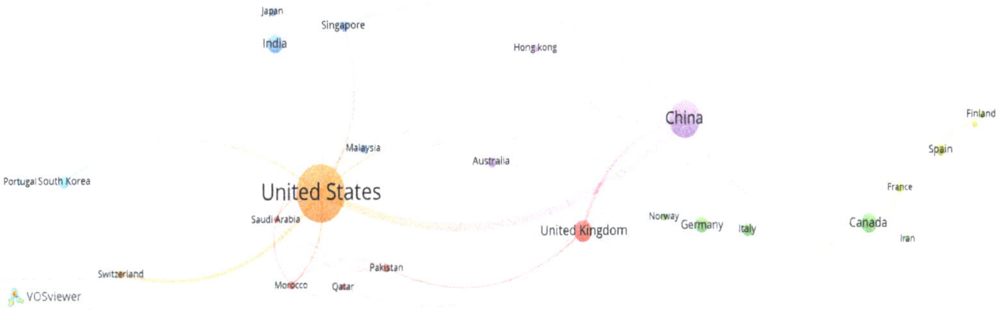

Figure 5. Co-authorship connections among countries.

There is a high volume of international collaborations between countries like the USA, China, and the UK. Different colors may be used to represent different regions or levels of collaboration. Among European countries, such as Germany, France, and the UK, connections could be displayed in one color. In contrast, connections between Asian countries, such as China, Japan, and Republic of Korea, could be displayed in another. There is likely a correlation between the thickness of the lines and the volume or frequency of collaborations, with thicker lines indicating stronger or more frequent collaborations. By identifying which countries are central in international research networks and the nature of their connections, the network effectively illustrates the global landscape of academic collaborations. The authors' varied backgrounds offer insightful local knowledge, guaranteeing that the study encompasses a broad spectrum of viewpoints on AI applications in HVAC systems. This study improves the generalizability of its findings across a range of climatic conditions and building types by integrating knowledge from diverse geographic locations.

3.5. AI-Based Method

Based on empirical research, various AI methods have been studied and are divided into two broad categories: the control and maintenance of HVAC systems. These methods aim to automate the necessary processes for HVAC system operation through the assistance of AI. By reducing manual steps in energy management, these methods have the potential to optimize energy consumption and mitigate energy waste. Specifically, an analysis of Figure 6 reveals that most articles (82.93%) examined the control method of HVAC systems, whereas a smaller percentage (17.07%) focused on their maintenance technique. This suggests that the control method has received more attention in the literature. Various AI techniques are employed to control and maintain HVAC systems. Control strategy, occupancy detection, consumption forecasting, and thermal comfort analysis are utilized in the control method. In contrast, fault detection and diagnosis (FDD) and maintenance management techniques are employed in the maintenance method. Figure 6 further illustrates that many articles (110) have explored control strategies for HVAC systems using AI, highlighting its prevalence as a control technique. Consumption forecasting, another commonly used control technique, is explored in 40 articles. Additionally, FDD, the principal maintenance method technique, is discussed in 37 articles.

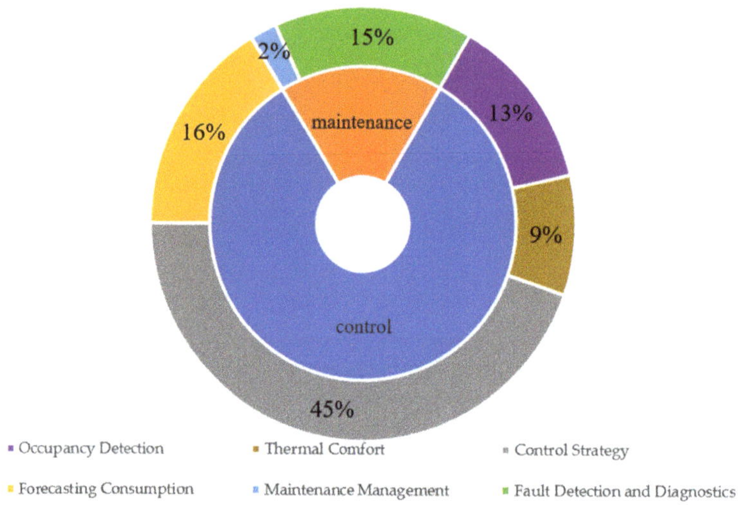

Figure 6. Number of articles per type of AI-based methods and technique.

This distribution of research focus indicates a clear trend favoring AI-driven control strategies over maintenance-related applications in HVAC systems. The dominance of control-focused studies suggests that researchers prioritize immediate energy efficiency improvements, adaptive control mechanisms, and real-time system optimizations, as these provide direct operational benefits. However, while fault detection and predictive maintenance remain less explored, they are crucial for long-term system reliability, cost reduction, and sustainability. The relatively lower number of studies on maintenance techniques highlights an opportunity for future research to develop more advanced AI-based maintenance frameworks, integrating real-time diagnostics, predictive analytics, and automated fault prevention to enhance the overall efficiency and longevity of HVAC systems.

3.6. AI Algorithms

Within the library, a combined number of 44 distinct algorithms are available. To depict the prevalence of these algorithms, Figure 7 presents the top 25 commonly utilized methods, categorized according to control and maintenance fields. Deep reinforcement learning (DRL) emerged as the most favored algorithm for control applications. Following suit, artificial neural network (ANN), Random Forest (RF), and long short-term memory (LSTM) algorithms ranked as the second, third, and fourth most commonly employed methods. Similarly, in maintenance approaches, Support Vector Machines (SVMs) and ANNs were identified as the predominant methods. Each algorithm has distinct advantages and disadvantages, making it suitable for specific tasks. A thorough examination of the primary benefits and drawbacks of several well-known algorithms used in HVAC control and maintenance applications can be found in Table 2. This detailed comparison serves as a valuable resource for selecting the most suitable algorithm for specific tasks across various domains. By understanding the strengths and limitations of each algorithm, practitioners can make informed decisions tailored to the unique requirements and constraints of their applications. This approach enables the effective and efficient implementation of AI models, enhancing the overall performance and management of HVAC systems.

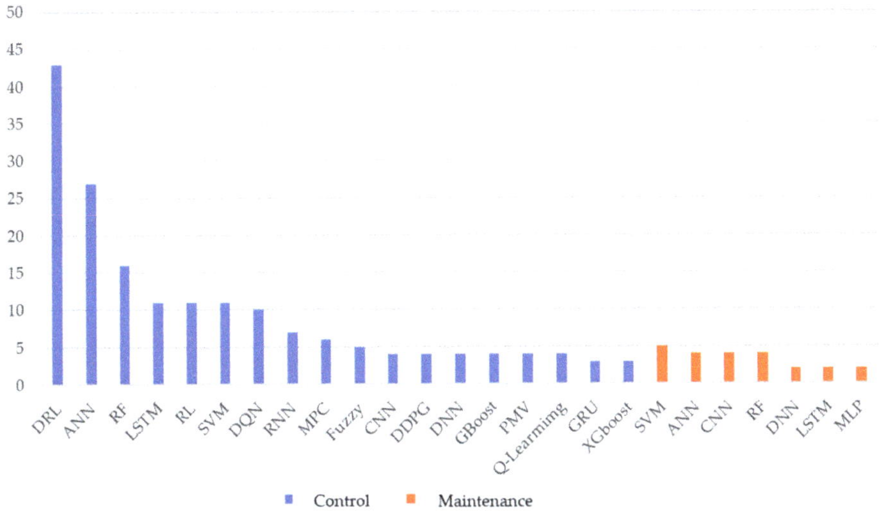

Figure 7. Number of articles per AI algorithm used.

Table 2. Advantages and disadvantages, associated techniques, and impacted components of each algorithm.

Algorithm	Advantages/Disadvantages	Technique	Component
ANN	Advantages: High accuracy in load forecasting; handles non-linear relationships. Disadvantages: Requires extensive training data; computationally intensive.	Forecasting Consumption Control Strategy Occupancy Detection Thermal Comfort	Chillers, Air Handlers [40] Compressors [41] Ventilation Systems [42] Duct Systems, Thermostats
SVM	Advantages: Effective for small datasets; robust to overfitting. Disadvantages: Not ideal for large datasets; kernel selection can be challenging.	Fault Detection and Diagnostics Occupancy Detection Maintenance Management	Airflow Systems [43] Air Quality Monitors [15] Compressors, Fans [44]
DRL	Advantages: Learns optimal control strategies without predefined models; adapts dynamically to uncertainties. Disadvantages: High complexity; slow convergence for large systems.	Control Strategy Forecasting Consumption Thermal Comfort	Entire HVAC System [45] Cooling Towers, Air Handlers [46] Thermostats, Sensors [47]
RF	Advantages: High accuracy for classification problems; resistant to overfitting. Disadvantages: May struggle with high-dimensional data; less interpretable.	Thermal Comfort Fault Detection and Diagnostics Forecasting Consumption	Thermostats [48] Ventilation Systems [49] Zone Equipment, Variable Air Volume (VAV) Systems [50]
LSTM	Advantages: Excellent for multivariate time series data; suitable for energy demand forecasting. Disadvantages: High memory requirements; slower training compared to traditional methods.	Forecasting Consumption Maintenance Management Thermal Comfort	Zone Temperature Controllers, Heat Pumps [40,51] Duct Systems, Heat Exchangers [44] Building Zones [52]

Figure 8, which presents a Sankey diagram, illustrates the flow and relationships between various AI methods, their techniques, and the specific algorithms used in HVAC systems for energy management. This visualization aids in understanding the predominant

AI strategies that have been applied to control and maintenance operations within the sector. It highlights the significant emphasis on control methods, particularly those utilizing deep reinforcement learning (DRL) and machine learning (ML) techniques, which are prominent in optimizing energy efficiency and operational effectiveness. The diagram also shows the linkage between these methods and key outcomes like energy consumption reduction and improved system reliability. By delineating these connections, Figure 8 provides crucial insights into how AI integration influences HVAC performance metrics and assists stakeholders in identifying areas of potential development or further research. This enhanced interpretation helps to bridge the gap between theoretical AI applications and practical HVAC system improvements, offering a clearer understanding of the strategic value of these technologies in building management.

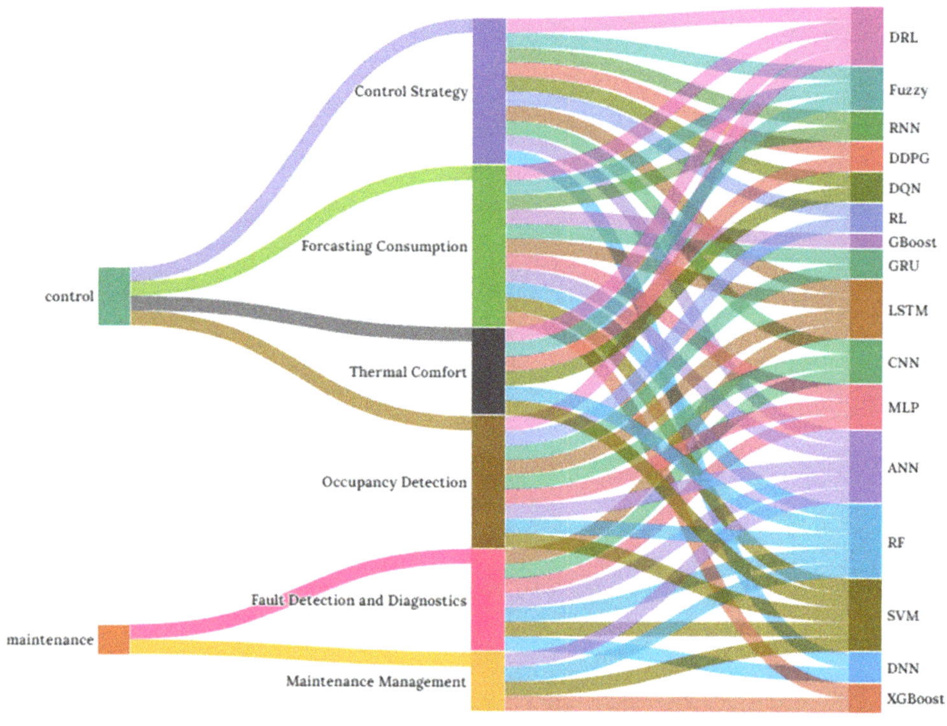

Figure 8. Sankey diagram between methods, techniques, and AI algorithms.

4. Systematic Literature Review

This section performs an exhaustive analysis utilizing the established library through a systematic literature review. After a comprehensive examination of these publications, significant subjects were clustered into primary informational classifications. These categories, their respective topics, and associated publications are exhibited in Table 3. AI models used for HVAC systems vary in complexity, precision, and computational needs. Model selection depends on data availability, real-time computing capacity, and system constraints [5,6]. Data quality issues, system integration, and fit to different building environments are also implementation challenges [7,8]. These challenges are key to delivering the successful implementation of AI-based HVAC systems [10,16].

Table 3. Categories and topics identified.

Categories	Topics	Publications
Control HVAC system	Control Strategy	[5,7,10,16,17,20–22,24,26,27,30,34,37–39,46,47,53–143]
	Occupancy Detection	[15,23,29,31,36,42,144–169]
	Forecasting Consumption	[14,18,28,33,45,50,51,170–202]
	Thermal Comfort	[25,35,41,49,203–220]
Maintenance HVAC system	Fault Detection and Diagnostics	[9,19,32,40,43,221–252]
	Maintenance Management	[12,44,253–255]

4.1. Control HVAC System

This category consolidates subjects about artificial intelligence methods employed in managing HVAC systems, which can potentially enhance energy efficiency within buildings. Within this category, the principal areas of focus include control strategy, occupancy detection, consumption forecasting, and thermal comfort assessment.

4.1.1. Control Strategy

The lack of intelligent and remote controls in managing chillers and air handlers within HVAC systems leads to significant energy consumption. However, the Internet of Things (IoT) has introduced technological advancements that facilitate remote control functionalities for these devices. Alternatively, the progress of machine learning (ML) and artificial intelligence (AI) allows devices to be trained to autonomously make informed decisions that influence interactions between humans and machines [56,57,106]. These advancements are supported by combining wireless sensor data or IoT with user feedback within an AI system. Model predictive control (MPC), which utilizes ML techniques to generate optimal schedules for HVAC systems, can be optimized and simulated using artificial neural networks (ANNs) [61,89,116]. Nonetheless, designing an optimal energy management algorithm for scheduling HVAC systems is challenging due to uncertainty in the models, parameters, and temporally linked operational constraints. The deep reinforcement learning (DRL) algorithm overcomes these challenges, as it does not require prior knowledge of uncertain parameters or the construction of a thermal dynamics model, making it suitable for energy management and improving user comfort [17,94,131,138]. A comparison between DRL and MPC is provided in a separate study [81]. The integration of IoT with AI presents promising advancements for HVAC systems. However, significant research gaps in scalability and user adaptability must be addressed to harness these technologies fully.

Modern HVAC systems rely on advanced control methods that incorporate machine learning (ML) techniques and fuzzy logic to achieve energy efficiency. For example, Elman neural networks have demonstrated significant potential in load forecasting by predicting HVAC system requirements based on dynamic environmental and occupancy data. These adaptive control strategies ensure systems respond promptly to changing conditions, reducing energy waste while maintaining performance levels. Additionally, hybrid models that combine ML and IoT functionalities allow more granular control of chillers and air handlers, ensuring improved coordination between system components [41,45].

AI-driven control strategies have demonstrated energy savings of up to 40% in various studies, depending on factors such as building type, climatic conditions, and AI model selection. For instance, reinforcement learning-based HVAC optimization has been shown to reduce energy consumption by 30–40% in smart buildings [16], while ANN-based predictive control strategies have achieved savings in the range of 20–35% [31].

4.1.2. Occupancy Detection

Building occupancy detection and estimation play a crucial role in enhancing building energy efficiency. Over the last ten years, numerous approaches have been suggested to enhance the precision of detecting and estimating occupancy [144,152,160,169]. Employing machine learning models and methodologies in predicting occupancy behavior and routines in buildings has spawned their application in the operational management of building systems [150,153]. In identified studies, machine learning techniques are employed to forecast patterns of occupancy and behavior, particularly emphasizing their potential implementation in building systems to enhance energy efficiency, indoor air quality, and thermal comfort. The model system and prediction timeframes are different, making it challenging to develop a perfect prediction model for building occupancy. In general, occupancy prediction models usually include data collection, occupancy prediction, and validation [145]. Despite the potentially higher prediction accuracy of ANNs, they may not possess the same level of transferability as traditional ML models [23,157,167]. In order to identify and classify real-time entities, a convolutional neural network (CNN) architecture was trained and implemented [146,151,165]. Long short-term memory (LSTM) can capture time series data. An attention model can accentuate their significance in energy prediction by allocating weights to the most pertinent inputs. Moreover, integrating the attention mechanism during the training of convolutional neural networks (CNNs) allows for decreased communication overhead. This mechanism effectively disregards extraneous information while amplifying pertinent data [154].

Advancements in AI have introduced innovative methods for precise occupancy detection in HVAC systems, such as using competitive learning and sensor fusion. These methods optimize the functionality of HVAC components like air handlers and ventilation systems by ensuring they align with actual occupancy needs. Furthermore, predictive models trained on historical occupancy and behavioral data can forecast usage patterns, enabling energy-efficient HVAC operation in both residential and commercial buildings [40,41].

4.1.3. Forecasting Consumption

Accurately estimating energy consumption in buildings, particularly within the HVAC system, which contributes significantly to electricity usage, is crucial for effective energy management [45,192]. In this regard, data-driven models have become extensively employed for precise energy consumption prediction [197].

Considering that HVAC accounts for a large portion of the electricity load in buildings, forecasting consumption is vital for energy management [42,187]. Data-driven models are widely used to predict energy consumption [197]. Knowing the properties of outdoor air and the number of individuals using a space enables the estimation of energy consumption of a cooling system in a building with a considerable level of precision [170]. In load forecasting, LSTM networks have demonstrated impressive outcomes [195]. These networks possess LSTM layers that explicitly grasp the temporal associations in forecasting multivariate time series, thus presenting significant potential in energy-efficient building technologies [198]. The recurrent neural network (RNN) model is employed to predict the thermal load of a building as well as the temperatures within its different zones [168,176,183,197].

Energy consumption forecasting is an essential aspect of modern HVAC systems, particularly for components like compressors and air conditioners, which contribute heavily to peak energy usage. Recurrent neural networks (RNNs) and predictive analytics platforms are increasingly utilized to forecast cooling and heating loads. These methods incorporate factors like weather patterns and occupancy schedules, enabling HVAC systems to pre-adjust their operations and optimize energy use in real time [40,45].

4.1.4. Thermal Comfort

Most individuals utilizing shared office spaces express discontent with their thermal comfort due to HVAC systems' inability to cater to individualized thermal environment needs [203,219]. By integrating considerations for both energy consumption and occupant thermal comfort, the regulation of thermal comfort can be approached as a problem of minimizing costs. One approach to predicting occupant thermal comfort involves the utilization of a Feedforward Neural Network (FNN) [217]. Additionally, certain studies employ an RF algorithm within HVAC systems to estimate the air quality index [202,204,217]. While AI significantly enhances HVAC efficiency, challenges such as system compatibility and data privacy remain prevalent. By using the environmental information, the HVAC system can be adjusted to predict thermal comfort in real time. Occupancy data, such as occupancy number and activity level, can be used to estimate indoor CO_2 and minimum ventilation levels. Thermal comfort prediction models also use occupancy and activity levels. This information can be used to optimize HVAC operations and minimize energy consumption [145].

Achieving optimal thermal comfort while minimizing energy consumption is a key challenge in HVAC design. Recent innovations involve using AI models, such as fuzzy logic controllers and reinforcement learning systems, to regulate key components like air ducts and chillers. These systems dynamically adjust HVAC parameters based on real-time environmental and user feedback, improving indoor air quality and occupant satisfaction [41,49].

4.2. Maintenance HVAC System

This classification aims to provide insights into the potential of AI and its algorithms in enhancing the energy efficiency of HVAC systems through maintenance practices. By exploring maintenance management concepts and FDD, it is feasible to reduce energy consumption while ensuring optimal comfort levels. This classification aims to illustrate some insights into the potential role of artificial intelligence and its algorithms in increasing the energy efficiency of HVAC systems by enhancing maintenance practices.

4.2.1. Fault Detection and Diagnostics

In recent years, scholars have intensified their efforts in advancing the study of automated fault detection and diagnostics, prompted by the rise in energy consumption attributable to building faults [228]. Such faults have led to suboptimal thermal comfort and reduced occupant productivity [231]. By investigating computing-based FDD methods, two distinct categories emerge: knowledge-based and data-driven approaches [18,19]. As a result of the insufficient number of fault samples, the imbalanced class classification problem, which is prominent in machine learning, makes it challenging to categorize faults [19,241]. Discrete Bayesian Networks (DisBNs) are employed to diagnose faults across different levels in commercial HVAC systems. The implementation of DisBNs addresses a common challenge faced by other FDD methods, namely the lack of sufficient fault data for training purposes. A DisBN considers the interactions between faults and fault symptoms (referred to as fault indicators) within HVAC systems, considering the impacts of faults at various levels [227]. To enhance the effectiveness of HVAC system detection and isolation, this study combines a multiscale representation, Principal Component Analysis (PCA), and ML classifiers [252]. By identifying faults and predicting future states, machine learning algorithms have the potential to aid in scheduling maintenance tasks [226].

Efficient fault detection and diagnostics are crucial for maintaining HVAC systems like compressors, chillers, and air handlers. AI-driven approaches, such as fuzzy inference systems and predictive maintenance frameworks, enable early identification of potential faults. These methods leverage historical maintenance data and real-time monitoring to reduce downtime and operational costs, ensuring components remain in optimal condition. For example, predictive models applied to chiller plants can detect anomalies linked to external factors like weather fluctuations [48,49].

AI-driven fault detection and diagnosis (FDD) has significantly enhanced HVAC reliability by identifying system anomalies before they result in failures. Machine learning models, such as Random Forests and Bayesian Networks, are extensively used to detect irregularities in temperature, airflow, and energy consumption patterns [223].

Studies have demonstrated that predictive maintenance frameworks leveraging AI can reduce HVAC downtime and extend equipment lifespan by providing real-time fault detection alerts. These systems proactively diagnose potential failures, allowing facility managers to address issues before they escalate into costly repairs [9].

4.2.2. Maintenance Management

The role of the HVAC system in facility management and maintenance operations is of paramount importance, as any malfunction can result in substantial financial drawbacks. The implementation of DL techniques can enhance the overall performance of buildings by reducing energy consumption, facilitating maintenance planning, and overseeing equipment operations. Such improvements are particularly notable in the domain of predictive maintenance. Notably, significant advancements have emerged in building operation and maintenance, as evident in recent studies [12,44]. Upon analysis, extreme gradient boosting (XGB) has been identified as the most optimal machine learning algorithm for predictive maintenance [253].

Effective maintenance management ensures the longevity and efficiency of HVAC components, such as heat exchangers, evaporators, and ducts. AI-based predictive models, such as extreme gradient boosting (XGB), have been shown to excel in anticipating equipment failures and optimizing maintenance schedules. Furthermore, integrating digital twins with predictive analytics provides a real-time simulation of HVAC operations, allowing for proactive interventions to prevent component degradation [41,48].

AI-driven predictive maintenance strategies play a critical role in minimizing HVAC operational costs. By utilizing sensor-based monitoring and real-time analytics, AI models can detect wear and tear in HVAC components, triggering maintenance alerts before failures occur [255]. Cloud-based AI models have been implemented for large-scale HVAC maintenance optimization, allowing for remote diagnostics, automated scheduling of repairs, and performance tracking across multiple buildings. These intelligent maintenance strategies not only enhance system efficiency but also reduce energy waste and carbon footprints in smart buildings [254].

4.3. Machine Learning Model Applicability in HVAC Applications

Several machine learning (ML) models are applicable to some extent based on the specific HVAC application:

- Supervised Learning (e.g., decision trees, Support Vector Machines, neural networks): Best suited for fault detection, energy consumption forecasting, and HVAC system diagnostics [14,19].

- Unsupervised Learning (e.g., Clustering, Principal Component Analysis): Suitable for anomaly detection and system optimization with unlabeled data [23].
- Reinforcement learning (e.g., deep Q-Networks, Actor–Critic Models): Most appropriate for adaptive HVAC control, real-time dynamic setpoint changes, and real-time energy optimization [26,32].

The model selection depends on the computational complexity, data availability, and requirement for real-time processing [30].

5. Discussion

This study elucidates the potential for AI to revolutionize energy management in HVAC systems globally, offering substantial benefits across various sectors. Even though this review is not focused on individual case studies, it combines major findings from various AI applications in HVAC systems, providing a systematic overview of their impact, challenges, and future directions. For the HVAC industry worldwide, the application of machine learning and deep learning highlights significant opportunities for energy savings and operational efficiencies crucial for commercial buildings. These adaptations enhance system performance and adjustment to diverse climatic conditions, making them versatile solutions, applicable globally. Further, by demonstrating AI's effectiveness in reducing energy consumption and emissions, this research supports establishing stricter energy efficiency standards and regulations internationally. Such findings can guide policy decisions to encourage or mandate the integration of intelligent systems in new and existing structures, aiding in achieving energy conservation and emission reduction targets. From a financial perspective, leveraging AI in building energy systems presents a significant advantage by decreasing operational costs and boosting the sustainability profile of real estate assets. This shift can draw investments, particularly under the increasing focus on Environmental, Social, and Governance (ESG) criteria within investment practices.

In assessing the effectiveness of AI technologies in HVAC systems, it is crucial to acknowledge several inherent limitations that may influence the generalizability and applicability of our findings. Our analysis depends heavily on the data presented in previously published research. The variability and limited availability of detailed operational data can restrict the extent to which conclusions can be generalized across different real-world scenarios. This limitation is significant as data deficiencies may skew the perceived effectiveness of AI techniques in HVAC systems. Furthermore, the review extensively covers various AI methods applicable to HVAC energy management. However, the practical deployment of these techniques is often hindered by existing infrastructure and the technological readiness of specific regions or industries. This mismatch between theoretical potential and practical feasibility can significantly limit the immediate applicability of our findings. Additionally, the impact of AI-driven solutions on energy efficiency and system performance can vary markedly depending on building architecture and climatic conditions. The studies reviewed may need to adequately represent this diversity, potentially limiting the reliability of AI solutions across different building types. These factors are critical as they not only influence the outcomes of our study but also affect the scalability and practical implementation of AI-driven energy management strategies.

Figure 9 presents a comprehensive overview of the current landscape, challenges, and future prospects of integrating AI in HVAC systems. The current landscape emphasizes the dominant role of AI-driven control methods and maintenance techniques in optimizing system performance and energy efficiency. However, significant challenges, such as infrastructure compatibility, scalability issues, data privacy concerns, integration with legacy systems, and the lack of long-term effectiveness studies, create barriers to wider adoption.

Despite these hurdles, we can envision a future landscape in which there is a transformative shift with advancements in IoT integration, adaptive algorithms, customization for regional climatic conditions, and zero-carbon energy management, highlighting AI's potential to redefine HVAC operations sustainably and effectively.

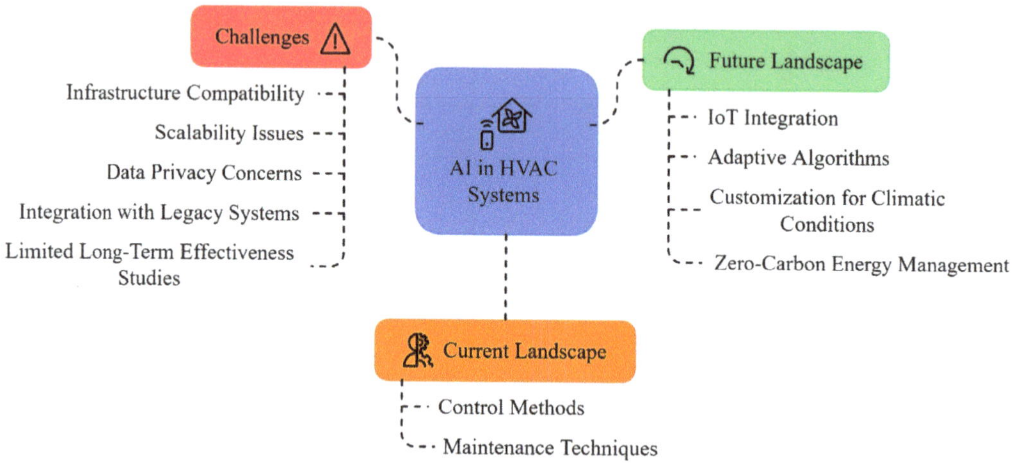

Figure 9. Overview of AI in HVAC Systems, highlighting the current landscape.

This review identifies several potential sources of error that could impact the validity and applicability of the findings. One such source is algorithmic bias, where AI models, heavily reliant on historical data, may not adequately capture the diversity of operational scenarios, building types, or climatic conditions. This can lead to biased predictions and management recommendations, affecting the accuracy of AI-driven solutions for HVAC systems. Another risk involves model overfitting, where complex AI algorithms such as deep learning models might overfit specific datasets used in studies. This tendency can hinder the algorithms' ability to generalize findings to other settings or adapt to evolving conditions due to climate change. Additionally, many studies utilize simulated data to train and validate AI algorithms. These simulations' assumptions can introduce errors, especially if they fail to accurately mimic real-world HVAC system behaviors and interactions with building occupants. These potential errors could undermine the effectiveness of the recommendations derived from this analysis, possibly leading to higher-than-expected energy usage or diminished occupant comfort. Such discrepancies have significant implications for policymakers and the scientific community, particularly in terms of meeting climate change targets. Erroneous predictions and suboptimal recommendations could impede the deployment of effective AI-driven energy management systems, thereby delaying progress toward crucial sustainability goals.

Even with the advancements made by AI-based HVAC systems, there are various gaps in the research area that need to be addressed. Apart from machine learning and deep learning approaches, scalability and implementation challenges hold back the widespread use of AI-based HVAC systems. Various AI models perform well in simulations but lack implementation capability due to computational complexity, integration problems, and varying building conditions. Furthermore, most existing AI solutions are designed for specific building classes or climate zones, making it difficult to develop generic models that can be implemented in various HVAC systems.

Another important challenge is the explainability and interpretability of AI models. Deep learning approaches, particularly neural networks, act as "black-box" models, reducing trust in AI-based HVAC control. Explainable AI (XAI) approaches need to be explored in future studies in order to improve model transparency and user acceptance.

Another limitation is real-time adaptation, as most AI-based HVAC models utilize historical data rather than learning in real time from varying environmental conditions. Developing AI systems with real-time learning and adaptation would make them more efficient in uncertain environments.

In addition, integration with smart grids and renewable energy systems is yet to be fully investigated. AI systems must be able to coordinate HVAC control with energy demand-response planning and renewable energy supplies in order to optimize efficiency.

Finally, the lack of standardized, high-quality datasets for training AI models is a primary limitation. Most HVAC datasets contain missing, biased, or inconsistent information, limiting the reliability of AI-based predictions. In the future, more efforts should be dedicated to creating standardized, high-quality datasets for developing AI-driven HVAC control.

Future AI-driven HVAC systems should incorporate adaptive models for better energy efficiency. The selection of ML models should be guided by the specific application, balancing accuracy, computational cost, and real-time adaptability.

Study Significance

Rising energy needs and demands for efficiency and intelligent HVAC systems have motivated researchers to take up AI-powered systems. In the present work, a critical review of applications of AI in optimizing HVAC systems has been provided, with perspectives on energy efficiency, preventive maintenance, and adaptive control methods.

The usefulness of this study emanates from the following contributions:

- Closing the theory and practice gap of AI in HVAC by categorizing AI techniques and determining their suitability in real situations.
- Identifying the major challenges in AI-based HVAC implementation, including data availability, model interpretability, and integration with existing building management systems.
- Highlighting the strengths of machine learning (ML) and deep learning (DL) models in improving HVAC performance, reducing operating costs, and improving occupant comfort.
- Providing future research opportunities by emphasizing the significance of hybrid AI models, IoT-based smart HVAC systems, and more transparent AI-driven decision-making.

With the integration of the latest advances in AI for HVAC, this research is a valuable reference work for researchers and industry experts, guiding future developments in smart and sustainable building control.

6. Conclusions

This paper critically assesses the current literature on the use of artificial intelligence (AI) in the improvement of the energy efficiency of HVAC systems. The conclusions indicate that AI provides an opportunity to decrease energy usage and costs and, at the same time, improve the thermal comfort of the occupants. In particular, the integration of the machine learning and deep learning approaches has been identified as being capable of cutting energy consumption by as much as 40%, which presents a strong solution for energy management in buildings. The scientometric analysis reflects the increasing number of studies that concentrate on the application of AI to HVAC systems; it also indicates a focus on real-world usability. Nevertheless, our research also points to a major drawback in the form of the lack of research on long-term effectiveness and scenarios in various climatic conditions, which are recommended for future research.

This paper thus concludes that, while AI has the potential to revolutionize the HVAC industry, there are still issues that need to be addressed, such as standardization, data privacy, and integration with the current infrastructure. For policymakers and industry leaders, these findings call for the development of legal and regulatory frameworks that will encourage the integration of AI technologies in the right manner and in a way that will facilitate the achievement of sustainable development goals and net-zero targets. In addition, the criticism in this article does more than amplify the importance of AI developments. Rather, it examines the real effects of these advancements and paves the way for future research that should concentrate on the implementation issues and costs of AI-based energy systems. By addressing these aspects, future research should aim to continue the exploration of AI capabilities and ensure these technologies are accessible, efficient, and effective across various regions and building types, thereby contributing to global efforts in energy sustainability.

The key findings are as follows:

- HVAC efficiency is improved significantly by the use of predictive control, supervision, and automated tuning thanks to the use of machine learning models.
- Some of the machine learning models skilled in fault diagnosis, predictive maintenance, and energy consumption shifting forecasting are artificial neural networks, decision trees, and reinforcement learning techniques.
- In comparison to other methods, deep reinforcement learning is more effective in the real-time controlling and regulating of HVAC systems with simultaneous adjustments in the environment and can lower the cost of operations.
- Even with the advanced technologies, data quality and the interpretability of the models in the context of integration of all system components is still a challenge to the dominance of AI in HVAC systems.
- Future research should focus on hybrid AI approaches, IoT integration, and explainable AI to enhance the practical application of AI-driven HVAC solutions.

Author Contributions: Conceptualization, A.H.M.R. and S.A.A.; Methodology, A.H.M.R. and S.A.A.; Software, A.H.M.R.; Validation, S.A.A. and M.T.; Formal analysis, S.A.A., A.H.M.R. and M.T.; Writing—original draft, A.H.M.R. and S.A.A.; Writing—Review and editing, M.K. and M.R.; Supervision and guidance throughout the project, M.K. and M.R. All authors have read and agreed to the published version of the manuscript.

Funding: This research received no external funding.

Conflicts of Interest: The authors declare no conflicts of interest.

Abbreviations

The following abbreviations are used in this manuscript:

HVAC	Heating, Ventilation, and Air Conditioning
SLR	systematic literature review
BEM	building energy management
DRL	deep reinforcement learning
ANN	artificial neural network
RF	Random Forest

SVM	Support Vector Machine
IoT	Internet of Things
FDD	fault detection and diagnosis
MPC	model predictive control
CNN	convolutional neural network
LSTM	long short-term memory
RNN	recurrent neural network
GAN	generative adversarial network
WoS	Web of Science
XGB	extreme gradient boosting
PCA	Principal Component Analysis
DisBN	Discrete Bayesian Network
DQN	deep Q-Network
DDPG	Deep Deterministic Policy Gradient
XAI	Explainable Artificial Intelligence
GRU	Gated Recurrent Unit
MLP	Multi-Layer Perceptron

References

1. Neri, E.; Coppola, F.; Miele, V.; Bibbolino, C.; Grassi, R. Artificial Intelligence: Who Is Responsible for the Diagnosis? *Radiol. Med.* **2020**, *125*, 517–521. [CrossRef]
2. Apanaviciene, R.; Vanagas, A.; Fokaides, P.A. Smart Building Integration into a Smart City (SBISC): Development of a New Evaluation Framework. *Energies* **2020**, *13*, 2190. [CrossRef]
3. Lee, S.; Choi, D.-H. Energy Management of Smart Home with Home Appliances, Energy Storage System and Electric Vehicle: A Hierarchical Deep Reinforcement Learning Approach. *Sensors* **2020**, *20*, 2157. [CrossRef]
4. Yousaf, S.; Bradshaw, C.R.; Kamalapurkar, R.; San, O. Investigating Critical Model Input Features for Unitary Air Conditioning Equipment. *Energy Build.* **2023**, *284*, 112823. [CrossRef]
5. Peng, Y.; Lei, Y.; Tekler, Z.D.; Antanuri, N.; Lau, S.-K.; Chong, A. Hybrid System Controls of Natural Ventilation and HVAC in Mixed-Mode Buildings: A Comprehensive Review. *Energy Build.* **2022**, *276*, 112509. [CrossRef]
6. Alanne, K.; Sierla, S. An Overview of Machine Learning Applications for Smart Buildings. *Sustain. Cities Soc.* **2022**, *76*, 103445. [CrossRef]
7. Sierla, S.; Ihasalo, H.; Vyatkin, V. A Review of Reinforcement Learning Applications to Control of Heating, Ventilation and Air Conditioning Systems. *Energies* **2022**, *15*, 3526. [CrossRef]
8. Tien, P.W.; Wei, S.; Darkwa, J.; Wood, C.; Calautit, J.K. Machine Learning and Deep Learning Methods for Enhancing Building Energy Efficiency and Indoor Environmental Quality—A Review. *Energy AI* **2022**, *10*, 100198. [CrossRef]
9. Nelson, W.; Culp, C. Machine Learning Methods for Automated Fault Detection and Diagnostics in Building Systems—A Review. *Energies* **2022**, *15*, 5534. [CrossRef]
10. Halhoul Merabet, G.; Essaaidi, M.; Ben Haddou, M.; Qolomany, B.; Qadir, J.; Anan, M.; Al-Fuqaha, A.; Abid, M.R.; Benhaddou, D. Intelligent Building Control Systems for Thermal Comfort and Energy-Efficiency: A Systematic Review of Artificial Intelligence-Assisted Techniques. *Renew. Sustain. Energy Rev.* **2021**, *144*, 110969. [CrossRef]
11. Miao, Y.; Yao, Y.; Hong, X.; Xiong, L.; Zhang, F.; Chen, W. Research on Optimal Control of HVAC System Using Swarm Intelligence Algorithms. *Build. Environ.* **2023**, *241*, 110467. [CrossRef]
12. Bouabdallaoui, Y.; Lafhaj, Z.; Yim, P.; Ducoulombier, L.; Bennadji, B. Predictive Maintenance in Building Facilities: A Machine Learning-Based Approach. *Sensors* **2021**, *21*, 1044. [CrossRef]
13. Xiong, L.; Yao, Y. Study on an Adaptive Thermal Comfort Model with K-Nearest-Neighbors (KNN) Algorithm. *Build. Environ.* **2021**, *202*, 108026. [CrossRef]
14. Touzani, S.; Granderson, J.; Fernandes, S. Gradient Boosting Machine for Modeling the Energy Consumption of Commercial Buildings. *Energy Build.* **2018**, *158*, 1533–1543. [CrossRef]
15. Peng, Y.; Rysanek, A.; Nagy, Z.; Schlüter, A. Using Machine Learning Techniques for Occupancy-Prediction-Based Cooling Control in Office Buildings. *Appl. Energy* **2018**, *211*, 1343–1358. [CrossRef]
16. Zhang, Z.; Chong, A.; Pan, Y.; Zhang, C.; Lam, K.P. Whole Building Energy Model for HVAC Optimal Control: A Practical Framework Based on Deep Reinforcement Learning. *Energy Build.* **2019**, *199*, 472–490. [CrossRef]

17. Yu, L.; Xie, W.; Xie, D.; Zou, Y.; Zhang, D.; Sun, Z.; Zhang, L.; Zhang, Y.; Jiang, T. Deep Reinforcement Learning for Smart Home Energy Management. *IEEE Internet Things J.* **2020**, *7*, 2751–2762. [CrossRef]
18. Deng, H.; Fannon, D.; Eckelman, M.J. Predictive Modeling for US Commercial Building Energy Use: A Comparison of Existing Statistical and Machine Learning Algorithms Using CBECS Microdata. *Energy Build.* **2018**, *163*, 34–43. [CrossRef]
19. Yan, K.; Chong, A.; Mo, Y. Generative Adversarial Network for Fault Detection Diagnosis of Chillers. *Build. Environ.* **2020**, *172*, 106698. [CrossRef]
20. Yu, L.; Sun, Y.; Xu, Z.; Shen, C.; Yue, D.; Jiang, T.; Guan, X. Multi-Agent Deep Reinforcement Learning for HVAC Control in Commercial Buildings. *IEEE Trans. Smart Grid* **2021**, *12*, 407–419. [CrossRef]
21. Brandi, S.; Piscitelli, M.S.; Martellacci, M.; Capozzoli, A. Deep Reinforcement Learning to Optimise Indoor Temperature Control and Heating Energy Consumption in Buildings. *Energy Build.* **2020**, *224*, 110225. [CrossRef]
22. Du, Y.; Zandi, H.; Kotevska, O.; Kurte, K.; Munk, J.; Amasyali, K.; Mckee, E.; Li, F. Intelligent Multi-Zone Residential HVAC Control Strategy Based on Deep Reinforcement Learning. *Appl. Energy* **2021**, *281*, 116117. [CrossRef]
23. Dai, X.; Liu, J.; Zhang, X. A Review of Studies Applying Machine Learning Models to Predict Occupancy and Window-Opening Behaviours in Smart Buildings. *Energy Build.* **2020**, *223*, 110159. [CrossRef]
24. Chen, B.; Cai, Z.; Bergés, M. Gnu-RL: A Precocial Reinforcement Learning Solution for Building HVAC Control Using a Differentiable MPC Policy. In Proceedings of the BuildSys 2019—Proceedings of the 6th ACM International Conference on Systems for Energy-Efficient Buildings, Cities, and Transportation, New York, NY, USA, 13–14 November 2019; Association for Computing Machinery, Inc.: New York, NY, USA, 2019; pp. 316–325.
25. Gao, G.; Li, J.; Wen, Y. DeepComfort: Energy-Efficient Thermal Comfort Control in Buildings Via Reinforcement Learning. *IEEE Internet Things J.* **2020**, *7*, 8472–8484. [CrossRef]
26. Zou, Z.; Yu, X.; Ergan, S. Towards Optimal Control of Air Handling Units Using Deep Reinforcement Learning and Recurrent Neural Network. *Build. Environ.* **2020**, *168*, 106535. [CrossRef]
27. Ding, X.; Du, W.; Cerpa, A. OCTOPUS: Deep Reinforcement Learning for Holistic Smart Building Control. In Proceedings of the BuildSys 2019—Proceedings of the 6th ACM International Conference on Systems for Energy-Efficient Buildings, Cities, and Transportation, New York, NY, USA, 13–14 November 2019; Association for Computing Machinery, Inc.: New York, NY, USA, 2019; pp. 326–335.
28. Ding, Y.; Zhang, Q.; Yuan, T.; Yang, K. Model Input Selection for Building Heating Load Prediction: A Case Study for an Office Building in Tianjin. *Energy Build.* **2018**, *159*, 254–270. [CrossRef]
29. Huchuk, B.; Sanner, S.; O'Brien, W. Comparison of Machine Learning Models for Occupancy Prediction in Residential Buildings Using Connected Thermostat Data. *Build. Environ.* **2019**, *160*, 106177. [CrossRef]
30. Zhang, C.; Kuppannagari, S.R.; Kannan, R.; Prasanna, V.K. Building HVAC Scheduling Using Reinforcement Learning via Neural Network Based Model Approximation. In Proceedings of the BuildSys 2019—Proceedings of the 6th ACM International Conference on Systems for Energy-Efficient Buildings, Cities, and Transportation, New York, NY, USA, 13–14 November 2019; Association for Computing Machinery, Inc.: New York, NY, USA, 2019; pp. 287–296.
31. Esrafilian-Najafabadi, M.; Haghighat, F. Occupancy-Based HVAC Control Systems in Buildings: A State-of-the-Art Review. *Build. Environ.* **2021**, *197*, 107810. [CrossRef]
32. Lee, K.-P.; Wu, B.-H.; Peng, S.-L. Deep-Learning-Based Fault Detection and Diagnosis of Air-Handling Units. *Build. Environ.* **2019**, *157*, 24–33. [CrossRef]
33. Liu, T.; Xu, C.; Guo, Y.; Chen, H. A Novel Deep Reinforcement Learning Based Methodology for Short-Term HVAC System Energy Consumption Prediction. *Int. J. Refrig.* **2019**, *107*, 39–51. [CrossRef]
34. Zhang, Z.; Lam, K.P. Practical Implementation and Evaluation of Deep Reinforcement Learning Control for a Radiant Heating System. In Proceedings of the BuildSys 2018—Proceedings of the 5th Conference on Systems for Built Environments, Shenzen, China, 7–8 November 2018; Association for Computing Machinery, Inc.: New York, NY, USA, 2018; pp. 148–157.
35. Taheri, S.; Razban, A. Learning-Based CO_2 Concentration Prediction: Application to Indoor Air Quality Control Using Demand-Controlled Ventilation. *Build. Environ.* **2021**, *205*, 108164. [CrossRef]
36. Carreira, P.; Costa, A.A.; Mansur, V.; Arsénio, A. Can HVAC Really Learn from Users? A Simulation-Based Study on the Effectiveness of Voting for Comfort and Energy Use Optimization. *Sustain. Cities Soc.* **2018**, *41*, 275–285. [CrossRef]
37. Ahn, K.U.; Park, C.S. Application of Deep Q-Networks for Model-Free Optimal Control Balancing between Different HVAC Systems. *Sci. Technol. Built Environ.* **2020**, *26*, 61–74. [CrossRef]
38. Javed, A.; Larijani, H.; Wixted, A. Improving Energy Consumption of a Commercial Building with IoT and Machine Learning. *IT Prof.* **2018**, *20*, 30–38. [CrossRef]

39. Zhang, Z.; Chong, A.; Pan, Y.; Zhang, C.; Lu, S.; Lam, K.P. A Deep Reinforcement Learning Approach to Using Whole Building Energy Model for HVAC Optimal Control. In Proceedings of the ASHRAE and IBPSA-USA Building Simulation Conference; American Society of Heating, Refrigerating and Air-Conditioning Engineers (ASHRAE), Chicago, IL, USA, 26–28 September 2018; pp. 675–682.
40. Cai, W.; Wen, X.; Li, C.; Shao, J.; Xu, J. Predicting the Energy Consumption in Buildings Using the Optimized Support Vector Regression Model. *Energy* **2023**, *273*, 127188. [CrossRef]
41. Qin, H.; Yu, Z.; Li, T.; Liu, X.; Li, L. Heating Control Strategy Based on Dynamic Programming for Building Energy Saving and Emission Reduction. *Int. J. Environ. Res. Public Health* **2022**, *19*, 14137. [CrossRef]
42. Esrafilian-Najafabadi, M.; Haghighat, F. Impact of Occupancy Prediction Models on Building HVAC Control System Performance: Application of Machine Learning Techniques. *Energy Build.* **2022**, *257*, 111808. [CrossRef]
43. Behravan, A.; Abboush, M.; Obermaisser, R. Deep Learning Application in Mechatronics Systems' Fault Diagnosis, a Case Study of the Demand-Controlled Ventilation and Heating System. In Proceedings of the 2019 Advances in Science and Engineering Technology International Conferences, ASET 2019, Dubai, United Arab Emirates, 26 March–10 April 2019; Institute of Electrical and Electronics Engineers Inc.: Piscataway, NJ, USA, 2019.
44. Sanzana, M.R.; Maul, T.; Wong, J.Y.; Abdulrazic, M.O.M.; Yip, C.-C. Application of Deep Learning in Facility Management and Maintenance for Heating, Ventilation, and Air Conditioning. *Autom. Constr.* **2022**, *141*, 104445. [CrossRef]
45. Meng, Q.; Xi, Y.; Ren, X.; Li, H.; Jiang, L.; Yang, L. Thermal Energy Storage Air-Conditioning Demand Response Control Using Elman Neural Network Prediction Model. *Sustain. Cities Soc.* **2022**, *76*, 103480. [CrossRef]
46. Sankaranarayanan, C.; Shaju, S.; Sukhwani, M. Actor-Critic Based Adaptive Control Strategy for Effective Energy Management. In Proceedings of the Proceedings of the 2022 5th International Conference on Advanced Systems and Emergent Technologies, IC_ASET 2022, Hammamet, Tunisia, 22–25 March 2022; Institute of Electrical and Electronics Engineers Inc.: Piscataway, NJ, USA, 2022; pp. 23–28.
47. Scarcello, L.; Cicirelli, F.; Guerrieri, A.; Mastroianni, C.; Spezzano, G.; Vinci, A. Pursuing Energy Saving and Thermal Comfort With a Human-Driven DRL Approach. *IEEE Trans. Hum. Mach. Syst.* **2022**, *53*, 707–719. [CrossRef]
48. Sanzana, M.R.; Abdulrazic, M.O.M.; Wong, J.Y.; Maul, T.; Yip, C.-C. Effects of External Weather on the Water Consumption of Thermal-Energy-Storage Air-Conditioning System. *Energy Nexus* **2023**, *10*, 100187. [CrossRef]
49. Isikdemir, Y.E.; Erturk, G.; Ates, H.; Tas, M.O. Fuzzy Inference and Machine Learning Based HVAC Control System for Smart Buildings. In Proceedings of the IEEE Global Energy Conference, GEC 2022, Batman, Turkey, 26–29 October 2022; Institute of Electrical and Electronics Engineers Inc.: Piscataway, NJ, USA, 2022; pp. 116–119.
50. Alden, R.E.; Gong, H.; Jones, E.S.; Ababei, C.; Ionel, D.M. Artificial Intelligence Method for the Forecast and Separation of Total and HVAC Loads with Application to Energy Management of Smart and NZE Homes. *IEEE Access* **2021**, *9*, 160497–160509. [CrossRef]
51. Xu, Y.; Gao, W.; Qian, F.; Li, Y. Potential Analysis of the Attention-Based LSTM Model in Ultra-Short-Term Forecasting of Building HVAC Energy Consumption. *Front. Energy Res.* **2021**, *9*, 730640. [CrossRef]
52. Fan, C.; He, W.; Liao, L. Real-Time Machine Learning-Based Recognition of Human Thermal Comfort-Related Activities Using Inertial Measurement Unit Data. *Energy Build.* **2023**, *294*, 113216. [CrossRef]
53. Naseem, T.; Javed, A.; Hamayun, M.T.; Jawad, M.; Ansari, E.A.; Fayyaz, M.A.B.; Ansari, A.R.; Nawaz, R. Design of an EnergyPlus Model-Based Smart Controller for Maintaining Thermal Comfortable Environment in Non-Domestic Building. *IEEE Access* **2023**, *11*, 33134–33147. [CrossRef]
54. Yang, S.; Wan, M.P. Machine-Learning-Based Model Predictive Control with Instantaneous Linearization—A Case Study on an Air-Conditioning and Mechanical Ventilation System. *Appl. Energy* **2022**, *306*, 118041. [CrossRef]
55. Xue, W.; Wang, H.; Li, K. PMV Inverse Model-Based Indoor Thermal Environment Control for Thermal Comfort and Energy Saving. In Proceedings of the Chinese Control Conference, CCC, Hefei, China, 25–27 July 2022; IEEE Computer Society: Washington, DC, USA, 2022; pp. 5294–5299.
56. Frassanito, R.; Buso, T.; Aumann, S.; Toniolo, J.; Albrici, P.; Canevari, P.; Iemmi, M.; Mapelli, F. How IoT and Artificial Intelligence Can Improve Energy Efficiency in Hospitals—A North Italian Case Study. *Proc. E3S Web Conf.* **2022**, *343*, 02001. [CrossRef]
57. Mahdi, M.N.; Bakare, T.A.; Ahmad, A.R.; Buhari, A.M.; Mohamed, K.S. Scalable Smartification of Commercial Buildings HVAC Systems Using the Internet of Things and Machine Learning. In Proceedings of the International Conference on Emerging Technologies and Intelligent Systems, Al Buraimi, Oman, 25–26 June 2021; Springer Science and Business Media Deutschland GmbH: Berlin/Heidelberg, Germany, 2022; Volume 322, pp. 165–174.
58. Du, Y.; Li, F.; Kurte, K.; Munk, J.; Zandi, H. Demonstration of Intelligent HVAC Load Management with Deep Reinforcement Learning: Real-World Experience of Machine Learning in Demand Control. *IEEE Power Energy Mag.* **2022**, *20*, 42–53. [CrossRef]

59. Aruta, G.; Ascione, F.; Bianco, N.; De Masi, R.F.; Mauro, G.M.; Vanoli, G.P. Model Predictive Control Based on Genetic Algorithm and Neural Networks to Optimize Heating Operation of a Real Low-Energy Building. In Proceedings of the 2022 7th International Conference on Smart and Sustainable Technologies, SpliTech 2022, Split, Croatia, 5–8 July 2022; Institute of Electrical and Electronics Engineers Inc.: Piscataway, NJ, USA, 2022.
60. Lin, X.; Guo, Q.; Yuan, D.; Gao, M. Bayesian Optimization Framework for HVAC System Control. *Buildings* **2023**, *13*, 314. [CrossRef]
61. Deng, Z.; Wang, X.; Jiang, Z.; Zhou, N.; Ge, H.; Dong, B. Evaluation of Deploying Data-Driven Predictive Controls in Buildings on a Large Scale for Greenhouse Gas Emission Reduction. *Energy* **2023**, *270*, 126934. [CrossRef]
62. Mohseni, S.-R.; Zeitouni, M.J.; Parvaresh, A.; Abrazeh, S.; Gheisarnejad, M.; Khooban, M.-H. FMI Real-Time Co-Simulation-Based Machine Deep Learning Control of HVAC Systems in Smart Buildings: Digital-Twins Technology. *Trans. Inst. Meas. Control* **2023**, *45*, 661–673. [CrossRef]
63. Andrés, E.; Cuéllar, M.P.; Navarro, G. On the Use of Quantum Reinforcement Learning in Energy-Efficiency Scenarios. *Energies* **2022**, *15*, 6034. [CrossRef]
64. Weinberg, D.; Wang, Q.; Timoudas, T.O.; Fischione, C. A Review of Reinforcement Learning for Controlling Building Energy Systems From a Computer Science Perspective. *Sustain. Cities Soc.* **2023**, *89*, 104351. [CrossRef]
65. Gehbauer, C.; Rippl, A.; Lee, E.S. Advanced Control of Dynamic Facades and HVAC with Reinforcement Learning Based on Standardized Co-Simulation. In Proceedings of the Building Simulation 2021: 17th Conference of IBPSA, Bruges, Belgium, 1–3 September 2021; International Building Performance Simulation Association: Toronto, ON, Canada, 2022; pp. 231–238.
66. Lavanya, R.; Murukesh, C.; Shanker, N.R. Development of Machine Learning Based Microclimatic HVAC System Controller for Nano Painted Rooms Using Human Skin Temperature. *J. Electr. Eng. Technol.* **2023**, *18*, 2343–2354. [CrossRef]
67. Deng, X.; Zhang, Y.; Qi, H. Towards Optimal HVAC Control in Non-Stationary Building Environments Combining Active Change Detection and Deep Reinforcement Learning. *Build. Environ.* **2022**, *211*, 108680. [CrossRef]
68. Abida, A.; Richter, P. HVAC Control in Buildings Using Neural Network. *J. Build. Eng.* **2023**, *65*, 105558. [CrossRef]
69. Elnour, M.; Himeur, Y.; Fadli, F.; Mohammedsherif, H.; Meskin, N.; Ahmad, A.M.; Petri, I.; Rezgui, Y.; Hodorog, A. Neural Network-Based Model Predictive Control System for Optimizing Building Automation and Management Systems of Sports Facilities. *Appl. Energy* **2022**, *318*, 119153. [CrossRef]
70. Liu, X.; Ren, M.; Yang, Z.; Yan, G.; Guo, Y.; Cheng, L.; Wu, C. A Multi-Step Predictive Deep Reinforcement Learning Algorithm for HVAC Control Systems in Smart Buildings. *Energy* **2022**, *259*, 124857. [CrossRef]
71. Darwazeh, D.; Gunay, B.; Duquette, J. Development of Inverse Greybox Model-Based Virtual Meters for Air Handling Units. *IEEE Trans. Autom. Sci. Eng.* **2021**, *18*, 323–336. [CrossRef]
72. Terzi, E.; Fagiano, L.; Farina, M.; Scattolini, R. Structured Modelling from Data and Optimal Control of the Cooling System of a Large Business Center. *J. Build. Eng.* **2020**, *28*, 101043. [CrossRef]
73. Murade, G.B.; Soni, B.; Mukherjee, A. HVAC Hybrid Control Methods for HEE in Buildings: Overview. In Proceedings of the 2021 7th IEEE International Conference on Advances in Computing, Communication and Control, ICAC3 2021, Mumbai, India, 3–4 December 2021; Institute of Electrical and Electronics Engineers Inc.: Piscataway, NJ, USA, 2021.
74. Faddel, S.; Tian, G.; Zhou, Q.; Aburub, H. On the Performance of Data-Driven Reinforcement Learning for Commercial HVAC Control. In Proceedings of the 2020 IEEE Industry Applications Society Annual Meeting, IAS 2020, Detroit, MI, USA, 10–16 October 2020; Institute of Electrical and Electronics Engineers Inc.: Piscataway, NJ, USA, 2020.
75. Mawson, V.J.; Hughes, B.R. Coupling Simulation with Artificial Neural Networks for the Optimisation of HVAC Controls in Manufacturing Environments. *Optim. Eng.* **2021**, *22*, 103–119. [CrossRef]
76. Talei, H.; Benhaddou, D.; Gamarra, C.; Benbrahim, H.; Essaaidi, M. Smart Building Energy Inefficiencies Detection through Time Series Analysis and Unsupervised Machine Learning. *Energies* **2021**, *14*, 6042. [CrossRef]
77. Gao, Y.; Li, S.; Fu, X.; Dong, W.; Lu, B.; Li, Z. Energy Management and Demand Response with Intelligent Learning for Multi-Thermal-Zone Buildings. *Energy* **2020**, *210*, 118411. [CrossRef]
78. Gan, V.J.L.; Luo, H.; Tan, Y.; Deng, M.; Kwok, H.L. Bim and Data-Driven Predictive Analysis of Optimum Thermal Comfort for Indoor Environment. *Sensors* **2021**, *21*, 4401. [CrossRef]
79. Gonçalves, D.; Sheikhnejad, Y.; Oliveira, M.; Martins, N. One Step Forward toward Smart City Utopia: Smart Building Energy Management Based on Adaptive Surrogate Modelling. *Energy Build.* **2020**, *223*, 110146. [CrossRef]
80. Lissa, P.; Schukat, M.; Barrett, E. Transfer Learning Applied to Reinforcement Learning-Based HVAC Control. *SN Comput. Sci.* **2020**, *1*, 127. [CrossRef]
81. Biswas, D. Reinforcement Learning Based HVAC Optimization in Factories. In Proceedings of the e-Energy 2020—Proceedings of the 11th ACM International Conference on Future Energy Systems, Online, 22–26 June 2020; Association for Computing Machinery, Inc.: New York, NY, USA, 2020; pp. 428–433.

82. Tien, P.W.; Wei, S.; Liu, T.; Calautit, J.; Darkwa, J.; Wood, C. A Deep Learning Approach towards the Detection and Recognition of Opening of Windows for Effective Management of Building Ventilation Heat Losses and Reducing Space Heating Demand. *Renew. Energy* **2021**, *177*, 603–625. [CrossRef]
83. Demirezen, G.; Fung, A.S.; Deprez, M. Development and Optimization of Artificial Neural Network Algorithms for the Prediction of Building Specific Local Temperature for HVAC Control. *Int. J. Energy Res.* **2020**, *44*, 8513–8531. [CrossRef]
84. Jang, Y.-E.; Kim, Y.-J.; Catalao, J.P.S. Optimal HVAC System Operation Using Online Learning of Interconnected Neural Networks. *IEEE Trans. Smart Grid* **2021**, *12*, 3030–3042. [CrossRef]
85. Sheikhnejad, Y.; Gonçalves, D.; Oliveira, M.; Martins, N. Can Buildings Be More Intelligent than Users? The Role of Intelligent Supervision Concept Integrated into Building Predictive Control. In Proceedings of the 6th International Conference on Energy and Environment Research, ICEER 2019, Aveiro, Portugal, 22–25 July 2019; Elsevier Ltd.: Amsterdam, The Netherlands, 2020; Volume 6, pp. 409–416.
86. Park, J.; Choi, H.; Kim, D.; Kim, T. Development of Novel PMV-Based HVAC Control Strategies Using a Mean Radiant Temperature Prediction Model by Machine Learning in Kuwaiti Climate. *Build. Environ.* **2021**, *206*, 108357. [CrossRef]
87. Aguilar, J.; Garces-Jimenez, A.; Gomez-Pulido, J.M.; Moreno, M.D.R.; De Mesa, J.A.G.; Gallego-Salvador, N. Autonomic Management of a Building's Multi-HVAC System Start-Up. *IEEE Access* **2021**, *9*, 70502–70515. [CrossRef]
88. Eiras-Franco, C.; Flores, M.; Bolón-Canedo, V.; Zaragoza, S.; Fernández-Casal, R.; Naya, S.; Tarrío-Saavedra, J. Case Study of Anomaly Detection and Quality Control of Energy Efficiency and Hygrothermal Comfort in Buildings. In Proceedings of the DATA 2019—Proceedings of the 8th International Conference on Data Science, Technology and Applications, Prague, Czech Republic, 26–28 July 2019; SciTePress: Setúbal, Portugal, 2019; pp. 145–151.
89. Pertzborn, A. Using Distributed Agents to Optimize Thermal Energy Storage. *J. Energy Storage* **2019**, *23*, 89–97. [CrossRef]
90. Gupta, A.; Badr, Y.; Negahban, A.; Qiu, R.G. Energy-Efficient Heating Control for Smart Buildings with Deep Reinforcement Learning. *J. Build. Eng.* **2021**, *34*, 101739. [CrossRef]
91. Ding, X.; Du, W.; Cerpa, A.E. MB2C: Model-Based Deep Reinforcement Learning for Multi-Zone Building Control. In Proceedings of the BuildSys 2020—Proceedings of the 7th ACM International Conference on Systems for Energy-Efficient Buildings, Cities, and Transportation, Online, 18–20 November 2020; Association for Computing Machinery, Inc.: New York, NY, USA, 2020; pp. 50–59.
92. Chen, Y.; Shi, Y.; Zhang, B. Optimal Control via Neural Networks: A Convex Approach. In Proceedings of the 7th International Conference on Learning Representations, ICLR 2019, New Orleans, LA, USA, 6–9 May 2019; International Conference on Learning Representations, ICLR: Appleton, WI, USA, 2019.
93. Jiang, Z.; Risbeck, M.J.; Ramamurti, V.; Murugesan, S.; Amores, J.; Zhang, C.; Lee, Y.M.; Drees, K.H. Building HVAC Control with Reinforcement Learning for Reduction of Energy Cost and Demand Charge. *Energy Build.* **2021**, *239*, 110833. [CrossRef]
94. Nagarathinam, S.; Menon, V.; Vasan, A.; Sivasubramaniam, A. MARCO—Multi-Agent Reinforcement Learning Based COntrol of Building HVAC Systems. In Proceedings of the e-Energy 2020—Proceedings of the 11th ACM International Conference on Future Energy Systems, Online, 22–26 June 2020; Association for Computing Machinery, Inc.: New York, NY, USA, 2020; pp. 57–67.
95. Xu, S.; Wang, Y.; Wang, Y.; O'Neill, Z.; Zhu, Q. One for Many: Transfer Learning for Building HVAC Control. In Proceedings of the BuildSys 2020—Proceedings of the 7th ACM International Conference on Systems for Energy-Efficient Buildings, Cities, and Transportation, Online, 22–26 June 2020; Association for Computing Machinery, Inc.: New York, NY, USA, 2020; pp. 230–239.
96. Liu, B.; Akcakaya, M.; McDermott, T.E. Automated Control of Transactive HVACs in Energy Distribution Systems. *IEEE Trans. Smart Grid* **2021**, *12*, 2462–2471. [CrossRef]
97. Wei, T.; Ren, S.; Zhu, Q. Deep Reinforcement Learning for Joint Datacenter and HVAC Load Control in Distributed Mixed-Use Buildings. *IEEE Trans. Sustain. Comput.* **2021**, *6*, 370–384. [CrossRef]
98. Coraci, D.; Brandi, S.; Piscitelli, M.S.; Capozzoli, A. Online Implementation of a Soft Actor-Critic Agent to Enhance Indoor Temperature Control and Energy Efficiency in Buildings. *Energies* **2021**, *14*, 997. [CrossRef]
99. Kurte, K.; Munk, J.; Kotevska, O.; Amasyali, K.; Smith, R.; McKee, E.; Du, Y.; Cui, B.; Kuruganti, T.; Zandi, H. Evaluating the Adaptability of Reinforcement Learning Based HVAC Control for Residential Houses. *Sustainability* **2020**, *12*, 7727. [CrossRef]
100. Zhao, H.; Zhao, J.; Shu, T.; Pan, Z. Hybrid-Model-Based Deep Reinforcement Learning for Heating, Ventilation, and Air-Conditioning Control. *Front. Energy Res.* **2021**, *8*, 610518. [CrossRef]
101. Brandi, S.; Fiorentini, M.; Capozzoli, A. Comparison of Online and Offline Deep Reinforcement Learning with Model Predictive Control for Thermal Energy Management. *Autom. Constr.* **2022**, *135*, 104128. [CrossRef]
102. Ala'raj, M.; Radi, M.; Abbod, M.F.; Majdalawieh, M.; Parodi, M. Data-Driven Based HVAC Optimisation Approaches: A Systematic Literature Review. *J. Build. Eng.* **2022**, *46*, 103678. [CrossRef]

103. Homod, R.Z.; Togun, H.; Kadhim Hussein, A.; Noraldeen Al-Mousawi, F.; Yaseen, Z.M.; Al-Kouz, W.; Abd, H.J.; Alawi, O.A.; Goodarzi, M.; Hussein, O.A. Dynamics Analysis of a Novel Hybrid Deep Clustering for Unsupervised Learning by Reinforcement of Multi-Agent to Energy Saving in Intelligent Buildings. *Appl. Energy* **2022**, *313*, 118863. [CrossRef]
104. Wei, T.; Chen, X.; Li, X.; Zhu, Q. Model-Based and Data-Driven Approaches for Building Automation and Control. In Proceedings of the IEEE/ACM International Conference on Computer-Aided Design, Digest of Technical Papers, ICCAD, San Diego, CA, USA, 5–8 November 2018; Institute of Electrical and Electronics Engineers Inc.: Piscataway, NJ, USA, 2018.
105. Chen, Y.; Zheng, Y.; Samuelson, H. Fast Adaptation of Thermal Dynamics Model for Predictive Control of HVAC and Natural Ventilation Using Transfer Learning with Deep Neural Networks. In Proceedings of the Proceedings of the American Control Conference, Denver, CO, USA, 1–3 July 2020; Institute of Electrical and Electronics Engineers Inc.: Piscataway, NJ, USA, 2020; pp. 2345–2350.
106. Park, S.; Park, S.; Choi, M.-I.; Lee, S.; Lee, T.; Kim, S.; Cho, K.; Park, S. Reinforcement Learning-Based Bems Architecture for Energy Usage Optimization. *Sensors* **2020**, *20*, 4918. [CrossRef]
107. Pathak, N.; Lachut, D.; Roy, N.; Banerjee, N.; Robucci, R. Non-Intrusive Air Leakage Detection in Residential Homes. In Proceedings of the ICDCN '18, 19th International Conference on Distributed Computing and Networking, Varanasi, India, 4–7 January 2018; Association for Computing Machinery: New York, NY, USA, 2018.
108. Mosaico, G.; Saviozzi, M.; Silvestro, F.; Bagnasco, A.; Vinci, A. Simplified State Space Building Energy Model and Transfer Learning Based Occupancy Estimation for HVAC Optimal Control. In Proceedings of the 5th International Forum on Research and Technologies for Society and Industry: Innovation to Shape the Future, RTSI 2019—Proceedings, Florence, Italy, 9–12 September 2019; Institute of Electrical and Electronics Engineers Inc.: Piscataway, NJ, USA, 2019; pp. 353–358.
109. He, T.; Jazizadeh, F.; Arpan, L. AI-Powered Virtual Assistants Nudging Occupants for Energy Saving: Proactive Smart Speakers for HVAC Control. *Build. Res. Inf.* **2022**, *50*, 394–409. [CrossRef]
110. Wei, S.; Tien, P.W.; Chow, T.W.; Wu, Y.; Calautit, J.K. Deep Learning and Computer Vision Based Occupancy CO_2 Level Prediction for Demand-Controlled Ventilation (DCV). *J. Build. Eng.* **2022**, *56*, 104715. [CrossRef]
111. Fang, X.; Gong, G.; Li, G.; Chun, L.; Peng, P.; Li, W.; Shi, X.; Chen, X. Deep Reinforcement Learning Optimal Control Strategy for Temperature Setpoint Real-Time Reset in Multi-Zone Building HVAC System. *Appl. Therm. Eng.* **2022**, *212*, 118552. [CrossRef]
112. Zhang, T.; Baasch, G.; Ardakanian, O.; Evins, R. On the Joint Control of Multiple Building Systems with Reinforcement Learning. In Proceedings of the e-Energy 2021—Proceedings of the 2021 12th ACM International Conference on Future Energy Systems, Online, 28 June–2 July 2021; Association for Computing Machinery, Inc: New York, NY, USA, 2021; pp. 60–72.
113. Fu, Q.; Chen, X.; Ma, S.; Fang, N.; Xing, B.; Chen, J. Optimal Control Method of HVAC Based on Multi-Agent Deep Reinforcement Learning. *Energy Build.* **2022**, *270*, 112284. [CrossRef]
114. Cicirelli, F.; Guerrieri, A.; Mastroianni, C.; Scarcello, L.; Spezzano, G.; Vinci, A. Balancing Energy Consumption and Thermal Comfort with Deep Reinforcement Learning. In Proceedings of the Proceedings of the 2021 IEEE International Conference on Human-Machine Systems, ICHMS 2021, Magdeburg, Germany, 8–10 September 2021; Institute of Electrical and Electronics Engineers Inc.: Piscataway, NJ, USA, 2021.
115. Martinez-Molina, A.; Alamaniotis, M. Enhancing Historic Building Performance with the Use of Fuzzy Inference System to Control the Electric Cooling System. *Sustainability* **2020**, *12*, 5848. [CrossRef]
116. Fu, C.; Zhang, Y. Research and Application of Predictive Control Method Based on Deep Reinforcement Learning for HVAC Systems. *IEEE Access* **2021**, *9*, 130845–130852. [CrossRef]
117. Mason, K.; Grijalva, S. Building HVAC Control via Neural Networks and Natural Evolution Strategies. In Proceedings of the 2021 IEEE Congress on Evolutionary Computation, CEC 2021—Proceedings, Kraków, Poland, 28 June–2 July 2021; Institute of Electrical and Electronics Engineers Inc.: Piscataway, NJ, USA, 2021; pp. 2483–2490.
118. Kurte, K.; Amasyali, K.; Munk, J.; Zandi, H. Comparative Analysis of Model-Free and Model-Based HVAC Control for Residential Demand Response. In Proceedings of the BuildSys 2021—Proceedings of the 2021 ACM International Conference on Systems for Energy-Efficient Built Environments, Coimbra, Portugal, 17–18 November 2021; Association for Computing Machinery, Inc: New York, NY, USA, 2021; pp. 309–313.
119. Ayadi, M.I.; Maizate, A.; Ouzzif, M.; Mahmoudi, C. Deep Learning in Building Management Systems over NDN: Use Case of Forwarding and HVAC Control. In Proceedings of the 2019 International Conference on Internet of Things (iThings) and IEEE Green Computing and Communications (GreenCom) and IEEE Cyber, Physical and Social Computing (CPSCom) and IEEE Smart Data (SmartData), Atlanta, GA, USA, 14–17 July 2019; IEEE: New York, NY, USA, 2019; pp. 1192–1198.
120. Kurte, K.; Munk, J.; Amasyali, K.; Kotevska, O.; Cui, B.; Kuruganti, T.; Zandi, H. Electricity Pricing Aware Deep Reinforcement Learning Based Intelligent HVAC Control. In Proceedings of the RLEM 2020—Proceedings of the 1st International Workshop on Reinforcement Learning for Energy Management in Buildings and Cities, Online, 17 November 2020; Association for Computing Machinery, Inc.: New York, NY, USA, 2020; pp. 6–10.

121. Tao, Y.; Qiu, J.; Lai, S. A Hybrid Cloud and Edge Control Strategy for Demand Responses Using Deep Reinforcement Learning and Transfer Learning. *IEEE Trans. Cloud Comput.* **2022**, *10*, 56–71. [CrossRef]
122. Blad, C.; Bøgh, S.; Kallesøe, C. A Multi-Agent Reinforcement Learning Approach to Price and Comfort Optimization in Hvac-Systems. *Energies* **2021**, *14*, 7491. [CrossRef]
123. Naug, A.; Quinones-Grueiro, M.; Biswas, G. A Relearning Approach to Reinforcement Learning for Control of Smart Buildings. In Proceedings of the Annual Conference of the Prognostics and Health Management Society, PHM, Online, 9–13 November 2020; Prognostics and Health Management Society: Rochester, NY, USA, 2020; Volume 12.
124. Zhang, C.; Zhang, Z.; Loftness, V. Bio-Sensing and Reinforcement Learning Approaches for Occupant-Centric Control. *ASHRAE Trans.* **2019**, *125*, 364–371.
125. Naug, A.; Quiñones-Grueiro, M.; Biswas, G. Continual Adaptation in Deep Reinforcement Learning-Based Control Applied to Non-Stationary Building Environments. In Proceedings of the RLEM 2020—Proceedings of the 1st International Workshop on Reinforcement Learning for Energy Management in Buildings and Cities, Online, 17 November 2020; Association for Computing Machinery, Inc.: New York, NY, USA, 2020; pp. 24–28.
126. Fang, X.; Gong, G.; Li, G.; Chun, L.; Peng, P.; Li, W.; Shi, X. Cross Temporal-Spatial Transferability Investigation of Deep Reinforcement Learning Control Strategy in the Building HVAC System Level. *Energy* **2023**, *263*, 125679. [CrossRef]
127. Murugesan, S.; Jiang, Z.; Risbeck, M.J.; Amores, J.; Zhang, C.; Ramamurti, V.; Drees, K.H.; Lee, Y.M. Less Is More: Simplified State-Action Space for Deep Reinforcement Learning Based HVAC Control. In Proceedings of the RLEM 2020—Proceedings of the 1st International Workshop on Reinforcement Learning for Energy Management in Buildings and Cities, Online, 17 November 2020; Association for Computing Machinery, Inc.: New York, NY, USA, 2020; pp. 20–23.
128. Liu, H.-Y.; Balaji, B.; Gao, S.; Gupta, R.; Hong, D. Safe HVAC Control via Batch Reinforcement Learning. In Proceedings of the Proceedings—13th ACM/IEEE International Conference on Cyber-Physical Systems, ICCPS 2022, Milano, Italy, 4–6 May 2022; Institute of Electrical and Electronics Engineers Inc.: Piscataway, NJ, USA, 2022; pp. 181–192.
129. Zenginis, I.; Vardakas, J.; Koltsaklis, N.E.; Verikoukis, C. Smart Home's Energy Management Through a Clustering-Based Reinforcement Learning Approach. *IEEE Internet Things J.* **2022**, *9*, 16363–16371. [CrossRef]
130. Dawood, S.M.; Hatami, A.; Homod, R.Z. Trade-off Decisions in a Novel Deep Reinforcement Learning for Energy Savings in HVAC Systems. *J. Build. Perform. Simul.* **2022**, *15*, 809–831. [CrossRef]
131. Masburah, R.; Sinha, S.; Jana, R.L.; Dey, S.; Zhu, Q. Co-Designing Intelligent Control of Building HVACs and Microgrids. In Proceedings of the 2021 24th Euromicro Conference on Digital System Design, DSD 2021, Online, 1–3 September 2021; Institute of Electrical and Electronics Engineers Inc.: Piscataway, NJ, USA, 2021; pp. 457–464.
132. Jneid, K.; Ploix, S.; Reignier, P.; Jallon, P. Deep Q-Network Boosted with External Knowledge for HVAC Control. In Proceedings of the BuildSys 2021—Proceedings of the 2021 ACM International Conference on Systems for Energy-Efficient Built Environments, Coimbra, Portugal, 17–18 November 2021; Association for Computing Machinery, Inc.: New York, NY, USA, 2021; pp. 329–332.
133. Zhong, X.; Zhang, Z.; Zhang, R.; Zhang, C. End-to-End Deep Reinforcement Learning Control for HVAC Systems in Office Buildings. *Designs* **2022**, *6*, 52. [CrossRef]
134. Chae, M.; Kang, K.; Koo, D.; Oh, S.; Chun, J.Y. Fuzzy Controller Algorithm for Automated HVAC Control. In Proceedings of the 37th International Symposium on Automation and Robotics in Construction, ISARC 2020: From Demonstration to Practical Use—To New Stage of Construction Robot, Kitakyushu, Japan, 27–28 October 2020; International Association on Automation and Robotics in Construction (IAARC): Oulu, Finland, 2020; pp. 566–570.
135. Xu, D. Learning Efficient Dynamic Controller for HVAC System. *Mob. Inf. Syst.* **2022**, *2022*, 157511. [CrossRef]
136. Deng, X.; Zhang, Y.; Zhang, Y.; Qi, H. Toward Smart Multizone HVAC Control by Combining Context-Aware System and Deep Reinforcement Learning. *IEEE Internet Things J.* **2022**, *9*, 21010–21024. [CrossRef]
137. Xu, S.; Fu, Y.; Wang, Y.; Yang, Z.; O'Neill, Z.; Wang, Z.; Zhu, Q. Accelerate Online Reinforcement Learning for Building HVAC Control with Heterogeneous Expert Guidances. In Proceedings of the BuildSys 2022—Proceedings of the 2022 9th ACM International Conference on Systems for Energy-Efficient Buildings, Cities, and Transportation, Boston, MA, USA, 9–10 November 2022; Association for Computing Machinery, Inc.: New York, NY, USA, 2022; pp. 89–98.
138. Scarcello, L.; Mastroianni, C. Cognitive Systems for Energy Efficiency and Thermal Comfort in Smart Buildings. In *IoT Edge Solutions for Cognitive Buildings*; Internet of Things; Springer Science and Business Media Deutschland GmbH: Berlin/Heidelberg, Germany, 2023; pp. 329–345, ISBN 21991073.
139. Homod, R.Z.; Yaseen, Z.M.; Hussein, A.K.; Almusaed, A.; Alawi, O.A.; Falah, M.W.; Abdelrazek, A.H.; Ahmed, W.; Eltaweel, M. Deep Clustering of Cooperative Multi-Agent Reinforcement Learning to Optimize Multi Chiller HVAC Systems for Smart Buildings Energy Management. *J. Build. Eng.* **2023**, *65*, 105689. [CrossRef]
140. Naug, A.; Quinones-Grueiro, M.; Biswas, G. Deep Reinforcement Learning Control for Non-Stationary Building Energy Management. *Energy Build.* **2022**, *277*, 112584. [CrossRef]

141. Kurte, K.; Amasyali, K.; Munk, J.; Zandi, H. Deep Reinforcement Learning with Online Data Augmentation to Improve Sample Efficiency for Intelligent HVAC Control. In Proceedings of the BuildSys 2022—Proceedings of the 2022 9th ACM International Conference on Systems for Energy-Efficient Buildings, Cities, and Transportation, Boston, MA, USA, 9–10 November 2022; Association for Computing Machinery, Inc.: New York, NY, USA, 2022; pp. 479–483.
142. Qin, H.; Yu, Z.; Li, T.; Liu, X.; Li, L. Energy-Efficient Heating Control for Nearly Zero Energy Residential Buildings with Deep Reinforcement Learning. *Energy* **2023**, *264*, 126209. [CrossRef]
143. Liao, Y.; Liu, Y.; Chen, C.; Zhang, L. Green Building Energy Cost Optimization with Deep Belief Network and Firefly Algorithm. *Front. Energy Res.* **2021**, *9*, 805206. [CrossRef]
144. Aliero, M.S.; Pasha, M.F.; Toosi, A.N.; Ghani, I. The COVID-19 Impact on Air Condition Usage: A Shift towards Residential Energy Saving. *Environ. Sci. Pollut. Res.* **2022**, *29*, 85727–85741. [CrossRef]
145. Zhang, W.; Wu, Y.; Calautit, J.K. A Review on Occupancy Prediction through Machine Learning for Enhancing Energy Efficiency, Air Quality and Thermal Comfort in the Built Environment. *Renew. Sustain. Energy Rev.* **2022**, *167*, 112704. [CrossRef]
146. Tien, P.W.; Wei, S.; Calautit, J.K.; Darkwa, J.; Wood, C. Real-Time Monitoring of Occupancy Activities and Window Opening within Buildings Using an Integrated Deep Learning-Based Approach for Reducing Energy Demand. *Appl. Energy* **2022**, *308*, 118336. [CrossRef]
147. Al Mohammad, A.; Hirmiz, H.; Perera, G.; Maleki, M. Motif-Based Occupancy Prediction for Energy Efficiency in HVAC. In Proceedings of the 2022 3rd International Conference on Pattern Recognition and Machine Learning, PRML 2022, Chengdu, China, 22–24 July 2022; Institute of Electrical and Electronics Engineers Inc.: Piscataway, NJ, USA, 2022; pp. 465–469.
148. Elkhoukhi, H.; NaitMalek, Y.; Bakhouya, M.; Berouine, A.; Kharbouch, A.; Lachhab, F.; Hanifi, M.; El Ouadghiri, D.; Essaaidi, M. A Platform Architecture for Occupancy Detection Using Stream Processing and Machine Learning Approaches. *Concurr. Comput.* **2020**, *32*, e5651. [CrossRef]
149. Jacoby, M.; Tan, S.Y.; Katanbaf, M.; Saffari, A.; Saha, H.; Kapetanovic, Z.; Garland, J.; Florita, A.; Henze, G.; Sarkar, S.; et al. Whisper: Wireless Home Identification and Sensing Platform for Energy Reduction. *J. Sens. Actuator Netw.* **2021**, *10*, 71. [CrossRef]
150. Acquaah, Y.; Steele, J.B.; Gokaraju, B.; Tesiero, R.; Monty, G.H. Occupancy Detection for Smart Hvac Efficiency in Building Energy: A Deep Learning Neural Network Framework Using Thermal Imagery. In Proceedings of the Applied Imagery Pattern Recognition Workshop, Washington, DC, USA, 13–15 October 2020; Institute of Electrical and Electronics Engineers Inc.: Piscataway, NJ, USA, 2020.
151. Acquaah, Y.T.; Gokaraju, B.; Tesiero, R.C.; Monty, G.H. Thermal Imagery Feature Extraction Techniques and the Effects on Machine Learning Models for Smart Hvac Efficiency in Building Energy. *Remote Sens.* **2021**, *13*, 3847. [CrossRef]
152. Khalil, M.; McGough, S.; Pourmirza, Z.; Pazhoohesh, M.; Walker, S. Transfer Learning Approach for Occupancy Prediction in Smart Buildings. In Proceedings of the 2021 12th International Renewable Engineering Conference, IREC 2021, Amman, Jordan, 14–15 April 2021; Institute of Electrical and Electronics Engineers Inc.: Piscataway, NJ, USA, 2021.
153. Issaraviriyakul, A.; Pora, W.; Panitantum, N. Cloud-Based Machine Learning Framework for Residential HVAC Control System. In Proceedings of the KST 2021—13th International Conference Knowledge and Smart Technology, Bangsaen, Thailand, 21–24 January 2021; Institute of Electrical and Electronics Engineers Inc.: Piscataway, NJ, USA, 2021; pp. 12–16.
154. Ye, Z.; O'neill, Z.; Hu, F. Hardware-Based Emulator with Deep Learning Model for Building Energy Control and Prediction Based on Occupancy Sensors' Data. *Information* **2021**, *12*, 499. [CrossRef]
155. McKee, E.; Du, Y.; Li, F.; Munk, J.; Johnston, T.; Kurte, K.; Kotevska, O.; Amasyali, K.; Zandi, H. Deep Reinforcement Learning for Residential Hvac Control with Consideration of Human Occupancy. In Proceedings of the IEEE Power and Energy Society General Meeting, Montreal, QC, Canada, 2–6 August 2020; IEEE Computer Society: Washington, DC, USA, 2020.
156. Momeni, M.; Wu, D.-C.; Razban, A.; Chen, J. Data-Driven Demand Control Ventilation Using Machine Learning CO_2 occupancy Detection Method. In Proceedings of the ECOS 2020—Proceedings of the 33rd International Conference on Efficiency, Cost, Optimization, Simulation and Environmental Impact of Energy Systems, Osaka, Japan, 29 June–3 July 2020; pp. 2060–2071.
157. Chitu, C.; Stamatescu, G.; Stamatescu, I.; Sgarciu, V. Assessment of Occupancy Estimators for Smart Buildings. In Proceedings of the 2019 10th IEEE International Conference on Intelligent Data Acquisition and Advanced Computing Systems: Technology and Applications, IDAACS 2019, Metz, France, 18–21 September 2019; Institute of Electrical and Electronics Engineers Inc.: Piscataway, NJ, USA, 2019; Volume 1, pp. 228–233.
158. Chandrasiri, A.P.; Geekiyanage, D. Real-Time Object Detection System for Building Energy Conservation: An IP Camera Based System. In Proceedings of the 34th Annual ARCOM Conference, ARCOM 2018, Belfast, Ireland, 3–5 September 2018; pp. 567–576.
159. Wang, C.; Lee, H. Economical and Non-Invasive Residential Human Presence Sensing via Temperature Measurement. In Proceedings of the ASME International Mechanical Engineering Congress and Exposition, Proceedings (IMECE), Pitsburgh, PA, USA, 9–15 November 2018; American Society of Mechanical Engineers (ASME): New York, NY, USA, 2018; Volume 6A-144113.

160. Tien, P.W.; Wei, S.; Calautit, J.K.; Darkwa, J.; Wood, C. A Vision-Based Deep Learning Approach for the Detection and Prediction of Occupancy Heat Emissions for Demand-Driven Control Solutions. *Energy Build.* **2020**, *226*, 110386. [CrossRef]
161. Choi, H.; Um, C.Y.; Kang, K.; Kim, H.; Kim, T. Application of Vision-Based Occupancy Counting Method Using Deep Learning and Performance Analysis. *Energy Build.* **2021**, *252*, 111389. [CrossRef]
162. Esrafilian-Najafabadi, M.; Haghighat, F. Occupancy-Based HVAC Control Using Deep Learning Algorithms for Estimating Online Preconditioning Time in Residential Buildings. *Energy Build.* **2021**, *252*, 111377. [CrossRef]
163. Lei, Y.; Zhan, S.; Ono, E.; Peng, Y.; Zhang, Z.; Hasama, T.; Chong, A. A Practical Deep Reinforcement Learning Framework for Multivariate Occupant-Centric Control in Buildings. *Appl. Energy* **2022**, *324*, 119742. [CrossRef]
164. Tien, P.W.; Wei, S.; Calautit, J.K.; Darkwa, J.; Wood, C. A Computer Vision Deep Learning Method for the Detection and Recognition of Manual Window Openings for Effective Operations of HVAC Systems in Buildings. In Proceedings of the International Conference of Architectural Science Association, Auckland, New Zeland, 26–28 November 2018; pp. 21–30.
165. Wei, S.; Tien, P.W.; Wu, Y.; Calautit, J.K. The Impact of Deep Learning–Based Equipment Usage Detection on Building Energy Demand Estimation. *Build. Serv. Eng. Res. Technol.* **2021**, *42*, 545–557. [CrossRef]
166. Esrafilian-Najafabadi, M.; Haghighat, F. Towards Self-Learning Control of HVAC Systems with the Consideration of Dynamic Occupancy Patterns: Application of Model-Free Deep Reinforcement Learning. *Build. Environ.* **2022**, *226*, 109747. [CrossRef]
167. Yayla, A.; Świerczewska, K.S.; Kaya, M.; Karaca, B.; Arayıcı, Y.; Ayözen, Y.E.; Tokdemir, O.B. Artificial Intelligence (AI)-Based Occupant-Centric Heating Ventilation and Air Conditioning (HVAC) Control System for Multi-Zone Commercial Buildings. *Sustainability* **2022**, *14*, 16107. [CrossRef]
168. Wei, S.; Tien, P.W.; Calautit, J.K.; Boukhanouf, R.; Wu, Y. Equipment Load Detection Using Deep Learning for Building Energy Management. *Proc. IOP Conf. Ser. Earth Environ. Sci.* **2020**, *463*, 012027.
169. Wei, S.; Tien, P.W.; Wu, Y.; Calautit, J.K. A Coupled Deep Learning-Based Internal Heat Gains Detection and Prediction Method for Energy-Efficient Office Building Operation. *J. Build. Eng.* **2022**, *47*, 103778. [CrossRef]
170. Zhou, Y. Advanced Renewable Dispatch with Machine Learning-Based Hybrid Demand-Side Controller: The State of the Art and a Novel Approach. In *Sustainable Developments by Artificial Intelligence and Machine Learning for Renewable Energies*; Elsevier: Amsterdam, The Netherlands, 2022; pp. 237–256, ISBN 978-032391228-0.
171. Kim, D.; Lee, Y.; Chin, K.; Mago, P.J.; Cho, H.; Zhang, J. Implementation of a Long Short-Term Memory Transfer Learning (LSTM-TL)-Based Data-Driven Model for Building Energy Demand Forecasting. *Sustainability* **2023**, *15*, 2340. [CrossRef]
172. Wang, B.; Wang, X.; Wang, N.; Javaheri, Z.; Moghadamnejad, N.; Abedi, M. Machine Learning Optimization Model for Reducing the Electricity Loads in Residential Energy Forecasting. *Sustain. Comput. Inform. Syst.* **2023**, *38*, 100876. [CrossRef]
173. Ma, H.; Xu, L.; Javaheri, Z.; Moghadamnejad, N.; Abedi, M. Reducing the Consumption of Household Systems Using Hybrid Deep Learning Techniques. *Sustain. Comput. Inform. Syst.* **2023**, *38*, 100874. [CrossRef]
174. Alawadi, S.; Mera, D.; Fernández-Delgado, M.; Alkhabbas, F.; Olsson, C.M.; Davidsson, P. A Comparison of Machine Learning Algorithms for Forecasting Indoor Temperature in Smart Buildings. *Energy Syst.* **2022**, *13*, 689–705. [CrossRef]
175. Rana, M.; Sethuvenkatraman, S.; Goldsworthy, M. A Data-Driven Approach Based on Quantile Regression Forest to Forecast Cooling Load for Commercial Buildings. *Sustain. Cities Soc.* **2022**, *76*, 103511. [CrossRef]
176. Metsä-Eerola, I.; Pulkkinen, J.; Niemitalo, O.; Koskela, O. On Hourly Forecasting Heating Energy Consumption of HVAC with Recurrent Neural Networks. *Energies* **2022**, *15*, 5084. [CrossRef]
177. Li, M. Optimizing HVAC Systems in Buildings with Machine Learning Prediction Models: An Algorithm Based Economic Analysis. In Proceedings of the—2020 Management Science Informatization and Economic Innovation Development Conference, MSIEID 2020, Guangzhou, China, 18–20 December 2020; Institute of Electrical and Electronics Engineers Inc.: Piscataway, NJ, USA, 2020; pp. 210–217.
178. Chen, Y.; Zhang, F.; Berardi, U. Day-Ahead Prediction of Hourly Subentry Energy Consumption in the Building Sector Using Pattern Recognition Algorithms. *Energy* **2020**, *211*, 118530. [CrossRef]
179. Borowski, M.; Zwolińska, K. Prediction of Cooling Energy Consumption in Hotel Building Using Machine Learning Techniques. *Energies* **2020**, *13*, 6226. [CrossRef]
180. Mawson, V.J.; Hughes, B.R. Optimisation of HVAC Control and Manufacturing Schedules for the Reduction of Peak Energy Demand in the Manufacturing Sector. *Energy* **2021**, *227*, 120436. [CrossRef]
181. Goyal, M.; Pandey, M. Extreme Gradient Boosting Algorithm for Energy Optimization in Buildings Pertaining to HVAC Plants. *EAI Endorsed Trans. Energy Web* **2020**, *21*, e1. [CrossRef]
182. Kajewska-Szkudlarek, J.; Bylicki, J.; Stańczyk, J.; Licznar, P. Neural Approach in Short-Term Outdoor Temperature Prediction for Application in Hvac Systems. *Energies* **2021**, *14*, 7512. [CrossRef]

183. Li, Y.; Peng, Y.; Zhang, D.; Mai, Y.; Ruan, Z.; Liang, S. Energy Usage Prediction Based on Multi-System Data for Public Buildings Using Machine Learning Methods. In Proceedings of the—2021 International Conference on Computers and Automation, CompAuto 2021, Paris, France, 7–9 September 2021; Institute of Electrical and Electronics Engineers Inc.: Piscataway, NJ, USA, 2021; pp. 7–13.
184. Chalapathy, R.; Khoa, N.L.D.; Sethuvenkatraman, S. Comparing Multi-Step Ahead Building Cooling Load Prediction Using Shallow Machine Learning and Deep Learning Models. *Sustain. Energy Grids Netw.* **2021**, *28*, 100543. [CrossRef]
185. Khamma, T.R.; Zhang, Y.; Guerrier, S.; Boubekri, M. Generalized Additive Models: An Efficient Method for Short-Term Energy Prediction in Office Buildings. *Energy* **2020**, *213*, 118834. [CrossRef]
186. Goyal, M.; Pandey, M. Ensemble-Based Data Modeling for the Prediction of Energy Consumption in HVAC Plants. *J. Reliab. Intell. Environ.* **2021**, *7*, 49–64. [CrossRef]
187. Zheng, J.; Ling, Z.; Kang, Y.; You, L.; Zhao, Y.; Xiao, Z.; Chen, X. HVAC Load Forecasting in Office Buildings Using Machine Learning. In Proceedings of the 2021 6th International Conference on Power and Renewable Energy, ICPRE 2021, Shanghai, China, 17–20 September 2021; Institute of Electrical and Electronics Engineers Inc.: Piscataway, NJ, USA, 2021; pp. 877–881.
188. Moayedi, H.; Mosavi, A. Double-Target Based Neural Networks in Predicting Energy Consumption in Residential Buildings. *Energies* **2021**, *14*, 1331. [CrossRef]
189. Kanewala, U.; Weerakoon, S.; Nawarathna, R. Exploring Deep Learning and Tree-Based Ensemble Models for Chiller Energy Consumption Predictions. In Proceedings of the 2021 IEEE 16th International Conference on Industrial and Information Systems, ICIIS 2021—Proceedings, Kandy, Sri Lanka, 9–11 December 2021; Institute of Electrical and Electronics Engineers Inc.: Piscataway, NJ, USA, 2021; pp. 306–311.
190. Kim, J.-H.; Seong, N.-C.; Choi, W. Modeling and Optimizing a Chiller System Using a Machine Learning Algorithm. *Energies* **2019**, *12*, 2860. [CrossRef]
191. Wang, J.; Li, G.; Chen, H.; Liu, J.; Guo, Y.; Sun, S.; Hu, Y. Energy Consumption Prediction for Water-Source Heat Pump System Using Pattern Recognition-Based Algorithms. *Appl. Therm. Eng.* **2018**, *136*, 755–766. [CrossRef]
192. Goyal, M.; Pandey, M. Towards Prediction of Energy Consumption of HVAC Plants Using Machine Learning. In Proceedings of the International Conference on Recent Developments in Science, Engineering and Technology, Gurugram, India, 15–16 November 2019; Springer: Berlin/Heidelberg, Germany, 2020; Volume 1229, pp. 254–265.
193. Sala-Cardoso, E.; Delgado-Prieto, M.; Kampouropoulos, K.; Romeral, L. Activity-Aware HVAC Power Demand Forecasting. *Energy Build.* **2018**, *170*, 15–24. [CrossRef]
194. Ahmad, M.W.; Mouraud, A.; Rezgui, Y.; Mourshed, M. Deep Highway Networks and Tree-Based Ensemble for Predicting Short-Term Building Energy Consumption. *Energies* **2018**, *11*, 3408. [CrossRef]
195. Bellatreche, L.; Garcia, F.; Pham, D.N.; Jiménez, P.Q. Sonder: A Data-Driven Methodology for Designing Net-Zero Energy Public Buildings. In Proceedings of the International Conference on Big Data Analytics and Knowledge Discovery, DaWaK 2020, Bratislava, Slovakia, 14–17 September 2020; Springer Science and Business Media Deutschland GmbH: Berlin/Heidelberg, Germany, 2020; Volume 12393, pp. 48–59.
196. Tian, W.; Lei, C.; Tian, M. Dynamic Prediction of Building HVAC Energy Consumption by Ensemble Learning Approach. In Proceedings of the—2018 International Conference on Computational Science and Computational Intelligence, CSCI 2018, Las Vegas, NV, USA, 12–14 December 2018; pp. 254–257.
197. Xiao, Z.; Gang, W.; Yuan, J.; Chen, Z.; Li, J.; Wang, X.; Feng, X. Impacts of Data Preprocessing and Selection on Energy Consumption Prediction Model of HVAC Systems Based on Deep Learning. *Energy Build.* **2022**, *258*, 111832. [CrossRef]
198. Liu, Z.; Jiang, G. Optimization of Intelligent Heating Ventilation Air Conditioning System in Urban Building Based on Bim and Artificial Intelligence Technology. *Comput. Sci. Inf. Syst.* **2021**, *18*, 1379–1394. [CrossRef]
199. Chen, Y.; Chandan, V.; Huang, Y.; Alam, M.J.E.; Ahmed, O.; Smith, L. Coordination of Behind-the-Meter Energy Storage and Building Loads: Optimization with Deep Learning Model. In Proceedings of the e-Energy 2019—Proceedings of the 10th ACM International Conference on Future Energy Systems, Phoenix, AZ, USA, 25–28 June 2019; Association for Computing Machinery, Inc.: New York, NY, USA, 2019; pp. 492–499.
200. Gangopadhyay, T.; Tan, S.Y.; Jiang, Z.; Sarkar, S. Interpretable Deep Attention Model for Multivariate Time Series Prediction in Building Energy Systems. In Proceedings of the International Conference on Dynamic Data Driven Applications Systems, DDDAS 2020, Boston, MA, USA, 2–4 October 2020; Springer Science and Business Media Deutschland GmbH: Berlin/Heidelberg, Germany, 2020; Volume 12312, pp. 93–101.
201. Runge, J.; Zmeureanu, R. Deep Learning Forecasting for Electric Demand Applications of Cooling Systems in Buildings. *Adv. Eng. Inform.* **2022**, *53*, 101674. [CrossRef]
202. Qiao, D.; Shen, B.; Dong, X.; Zheng, H.; Song, W.; Wu, S. MTL-Deep-STF: A Multitask Learning Based Deep Spatiotemporal Fusion Model for Outdoor Air Temperature Prediction in Building HVAC Systems. *J. Build. Eng.* **2022**, *62*, 105364. [CrossRef]

203. Singh, A. Optimal Thermal Comfort Based Energy Efficient Strategy of HVAC System Using Supervised Learning Based Classifier with Demand Response. In Proceedings of the 10th IEEE International Conference on Power Electronics, Drives and Energy Systems, PEDES 2022, Jaipur, India, 14–17 December 2022; Institute of Electrical and Electronics Engineers Inc.: Piscataway, NJ, USA, 2022.
204. Metzmacher, H.; Syndicus, M.; Warthmann, A.; van Treeck, C. Exploratory Comparison of Control Algorithms and Machine Learning as Regulators for a Personalized Climatization System. *Energy Build.* **2022**, *255*, 111653. [CrossRef]
205. Yu, L.; Xu, Z.; Zhang, T.; Guan, X.; Yue, D. Energy-Efficient Personalized Thermal Comfort Control in Office Buildings Based on Multi-Agent Deep Reinforcement Learning. *Build. Environ.* **2022**, *223*, 109458. [CrossRef]
206. Boutahri, Y.; Tilioua, A. Artificial Intelligent-Based System for Thermal Comfort Control in Smart Building. In Proceedings of the International Conference on Artificial Intelligence and Smart Environment, Errachidia, Morocco, 24–26 November 2022; Springer Science and Business Media Deutschland GmbH: Berlin/Heidelberg, Germany, 2023; Volume 635, pp. 240–246.
207. Acquaah, Y.T.; Gokaraju, B.; Tesioro, R.C.; Monty, G.; Roy, K. Occupancy and Thermal Preference-Based HVAC Control Strategy Using Multisensor Network. *IEEE Sens. J.* **2023**, *23*, 11785–11795. [CrossRef]
208. Hou, F.; Ma, J.; Kwok, H.H.L.; Cheng, J.C.P. Prediction and Optimization of Thermal Comfort, IAQ and Energy Consumption of Typical Air-Conditioned Rooms Based on a Hybrid Prediction Model. *Build. Environ.* **2022**, *225*, 109576. [CrossRef]
209. Alsaleem, F.; Tesfay, M.K.; Rafaie, M.; Sinkar, K.; Besarla, D.; Arunasalam, P. An IoT Framework for Modeling and Controlling Thermal Comfort in Buildings. *Front. Built Environ.* **2020**, *6*, 87. [CrossRef]
210. Qian, J.; Cheng, X.; Yang, B.; Li, Z.; Ren, J.; Olofsson, T.; Li, H. Vision-Based Contactless Pose Estimation for Human Thermal Discomfort. *Atmosphere* **2020**, *11*, 376. [CrossRef]
211. Ghahramani, A.; Galicia, P.; Lehrer, D.; Varghese, Z.; Wang, Z.; Pandit, Y. Artificial Intelligence for Efficient Thermal Comfort Systems: Requirements, Current Applications and Future Directions. *Front. Built Environ.* **2020**, *6*, 49. [CrossRef]
212. Gao, N.; Shao, W.; Rahaman, M.S.; Zhai, J.; David, K.; Salim, F.D. Transfer Learning for Thermal Comfort Prediction in Multiple Cities. *Build. Environ.* **2021**, *195*, 107725. [CrossRef]
213. Somu, N.; Sriram, A.; Kowli, A.; Ramamritham, K. A Hybrid Deep Transfer Learning Strategy for Thermal Comfort Prediction in Buildings. *Build. Environ.* **2021**, *204*, 108133. [CrossRef]
214. Abdulgader, M.; Lashhab, F. Energy-Efficient Thermal Comfort Control in Smart Buildings. In Proceedings of the 2021 IEEE 11th Annual Computing and Communication Workshop and Conference, CCWC 2021, Online, 27–30 January 2021; Institute of Electrical and Electronics Engineers Inc.: Piscataway, NJ, USA, 2021; pp. 22–26.
215. Lu, S.; Cochran Hameen, E. An Interactive Task Conditioning System Featuring Personal Comfort Models and Non-Intrusive Sensing Techniques: A Field Study in Shanghai. *Technologies* **2021**, *9*, 90. [CrossRef]
216. Chen, Z.; Tao, Z.; Chang, A. A Data-Driven Approach to Optimize Building Energy Performance and Thermal Comfort Using Machine Learning Models. In Proceedings of the ICCIR '21: Proceedings of the 2021 1st International Conference on Control and Intelligent Robotics, Guangzhou, China, 18–20 June 2021; Association for Computing Machinery: New York, NY, USA, 2021; pp. 464–469.
217. Ding, Z.; Fu, Q.; Chen, J.; Wu, H.; Lu, Y.; Hu, F. Multi-Zone Residential HVAC Control with Satisfying Occupants' Thermal Comfort Requirements and Saving Energy via Reinforcement Learning. In Proceedings of the International Conference on Parallel and Distributed Computing: Applications and Technologies, PDCAT 2021, Guangzhou, China, 17–19 December 2021; Springer Science and Business Media Deutschland GmbH: Berlin/Heidelberg, Germany, 2022; Volume 13148, pp. 441–451.
218. Syed Ahmad, S.S.; Yung, S.M.; Kausar, N.; Karaca, Y.; Pamucar, D.; Al Din Ide, N. Nonlinear Integrated Fuzzy Modeling to Predict Dynamic Occupant Environment Comfort for Optimized Sustainability. *Sci. Program.* **2022**, *2022*, 1–13. [CrossRef]
219. Qin, S.; Yu, L.; Yue, D.; Shen, C. Optimal HVAC Control in Shared Office Spaces Based on Deep Reinforcement Learning. In *Proceedings—2021 China Automation Congress, CAC 2021*; Beijing, China, 22–24 October 2021, Institute of Electrical and Electronics Engineers Inc.: Piscataway, NJ, USA, 2021; pp. 1599–1604.
220. Cho, S.; Nam, H.J.; Shi, C.; Kim, C.Y.; Byun, S.-H.; Agno, K.-C.; Lee, B.C.; Xiao, J.; Sim, J.Y.; Jeong, J.-W. Wireless, AI-Enabled Wearable Thermal Comfort Sensor for Energy-Efficient, Human-in-the-Loop Control of Indoor Temperature. *Biosens. Bioelectron.* **2023**, *223*, 115018. [CrossRef]
221. Bezyan, B.; Zmeureanu, R. Detection and Diagnosis of Dependent Faults That Trigger False Symptoms of Heating and Mechanical Ventilation Systems Using Combined Machine Learning and Rule-Based Techniques. *Energies* **2022**, *15*, 1691. [CrossRef]
222. Hosamo, H.H.; Svennevig, P.R.; Svidt, K.; Han, D.; Nielsen, H.K. A Digital Twin Predictive Maintenance Framework of Air Handling Units Based on Automatic Fault Detection and Diagnostics. *Energy Build.* **2022**, *261*, 111988. [CrossRef]
223. Chen, Y.; Wen, J.; Pradhan, O.; Lo, L.J.; Wu, T. Using Discrete Bayesian Networks for Diagnosing and Isolating Cross-Level Faults in HVAC Systems. *Appl. Energy* **2022**, *327*, 120050. [CrossRef]

224. Martinez-Viol, V.; Urbano, E.M.; Torres Rangel, J.E.; Delgado-Prieto, M.; Romeral, L. Semi-Supervised Transfer Learning Methodology for Fault Detection and Diagnosis in Air-Handling Units. *Appl. Sci.* **2022**, *12*, 8837. [CrossRef]
225. Yan, K.; Zhou, X. Chiller Faults Detection and Diagnosis with Sensor Network and Adaptive 1D CNN. *Digit. Commun. Netw.* **2022**, *8*, 531–539. [CrossRef]
226. Zhang, C.; Zhao, Y.; Zhao, Y.; Li, T.; Zhang, X. Causal Discovery and Inference-Based Fault Detection and Diagnosis Method for Heating, Ventilation and Air Conditioning Systems. *Build. Environ.* **2022**, *212*, 108760. [CrossRef]
227. Du, Z.; Chen, S.; Li, P.; Chen, K.; Liang, X.; Zhu, X.; Jin, X. Knowledge-Extracted Deep Learning Diagnosis and Its Cloud-Based Management for Multiple Faults of Chiller. *Build. Environ.* **2023**, *235*, 110228. [CrossRef]
228. Mertens, N.; Wohlfahrt, T.; Hartmann, N.; Reddy, C.B.V. Automatic Transformation of HVAC Diagrams into Machine-Readable Format. In *Product Lifecycle Management. PLM in Transition Times: The Place of Humans and Transformative Technologies, Proceedings of the 19th IFIP WG 5.1 International Conference, PLM 2022, Grenoble, France, 10–13 July 2022*; Noël, F., Nyffenegger, F., Rivest, L., Bouras, A., Eds.; Springer Science and Business Media Deutschland GmbH: Berlin/Heidelberg, Germany, 2023; Volume 667, pp. 410–419.
229. Chen, J.; Zhang, L.; Li, Y.; Shi, Y.; Gao, X.; Hu, Y. A Review of Computing-Based Automated Fault Detection and Diagnosis of Heating, Ventilation and Air Conditioning Systems. *Renew. Sustain. Energy Rev.* **2022**, *161*, 112395. [CrossRef]
230. Albayati, M.G.; Faraj, J.; Thompson, A.; Patil, P.; Gorthala, R.; Rajasekaran, S. Semi-Supervised Machine Learning for Fault Detection and Diagnosis of a Rooftop Unit. *Big Data Min. Anal.* **2023**, *6*, 170–184. [CrossRef]
231. Yang, C.; Gunay, B.; Shi, Z.; Shen, W. Machine Learning-Based Prognostics for Central Heating and Cooling Plant Equipment Health Monitoring. *IEEE Trans. Autom. Sci. Eng.* **2021**, *18*, 346–355. [CrossRef]
232. Li, G.; Zheng, Y.; Liu, J.; Zhou, Z.; Xu, C.; Fang, X.; Yao, Q. An Improved Stacking Ensemble Learning-Based Sensor Fault Detection Method for Building Energy Systems Using Fault-Discrimination Information. *J. Build. Eng.* **2021**, *43*, 102812. [CrossRef]
233. Zhu, X.; Chen, K.; Anduv, B.; Jin, X.; Du, Z. Transfer Learning Based Methodology for Migration and Application of Fault Detection and Diagnosis between Building Chillers for Improving Energy Efficiency. *Build. Environ.* **2021**, *200*, 107957. [CrossRef]
234. Parzinger, M.; Wellisch, U.; Hanfstaengl, L.; Sigg, F.; Wirnsberger, M.; Spindler, U. Identifying Faults in the Building System Based on Model Prediction and Residuum Analysis. *Proc. E3S Web Conf.* **2020**, *172*, 22001.
235. Alghanmi, A.; Yunusa-Kaltungo, A.; Edwards, R. A Comparative Study of Faults Detection Techniques on HVAC Systems. In Proceedings of the 2021 IEEE PES/IAS PowerAfrica, PowerAfrica 2021, Nairobi, Kenya, 23–27 August 2021; Institute of Electrical and Electronics Engineers Inc.: Piscataway, NJ, USA, 2021.
236. Dey, M.; Rana, S.P.; Dudley, S. Smart Building Creation in Large Scale HVAC Environments through Automated Fault Detection and Diagnosis. *Future Gener. Comput. Syst.* **2020**, *108*, 950–966. [CrossRef]
237. Dey, M.; Rana, S.P.; Dudley, S. A Case Study Based Approach for Remote Fault Detection Using Multi-Level Machine Learning in a Smart Building. *Smart Cities* **2020**, *3*, 401–419. [CrossRef]
238. Parzinger, M.; Hanfstaengl, L.; Sigg, F.; Spindler, U.; Wellisch, U.; Wirnsberger, M. Residual Analysis of Predictive Modelling Data for Automated Fault Detection in Building's Heating, Ventilation and Air Conditioning Systems. *Sustainability* **2020**, *12*, 6758. [CrossRef]
239. Dowling, C.P.; Zhang, B. Transfer Learning for HVAC System Fault Detection. In Proceedings of the American Control Conference, Denver, CO, USA, 1–3 July 2020; Institute of Electrical and Electronics Engineers Inc.: Piscataway, NJ, USA, 2020; pp. 3879–3885.
240. Gharsellaoui, S.; Mansouri, M.; Trabelsi, M.; Harkat, M.-F.; Refaat, S.S.; Messaoud, H. Interval-Valued Features Based Machine Learning Technique for Fault Detection and Diagnosis of Uncertain HVAC Systems. *IEEE Access* **2020**, *8*, 171892–171902. [CrossRef]
241. Zhong, C.; Yan, K.; Dai, Y.; Jin, N.; Lou, B. Energy Efficiency Solutions for Buildings: Automated Fault Diagnosis of Air Handling Units Using Generative Adversarial Networks. *Energies* **2019**, *12*, 527. [CrossRef]
242. McHugh, M.K.; Isakson, T.; Nagy, Z. Data-Driven Leakage Detection in Air-Handling Units on a University Campus. *ASHRAE Trans.* **2019**, *125*, 381–388.
243. Gharsellaoui, S.; Mansouri, M.; Refaat, S.S.; Abu-Rub, H.; Messaoud, H. Multivariate Features Extraction and Effective Decision Making Using Machine Learning Approaches. *Energies* **2020**, *13*, 609. [CrossRef]
244. Yang, C.; Shen, W.; Gunay, B.; Shi, Z. Toward Machine Learning-Based Prognostics for Heating Ventilation and Air-Conditioning Systems. *ASHRAE Trans.* **2019**, *125*, 106–115.
245. Gharsellaoui, S.; Mansouri, M.; Refaat, S.S.; Abu-Rub, H.; Messaoud, H. Enhanced Machine Learning Approaches for Diagnosing Building Systems. In Proceedings of the 2019 International Conference on Internet of Things, Embedded Systems and Communications, IINTEC 2019—Proceedings, Tunis, Tunisia, 20–22 December 2019; Institute of Electrical and Electronics Engineers Inc.: Piscataway, NJ, USA, 2019; pp. 136–141.

246. Li, G.; Yao, Q.; Fan, C.; Zhou, C.; Wu, G.; Zhou, Z.; Fang, X. An Explainable One-Dimensional Convolutional Neural Networks Based Fault Diagnosis Method for Building Heating, Ventilation and Air Conditioning Systems. *Build. Environ.* **2021**, *203*, 108057. [CrossRef]
247. Li, Y.; O'Neill, Z. An Innovative Fault Impact Analysis Framework for Enhancing Building Operations. *Energy Build.* **2019**, *199*, 311–331. [CrossRef]
248. Tasfi, N.L.; Higashino, W.A.; Grolinger, K.; Capretz, M.A.M. Deep Neural Networks with Confidence Sampling for Electrical Anomaly Detection. In Proceedings of the—2017 IEEE International Conference on Internet of Things, IEEE Green Computing and Communications, IEEE Cyber, Physical and Social Computing, IEEE Smart Data, iThings-GreenCom-CPSCom-SmartData 2017, Exeter, UK, 21–23 June 2017; Institute of Electrical and Electronics Engineers Inc.: Piscataway, NJ, USA, 2018; pp. 1038–1045.
249. Xu, S.; Fu, Y.; Wang, Y.; O'Neill, Z.; Zhu, Q. Learning-Based Framework for Sensor Fault-Tolerant Building HVAC Control with Model-Assisted Learning. In Proceedings of the BuildSys 2021—Proceedings of the 2021 ACM International Conference on Systems for Energy-Efficient Built Environments, Coimbra, Portugal, 17–18 November 2021; Association for Computing Machinery, Inc.: New York, NY, USA, 2021; pp. 1–10.
250. Wang, D.; Li, X. A Novel Virtual Sensor Modeling Method Based on Deep Learning and Its Application in Heating, Ventilation, and Air-Conditioning System. *Energies* **2022**, *15*, 5743. [CrossRef]
251. Li, G.; Chen, L.; Liu, J.; Fang, X. Comparative Study on Deep Transfer Learning Strategies for Cross-System and Cross-Operation-Condition Building Energy Systems Fault Diagnosis. *Energy* **2023**, *263*, 125943. [CrossRef]
252. Du, Z.; Chen, S.; Anduv, B.; Zhu, X.; Jin, X. IoT Intelligent Agent Based Cloud Management System by Integrating Machine Learning Algorithm for HVAC Systems. *Int. J. Refrig.* **2023**, *146*, 158–173. [CrossRef]
253. Hosamo, H.H.; Nielsen, H.K.; Kraniotis, D.; Svennevig, P.R.; Svidt, K. Improving Building Occupant Comfort through a Digital Twin Approach: A Bayesian Network Model and Predictive Maintenance Method. *Energy Build.* **2023**, *288*, 112992. [CrossRef]
254. Haque, N.I.; Rahman, M.A.; Shahriar, H. Ensemble-Based Efficient Anomaly Detection for Smart Building Control Systems. In Proceedings of the—2021 IEEE 45th Annual Computers, Software, and Applications Conference, COMPSAC 2021, Madrid, Spain, 12–16 July 2021; Institute of Electrical and Electronics Engineers Inc.: Piscataway, NJ, USA, 2021; pp. 504–513.
255. Alonso, S.; Morán, A.; Pérez, D.; Prada, M.A.; Díaz, I.; Domínguez, M. Estimating Cooling Production and Monitoring Efficiency in Chillers Using a Soft Sensor. *Neural Comput. Appl.* **2020**, *32*, 17291–17308. [CrossRef]

Disclaimer/Publisher's Note: The statements, opinions and data contained in all publications are solely those of the individual author(s) and contributor(s) and not of MDPI and/or the editor(s). MDPI and/or the editor(s) disclaim responsibility for any injury to people or property resulting from any ideas, methods, instructions or products referred to in the content.

buildings

MDPI

Article

Strategies to Optimise Project Management Implementation in the Delivery of Renewable Energy Projects in Indonesia

Yohanes Arwadi [1], Alex Torku [1,*], Mershack O. Tetteh [2] and Francis K. Bondinuba [3,4,5]

1 School of Civil Engineering, University of Leeds, Leeds LS2 9JT, UK; cn23ya@leeds.ac.uk
2 Department of Civil Engineering, College of Engineering and Physical Sciences, Aston University, Birmingham B4 7ET, UK; m.tetteh@aston.ac.uk
3 Department of Construction Technology and Quantity Surveying, Faculty of Built and Natural Environment, Kumasi Technical University, Kumasi P.O. Box 854, Ghana; francis.kbondinuba@kstu.edu.gh
4 Faculty of Engineering and the Built Environment, Durban University of Technology, P.O. Box 1334, Durban 4000, South Africa
5 The Urban Institute, School of Energy, Geoscience, Infrastructure and Society, Heriot-Watt University, Edinburgh EH14 4AS, UK
* Correspondence: a.torku@leeds.ac.uk

Abstract: One of Indonesia's sustainability targets is achieving full energy transitions and net zero emissions by 2060 or sooner. Thus, significant strides and massive investments are being made to revolutionise the development and delivery of infrastructure projects, particularly in the energy sector. The State Electricity Company is driving these advances, targeting a surge in renewable energy usage to 23% by 2025 and 31% by 2050. Despite these efforts, progress remains critically low, highlighting the urgent need for more effective strategies and execution. A comprehensive understanding of the critical success factors (CSFs) and barriers—and how they influence these projects—is essential to ensuring their successful delivery. This study aims to identify and model the influences of various CSFs and barriers to renewable energy projects' success in Indonesia. Using a questionnaire survey with 182 respondents, this study explored the critical success factors, key challenges, and applicable strategies. The data were analysed using exploratory factor analysis and structural equation modelling. The analysis revealed four critical success factors, including stakeholders' collaboration and project supervision, and 10 key challenges, including insufficient planning, poor stakeholder communication, and unrealistic cost and time estimation. The strategies involved a cross-functional framework for effective collaboration, early consultation for integrating collaboration and supervision, and frequent reporting for enhanced supervision. These strategies are crucial to addressing emerging challenges in renewable energy projects and achieving sustainable development goals in Indonesia.

Keywords: challenges; critical success factors; Indonesia; project management; renewable energy; strategies

1. Introduction

Indonesia's energy consumption is mainly from fossil fuels, with nearly 100% electricity access and renewable energy making up 14.5% of the mix [1]. Energy demand is expected to rise by 80% by 2030, with targets of 17–19% renewable energy by 2025 and 44% by 2030 [1]. The country has significant renewable energy potential. Transitioning to renewable energy is crucial for reducing emissions and ensuring sustainability [1,2].

Indonesia has committed to achieving net zero emissions (NZEs) and sustainable development goals (SDGs) by 2060 or sooner [3]. According to the electricity supply business plan (RUPTL) 2021–2030, the State Electricity Company (PLN) is planning to boost the utilisation of renewable energy to 23% or higher in 2025 and 31% or higher in 2050 of electricity provision throughout the country [4]. However, PLN mainly relies on fossil fuel energy for electricity provision in Indonesia, where approximately 80% of the generated energy is from coal firing [5]. It shows that renewable energy projects (REPs) are unpopular in Indonesia's SDGs.

It is challenging for developing countries to transition to renewable energy [6]. Several projects involving renewable energy technologies in developing countries have failed [7]. Indonesia—one of the developing countries in Asia—is likely to face similar challenges in achieving its renewable energy development goals if appropriate project management (PM) strategies are not in place [8]. The nature of power plant projects often influences their implementation and demands unique solutions throughout their execution [9]. Projects have an embedded characteristic of being unique; various projects encounter different challenges in reaching their objectives [10]. Different levels of complexity across the projects produce various levels of uncertainty, which leads to distinct challenges in their execution [11].

Furthermore, the uniqueness of the project creates various dimensions for project success—project success has different criteria depending on its uniqueness [12]. Critical success factors (CSFs) are essential to effective project management (PM) and have a direct relationship with project success [13]. While Indonesia is committed to achieving ambitious SDGs, it is anticipated that the country will face significant challenges in executing complex renewable energy projects. This underscores the crucial role of CSFs in the development of renewable energy in Indonesia, providing essential guidance for effective PM implementation. Therefore, identifying CSFs for REPs could help improve PM practice in delivering renewable energy projects in Indonesia.

PLN monopolises the energy sector in Indonesia, and it is mandated by the government to achieve the target of 23% renewable energy in 2025, according to RUPTL 2021–2030 [4]. This implies that the implementation of PM in PLN and their related organisations has a vital role in REPs in Indonesia. Therefore, identifying the key challenges and CSFs from the perspective of PLN and their vendors is vital to determining the PM enhancement strategies in REPs in Indonesia. Specifically, this research aims to (1) identify the CSFs for Indonesia's renewable energy projects, (2) identify the key challenges in Indonesia's energy sector development, and (3) identify applicable project management strategies to deliver renewable energy projects in Indonesia.

The outcome of this study is a set of strategies to optimise PM implementation by addressing CSFs and challenges specific to Indonesia's renewable energy projects. This research offers both practical implications, by identifying effective strategies for enhancing PM in the delivery of renewable energy projects, and empirical implications, by demonstrating the benefits of incorporating diverse perspectives to identify key CSFs and challenges. This approach helps determine the appropriate management style based on the project's type and the organisation's characteristics.

1.1. Critical Success Factors (CSFs)

Many studies have explored the definition of project success in various aspects and how it is measured. For instance, Kerzner [14–16] discovered that project success is determined by the ability to satisfy technical goals involving cost, time, and quality, the benefits and values gained by the stakeholders, and how the project is performed. However,

Boyd [17] categorises five maxims of project satisfaction, including delivering the products for customers' needs, delivering quality while maintaining costs, delivering the project consistent with the customers' timeframe, delivering the desired degree of feedback for customers' desires, and having a fair conflict resolution system for both the customers and the team. This clarifies that a wide range of definitions for project success differ regarding the individual's perspectives. Nonetheless, Rockart [18] suggested that CSFs are variables in PM whose fulfilment will ensure successful delivery for stakeholders. Moreover, Ofori [19] defines CSFs as a set of project areas that must be managed effectively based on the project's environment to achieve project success. These studies show that CSF can boost project performance to achieve its success. Thus, exploring CSF is essential to understand which key areas in PM can contribute effectively to project success.

According to Pinto and Slevin [20], understanding CSF can give a guideline to execute better and monitor the PM approach for achieving project success. Moreover, Rockart [21] suggested that CSF identification provides significant information to prioritise several areas aligned with the expected success. CSF can also align resource allocation towards favourable criteria in the project to enhance its value with minimum resource utilisation [22]. In addition, Shenhar and Dvir [23] implied that CSF is highlighted as a strategic plan to achieve strategic objectives with effective planning. In summary, identifying CSF helps deliver the project and strategic objectives in various dimensions, such as effective control and monitoring, resource utilisation, and strategic goals.

Meanwhile, studies believed that the determination of CSFs in the projects is influenced by various factors, such as the context of the industry [24], specific stakeholders [25], and the project's concept and complexity [12]. This shows that multiple aspects influence the required CSFs in the project. For instance, Pinto and Slevin [26] emphasised that clear objectives, competent teams, realistic cost and time estimates, and adequate resources are part of CSFs in a more general study. However, a more recent study in a developing country in South America argued that a competent project team and adequate resources are not involved in CSFs [27]. Meanwhile, a study in Thailand, which is more similar to Indonesia, clarified that a competent team and adequate resources should be prioritised as CSF, excluding realistic cost and time estimates. These studies explain that various project contexts result in different priorities, which should be considered in CSF selection. Therefore, a detailed evaluation of the project, such as its context and stakeholders' perspectives, is required to identify CSFs correctly. This study has collected and established a breakdown list of several relevant success factors (SFs) (Table 1) based on various studies from 1987 to 2024.

Table 1. List of success factors based on the literature.

Success Factor	Reference													
	[20,26]	[14–16]	[17]	[28]	[29]	[30]	[31]	[25,32]	[33]	[19]	[27]	[34]	[35]	[36]
Clear project management objectives (SF1)	√			√	√	√			√		√	√		√
Top management support (SF2)	√			√	√	√	√	√	√	√	√	√	√	
Information/communication (SF3)	√		√		√	√	√			√			√	√
Clear involvement (SF4)	√	√	√			√	√		√	√	√	√		√
Competent project team (SF5)	√						√	√	√	√		√	√	√
Authority of project manager/leader (SF6)	√					√				√				√
Realistic cost and time estimates (SF7)	√	√	√	√						√	√		√	
Adequate project control (SF8)	√				√	√						√	√	√
Problem-solving abilities (SF9)	√						√				√			√
Project performance and quality (SF10)		√	√							√	√	√	√	
Adequate resources (SF11)	√	√				√	√		√	√				√
Planning and controlling (SF12)	√	√		√	√	√		√	√		√	√		√
Monitor performance and feedback (SF13)	√		√	√			√	√		√				√
Project mission/common goals (SF14)	√					√	√			√	√	√	√	
Project ownership (SF15)	√	√			√			√	√			√		

1.2. Project Management (PM) Challenges in Developing Countries

Based on Kerzner's [37] study, PM has experienced significant evolution over the years and has become a strategic necessity to create business value using technology to ensure its effectiveness. However, PM in developing countries has encountered several drawbacks over the years [38–40]. Furthermore, studies have found that PM maturity levels in developing countries have fallen from 2010 [41] to 2020 [42]. However, as the author mentioned, this research is conducted in South Africa, which may have a specific situation that needs to be clarified. Thus, PM implementation in developing countries needs improvement strategies for project challenges.

Meanwhile, studies discovered that challenges produce adversarial impacts on the project, such as defects and incomplete deliverables [43] and dynamic changes that require responsive management practice [44]. Specifically in infrastructure megaprojects, Flyvbjerg et al. [45] mentioned that challenges make projects suffer from cost overruns and scope changes. These studies indicate that challenges in PM result in various negative impacts on delivering project objectives. Pinto and Slevin [26] also believe that identifying challenges in PM practice becomes the key to improving project success. Other studies found that the identification of challenges provides a few benefits, including clarifying potential risks and their mitigation [46,47], generating an enhanced management approach [48], and determining improved stakeholder collaboration [49]. Hence, identifying challenges is important to address emerging challenges and minimise their impacts on project results.

Also, studies found several sources for challenges that arise in the project, such as organisational factors [10,26], environmental factors [45,50], technical factors [51,52], and the human factor [53,54]. Nonetheless, Flyvbjerg et al. [45] suggested that PM challenges are likely context-specific. For instance, research and development projects are struggling to manage innovation and transfer knowledge [55], while information technology projects encounter rapid changes and customer involvement issues [56], and construction projects regularly deal with strict budgets and timelines and team integration issues [10]. Specifically for developing countries, Ofori [24] found that construction projects have common challenges, including resource limitations, regulatory barriers, and inadequate stakeholder communication. Yanwen [40] added that there are several challenges in PM in developing countries, such as a lack of PM infrastructure and PM awareness at the top management level. However, Munns and Bjermi [57] believe that challenges in developing countries emerge between medium- and large-scale projects. Meanwhile, Flyvbjerg et al. [45] suggested megaprojects in construction tend to face challenges such as underestimation of cost, optimism bias, and stakeholders' complexity. Hence, it is inevitable for REPs in Indonesia to encounter challenges since REPs are mostly large projects that require an advanced management approach. In summary, the literature indicates that many factors influence the various challenges arising from the project. Therefore, collecting perspectives from professionals and practitioners with relevant experience can help evaluate specific challenges in this study.

While tailoring this literature review, distinct challenges are identified in PM practice in infrastructure projects in developing countries, namely insufficient planning [58–60], design and cost changes [61–63], and lack of appropriate skilled workers [64,65]. Moreover, the recent studies by McDermot et al. [27] and Welde and Bukkestein [66] found similar challenges, such as poor communication, unclear scopes, and insufficient performance and quality control. This literature review shows a wide variety of PM challenges. Thus, a thorough assessment must determine which challenges suit specific projects. To summarise, the list of collected PM challenges relevant to the context of REPs in developing countries is provided in this literature review (Table 2).

According to Boonstra and Reezigt [67], adopting appropriate strategies relevant to the specific project involves the combination of overviewing PM strategies and understanding the project's complexity. Although the project's complexity may differ for each project, the energy sector's projects in developing countries tend to have higher complexity [40,57]. This leads to the urgency of improving strategies in PM practice to deliver successful execution in REPs in Indonesia. Nonetheless, despite several studies examining PM strategies [68–70], it is essential to examine the appropriate PM strategies by addressing the project goals and the issues for reaching those goals [67]. Furthermore, finding the relationship between CSFs and challenges is essential to minimise the adversarial impacts of the challenges [22,26] and enable effective management in the project [46]. Hence, the generation of PM strategies is influenced by the relevant CSFs, key challenges and the relationship of both to ensure its alignment with the project's requirements.

Table 2. List of challenges based on the literature.

Challenges	Reference													
	[38]	[39]	[71]	[22]	[64]	[72]	[62]	[65]	[59]	[27]	[66]	[73]	[34]	[36]
Insufficient planning (C1)	√			√	√	√			√		√		√	√
Lack of skilled workers (C2)	√			√	√	√	√	√	√	√	√	√		
Poor stakeholders' communication (C3)	√		√		√	√	√			√		√	√	√
Unrealistic cost and time estimation (C4)	√	√	√		√		√		√	√	√			√
Unclear project scope (C5)	√						√	√	√	√		√	√	
Underestimation of risk control and management (C6)	√				√					√		√	√	√
Insufficient resources (C7)	√	√	√	√						√	√	√		
Inadequate project delivery system (C8)	√				√	√							√	√
Lack of technology knowledge (C9)	√						√				√	√	√	
Poor top management support (C10)		√	√							√	√			√
Insufficient cash flow (C11)	√	√			√		√		√	√		√		√
Inappropriate analysis and design (C12)	√	√		√	√	√		√	√		√	√	√	√
Inadequate performance and quality control (C13)	√		√	√			√	√		√		√	√	√
Lack of accountability in decision-making (C14)	√				√		√			√	√		√	√
Inefficient bureaucracy system (C15)	√	√				√		√	√			√		

2. Research Methods

To observe social reality for objective findings, positivism philosophy is beneficial in measuring phenomena amidst distinct perspectives [74]. Further, the literature explains that positivist research involves a deductive approach by collecting facts for hypothesis testing. Moreover, Cresswell [75] suggested that positivism philosophy is a quantitative study exploring the research's aim through identification, measurement, and assessment. The quantitative study also allows the validation of the preliminary assumptions through numeric analysis to produce the data results for attaining research objectives [76]. In other studies about the construction projects, this method enables an in-depth knowledge of infrastructure projects, which can provide valuable objective opinions on the project's complexities [77,78]. Therefore, a conceptual model for quantitative study is established with a positivism philosophy and a deductive approach utilising theories exploration in the literature review and professionals' perspectives collection for analysis (Figure 1).

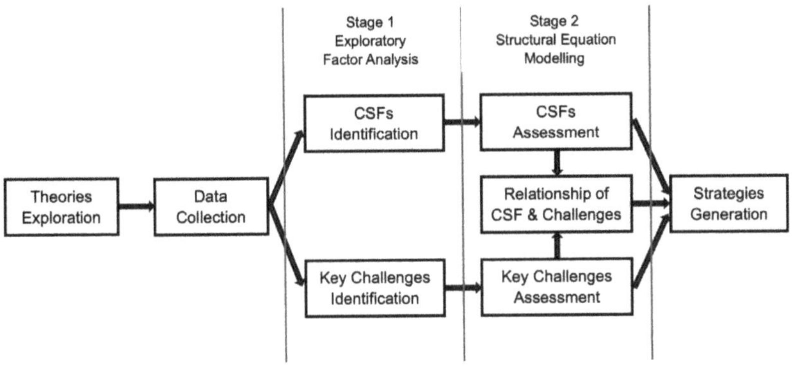

Figure 1. Conceptual model of the research.

2.1. Data Collection

The population of this research comprised practitioners and professionals who contributed to renewable energy projects in Indonesia—including the client, contractor, and consultant. These different roles are important to produce multiple perspectives to achieve the aim of this research. Although PLN [79] recorded 820 renewable energy projects in Indonesia in 2023, there was no reliable central database for practitioners and professionals who contributed to these renewable energy projects (sampling frame). Therefore, nonprobability, purposive sampling, and snowball sampling techniques were used to select relevant participants for this study. In purposive sampling, the selection of participants is based on a specific purpose [80], focusing on individuals who possess relevant knowledge and experience regarding the topic being studied. Snowball sampling entails identifying initial participants who then assist in locating additional participants in the target population [81]. Given the challenges associated with recruiting a large and diverse pool of experts in the construction field, purposive and snowball sampling have frequently been employed in construction research [82–84]. The questionnaire was disseminated through personalised emails, which included a Microsoft Word file attachment and a web link generated to facilitate online responses. The total number of distributions is indeterminate because potential respondents were kindly requested to forward the questionnaire to other practitioners and professionals deemed appropriate. As a result of various limitations, including the demanding schedules of experts, a total of 182 responses were collected all around Indonesia. This sample size could be considered both representative and acceptable. Moreover, the response rate is comparable to those of similar surveys in the field of construction [83–85]. Similarly, this meets the central limit theorem's requirement of a minimum sample size of 30 for any group, as recommended by Sproull [86] and Ott and Longnecker [87]. A sample of the questionnaire is provided in Appendix A to have a better understanding of the survey. This study was reviewed and approved by the University of Leeds Research Ethics Committee (Approval Code: MEEC 22-005).

2.2. Data Analysis Stage 1 (EFA)

First, this study tested the accuracy of the data by calculating Cronbach's coefficient alpha to measure the correlation between two parallel tests, those in the initial tests [88]. The coefficient alpha was calculated as follows [88]:

$$\alpha = \frac{k}{k-1}\left(1 - \frac{\sum_i S_i^2}{S_t^2}\right)$$

where

α = Cronbach's coefficient alpha.

k = the number of items.

S_i^2 = the variance of scores on the item.

S_t^2 = the variance of total test scores.

Cronbach and Shavelson [88] classified the coefficient alpha value proceeding analysis.

Further, the data are analysed using EFA with the Principal Component Analysis (PCA) method and the varimax rotation method at two levels utilising IBM SPSS Statistics 29 to select which indicators of CSFs and challenges fit the PM implementation. EFA with PCA and varimax rotation provides several benefits for research, including dimension reduction to focus on the most important options [89], identification of underlying constructs as unobserved latent variables [90], and a factor solution with higher accuracy to represent the true structure of the data [91]. These benefits of EFA allow for effective factor reduction and structure development from the collected elements to ensure the validity of the research. Several studies describe the required thresholds in EFA with PCA and varimax rotation method analysis, including the value of Kaiser–Meyer–Olkin (KMO) above 0.700 [92], p-value in Bartlett's test below 0.050 [91], factor loadings in communalities greater than 0.500, and cumulative loadings greater than 60% [93]. Additionally, in evaluating the variable's group, the factor loading for each indicator should be greater than 0.500 in EFA varimax to ensure the significance among indicators [93]. Also, Hair [93] suggested that the cumulative loadings should be above 60% to ensure that the factor solution captures a substantial portion of the data's variability. Hence, the EFA analysis in this study complied with these requirements to generate an accurate factor solution and unobserved latent variables with relevant underlying constructs.

2.3. Data Analysis Stage 2 (SEM)

The second analysis in this research involves SEM with Maximum Likelihood Estimation (MLE) as an integrated analysis from EFA. Based on Hoyle [94], SEM with MLE has a strong capability to model the relevancy of the latent variables as the unobserved constructs from the observed indicators, which allows the transfer of the analysis from the indicators to constructs. This method can also facilitate the analysis of new unobserved constructs and generate the correlations between variables as a relationship assessment [95]. SEM with MLE can provide several benefits, such as reliable correlation estimates and an accurate relationship model [95], a more holistic approach to model exploration and a description of significant correlations [96], and accurate assessment and model refinement through the calculation of various model fit indices [97]. This results in using SEM with MLE in this research to evaluate established model fitness and calculate the variables' estimates and their correlation.

In the first stage of SEM with MLE, it is necessary to assess the normality of the data with kurtosis and skewness measurement [95]. The normality of the data should be obtained with the skewness value less than absolute $|2.000|$ and kurtosis less than absolute $|7.000|$ [98,99]. Furthermore, Barret [100] clarified that the evaluation of model fit could utilise the comparison of chi-square over the degree of freedom (x^2/df), and the value should be below 5.000 to be acceptable [95]. Nonetheless, the assessment using chi-square is highly sensitive to sample size, which often causes the larger size with the appropriate model to become unfit [95]. To avoid this inaccurate indication, evaluating the model fit in SEM with other parameters such as the second stage of SEM is recommended. Several studies discovered additional parameters can be used to confirm the model fit, such as SRMR should be below 0.080 to be an acceptable fit [97], TLI is above 0.900 to be acceptable [97,101]. Moreover, CFI is required to be over 0.900 [101,102], and RMSEA should be less than 0.080 to be accepted [97,103]. Finally, the impact of the independent variables on the dependent variable is evaluated with the requirement of a p-value below 0.050 and a t-value greater than absolute $|1.960|$ for a p-value of 0.050 [95]. The author added that the standardised estimates in SEM indicate the interpretation and the strength of the relationship between two variables [95]. Therefore, these stages were carried out in the analysis of SEM with the MLE method to provide a full description of the model sufficiency and the relationship between the variables in this study.

This research integrates EFA and SEM to facilitate an in-depth analysis of indicators identified in the previous literature. EFA is advantageous for filtering indicators, enabling the identification of key indicators within variables and the formation of indicator groups within each variable that share a common construct as latent variables. Conversely, SEM examines the relationships among all variables to elucidate their mutual impacts. This combination yields a more comprehensive assessment of existing data by concentrating on significant factors and revealing their interrelationships in PM practices to achieve project success. Consequently, SEM analysis complements EFA by generating more accurate strategies in this research, illustrating the connections between key CSFs and challenges previously identified by EFA.

3. Results and Findings

This study's questionnaire survey collected 182 responses from various professionals and practitioners contributing to Indonesia's energy sector development. The survey results contain categories, including respondents' backgrounds, awareness of Indonesia's SDG commitment, and perspectives about CSFs and challenges in REPs in Indonesia.

3.1. Respondents' Background

The result indicates that most participants have 5–10 years of experience (Figure 2a). The result for the working environment is approximately equal between the two groups, with 51% of respondents working on-site and 49% working in the office (Figure 2b). Most participants work in non-managerial positions within their organisation (Figure 2c). Most participants have an owner/client role in the project (Figure 2d). These results imply that all participants have distinct experiences in energy sector development. Thus, this diversity enables the data collection to represent the target population for this research.

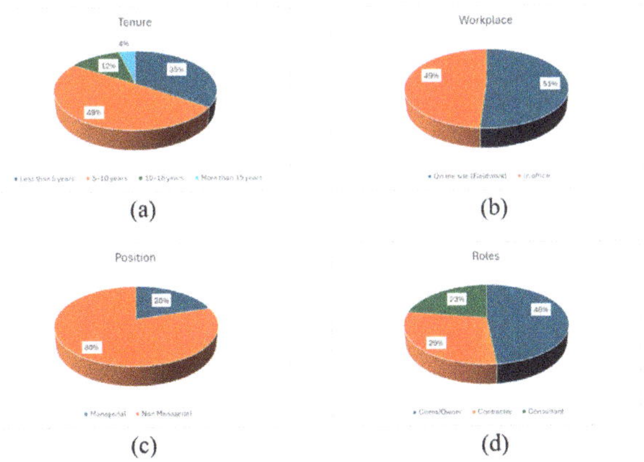

Figure 2. Respondents' background: (**a**) level formatting of experience result, (**b**) respondents' workplace result, (**c**) organisational position result, and (**d**) project roles result.

3.2. Respondents' Awareness

First, most participants know Indonesia's SDG commitment, with 54% agreeing and 25% strongly agreeing that they were aware (Figure 3a). Second, out of 182 participants, most believe Indonesia will encounter potential issues in performing REPs, with approximately 55% agreeing and 18% strongly agreeing (Figure 3b). Lastly, most participants agree that Indonesia can achieve the SDG in 2060 (Figure 3c). This leads to the description of the respondents' level of awareness in this research and their distinct opinions in this case.

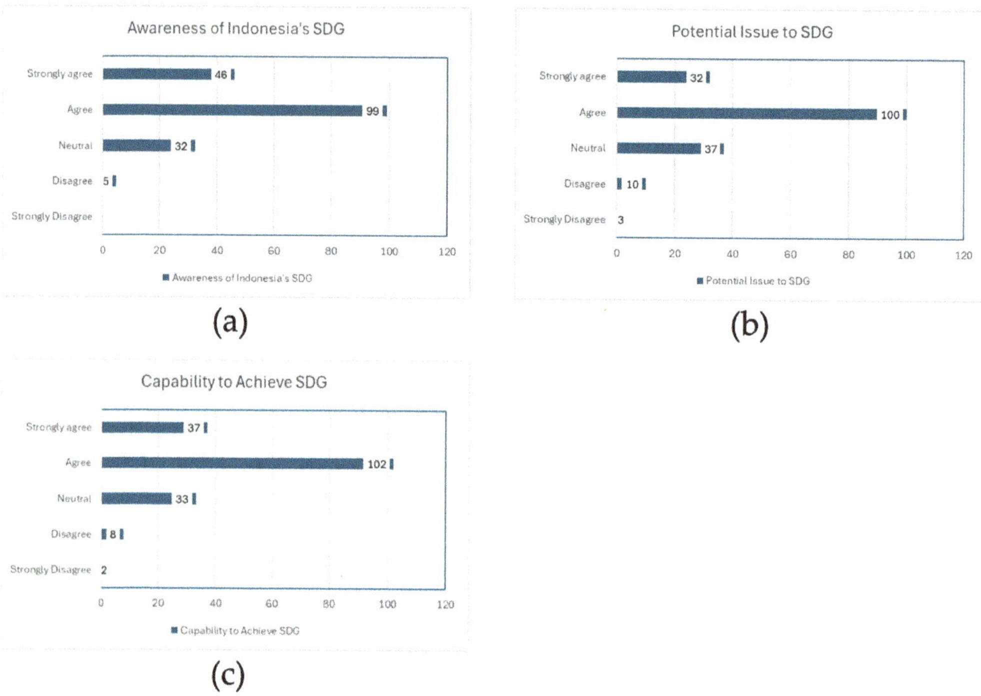

Figure 3. Respondents' awareness: (**a**) awareness of SDG result, (**b**) awareness of issue result, and (**c**) awareness of capability result.

3.3. Results and Findings of CSFs

This survey involved 15 SFs from the literature review, and respondents were asked using a Likert scale from 1 to 5 (from strongly disagree to strongly agree) to measure their level of agreement. Based on the survey, Clear stakeholders' involvement (SF 4) obtained the most responses for strongly agreeing with 99 responses, and Leader's authority (SF6) obtained the most responses for agreeing with 114. Despite these SFs collected from the literature, several respondents disagree with some factors. For instance, project ownership (SF 15) obtained the most disagreement from the respondents, with six responses, and two respondents selected "strongly disagree" for information/communication (SF 2). However, most respondents believed that these factors significantly influence PM implementation, which implies the validity of these factors since they are collected from the previous literature. The graphical description of the results of CSF can be seen in Figure 4.

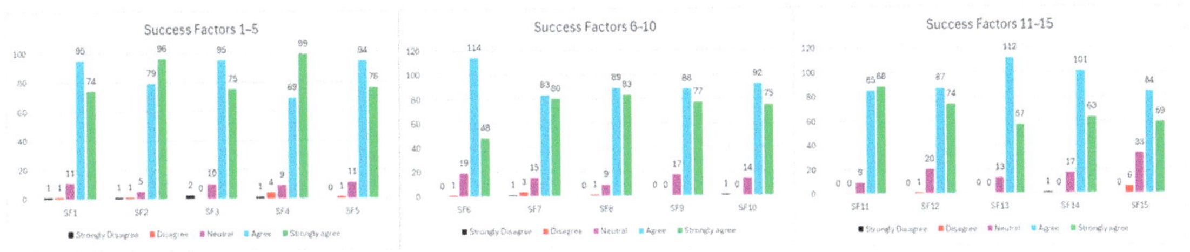

Figure 4. Respondents' perspectives on SFs.

3.4. Results and Findings of Challenges

This survey also utilised a Likert scale similar to CSFs to capture the significant challenges for this research. According to the results, insufficient planning (C1) obtained the most strongly agreed responses from 109 respondents, while inadequate project system (C8) had the most agreed responses from 94 respondents. Although the number is relatively low, some respondents had different perspectives on particular challenges. For instance, insufficient bureaucracy (C2) obtained the highest number of 7 responses with disagree, while there were no strongly disagree responses from the respondents in the challenges questions. In summary, the respondents showed diversity in their responses in this survey for the significant challenges in PM implementation. The interpretation of this result is that the challenges collected from the literature are relevant and valid to the recent situation in Indonesia. The graphical description of the results of the challenges is shown in Figure 5.

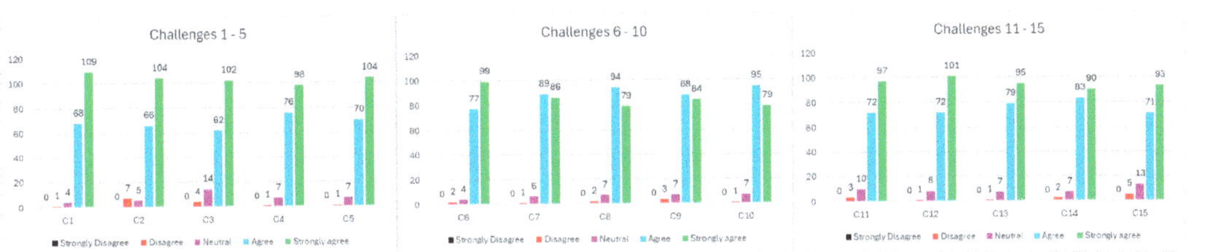

Figure 5. Respondents' perspectives on challenges.

4. Discussion

4.1. Identification of CSFs

First, the collected data of SFs was analysed with SPSS to find Cronbach's coefficient alpha. The calculation resulted in a value of 0.863, which is indicated as very good [88]. This leads to a strong covariance between the observed score and the true score in the data since the reliability test evaluates random measurement error [88], which is only less than 14% in this case.

Second, survey data regarding SF was analysed using EFA with PCA and the varimax rotation method at two levels to identify appropriate CSFs for this research. In the first level of the EFA analysis, the overall calculations yielded acceptable values, with a KMO statistic of 0.852 and Bartlett's p-value less than 0.010. However, the analysis of all SFs resulted in varied factor loadings ranging from 0.419 to 0.767 (Table 3), encompassing four latent variables with cumulative loadings of 57%. Consequently, the SFs with factor loading values below 0.500 were removed from consideration, including SF9 (problem-solving ability) and SF11 (adequate resources). According to Pinto and Slevin [26], CSFs in projects differ based on several contexts, including specifications, environment, and objectives. Furthermore, multiple studies support the notion that the diversity of CSFs is influenced by specific factors, such as project context [19], stakeholder expectations [25], and levels of complexity and strategic concepts [12]. Additionally, de Wit [104] noted that the determination of project success depends on various perspectives, which can impact the relevant CSFs for the project. This indicates that the SFs eliminated in this study are not deemed significant, as the experienced respondents did not regard them as relevant factors for Indonesia's REPs.

Table 3. Detailed results of communalities' extraction for CSFs in the first EFA analysis.

Indicators	Extraction 1	Description
SF1	0.767	Significant
SF2	0.613	Significant
SF3	0.541	Significant
SF4	0.614	Significant
SF5	0.578	Significant
SF6	0.536	Significant
SF7	0.539	Significant
SF8	0.510	Significant
SF9	0.475	Non-significant
SF10	0.537	Significant
SF11	0.419	Non-significant
SF12	0.683	Significant
SF13	0.585	Significant
SF14	0.572	Significant
SF15	0.621	Significant

Finally, the second EFA analysis, similar to the previous method, was undertaken with 13 indicators of SF by excluding SF9 and SF11. This analysis resulted in four unobserved variables and higher cumulative covariance (61%) than the first analysis. The factor loading value for all SFs also increased compared to the previous analysis, ranging from 0.605 to 0.826; the detailed breakdown can be seen in Table 4. These results lead to the model description for further analysis in SEM as latent variables in CSFs.

Table 4. Detailed result of variance loading and component matrix for CSFs in EFA.

Indicators	Factor CSF 1	Factor CSF 2	Factor CSF 3	Factor CSF 4
SF5	0.765			
SF2	0.713			
SF4	0.707			
SF6	0.650			
SF3	0.635			
SF7		0.682		
SF10		0.665		
SF14		0.605		
SF13			0.762	
SF15			0.689	
SF1			0.585	
SF12				0.826
SF8				0.637
Rotation total	2.945	1.766	1.704	1.533
Sums of squared loadings % of variance (cumulative %)				
	22.7 (22.7)	13.6 (36.2)	13.1 (49.3)	11.8 (61.1)

4.1.1. CSF1 (Stakeholder Collaboration)

Based on the analysis, group 1 (CSF1) consists of five indicators, namely, SF5 (team competency), SF2 (top management support), SF4 (clear stakeholder involvement), SF6 (leader authority), and SF3 (information and communication). According to Mitchell et al. [105], several areas are connected to building effective collaboration with the stakeholders, including the leader's authority, management support, stakeholder involvement, sufficient team competency, and proper communication. Moreover, PMI [52] emphasises that stakeholder management, as a key framework, is established by some essential aspects such as identifying the competencies and proper communication within the project team, and the leader's authority and management support play a vital role in enabling stakeholder management. There is a dynamic process where stakeholders interact and engage within the project team as stakeholders collaborate, which allows building the relationship and maintaining communication to align stakeholders' involvement with project objectives [106]. This leads to the construct of all indicators in group 1 interpreting stakeholder collaboration for its latent variable.

In addition, Ofori [24] suggested that stakeholder collaboration involving capacity building and stakeholder involvement in infrastructure projects in developing countries is beneficial to ensure sustainable development. Müller and Turner [32] also highlighted that stakeholder collaboration can effectively navigate the project environment in developing countries by providing clear direction and facilitating good communication. Therefore, stakeholder collaboration, as the construct in group 1, is appropriate to be one of the CSFs in REPs in Indonesia to support the achievement of project success.

4.1.2. CSF2 (Delivery Management)

The second group (CSF2) is composed of three indicators, including SF7 (time and cost estimation), SF10 (performance and quality), and SF14 (project mission/common goals). This group focuses on delivery management since it is a process to ensure that the project deliverables are performed in the specified requirements according to the designed schedule and budget plan to achieve its goals [44]. Furthermore, Baccarini [11] states that delivery management involves cost and time estimation to properly plan the schedule and financial resources for completing the project. de Wit [104] also highlighted that performance and quality management are integrated into the delivery management to meet the project goals expected by stakeholders. Furthermore, clear project goals can guide delivery management, such as providing direction in decision-making and ensuring the project contributes to strategic values [23]. In addition, Kerzner [16] also implied that the determination of project delivery requires a holistic approach involving cost and time estimation, performance and quality management, and project goals. This indicates that the interaction among indicators in group CSF2 is aligned with delivery management.

Additionally, delivery management is critical in projects in developing countries to identify the timeline and schedule to meet the requirements and avoid potential negative consequences since they tend to be cost- and time-sensitive [107]. Moreover, delivery management provides a more adaptive and flexible project, essential to cope with changes and maintain the quality of complex projects [108]. Thus, delivery management is recommended as the relevant CSF in this case study since REPs in Indonesia tend to be megaprojects with high complexities.

4.1.3. CSF3 (Project Supervision)

Meanwhile, group 3 (CSF3) is also formed from three indicators, which are SF13 (monitor and feedback), SF15 (ownership), and SF1 (clear objective). Monitoring involves continuous assessment, and feedback involves timely information, which is crucial in project supervision to maintain alignment with project objectives [46]. Ownership refers to project supervision, which drives accountability and influences the project's governance to meet the objectives [22]. Moreover, Lock [109] emphasised that well-defined objectives are a foundation for successful project supervision to enable a shared understanding of the expected outcomes among stakeholders and utilise them to evaluate the project efforts. Kerzner [46] also defined project supervision as an aspect of the PM that involves monitoring and guidance provision for the project aligned with its governance and objectives. These studies imply that the indicators in CSF3 work together to develop effective project supervision to describe the stakeholder's responsibility and proper inspection measured with clear expected objectives.

Furthermore, effective supervision can ensure accountability and adherence to its objectives while reducing the potential corruption as a key concern in developing countries [110]. Project supervision can enhance project performance and outcomes with regular monitoring and feedback as guidelines to meet the objectives by facilitating timely decisions to address changes and emerging issues often encountered in developing countries [107]. Hence, project supervision can positively impact REPs in Indonesia as a relevant CSF in this research.

4.1.4. CSF4 (Integration Management)

Lastly, the fourth group (CSF4) comprises the rest of the indicators, which are SF12 (planning and controlling) and SF8 (project control). Planning and controlling are continuous cycles in integration management to allow plan updates and realignment of the progress to stay on track [44]. Another study found that project control is essential in integration management to provide adjustment plans and actions related to the impact of changes in the project [48]. Based on PMI [47], integration management is a key aspect of PM that coordinates planning and controlling to manage conflicts and interdependencies effectively. These studies indicate that integration management involves coordinating planning and control processes, adjusting plans based on changes, and maintaining the project aligned with its goals.

According to Turner and Müller [54], projects in developing countries often have multiple interlinked components, making integration management essential for aligning various project elements and adjusting appropriate actions for each process. In addition, Kerzner [46] implied that integration management provides alignment between project activities and strategic goals, which are more important in developing countries where projects have strategic initiatives related to broader issues. Therefore, integration management can be suggested as part of the CSFs to provide more aligned processes in REPs to ensure the achievement of broader strategic objectives in Indonesia.

4.2. *Identification of Key Challenges*

Studies found that identifying challenges in PM can contribute to the project's success in various aspects [26,47,48]. This identification of challenges was similar to SF's analysis. Cronbach's coefficient alpha was calculated, resulting in an excellent value of 0.938. This means the survey data are valid and reliable for further quantitative analysis in this study.

While carrying a similar EFA test with SFs, 'challenges' data achieved an excellent KMO value of 0.946. For Barlett's test, the resulting value is lower than 0.010, which means that the data in the variable is correlated and acceptable for EFA. These results show the reliability of the research, which proves the validity of the collected data. However, the variable group result generated lower indicator values, ranging from 0.370 to 0.676 (Table 5).

Hence, some indicators with factor loading below 0.500 are eliminated, which are C2 (lack of skilled workers), C7 (insufficient resources), C9 (lack of technology knowledge), C10 (poor management support), and C11 (insufficient cash flow). PM challenges may vary in specific projects since they are context-specific and affected by several factors [45]. Several studies also discussed the various emerging challenges in the project influenced by distinct factors such as project type [10,56] and industry and environment [24]. Also, other studies suggested that existing challenges are affected by specific factors in the project, such as its scope and complexity [111], organisational and cultural context [45], and stakeholders [112]. These studies imply that the identification of challenges requires the details of the project since it can be affected by various factors, especially for REPs in Indonesia, which may have specific characteristics.

Table 5. Detailed results of communalities' extraction for challenges in the first EFA analysis.

Indicators	Extraction 1	Description
C1	0.649	Significant
C2	0.452	Non-significant
C3	0.567	Significant
C4	0.639	Significant
C5	0.627	Significant
C6	0.602	Significant
C7	0.481	Non-significant
C8	0.571	Significant
C9	0.370	Non-significant
C10	0.445	Non-significant
C11	0.478	Non-significant
C12	0.511	Significant
C13	0.676	Significant
C14	0.606	Significant
C15	0.526	Significant

The second EFA analysis was undertaken with 10 remaining indicators to enhance their correlation. This analysis resulted in a greater cumulative loadings value of 61%, with factor loadings ranging from 0.725 to 0.840, with the breakdown results, as shown in Table 6. Tabachnick and Fidell [89] suggested that EFA may extract a single component in the variable when the elements are highly intercorrelated, which leads to an inability to distinguish between factors. Nonetheless, a single-factor solution may result from an unseparated homogeneous dataset across multiple elements [91]. Moreover, the elements collected from the previous literature contain different aspects of challenges, such as poor communication from human aspects [53], inadequate project delivery system from technical aspects [47], and unclear project scopes from organisational aspects [10]. Hence, the dataset in challenges can have a high correlation rather than homogeneous elements. This can also be seen from the high values of Cronbach's coefficient alpha [113] and factor loadings [114].

The challenge with the highest factor loading, C13 (inadequate project control), is aligned with a study by Sovacool and Cooper [115], who found that megaprojects in the energy sector involve higher complexity, which leads to difficulties in their control. Furthermore, Sambasivan and Soon [116] mentioned that poor control is indicated as a key challenge in developing countries, resulting in ineffective PM. In addition, inadequate control is also found in Vietnam, which leads to delays and cost overruns [62]. Therefore, C13 is supported by similar studies to be an acceptable key challenge in this research.

Meanwhile, C5 (unclear project scopes) is also supported by Pinto and Slevin [26], who mentioned that unclear scopes result in misaligned objectives and failure to meet stakeholders' expectations. Moreover, McDermot et al. [27] believed that ill-defined scope is one of the most common challenges in developing countries, manifesting in errors and changes. Additionally, Vindi et al. [34] found that REP in Indonesia required contractual renegotiation due to design and scope changes. These studies indicate that C5 is often encountered by projects in developing countries, particularly in Indonesia's REPs. Therefore, the analysis of C5 is proven to be a significant challenge that adversely impacts the project's success.

Table 6. Detailed result of variance loading and component matrix for key challenges in EFA.

Indicators	Factor Challenges
C13	0.840
C5	0.815
C14	0.802
C4	0.800
C1	0.796
C6	0.780
C3	0.762
C8	0.760
C15	0.754
C12	0.725
Rotation total	1.001
Sums of squared loadings % of variance (cumulative %)	
	6.148 (61.477)

For C14 (lack of accountability in decision-making), Yanwen [40] found that lack of accountability happened in the form of bribery and corruption in complex project management in developing countries. Poor accountability in developing countries is influenced by distinct cultural habits not established with structured management [117]. Sustainable development projects in Indonesia need to adopt legal frameworks to enhance accountability in decision-making [118]. Thus, specific cultures in developing countries may lead to C14, including in sustainable development projects in Indonesia, which can influence the achievement of project success.

Furthermore, C4 (unrealistic cost and time estimation) is clarified by Welde and Bukkestein [66], who mentioned that inaccurate cost and time estimation influenced large government project delays. The estimation issue involves optimism bias and prospect theory, leading to underestimating uncertainties and errors in the project [119]. This is common in megaprojects since they are driven by inadequate information but higher ambition [45], especially in public facility projects that involve social values [120]. Therefore, C4 is considered an important challenge in REPs in Indonesia since REPs are not very common and are associated with more social ambition.

Meanwhile, C1 (insufficient planning) is supported by Zidane and Andersen [121], who found that poor planning is the most frequent issue in construction projects in Norway. Also, a study in Malaysia found that improper planning is the most impactful issue in construction projects, leading to delays [116]. Moreover, Ofori [122] believed that inadequate planning is often encountered in construction projects in developing countries and is associated with high complexity. Megaprojects in the energy sector are complex, especially because they often interfere with political uncertainties, regulatory matters, and environmental concerns [115]. Hence, the characteristics of REPs enhance the impact of C1 on project management practice, especially in developing countries, including Indonesia.

For C6 (underestimation of risk management and control), large construction projects are exposed to risks embedded with several issues [123]. The common mistake in risk management practice in megaprojects is that it is often reactive rather than proactive, resulting in poor mitigation [124]. Moreover, a study found that insufficient risk management often happens in developing countries, which leads to delays and demands for changes in infrastructure projects [27]. Especially for megaprojects, higher complexity and uncertainty require robust risk management and control to ensure that project objectives are met [45]. Therefore, C6 is a key challenge in REPs in Indonesia since developing countries fail to deal with the scale of the complexity and uncertainty in megaprojects.

Furthermore, C8 (inadequate project delivery system) is supported by Duy Nguyen et al. [72], clarifying that improper project delivery systems are a common phenomenon in large construction projects in Vietnam, resulting in inefficient execution. The degree of project delivery shows a significant relationship with project management maturity, which is decreasing in developing countries [42,125]. The poor delivery system in REP in Indonesia leads to its failure to meet its objectives [34]. Therefore, C8 is a significant challenge for this research, especially when project management maturity in developing countries has fallen recently.

For C15 (insufficient bureaucracy), Merrow [126] believed that megaprojects often encounter bureaucracy issues involving poor administrative capacity to manage their complexities. Insufficient bureaucracy is an issue in large governance-related projects in the form of ineffective coordination with superior bodies [66]. For instance, large construction companies in Vietnam also encounter poor bureaucracy associated with bribery, which escalates the cost [72]. Governance structures influence the bureaucracy issue, leading to inefficiencies in the project [127]. Additionally, Ahsan and Gunawan [128] implied that bureaucratic problems caused policy changes and slow decision-making processes in developing countries. Therefore, C15 is proven to significantly impact megaprojects and developing countries, which is relevant to this research.

Finally, C12 (inappropriate analysis and design) is aligned with research by McDermot et al. [27], which suggested that the design process for infrastructure projects in developing countries includes technical difficulties and complexities requiring involvement from all parties. For example, large projects in Vietnam often encounter design changes caused by poor analysis [62]. In addition, REP in Indonesia encountered issues in analysis and design related to its compliance with technical standards [34]. Thus, the significance of C12 in this research is evident—REPs in Indonesia involve high complexities, which increases the possibility of encountering such issues.

4.3. CSFs and Key Challenges Assessment and Relationship

This research established a structured model based on the variables from previous analysis, including five independent variables (Figure 6), four latent variables for CSFs, one variable for key challenges, and one dependent variable for project success. Afterwards, the normality requirements are evaluated to check if the model fits for SEM with the MLE method. This analysis produced acceptable values ranging from −1.547 to 0.182 for skewness and −0.629 to 4.545 for kurtosis. In summary, SEM with the MLE method can be carried out with the data collection using the model. Further, the model's fitness was analysed by assessing several requirements for SEM with MLE. Based on the calculation, the generated model for this research was fit since all resulting values met the requirements

(Table 7). Therefore, using SEM with the MLE method, this study had sufficient data and an appropriate model to investigate the covariance for all variables.

Table 7. Comparison of the results and the requirements for model fit in SEM.

	(x^2/df)	SRMR	TLI	CFI	RMSEA
Result value	1.674	0.023	0.909	0.921	0.061
Requirement	<5.000	<0.080	>0.900	>0.900	<0.080

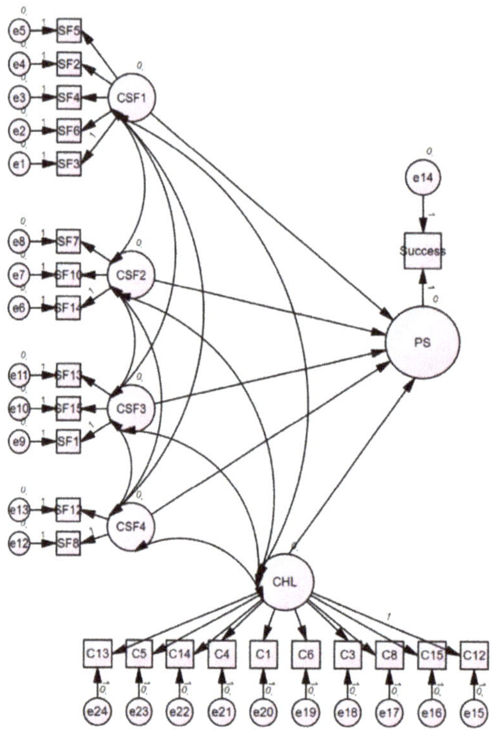

Figure 6. Research covariance model for SEM.

This analysis showed that the established model can capture the relationship between CSFs and challenges and their impact on project success. Based on Jugdev and Muller [129], managing CSF allows effective mitigation for the impact of challenges, leading to a higher chance of reaching project success. Also, CSFs are interdependent with challenges since addressing challenges can enhance the effectiveness of CSF implementation to achieve project success [22]. Thus, the established model provides fitness and significant relationships between variables supported by the SEM analysis and the previous literature.

Furthermore, SEM was conducted to find the correlation of all variables in the model at the following stage of evaluation. For the covariances of CSF1, CSF3, and challenges to project success, the resulting values were indicated to meet the requirement with p-values of 0.002 and 0.01, respectively, and t-values of 3.059, 3.597, and 2.568, respectively, leading to their significant impact on project success. However, CSF2 and CSF4 indicated higher p-values of 0.433 and 0.202 and lower t-values of 0.784 and 1.275, implying their insignificant correlation to the project's success. This means that both CSF2 and CSF4 have failed to show their significant correlation to project success. Hence, this analysis indicated that only three variables, CSF1, CSF3, and key challenges, are accepted for further analysis to respond to the research aim. Nonetheless, all independent variables indicate various values for standardised estimates from −2.70 to 0.966 in the analysis model, which measures the correlation between each independent and dependent variable (Figure 6). Therefore, the strategy generation will involve standardised estimates, *p*-values, and t-values for all variables in the summary result (Table 8) to direct its consideration.

Table 8. Summary covariance for independent variables to dependent variable.

Hypothesised Relationship	Standardised Estimates	T-Value	*p*-Value	Decision
CSF 1 to project success	0.499	3.059	0.002	Accepted
CSF 2 to project success	−0.179	−0.784	0.433	Rejected
CSF 3 to project success	0.966	3.597	<0.001	Accepted
CSF 4 to project success	0.236	1.275	0.202	Rejected
Challenges to project success	−0.27	−2.568	0.01	Accepted

Specific CSFs can enhance the opportunity to successfully deliver specific megaprojects in developing countries [24]. Strong stakeholder collaboration and effective supervision in megaprojects help deal with their complexities, improving project outcomes [119]. Hence, focusing the strategies on CSF1 and CSF3 can contribute to advancing PM implementation in Indonesia's REPs. Meanwhile, Flyvbjerg et al. [45] specified the challenges for megaprojects that failed to reach success and suffered from cost escalation and scope changes. Moreover, similar studies in developing countries have revealed adverse impacts of the key challenges on project success, such as C1, C3 [116], C4, C6, C15 [27], C5, C8 [34], C12, C13, and C14 [62]. Thus, the relationship between key challenges and project success in this research is supported by the previous literature.

Moreover, this research also seeks the correlation between CSFs and key challenges with correlation estimates, which can describe the strength and direction of the relationship across variables [95]. The analysis showed various values of correlation estimates between all latent variables of CSFs and the challenges from 0.546 to 0.590. Nonetheless, the correlation estimates were focused on the relationships between CSF 1 and CSF 3 with the challenges. Note that CSF 2 and 4 were rejected in the previous discussion. The correlation estimates showed a positive linear relationship between the variables, resulting in an estimated value above 0.56 for all accepted variables, as seen in Table 9. Therefore, SEM analysis proved the correlation between key CSF and challenges in this research from the main variables significant to the project's success.

Table 9. Summary result of the correlation between independent variables.

Variable Correlation	Estimates Value	Decision
CSF 1 to challenges	0.590	Accepted
CSF 2 to challenges	0.587	Rejected
CSF 3 to challenges	0.569	Accepted
CSF 4 to challenges	0.546	Rejected
CSF 1 to CSF 3	0.673	Accepted

Zwikael and Ahn [130] found the relationship between CSFs and challenges that navigate the determination of CSFs to tackle emerging challenges in the project. Especially for CSF1, studies indicated that stakeholders' collaboration plays a significant role in addressing specific challenges such as building communication [131], ensuring the alignment of project scopes, enhancing overall accountability in decision-making [112], and minimising complicated bureaucracy [72]. For CSF3, studies found that effective supervision has an important influence on several challenges, for instance, enforcing detailed planning [46], validating accurate schedule and cost estimation [132], ensuring effective risk management [130], improving delivery system [133], developing detailed analysis and design [112], and enhancing quality control process [32]. Turner and Müller [54] also highlighted that project supervision can facilitate team communication and involvement rather than just monitoring process, which enhances stakeholders' collaboration. Also, Karlsen [134] mentioned that project supervision helps to reduce conflicts and align the involvement of stakeholders, resulting in better collaboration. Hence, CSF1, CSF3, and key challenges in this research provided a strong relationship between each other, which is proven by existing knowledge.

4.4. PM Strategies

4.4.1. Cross-Functional Framework (S1)

REPs involve stakeholders with various expectations for the project, especially when the expectation is associated with social values. This leads to interdisciplinary requirements in project execution simultaneously, which could result in conflicts. Utilising S1 can facilitate relationship building and extensive consideration to determine project scopes and stakeholders' involvement. S1 enables a more detailed assessment of the decision-making process by gathering input from all stakeholders. This assessment can navigate management support and the leader's authority to ensure that the decision is aligned with the strategic objectives. Therefore, this framework helps to develop effective coordination among various backgrounds to enhance the bureaucratic system and accountability in the decision-making process.

Literature evidence: According to Yang and Shen [135], a framework involving various stakeholders fosters better collaboration in diverse environments where conflicts may emerge. The cross-functional team also enables the integration of different disciplines in the decision-making process to ensure comprehensive problem-solving, which promotes accountability [136]. Other studies also mentioned that cross-functional teams can simplify bureaucratic processes to avoid delays [137] and provide rich learning spaces to develop team competency [138]. Hence, S1 allows the management of CSF1 by integrating stakeholders' multidisciplinary approaches to optimise their involvement in achieving project success.

4.4.2. Early Consultation (S2)

The potential impact of early consultation should be considered in PM implementation, particularly in projects associated with high complexity. The impact of the challenges in complex projects can be minimised and mitigated properly with S2. Another impact of this strategy is associated with the team's diversity in REPs. The consultation process is essential for information distribution about diverse aspects to ensure that everyone keeps track of project updates. Also, it can increase their involvement to provide effective monitoring for the project. Therefore, early consultation can contribute to managing stakeholders' collaboration and project supervision as key CSFs to overcome a project's complexity.

Literature evidence: Intensive consultation encourages stakeholders to develop a greater sense of ownership, foster strong collaboration, and increase supervision throughout the project [106]. According to Rowlinson and Cheung [139], consulting with stakeholders creates a platform for knowledge sharing when facing challenges. It empowers them to become more involved, resulting in a shared perspective on project implementation.

Furthermore, Jepsen and Eskerod [140] suggested that early consultation provides effective control and guidance in the decision-making process, aligning with project objectives. Thus, these studies illustrate how the correlation between CSF1 and CS2 manifests in S2, allowing for effective resolution to address the key challenges.

4.4.3. Frequent Reporting (S3)

S3 allows detailed supervision associated with effective monitoring to ensure that the project meets its objectives. It can resolve issues in various project phases, such as improper planning and poor project delivery. Furthermore, regular reporting can support risk management implementation by providing early warnings to stakeholders to promptly mitigate the risk's impact. In addition, S3 helps to reinforce the clarity of the project's objectives among stakeholders, ensuring mutual understanding within the team. Afterwards, the stakeholders can feel more involved with the regular updates, increasing their sense of belonging and level of contribution. Therefore, this strategy can effectively manage CSF2, leading to a more responsible environment among stakeholders and maintaining control to achieve project success.

Literature evidence: A study mentioned that routine reporting enables continuous monitoring to support timely corrective actions, improving control [46]. Hillson [141] emphasised that regular reporting is a proactive approach to risk management to prevent the escalation of adverse impacts from the issues and maintain project stability. Based on Locke and Latham [142], project reports can provide constant clarification about project objectives for the project team and support aligning planning and delivery with the expected outcomes. Lastly, an increased sense of ownership can be obtained by consistently reminding relevant stakeholders, which encourages them to meet their commitments for their role in the project [131]. Hence, the implementation of S2 is aligned with CSF2 by allowing immediate signals for necessary adjustments and ensuring the project team's contribution to meeting project objectives.

Based on the discussion of PM strategies in this research, the integration arrangement of PM strategies with the key CSF, challenges, and their correlations are summarised in Table 10.

Table 10. Integration of PM strategies and key CSFs, challenges, and their correlations.

Key CSF	PM Strategies	Key Challenges
CSF 1	Cross-functional framework (S1)	C3, C5, C14, and C15
	Early consultation (S2)	C1, C3, C4, C5, C6, C8, C12, C13, C14, and C15
CSF 2	Frequent reporting (S3)	C1, C4, C6, C8, C12, and C13

5. Conclusions

This research aims to identify applicable project management strategies to deliver renewable energy projects in Indonesia. To achieve this aim, the research (1) identified the CSFs for Indonesia's renewable energy projects, (2) identified the key challenges in Indonesia's energy sector development, and (3) the identified CSFs and challenges were used to inform project management strategies to deliver renewable energy projects in Indonesia. A questionnaire survey collected data from 182 respondents involved in Indonesia's energy sector development. The data were analysed using exploratory factor analysis and structural equation modelling. The result revealed 13 CSFs grouped into stakeholder collaboration, delivery management, project supervision, and integration management. The analysis identified 10 key challenges in Indonesia's energy sector development. These challenges included insufficient planning, poor stakeholder communication, and unrealistic cost and time estimation. Three main strategies were recommended to address emerging challenges in renewable energy projects and achieve sustainable development goals in Indonesia. The strategies included a cross-functional framework for effective collaboration, early consultation for integrating collaboration and supervision, and frequent reporting for enhanced supervision. Given Indonesia's well-documented issues with corruption, collusion, and nepotism (KKN), these strategies can also address these concerns by ensuring regulatory compliance and enhancing the monitoring system throughout the process.

While this study successfully achieved its aim, it is essential to acknowledge certain limitations and propose future research directions. The relatively low number of responses may impact the generalisability of the findings. The adopted methodology is cross-sectional, concentrating data collection on a specific moment in time. Future research could benefit from larger sample sizes to validate the findings. Moreover, employing a multiple-case design using secondary data from the existing literature could enhance both the internal and external validity of the research. This study is specifically limited to renewable energy projects in Indonesia, as the considerations for these projects are influenced by the surrounding context and the various types of projects. Further studies are recommended to verify how the strategies impact projects with typical characteristics. This study focused on renewable energy projects in Indonesia within the framework of committing to SDGs, recognising that different project types can generate distinct considerations for CSFs and challenges. Additionally, this study recommended innovative strategies, necessitating further research to verify the impact of these strategies on projects with typical characteristics. Given the high scale and complexity of renewable energy projects, a more detailed study is required to assess the implementation of these strategies for enhancing PM practices in managing uncertainties. A broader range of case studies from various industries and developing countries should further clarify the specific indicators and their relevance to project success in their contexts.

This study develops both practical and empirical implications for the industry. Regarding practical implications, the method of identifying CSFs and challenges can provide valuable lessons and help determine appropriate strategies for projects. These strategies can be integrated into the project delivery system at PLN to ensure comprehensive consideration in the decision-making process, utilising a cross-functional framework. This approach facilitates proper management of changes through early consultation and frequent reporting while maintaining project performance. As for empirical implications, this research highlights the benefits of incorporating diverse perspectives to gain extensive insights into the project. Since project managers must determine the appropriate PM style based on the project and the organisation, identifying distinct expectations from multiple stakeholders can enhance PM performance by addressing their requirements effectively.

Author Contributions: Y.A.: conceptualisation, formal analysis, data curation, methodology, writing—original draft; A.T.: conceptualisation, supervision, writing—review and editing; M.O.T.: methodology, writing—review and editing; F.K.B.: methodology, writing—review and editing. All authors have read and agreed to the published version of the manuscript.

Funding: This research received no external funding.

Data Availability Statement: The data supporting this study's findings are available from the corresponding author upon reasonable request.

Conflicts of Interest: The authors declare no conflicts of interest.

Abbreviations

The following abbreviations are used in this manuscript:

CFA	confirmatory factor analysis
CFI	comparative fit index
CSF	critical success factor
EFA	exploratory factor analysis
ENDC	Enhanced Nationally Determined Contribution
KKN	corruption, collusion, and nepotism
KMO	Kaiser–Meyer–Okin
MLE	maximum likelihood estimation
NZE	net zero emission
PCA	principal component analysis
PLN	State Electric Company
PM	project management
PMI	project management institute
REP	renewable energy project
RMSEA	root mean squared error of approximation
RUPTL	electricity supply business plan
SDG	sustainable development goals
SEM	structural equation modelling
SF	success factor
SRMR	standardised root mean square residual
TLI	Tucker–Lewis index
UN	United Nations

Appendix A. Sample of the Survey Questionnaire

You are being invited to take part in a research project. Before you decide, you need to understand why the research is being performed and what it will involve. Please take time to read the following information carefully. Ask us if anything could be clarified or if you want more information. Take time to decide whether you wish to take part.

This research study aims to explore the most suitable project management strategies to deliver renewable energy projects in Indonesia. The participants in this research study are professionals and practitioners who are participating in energy sector development in Indonesia. If you meet this criterion, we invite you to participate in the survey. For those who do not meet this specific category, we sincerely appreciate the time you have taken to read this questionnaire.

What is this document? This document serves as an information sheet for this research study. It explains the nature of this study, your rights, and what will be performed with your data.

Nature of this study. This research study involves completing the questionnaire, which consists of statements. You will be asked to choose one of the answers from the options provided based on your experiences and perceptions. Before you start, we may have questions about you (e.g., tenures, positions, and role in the project). You need to answer every question in the questionnaire. It will take you approximately 10–15 min to complete the questionnaire. You will be given full instructions shortly.

Risks and benefits. There are no known risks associated with this study. While there is no direct benefit to participants, your participation will contribute to our knowledge. The results of this study are expected to provide valuable insights that can be used to enhance the effectiveness of future delivery performance in renewable energy projects in Indonesia.

Confidentiality and use of data. To safeguard your privacy, we will never share personal information with anyone outside the researcher, his supervisor, and assessment staff at the University of Leeds as required for assessment purposes related to the researcher's degree. The data will be securely stored in encrypted files to prevent unauthorised access. Responses will be made anonymous to protect your privacy, and any data will be presented in an aggregated form to prevent the identification of individuals.

Voluntary participation and right to withdraw. Your participation is voluntary, and you may withdraw from this study at any time and for any reason. If you have any questions or need further information, please contact the researcher by email. Please visit the University of Leeds Privacy Notice for more information regarding ethics and data protection. By completing and returning this questionnaire, you imply your consent to participate in this research study. We sincerely appreciate your time, effort, and valuable insights in this research study. Thank you for your participation.

You may continue to the next page to give consent for your participation.

Consent Form

Please give a checklist next to each statement if you agree:

- ___ I agree to participate in this study voluntarily, and I am free to withdraw at any time without giving any reason.
- ___ I confirm that I have read and understood the information explaining the research project and how my data will be stored and used.
- ___ I understand that my responses will be kept strictly confidential.
- ___ I understand that relevant sections of the data collected during this study may be reviewed by the researchers.

By clicking "I agree," you indicate your consent to participate in this research.
I agree (button to proceed)

Section 1

First, you will be asked some demographic information:

1. Have you participated in energy sector projects in Indonesia?
    - Yes.
    - No.
2. What is your gender?
    - Male.
    - Female.
    - Prefer not to say.
3. How long have you been participating in energy sector's projects?
    - Less than 5 years.
    - 5–10 years.
    - 10–15 years.
    - More than 15 years.
4. What is the environment of your daily job?
    - On site.
    - In the office.
5. What is your current position in the job?
    - Managerial.
    - Non-managerial.

6. What is the role of the organisation in the projects that you are working for?
    - Client.
    - Contractor.
    - Consultant.

Instructions

Please read the following instructions carefully before starting the questionnaire:

1. Please answer each question honestly based on your personal experiences and perceptions.
2. There are no right or wrong answers.
3. For the questions, you will be asked to indicate your level of agreement on a 5-point scale:

    1 = Strongly Disagree.
    2 = Disagree.
    3 = Neutral.
    4 = Agree.
    5 = Strongly Agree.
    After completing the questionnaire, please click the "Submit" button to send your responses.

Section 2

Table A1. Samples questions in Section 2.

No	Question	Answer (Please Select One)				
		1	2	3	4	5
1	Do you think you know about Indonesia's Sustainable Development Goals (SDGs) to reach Net Zero Emissions and Energy Transitions in 2060?					
2	Do you think Indonesia, as a developing country, has barriers to achieve SDG in 2060?					
3	Do you agree that Indonesia has the capability to achieve the SDG in 2060?					
	CRITICAL SUCCESS FACTORS (CSFs)					
4	Do you agree that Indonesia needs to prioritise several particular project management areas to reach project success in renewable energy projects for SDG?					
5	Do you agree that clear project management objectives can significantly determine project success while carrying out the renewable energy projects for SDG?					
6	Do you agree that top management support can significantly determine project success while carrying out the renewable energy projects for SDG?					
7	Do you agree that clear information and communication can significantly determine project success while carrying out the renewable energy projects for SDG?					
8	Do you agree that clear stakeholder involvement can significantly determine project success while carrying out the renewable energy projects for SDG?					
9	Do you agree that the competency of the project team can significantly determine project success while carrying out the renewable energy projects for SDG?					
10	Do you agree that the authority of project manager/leader can significantly determine project success while carrying out the renewable energy projects for SDG?					
11	Do you agree that problem-solving abilities can significantly determine project success while carrying out the renewable energy projects for SDG?					
12	Do you agree that project performance and quality can significantly determine project success while carrying out the renewable energy projects for SDG?					
13	Do you agree that realistic cost and time estimates can significantly determine project success while carrying out the renewable energy projects for SDG?					
14	Do you agree that adequate project control can significantly determine project success while carrying out the renewable energy projects for SDG?					
15	Do you agree that adequate resources can significantly determine project success while carrying out the renewable energy projects for SDG?					
16	Do you agree that planning and controlling can significantly determine project success while carrying out the renewable energy projects for SDG?					
17	Do you agree that performance monitoring and feedback can significantly determine project success while carrying out the renewable energy projects for SDG?					
18	Do you agree that clear project missions/common goals can significantly determine project success while carrying out the renewable energy projects for SDG?					
19	Do you agree that project ownership can significantly determine project success while carrying out the renewable energy projects for SDG?					
20	Do you think Indonesia will face challenges while carrying out the renewable energy projects for SDG?					
21	Do you agree that insufficient planning will have a strong impact on project performance while carrying out the renewable energy projects for SDG?					
22	Do you agree that lack of skilled workers will have a strong impact on project performance while carrying out the renewable energy projects for SDG?					
23	Do you agree that poor stakeholder communication will have a strong impact on project performance while carrying out the renewable energy projects for SDG?					
24	Do you agree that unrealistic cost and time estimation will have a strong impact on project performance while carrying out the renewable energy projects for SDG?					
25	Do you agree that unclear project scopes will have a strong impact on project performance while carrying out the renewable energy projects for SDG?					
26	Do you agree that underestimation of risk control and management will have a strong impact on project performance while carrying out the renewable energy projects for SDG?					
27	Do you agree that insufficient resources will have a strong impact on project performance while carrying out the renewable energy projects for SDG?					

Table A1. *Cont.*

No	Question	Answer (Please Select One)				
		1	2	3	4	5
	CHALLENGES					
28	Do you agree that an inadequate project delivery system will have a strong impact on project performance while carrying out the renewable energy projects for SDG?					
29	Do you agree that lack of technology knowledge will have a strong impact on project performance while carrying out the renewable energy projects for SDG?					
30	Do you agree that poor top management support will have a strong impact on project performance while carrying out the renewable energy projects for SDG?					
31	Do you agree that insufficient cash flow will have a strong impact on project performance while carrying out the renewable energy projects for SDG?					
32	Do you agree that inappropriate analysis and design will have a strong impact on project performance while carrying out the renewable energy projects for SDG?					
33	Do you agree that an inadequate project performance and quality control will have a strong impact on project performance while carrying out the renewable energy projects for SDG?					
34	Do you agree that lack of accountability in decision-making will have a strong impact on project performance while carrying out the renewable energy projects for SDG?					
35	Do you agree that insufficient bureaucracy system will have a strong impact on project performance while carrying out the renewable energy projects for SDG?					
	IMPACT OF CSF AND CHALLENGES ON PROJECT SUCCESS					
36	Do you agree that the critical success factors and challenges have a strong impact on Project Success in renewable energy projects for SDG?					

References

1. International Energy Agency (IEA). Indonesia Energy Outlook. 2024. Available online: https://www.iea.org/reports/indonesia-energy-outlook-2024 (accessed on 19 March 2025).
2. International Renewable Energy Agency (IRENA). Renewable Energy Statistics 2024. 2024. Available online: https://www.irena.org/publications/2024/Jan/Renewable-Energy-Statistics-2024 (accessed on 19 March 2025).
3. Indonesia Government. *G20pedia 2022*; Indonesia Government: Bali, Indonesia, 2022. Available online: https://indonesia.go.id/g20/kategori/g20/4177/g20pedia (accessed on 10 February 2024).
4. Ministry of Energy and Mineral Resources of Indonesia. Rencana Usaha Penyediaan Tenaga Listrik (RUPTL). 2018–2027. Jakarta: Ministry of Energy and Mineral Resources. 2021. Available online: https://gatrik.esdm.go.id/assets/uploads/download_index/files/38622-ruptl-pln-2021-2030.pdf (accessed on 14 February 2024).
5. PLN. *Statistik PLN 2022*; State Electricity Company of Indonesia: Jakarta, Indonesia, 2023. Available online: https://web.pln.co.id/statics/uploads/2023/05/Statistik-PLN-2022-Final.pdf (accessed on 27 February 2024).
6. Donastorg, A.; Renukappa, S.; Suresh, S. Financing Renewable Energy Projects in Developing Countries: A Critical Review. *IOP Conf. Ser. Earth Environ. Sci.* **2017**, *83*, 012012.
7. Stapleton, G.J. Successful implementation of renewable energy technologies in developing countries. *Desalination* **2009**, *248*, 595–602. [CrossRef]
8. Kim, K.; Park, H.; Kim, H. Real options analysis for renewable energy investment decisions in developing countries. *Renew. Sustain. Energy Rev.* **2017**, *75*, 918–926. [CrossRef]
9. Alsakini, W.; Wikström, K.; Kiiras, J. Proactive schedule management of industrial turnkey projects in developing countries. *Int. J. Proj. Manag.* **2004**, *22*, 75–85. [CrossRef]
10. Morris, P.W.G. *The Management of Projects*; Thomas Telford: London, UK, 1994.
11. Baccarini, D. The concept of project complexity—A review. *Int. J. Proj. Manag.* **1996**, *14*, 201–204.
12. Shenhar, A.J.; Dvir, D.; Levy, O.; Maltz, A.C. Project Success: A Multidimensional Strategic Concept. *Long Range Plan.* **2001**, *34*, 699–725. [CrossRef]
13. Pinto, J.K.; Slevin, D.P. Critical Success Factors across the Project Life Cycle. *Proj. Manag. J.* **1988**, *19*, 67–75.
14. Kerzner, H. *Project Management: A Systems Approach to Planning, Scheduling and Controlling*; Van Nos-trand Reinhold: New York, NY, USA, 1992.
15. Kerzner, H. *Project management: A Systems Approach to Planning, Scheduling and Controlling*, 7th ed.; John Wiley and Sons: New York, NY, USA, 2001.
16. Kerzner, H. *Project Management: A Systems Approach to Planning, Scheduling and Controlling*, 8th ed.; John Wiley & Sons, Inc.: Hoboken, NJ, USA, 2003.
17. Boyd, A. The five maxims of project satisfaction. *Aslib Proc.* **2001**, *53*, 423–430. [CrossRef]
18. Rockart, J.F. The Changing Role of the Information Systems Executive: A Critical Success Factors Perspective. *Sloan Manag. Rev.* **1982**, *24*, 3–13.
19. Ofori, D.F. Project Management Practices and Critical Success Factors–A Developing Country Perspective. *Int. J. Bus. Manag.* **2013**, *8*, 14. [CrossRef]
20. Pinto, J.K.; Slevin, D.P. Critical Success Factors in R&D Projects. *Res. Manag.* **1989**, *32*, 31–35. [CrossRef]

21. Rockart, J.F. *Chief Executives Define Their Own Data Needs*; Harvard Business Review: Brighton, MA, USA, 1979.
22. Belassi, W.; Tukel, O.I. A new framework for determining critical success/failure factors in projects. *Int. J. Proj. Manag.* **1996**, *14*, 141–151. [CrossRef]
23. Shenhar, A.J.; Dvir, D. *Reinventing Project Management: The Diamond Approach to Successful Growth and Innovation*; Harvard Business Review Press: Boston, MA, USA, 2007.
24. Ofori, G. Challenges facing construction industries in developing countries. *Build. Res. Inf.* **2002**, *30*, 149–151.
25. Turner, R.; Zolin, R. Forecasting Success on Large Projects: Developing Reliable Scales to Predict Multiple Perspectives by Multiple Stakeholders over Multiple Time Frames. *Proj. Manag. J.* **2012**, *43*, 87–99. [CrossRef]
26. Pinto, J.K.; Slevin, D.P. Critical Success Factors in Effective Project Implementation. In *Project Management Handbook*; Cleveland, D.I.E., King, W.R., Eds.; Van Nostrand Reinhold: New York, NY, USA, 1987; pp. 512–719.
27. McDermot, E.; Agdas, D.; Díaz, C.R.R.; Rose, T.; Forcael, E. Improving performance of infrastructure projects in developing countries: An Ecuadorian case study. *Int. J. Constr. Manag.* **2020**, *22*, 2469–2483. [CrossRef]
28. Yeo, K. Critical failure factors in information system projects. *Int. J. Proj. Manag.* **2002**, *20*, 241–246. [CrossRef]
29. Frese, R.; Sauter, V. Project Success and Failure: What Is Success, What Is Failure, And How Can You Improve Your Odds For Success? 2003. Available online: http://www.umsl.edu/~sauterv/analysis/6840_f03_papers/frese/ (accessed on 7 April 2024).
30. Andersen, E.S.; Birchall, D.; Jessen, S.A.; Money, A.H. Exploring Project Success. *Balt. J. Manag.* **2006**, *1*, 127–147.
31. Hyväri, I. Success of Projects in Different Organizational Conditions. *Proj. Manag. J.* **2006**, *37*, 31–41. [CrossRef]
32. Müller, R.; Turner, R. Leadership competency profiles of successful project managers. *Int. J. Proj. Manag.* **2010**, *28*, 437–448. [CrossRef]
33. Khang, D.B.; Moe, T.L. Success Criteria and Factors for International Development Projects: A Life-Cycle-Based Framework. *Proj. Manag. J.* **2008**, *39*, 72–84. [CrossRef]
34. Kurniawan, V.A.; Pradheksa, P.Y.; Saleh, R. The failure of micro-hydro technology: A case study of the Banyubiru project in Central Java, Indonesia. *Renew. Sustain. Energy Transit.* **2024**, *5*, 100081. [CrossRef]
35. Alqahtani, F.K.; Al-Jadhai, S.; Alromihy, H.; Alsaud, M.; Sherif, M.; Mohamed, A.G. Appraising critical success factors in sustainable housing projects: A comparative study of PPP modalities in Saudi Arabia. *Heliyon* **2024**, *10*, e32854. [CrossRef]
36. Roux, E.; Argyropoulou, M.; Argyropoulou, R. Critical success factors influencing project success in turnkey mechanical engineering projects within a project-based organisation. *Procedia Comput. Sci.* **2024**, *239*, 693–700. [CrossRef]
37. Kerzner, H. *Project Management: A Systems Approach to Planning, Scheduling, and Controlling*, 13th ed.; Wiley-Blackwell: Hoboken, NJ, USA, 2022.
38. Avots, I. Projects in developing countries. *Bus. Horiz.* **1972**, *15*, 69–73. [CrossRef]
39. Rondinelli, D.A. Why development projects fail: Problems of project management in developing countries. *Proj. Manag. Q.* **1976**, *7*, 10–15.
40. Yanwen, W. The Study on Complex Project Management in Developing Countries. *Phys. Procedia* **2012**, *25*, 1547–1552. [CrossRef]
41. Pretorius, S.; Steyn, H.; Jordaan, J. Project management maturity and project management success in the engineering and construction industries in Southern Africa. *S. Afr. J. Ind. Eng.* **2012**, *23*, 1–12. [CrossRef]
42. Pretorius, S.; Steyn, H.; Bond-Barnard, T.J. Project management maturity and project management success in de-veloping countries. *S. Afr. J. Ind. Eng.* **2023**, *34*, 36–48.
43. Atkinson, R. Project management: Cost, Time and quality, Two Best Guesses and a phenomenon, Its Time to Accept Other Success Criteria. *Int. J. Proj. Manag.* **1999**, *17*, 337–342.
44. Turner, R.J. *The Handbook of Project-based Management*; McGraw-Hill Education: New York, NY, USA, 2009.
45. Flyvbjerg, B.; Bruzelius, N.; Rothengatter, W. *Megaprojects and Risk: An Anatomy of Ambition*; Cambridge University Press: Cambridge, UK, 2003.
46. Kerzner, H. *Project Management: A Systems Approach to Planning, Scheduling, and Controlling*, 12th ed.; Wiley: Hoboken, NJ, USA, 2017.
47. PMI. *A Guide to the Project Management Body of Knowledge (PMBOK Guide)*, 6th ed.; Project Management Institute, Inc.: Newtown Square, PA, USA, 2017.
48. Meredith, J.R.; Mantel, S.J. *Project Management: A Managerial Approach*, 9th ed.; International Student Version; John Wiley: Hoboken, NJ, USA, 2012.
49. Freeman, R.E. *Strategic Management: A Stakeholder Approach*; Pitman: Boston, MA, USA, 1984. [CrossRef]
50. Zwikael, O.; Smyrk, J. *Project Management for the Creation of Organisational Value*; Springer: London, UK; New York, NY, USA, 2011.
51. Boehm, B.W. Software Engineering Economics. *IEEE Trans. Softw. Eng.* **1984**, *10*, 4–21.
52. PMI. *A Guide to the Project Management Body of Knowledge (PMBOK Guide)*, 5th ed.; Project Management Institute, Inc.: Newtown Square, PA, USA, 2013.

53. Katz, D. The motivational basis of organizational behavior. *Syst. Res. Behav. Sci.* **1964**, *9*, 131–146. [CrossRef] [PubMed]
54. Turner, J.R.; Müller, R. The Project Manager's Leadership Style as a Success Factor on Projects: A Literature Review. *Proj. Manag. J.* **2005**, *36*, 49–61. [CrossRef]
55. Cohen, W.M.; Levinthal, D.A. Absorptive Capacity: A New Perspective on Learning and Innovation. *Adm. Sci. Q.* **1990**, *35*, 128–152. [CrossRef]
56. Boehm, B.; Turner, R. Management challenges to implementing agile processes in traditional development organizations. *IEEE Softw.* **2005**, *22*, 30–39.
57. Munns, A.; Bjeirmi, B. The role of project management in achieving project success. *Int. J. Proj. Manag.* **1996**, *14*, 81–87. [CrossRef]
58. Akogbe, R.-K.T.M.; Feng, X.; Zhou, J. Importance and ranking evaluation of delay factors for development construction projects in Benin. *KSCE J. Civ. Eng.* **2013**, *17*, 1213–1222. [CrossRef]
59. Gardezi, S.S.S.; Manarvi, I.A. Time Extension Factors in Construction Industry of Pakistan. *Procedia Eng.* **2014**, *77*, 196–204. [CrossRef]
60. Ballesteros Pérez, P.; del Campo-Hitschfeld, M.L.; Gonzalez-Naranjo, M.A.; González-Cruz, M.C. Climate and construction delays: Case study in Chile. *Eng. Constr. Archit. Manag.* **2015**, *22*, 596–621.
61. Assaf, S.A.; Al-Hejji, S. Causes of delay in large construction projects. *Int. J. Proj. Manag.* **2006**, *24*, 349–357. [CrossRef]
62. Le-Hoai, L.; Lee, Y.D.; Lee, J.Y. Delay and cost overruns in Vietnam large construction projects: A comparison with other selected countries. *KSCE J. Civ. Eng.* **2008**, *12*, 367–377. [CrossRef]
63. Sweis, G.; Sweis, R.; Abu Hammad, A.; Shboul, A. Delays in construction projects: The case of Jordan. *Int. J. Proj. Manag.* **2008**, *26*, 665–674. [CrossRef]
64. Odeh, A.M.; Battaineh, H.T. Causes of construction delay: Traditional contracts. *Int. J. Proj. Manag.* **2002**, *20*, 67–73. [CrossRef]
65. Emuze, F.A.; Smallwood, J.J. Reducing 'non-value-adding activities' when building infrastructure in South Africa. *Proc. Inst. Civ. Eng.—Manag. Procure. Law* **2013**, *166*, 121–127.
66. Welde, M.; Bukkestein, I. Over time or on time? A study of delays in large government projects. *Procedia Comput. Sci.* **2022**, *196*, 772–781. [CrossRef]
67. Boonstra, A.; Reezigt, C. A Complexity Framework for Project Management Strategies. *Proj. Manag. J.* **2023**, *54*, 253–267. [CrossRef]
68. Floricel, S.; Michela, J.L.; Piperca, S. Complexity, uncertainty-reduction strategies, and project performance. *Int. J. Proj. Manag.* **2016**, *34*, 1360–1383. [CrossRef]
69. Pellerin, R.; Perrier, N. A review of methods, techniques and tools for project planning and control. *Int. J. Prod. Res.* **2018**, *57*, 2160–2178. [CrossRef]
70. Shenhar, A.; Holzmann, V. The Three Secrets of Megaproject Success: Clear Strategic Vision, Total Alignment, and Adapting to Complexity. *Proj. Manag. J.* **2017**, *48*, 29–46. [CrossRef]
71. Hughes, M.W. Why Projects Fail: The Effects of Ignoring Obvious Management Principles. *Ind. Eng. Mag.* **1986**, *18*, 12–18.
72. Nguyen, L.D.; Ogunlana, S.O.; Lan, D.T.X. A study on project success factors in large construction projects in Vietnam. *Eng. Constr. Arch. Manag.* **2004**, *11*, 404–413. [CrossRef]
73. Nyarko, K.; Whale, J.; Urmee, T. Drivers and challenges of off-grid renewable energy-based projects in West Africa: A review. *Heliyon* **2023**, *9*, e16710. [CrossRef] [PubMed]
74. Saunders, M.N.K.; Lewis, P.; Thornhill, A. *Research Methods for Business Students*, 8th ed.; Pearson: New York, NY, USA, 2019.
75. Creswell, J.W. *Qualitative Inquiry & Research Design: Choosing Among Five Approaches*, 3rd ed.; SAGE Publications: Los Angeles, CA, USA, 2013.
76. Hair, J.F. *Multivariate Data Analysis*, 6th ed.; Pearson Prentice Hall PTR: Upper Saddle River, NJ, USA, 2006.
77. Choi, H.-H.; Cho, H.-N.; Seo, J.W. Risk Assessment Methodology for Underground Construction Projects. *J. Constr. Eng. Manag.* **2004**, *130*, 258–272. [CrossRef]
78. Forcael, E.; Vargas, S.; Opazo, A.; Medina, L. Role of the Civil Engineer in the Contemporary Chilean Society. *Rev. Construcción (Univ. Católica Chile)* **2013**, *12*, 74–89.
79. PLN. Collaboration for Sustainability. Jakarta: State Electricity Company of Indonesia. 2024. Available online: https://eppid.pln.co.id/wp-content/uploads/2024/07/Laporan%20TJSL%202023.pdf (accessed on 10 February 2025).
80. Braimah, N.; Ndekugri, I. Consultants' Perceptions on Construction Delay Analysis Methodologies. *J. Constr. Eng. Manag.* **2009**, *135*, 1279–1288. [CrossRef]
81. Hair, J.F.; Hult, T.M.; Ringle, C.; Sarstedt, M. *A Primer on Partial Least Squares Structural Equation Modeling (PLS-SEM)*; SAGE Publications: Los Angeles, CA, USA, 2017.
82. Choi, B.; Ahn, S.; Lee, S. Role of social norms and social identifications in safety behavior of construction workers. I: Theoretical model of safety behavior under social influence. *J. Constr. Eng. Manag.* **2017**, *143*, 04016124.

83. Chan, A.P.; Darko, A.; Ameyaw, E.E.; Owusu-Manu, D.G. Barriers affecting the adoption of green building tech-nologies. *J. Manag. Eng.* **2017**, *33*, 04016057.
84. Tetteh, M.O.; Chan, A.P.C.; Darko, A.; Torku, A.; Nani, G. Critical Barriers to International Construction Joint Ventures Success: Multiexpert Views and Contextual Disparities. *J. Constr. Eng. Manag.* **2021**, *147*, 04021081. [CrossRef]
85. Owusu-Manu, D.; Torku, A.; Pärn, E.; Addy, M.N.; Edwards, D.J. An empirical assessment of innovation practices of quantity surveying firms in Ghana. *J. Constr. Proj. Manag. Innov.* **2017**, *7* (Suppl. S1), 1843–1858.
86. Sproull, N. *Handbook of Research Methods*; Scarecrow: London, UK, 1995.
87. Ott, R.L.; Longnecker, M. *An Introduction to Statistical Methods and Data Analysis*; Cengage Learning Inc.: Boston, MA, USA, 2010.
88. Cronbach, L.J.; Shavelson, R.J. My Current Thoughts on Coefficient Alpha and Successor Procedures. *Educ. Psychol. Meas.* **2004**, *64*, 391–418. [CrossRef]
89. Tabachnick, B.G.; Fidell, L.S. *Using Multivariate Statistics*, 5th ed.; Pearson/Allyn & Bacon: Boston, MA, USA, 2007.
90. Fabrigar, L.R.; Wegener, D.T.; MacCallum, R.C.; Strahan, E.J. Evaluating the Use of Exploratory Factor Analysis in Psychological Research. *Psychol. Methods* **1999**, *4*, 272–299.
91. Field, A. *Discovering Statistics Using IBM SPSS Statistics*, 5th ed.; SAGE Publications: London, UK, 2017.
92. Hutcheson, G.D.; Sofroniou, N. *The Multivariate Social Scientist: Introductory Statistics Using Generalized Linear Models*, 1st ed.; SAGE: London, UK, 1999.
93. Hair, J.F. *Multivariate Data Analysis: A Global Perspective*, 7th ed.; Pearson: Upper Saddle River, NJ, USA, 2010.
94. Hoyle, R.H. *Structural Equation Modeling for Social and Personality Psychology*, 1st ed.; Sage Publications: London, UK, 2011. [CrossRef]
95. Kline, R.B. *Principles and Practice of Structural Equation Modeling*, 4th ed.; The Guilford Press: New York, NY, USA, 2016.
96. Byrne, B.M. *Structural Equation Modeling With AMOS: Basic Concepts, Applications, and Programming*, 2nd ed.; Taylor and Francis: Hoboken, NJ, USA, 2013.
97. Hu, L.T.; Bentler, P.M. Cutoff criteria for fit indexes in covariance structure analysis: Conventional criteria versus new alternatives. *Struct. Equ. Model. Multidiscip. J.* **1999**, *6*, 1–55. [CrossRef]
98. West, S.G.; Taylor, A.B.; Wu, W. Model it and model selection in structural equation modeling. *Handb. Struct. Equ. Model.* **2012**, *1*, 209–231.
99. Tabachnick, B.G.; Fidell, L.S. *Using Multivariate Statistics*, 6th ed.; International Edition; Pearson: Boston, MA, USA, 2013.
100. Barrett, P. Structural equation modelling: Adjudging model fit. *Personal. Individ. Differ.* **2006**, *42*, 815–824. [CrossRef]
101. Kunnan, A.J. An introduction to structural equation modelling for language assessment research. *Lang. Test.* **1998**, *15*, 295–332.
102. Bentler, P.M. Comparative Fit Indexes in Structural Models. *Psychol. Bull.* **1990**, *107*, 238–246. [CrossRef] [PubMed]
103. Steiger, J.H. Structural Model Evaluation and Modification: An Interval Estimation Approach. *Multivar. Behav. Res.* **1990**, *25*, 173–180. [CrossRef]
104. de Wit, A. Measurement of project success. *Int. J. Proj. Manag.* **1988**, *6*, 164–170.
105. Mitchell, R.K.; Agle, B.R.; Wood, D.J. Toward a Theory of Stakeholder Identification and Salience: Defining the Principle of Who and What Really Counts. *Acad. Manag. Rev.* **1997**, *22*, 853–886. [CrossRef]
106. Bourne, L.; Walker, D.H. Project relationship management and the Stakeholder Circle™. *Int. J. Manag. Proj. Bus.* **2008**, *1*, 125–130. [CrossRef]
107. Ika, L.A.; Diallo, A.; Thuillier, D. Critical success factors for World Bank projects: An empirical investigation. *Int. J. Proj. Manag.* **2012**, *30*, 105–116. [CrossRef]
108. Turner, J.; Cochrane, R. Goals-and-methods matrix: Coping with projects with ill defined goals and/or methods of achieving them. *Int. J. Proj. Manag.* **2002**, *11*, 93–102. [CrossRef]
109. Lock, D. *Project Management*, 10th ed.; Routledge: London, UK, 2019.
110. Diallo, A.; Thuillier, D. The success dimensions of international development projects: The perceptions of African project coordinators. *Int. J. Proj. Manag.* **2004**, *22*, 19–31. [CrossRef]
111. Williams, T. The need for new paradigms for complex projects. *Int. J. Proj. Manag.* **1999**, *17*, 269–273. [CrossRef]
112. Olander, S.; Landin, A. Evaluation of stakeholder influence in the implementation of construction projects. *Int. J. Proj. Manag.* **2005**, *23*, 321–328. [CrossRef]
113. Cortina, J.M. What is coefficient alpha? An examination of theory and applications. *J. Appl. Psychol.* **1993**, *78*, 98–104.
114. Costello, A.; Osborne, J. Best Practices in Exploratory Factor analysis: Four Recommendations for Getting the Most from Your Analysis Recommendations for Getting the Most from Your Analysis. *Res. Eval. Pract. Assess. Res. Eval.* **2005**, *10*, 7.
115. Sovacool, B.K.; Cooper, C.J. *The Governance of Energy Megaprojects*; Edward Elgar Publishing: Cheltenham, UK, 2019.
116. Sambasivan, M.; Soon, Y.W. Causes and effects of delays in Malaysian construction industry. *Int. J. Proj. Manag.* **2007**, *25*, 517–526. [CrossRef]

117. Muriithi, N.; Crawford, L. Approaches to project management in Africa: Implications for international development projects. *Int. J. Proj. Manag.* **2003**, *21*, 309–319. [CrossRef]
118. Patlis, J.M. The role of law and legal institutions in determining the sustainability of integrated coastal management projects in Indonesia. *Ocean Coast. Manag.* **2005**, *48*, 450–467. [CrossRef]
119. Flyvbjerg, B. What you Should Know about Megaprojects and Why: An Overview. *Proj. Manag. J.* **2014**, *45*, 6–19. [CrossRef]
120. Wachs, M. When planners lie with numbers. *J. Am. Plan. Assoc.* **1989**, *55*, 476–479.
121. Zidane, Y.J.-T.; Andersen, B. The top 10 universal delay factors in construction projects. *Int. J. Manag. Proj. Bus.* **2018**, *11*, 650–672. [CrossRef]
122. Ofori, G. Construction in Developing Countries. *Constr. Manag. Econ.* **2007**, *25*, 1–6.
123. Guo, F.; Chang-Richards, Y.; Wilkinson, S.; Li, T.C. Effects of project governance structures on the management of risks in major infrastructure projects: A comparative analysis. *Int. J. Proj. Manag.* **2014**, *32*, 815–826. [CrossRef]
124. Akhavan, P.; Zahedi, M.R. Critical Success Factors in Knowledge Management Among Project-Based Organizations: A Multi-Case Analysis. *ICFAI Univ. J. Knowl. Manag.* **2014**, *12*, 20.
125. Busse, R.; Zafer, H.; Warner, M. Rethinking the Roles of Project Management Maturity and Organisational Culture for Perceived Performance: An Empirical Study Based on German Evidence. *Eur. J. Int. Manag.* **2020**, *14*, 730–752. [CrossRef]
126. Merrow, E.W. *Industrial Megaprojects: Concepts, Strategies, and Practices for Success*, 1st ed.; Wiley: Hoboken, NJ, USA, 2011.
127. Miller, R.; Lessard, D.R.; Floricel, S. *The Strategic Management of Large Engineering Projects: Shaping Institutions, Risks, and Governance*; MIT Press: Cambridge, MA, USA, 2000.
128. Ahsan, K.; Gunawan, I. Analysis of cost and schedule performance of international development projects. *Int. J. Proj. Manag.* **2010**, *28*, 68–78. [CrossRef]
129. Jugdev, K.; Müller, R. A Retrospective look at our Evolving Understanding of Project Success. *Proj. Manag. J.* **2005**, *36*, 19–31. [CrossRef]
130. Zwikael, O.; Ahn, M. The Effectiveness of Risk Management: An Analysis of Project Risk Planning Across Industries and Countries. *Risk Anal.* **2010**, *31*, 25–37. [CrossRef]
131. Müller, R.; Turner, R. The Influence of Project Managers on Project Success Criteria and Project Success by Type of Project. *Eur. Manag. J.* **2007**, *25*, 298–309. [CrossRef]
132. Meredith, J.R.; Mantel, S.J. *Project Management: A Managerial Approach*, 8th ed.; International Student Version; John Wiley: Hoboken, NJ, USA, 2011.
133. Kerzner, H. *Project Management: A Systems Approach to Planning, Scheduling and Controlling*, 9th ed.; John Wiley: Hoboken, NJ, USA, 2006.
134. Karlsen, J.T. Project Stakeholder Management. *Eng. Manag. J.* **2002**, *14*, 19–24. [CrossRef]
135. Yang, R.J.; Shen, G.Q.P. Framework for Stakeholder Management in Construction Projects. *J. Manag. Eng.* **2015**, *31*, 04014064. [CrossRef]
136. Hitt, M.A.; Ireland, R.D.; Hoskisson, R.E.; Agarwal, Y. Strategic management competitiveness and globalization. *Financ. India* **2003**, *XVII*, 321–323.
137. Katzenbach, J.R.; Smith, D.K. *The Wisdom of Teams: Creating the High-Performance Organization*; Harvard Business School Press: Boston, MA, USA, 1993.
138. Zhang, L.; Guo, H. Enabling knowledge diversity to benefit cross-functional project teams: Joint roles of knowledge leadership and transactive memory system. *Inf. Manag.* **2019**, *56*, 103156. [CrossRef]
139. Rowlinson, S.; Cheung, Y.K.F. Stakeholder management through empowerment: Modelling project success. *Constr. Manag. Econ.* **2008**, *26*, 611–623. [CrossRef]
140. Jepsen, A.L.; Eskerod, P. Stakeholder analysis in projects: Challenges in using current guidelines in the real world. *Int. J. Proj. Manag.* **2008**, *27*, 335–343. [CrossRef]
141. Hillson, D. Extending the risk process to manage opportunities. *Int. J. Proj. Manag.* **2002**, *20*, 235–240. [CrossRef]
142. Locke, E.A.; Latham, G.P. Building a practically useful theory of goal setting and task motivation: A 35-year odyssey. *Am. Psychol.* **2002**, *57*, 705–717. [CrossRef]

Disclaimer/Publisher's Note: The statements, opinions and data contained in all publications are solely those of the individual author(s) and contributor(s) and not of MDPI and/or the editor(s). MDPI and/or the editor(s) disclaim responsibility for any injury to people or property resulting from any ideas, methods, instructions or products referred to in the content.

buildings

MDPI

Article

Comparison of Energy Demand for Heating and CO_2 Emissions in Urban and Rural Areas, the Case of Hajdú-Bihar County, Hungary

Ferenc Kalmár *, Béla Bodó and Tünde Kalmár

Department of Building Services and Building Engineering, Faculty of Engineering, University of Debrecen, Hungary, Otemeto Str. 2-4, 4028 Debrecen, Hungary; bela.bodo@karakalterv.hu (B.B.); kalmar_tk@eng.unideb.hu (T.K.)

* Correspondence: fkalmar@eng.unideb.hu; Tel.: +36-52-415155 (ext. 77774)

Abstract: Mitigating carbon dioxide emissions in the building sector is a primary global goal. This paper compares different residential buildings in urban and rural regions of Hajdú-Bihar County (Hungary). Significant differences were found between urban and rural single-family houses concerning their energy performance; however, the differences in CO_2 emissions were not significant. Only the differences in specific heat losses were significant between urban single-family and masonry-structured multifamily buildings. Panel buildings demonstrate the best energy performance from their construction period, but due to high investment costs and the inability to change the heat source, the CO_2 emissions from these buildings have a lower limit today. In both single-family houses and masonry-structured multifamily buildings, meeting the heat demand can be achieved with zero CO_2 emissions using existing technologies.

Keywords: urban residential buildings; rural residential buildings; multifamily buildings; heat losses; heat demand; CO_2 emissions

1. Introduction

1.1. Energy-Saving Efforts in the Building Sector

Mitigating greenhouse gas emissions is one of the main goals of each country. In the European Union, Directive 2018/844 of the European Parliament and the Council stated that the Union is committed to developing a sustainable, competitive, secure and decarbonised energy system by 2050 [1]. Member states are required to set out a roadmap with measures, aimed towards the long-term 2050 goal of reducing greenhouse gas emissions in the Union by 80–95% compared to 1990. This is to ensure a highly energy-efficient and decarbonised national building stock and to facilitate the cost-effective transformation of existing buildings into nearly zero-energy buildings [1]. Moreover, in line with the scientific advice from the European Scientific Advisory Board on Climate Change and based on a detailed Impact Assessment, in 2024, the communication of the Commission to the European Parliament presents a 90% net GHG emission reduction compared to 1990 levels as the recommended target for 2040 [2]. In the Union, building stock is responsible for 36% of CO_2 emissions [1]. Ürge-Vorsatz et al. stated that globally, over 60% of residential energy is demanded for thermal uses [3]. Within this, space heating comprises 32–33% of the global total building energy use. They found that biomass is still by far the dominant fuel when a global picture is considered, covering most of the energy needs of the poor in the least developed countries, used with very low efficiencies. Berardi stated that the building sector deserves to be considered alone, without being mixed with other subsectors within a "third sector" besides transportation and industry [4]. Furthermore, he drew attention to the importance of data accountability and the reliability of buildings' energy consumption. He proposed that large countries should provide data on the energy consumption in buildings, dividing it into regions or provinces with the same climatic conditions [4]. Parshall et al. had the same opinion since they suggested counties as the best spatial scale for a local energy data inventory [5]. They argued that raw energy data are available at the county scale, and counties are a recognised political unit with some authority to formulate local policy.

1.2. CO_2 Emissions in the Building Sector

Zhou et al. evaluated trajectories for China's building sectors to 2050 and the potential impact of policies on fully deploying today's maximum techno-economically feasible efficiency and renewables measures for Chinese buildings using four scenarios [6]. They pointed out that by developing and implementing energy efficiency and renewable energy policies, China can achieve CO_2 emissions from the energy use in buildings that are just 20% above 2010 levels in 2050 and are likely to decline rapidly thereafter. Akan and Akan determined the optimum insulation thickness of buildings depending on the life cycle cost, considering the heating and cooling energy demands separately. They found that the CO_2 emissions can be reduced by approximately 66–76% during the heating season and by 46–69% during the cooling season using the optimal insulation thickness [7]. Canbolat and Albak performed a multi-objective optimisation of the whole external wall structure of buildings, considering the heat source as well [8]. The multi-objective optimisation aimed to obtain minimum CO_2 emissions with minimum costs. They obtained different structural materials and layers depending on the main goal of the optimisation. The study pointed out that the boundary parameters and the optimisation objectives have to be well defined to obtain appropriate results. Zhu et al. drew attention to the fact that besides the operational CO_2 emissions of buildings, the embodied CO_2 has to be considered [9]. They found that in China, the annual embodied carbon dioxide accounts for roughly one half of the total carbon dioxide emissions in the building sector in 2015.

1.3. Urban vs. Rural Buildings

Several authors studied the energy consumption of residential buildings built in urban and rural regions. Heinonen and Junnila showed that in Finland, cities are more energy intensive than the surrounding suburban and rural areas [10]. The authors explained this urban–rural difference with the higher incomes of urban residents. The same authors analysed holistically the residential energy consumption patterns and the overall housing energy requirements of urban and rural households in Finland [11]. They studied three of the most common housing types in Finland and found that each housing mode was less energy intensive in rural areas, the energy mixes vary significantly between the studied building types, and the behavioural differences are significant between the different housing modes. This finding suggests that not only are the construction technology and thermal requirements essential when the decarbonisation potential is analysed, but also the settlement type where the building was built. Nematchoua et al. claimed that consistent data at the neighbourhood level are crucial to understanding the origins of climate change at regional, urban and rural scales [12]. Cai and Jiang, through the example of Shaanxi province, found substantial differences between the energy sources used in rural and urban areas [13]. In rural areas, the place of coal was taken by biomass, while in urban areas, coal was replaced mainly by natural gas. Guan et al. established a bottom-up Adaptive Weighting Divisia decomposition model to analyse the differences in the per capita CO_2 emissions across urban and rural areas of different provinces in China [14]. Their results showed that the average household CO_2 emissions in urban areas consistently exceeded those in rural areas. Zhang et al. in their study concluded that residential energy consumption both in urban and rural regions rapidly increased in recent years [15]. However, the growth rate of urban residential energy consumption was faster than that of rural areas; hence, the gap in residential energy consumption per capita between urban and rural areas became narrower over the study period. In contrast, Cheshmehzangi concluded that urban housing has benefited much from China's green strategies and energy housing plans, while rural housing has experienced minimal change in the past few decades [16]. Zhou et al., taking the Shaanxi province as an example, developed a simulation model to predict the trends in carbon dioxide emission changes in urban and rural buildings [17]. They concluded that structural adjustments to the building energy consumption have the greatest potential for carbon reduction and play a central role in achieving a green, low-carbon pathway. Du et al. found that the rebound effect might impact the expected results in saving energy and reducing CO_2 emissions [18]. Differences can be observed between the rebound effect in urban and rural areas. In China, the magnitudes of the rebound effect for residential buildings in urban areas fluctuate between 79.43% and 110.00%, while that in rural areas increases from 115.28% to 120.40%. The rebound effect is mainly because the energy demand for residential buildings would increase when the energy efficiency improves [18]. Wang et al. developed a model to analyse the movement of CO_2 emissions gravity centres in China [19]. They found that per capita disposable income is one of the primary factors affecting the gravity centres. The findings of Zhou and Shi (obtained for the 2000–2015 period) contradict this statement. They concluded that urbanisation is an emollient for the inequality of urban–rural energy consumption because it could improve the rural residents' accessibility to modern energy services and the affordability of clean energy [20]. Based on balanced panel data covering 30 provinces of China for the period 1990–2012, Zhou et al. found that the impact of employment structural transformation on the energy consumption and CO_2 emissions across China was generally positive and statistically significant but differed considerably across regions [21]. He et al. concluded that in China, the lack of knowledge about building energy efficiency and the energy consumption habits of rural residents in comparison with urban residents is poor [22]. The effects of urban and rural residents' lifestyles on the energy consumption of buildings are also emphasised in [23–26]. The demand of urban residents for more electronics and improved indoor comfort drives energy consumption higher, while the traditional ways of using natural options are still dominant [25]. The differences between urban and rural energy consumption are also accentuated by the urban microclimate. Several studies have demonstrated that the urban heat island affects buildings' energy consumption [27–29]. Li et al. showed that the urban heat island effect can increase cooling energy consumption from 10% to 120% and decrease heating energy consumption from 3% to 45% [30]. The urban night-time air temperatures can be higher by 3–4 °C than adjacent non-urban areas [31]. Climate change also affects the energy consumption in buildings (decreased energy for heating and increased energy for cooling); however, the impact is different in urban and rural areas [32]. Madlener and Sunak showed that the effects of urbanisation differ considerably between developed and developing countries as well as within the group of developing countries [33]. Hence, this process should be well assessed in different countries to choose the optimal local building energy solutions. Myszczyszyn and Supron analysed data between 1996 and 2020 and found that in central European countries, the urbanisation rate affected CO_2 emissions, influencing the enhanced energy consumption [34].

1.4. Research Goal

Summarising the references discussed above, it can be stated that the energy performance characteristics of the existing building stock have to be known in detail to meet the ambitious decarbonisation goals. Furthermore, the differences between rural and urban buildings have to be accounted for when decarbonisation strategies are elaborated. Employment, income, energy mix, and lifestyle strongly affect energy consumption. An unfortunate combination of these can lead to energy poverty [35]. Because of similar building policies implemented in the last 80 years, common features can be observed in the Central and Eastern European countries' building stocks. Different building typologies have already been created for these countries [36,37]. However, the urban–rural differences have not been researched so far from the point of view of energy consumption and CO_2 emissions. It is well known that the A/V ratio of the building affects the heat demand. Hence, the specific heat losses of single-family houses are expected to exceed those of multifamily buildings. Our research aimed to show the energy and CO_2 emission differences between urban and rural areas in Central and Eastern European countries through the example of Debrecen and Hajdú-Bihar county (Hungary).

2. Methodology

First, the available data on the building stock provided by the Hungarian Central Statistical Office were assessed and analysed. Thereafter, the energy assessment of 1200 flats/houses in Debrecen and Hajdú-Bihar county was carried out. Debrecen is the second largest city in Hungary. Out of 1200 energy assessments, 841 were suitable for the planned research: 236 were single-family houses in Debrecen, 372 were single-family houses in small towns and villages in rural regions of the county, 115 were flats in panel buildings in Debrecen, and 118 were flats in masonry-structured multifamily buildings in Debrecen. The building envelope and the building service systems were assessed on-site, and the calculations were carried out using Winwatt32 7.63 software, which integrates the energy-performance-calculation methodology used in Hungary. Based on the building energy data, statistical analyses were performed using the ORIGIN LAB 9.55 software. This software for data analysis and graphing, available at the University of Debrecen, offers all the features needed to analyse the data gathered in this study. The results are presented using box-plot diagrams. The linear fit method was used to see the variation trends over the years of different parameters [38]. To compare different groups of buildings, the paired sample t-test was performed at a $p = 0.05$ significance level [39]. The energy performance analysis of detached houses in Debrecen was already conducted [40]. Buildings built between 1930 and 2023 were classified into eight categories. In this manuscript, single-family houses from the rural region and masonry-structured multifamily buildings were classified similarly into eight categories: buildings built before 1959, followed by buildings built between 1960 and 1969, 1970 and 1979, 1980 and 1989, 1990 and 1999, 2000 and 2009, 2010 and 2019, and 2020 and 2023. The assessed panel buildings (insulated concrete sandwich panels) were built in Debrecen only in two decades: 1960–1969 and 1969–1970. The A/V values, the net heated area, the specific heat loss coefficient, the net energy demand for heating and the specific CO_2 emissions were statistically analysed for all buildings. Using the paired sample t-test method at a $p = 0.05$ significance level, the differences between rural and urban single-family houses were analysed for each category; thereafter, the significance analysis of the differences between urban single-family houses and masonry-built multifamily houses was conducted.

Finally, through some good examples, rational envelope refurbishments, energy sources and heat generation modes were proposed for the analysed typical buildings to obtain minimum CO_2 emissions. The steps of the performed research are presented in Figure 1.

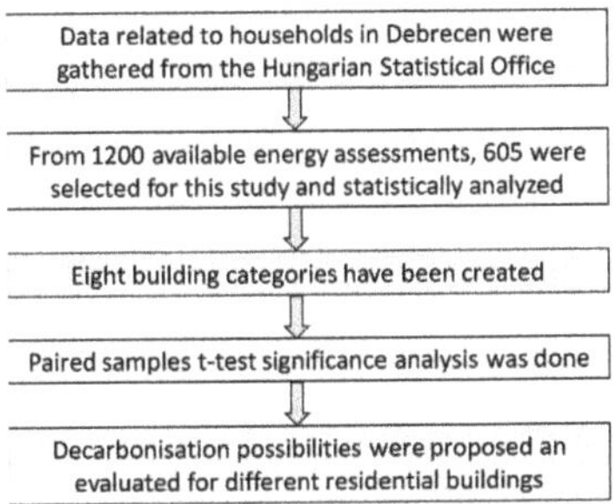

Figure 1. Flowchart of research algorithm.

3. Features of Building Stock

In Hungary, heating accounts for approximately 71% of households' total final energy consumption [41]. Obviously, single-family houses predominantly characterise rural areas, while urban areas feature a significant number of flats in multifamily houses. According to the Hungarian Central Statistical Office, the number of single-family houses in Debrecen is 37,615, while the number of flats placed in multifamily houses is 50,008, of which 26,233 are flats in panel buildings (26,776 are flats in masonry-structured multifamily buildings). In Hajdú-Bihar county, there are 208,075 households in total [42].

When analysing the building structures, it can be observed that the building stock is extremely diverse. However, specific building materials have been used for external building elements in different decades over the last century. The differences can primarily be shown through the structure of the external walls. Until 1960, masonry blocks were mainly made of adobe mixed with grain crumbs (Figure 2a). The thermal conductivity of this mixture was between 0.40 and 0.65 W/mK, depending on the density of the material (1100–1400 kg/m^3, respectively). The wall thickness ranged from 40 to50 cm resulting in an overall heat transfer coefficient of 0.70 to 1.28 W/m^2K. The ground level floor consists of concrete without an insulation layer. The beam-framed slab under the attic is covered again with adobe mixed with grain crumbs.

After the 1960s, the adobe was gradually replaced by burnt clay bricks (Figure 2b). The thermal conductivity of burnt clay bricks is 0.78 W/mK, making the overall heat transfer coefficient of the 38 cm walls 1.43 W/m^2K.

From the years 1970 to 1980, the burnt clay bricks with several vertical holes became widely used. These bricks were called B30, as their length and the thickness of the walls built from them was 30 cm (Figure 2c). They were also commonly used for multifamily buildings. The thermal conductivity of these bricks is 0.64 W/mK, leading to an overall heat transfer coefficient of 1.47 W/m^2K for a 30 cm thick wall.

Later, the B30 brick was enhanced by increasing its length to 38 cm and increasing the number of vertical holes in a special arrangement (HB38 brick). However, this brick was not widely used. Instead, a similar brick but with better thermal properties, PTH38, saw extensive use after the year 2000 (Figure 2d). The thermal conductivity of this brick was 0.2 W/mK, and the length (thickness of the wall) could be either 30 cm, 38 cm or sometimes

44 cm. Consequently, the overall heat transfer coefficient of the wall was 0.58 W/m^{2}K (30 cm), 0.47 W/m^{2}K (38 cm) or 0.41 W/m^{2}K (44 cm). In the latter two cases, even the requirements of the 7/2006 Building energy performance regulation (0.45 W/m^{2}K) were satisfied.

(**a**) External wall from adobe bricks (**b**) Burnt clay brick (**c**) B30 brick (**d**) PTH 30 brick

Figure 2. Building materials used for residential building walls.

After the year 2000, PTH bricks were extensively used even in multifamily buildings. Moreover, these bricks (or some improved versions) are used today, insulated with 15–20 cm of expanded polystyrene (EPS). These external walls have an overall heat transfer coefficient of 0.2 W/m ^{2} K (15 cm thick EPS) or 0.16 W/m ^{2} K (20 cm thick EPS).

It can be observed that shifting from adobe bricks to burnt clay bricks and B30 bricks led to an increase in the overall heat transfer coefficient of the walls. Subsequently, a significant reduction in the overall heat transfer coefficient was achieved in the year 2000 with PTH bricks, and again after 2020 (imposed by the requirements of nearly zero energy buildings), the overall heat transfer coefficient was reduced nearly by half.

It should be mentioned that in the 1970s and 1980s, silicate bricks were also used in considerable quantities. When producing silicate, aluminium paste is added to water, cement and lime, which reacts with the water to form hydrogen gas. In the silicate structure, pores form in place of the hydrogen, improving the material's heat conductivity (0.22–0.36 W/mk depending on the density). This is what we call aerated concrete today (heat conductivity 0.13–0.15 W/mk). The wall thickness was primarily 30 cm.

Currently, several single-family houses are built from light-frame (wood or metal) structures (plasterboard, OSB, 16 cm mineral wool, OSB, 10 cm EPS, plaster).

During the years 1970–1980, buildings constructed with industrialised technology gained popularity in Central and Eastern European countries. The prefabricated panels already included a polystyrene layer of 6–8 cm. These blocks of flats may contain hundreds of apartments. Due to thermal bridges, the walls of these buildings have a mean overall heat transfer coefficient of 1. 04–1. 96 W/m ^{2} K (even reaching 1. 98 in some cases). These buildings are mostly connected to the district heating system. Despite the advantageous A/V parameter, these buildings exhibit relatively high energy consumption (due to the high overall heat transfer coefficient of the external walls).

4. Results and Discussion

4.1. Rural Single-Family Houses

The analysed buildings were categorised into eight groups. The thermal performance requirements related to the external building elements have changed several times in the last decades, but were set out in a decree only in 2006. A building pocket book mentioned first the overall heat transfer coefficient (U) of external walls in 1934 [43]. The authors recommended U values lower than 1.68 W/m^{2}K. Later, lower overall heat transfer coefficients were recommended in standards or prescriptions, but these were not compulsory. Hence, the overall heat transfer coefficients of the external building elements (especially the walls) were determined mainly by the construction materials and technology. Consequently, it was decided to carry out an energy performance analysis, grouping the buildings depending on the year of construction. It was shown in the previous section that the development and utilisation of different building blocks strongly affect the thermal properties of external walls. The same grouping methodology was used in the case of urban single-family houses [38].

The A/V parameter and the net heated area for rural single-family houses in Hajdú-Bihar county, based on 372 assessments, are presented in Figure 3. It can be observed that the decade-by-decade variation in the A/V parameter is not significant, but the decrease is significant between the oldest and newest buildings. The net floor area is increasing, but the decade-by-decade differences are not significant. However, the difference between the oldest and the recently built buildings is significant. The specific heat loss coefficient and the net heat demand by groups are presented in Figure 4. A continuous decrease in these parameters can be observed over the years.

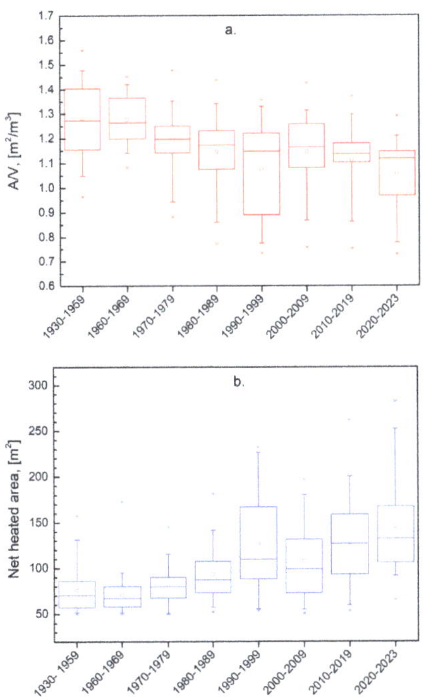

Figure 3. A/V ratio (**a**) and net heated area (**b**) of the analysed rural single-family houses.

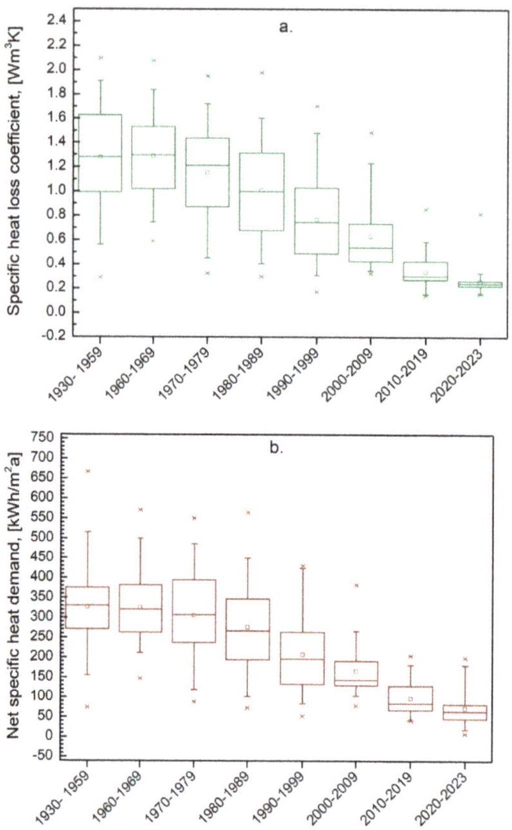

Figure 4. Specific heat loss coefficient (**a**) and net specific heat demand (**b**) of the analysed rural single-family houses.

The specific heat loss coefficient and the specific heat demand are shown in Figure 4.

The specific CO_2 emissions related to the energy consumption are presented in Figure 5.

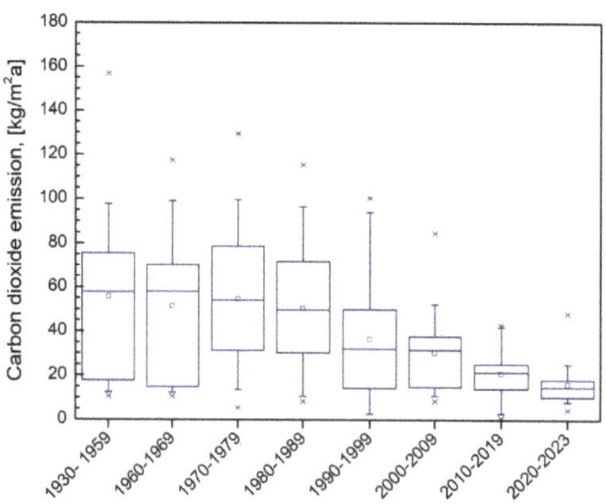

Figure 5. Specific CO_2 emissions of the analysed rural single-family houses.

It can be seen that there is practically no difference between the specific CO_2 emissions of rural single-family houses until 1989. Thereafter, a decrease can be observed caused by the thermally efficient masonry blocks and strict energy performance requirements set out in the 7/2006 decree [44]. The extremely high emission value was observed in the case of a building with poor thermal properties of the envelope, and coal was used for heating and electricity for domestic hot water production. However, it can be observed that in each analysed decade, there are buildings with specific CO_2 emissions below 10 kg/m^{2}a. Currently, the requirement is 20 kg/m^{2}a. This means that there are lots of already refurbished single-family houses.

4.2. Masonry-Structured Multifamily Buildings

A total of 118 analysed flats were placed in masonry-structured multifamily buildings in Debrecen. Figures 6–8 show the results obtained for these buildings.

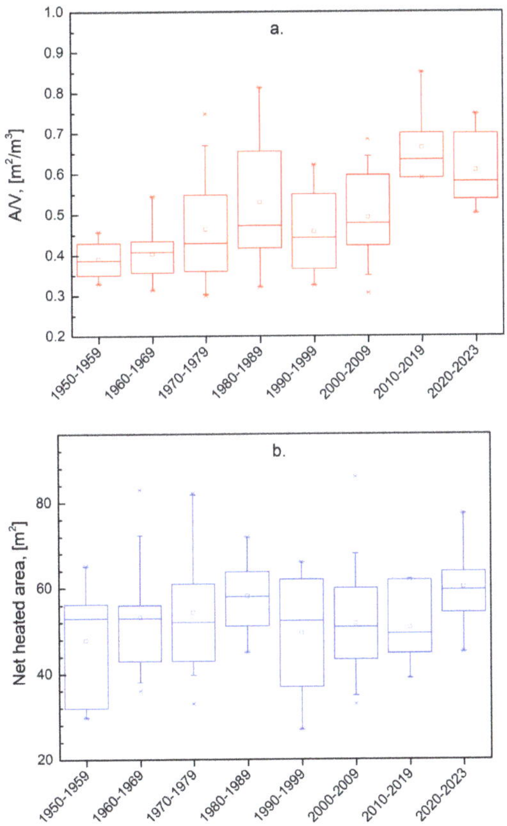

Figure 6. A/V ratio (**a**) and net heated area (**b**) of the analysed masonry-structured multifamily buildings.

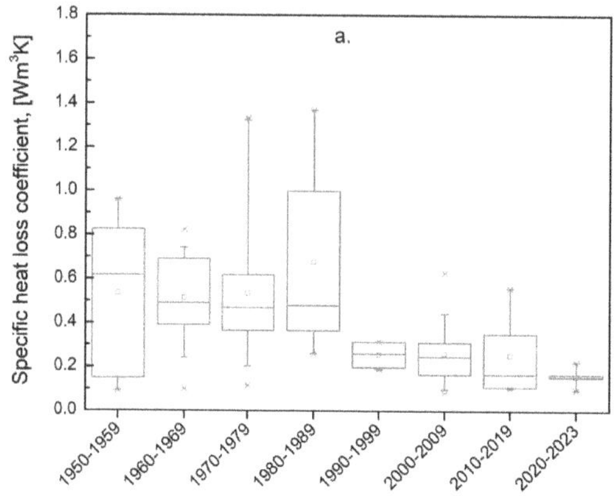

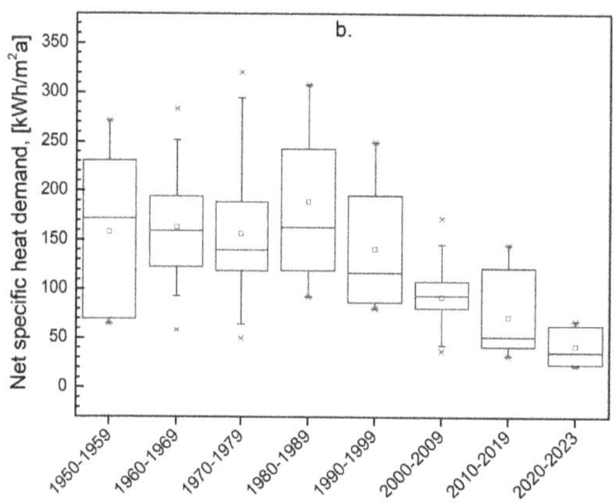

Figure 7. Specific heat loss coefficient (**a**) and net specific heat demand (**b**) of the analysed masonry-structured multifamily buildings.

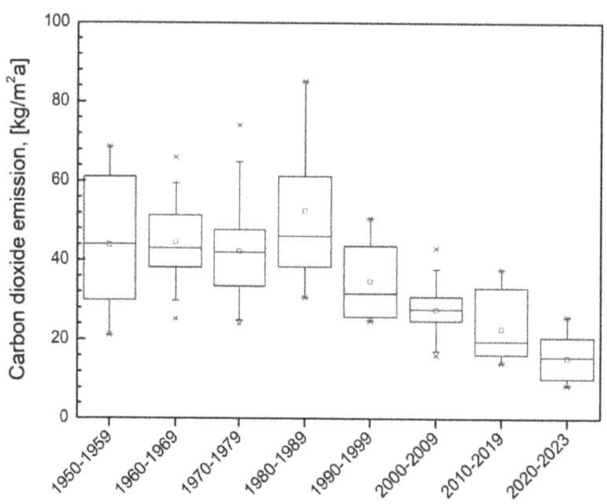

Figure 8. Specific CO_2 emissions of the analysed masonry-structured multifamily buildings.

It can be observed that the A/V parameter of masonry-structured multifamily buildings increased over the years.

The differences are not significant except for the decades 2000–2009 and 2010–2020. Practically, this means that the number of flats in a building decreased in the last decades.

The net heated area of a flat shows a small variation during the analysed decades, but only between the 1980–1989 and 1990–1999 decades (the political change) can a significant decrease be observed.

Analysing the specific CO_2 emissions, it can be affirmed that before 2000, there were no buildings with emissions below 20 kg/m^{2}a. Most buildings with emissions below 20 kg/m^{2}a were found only after 2020.

4.3. Panel Buildings

In the framework of this research, 115 apartments placed in industrialised panel buildings were assessed. These buildings were built only between the years 1970 and1989, so the obtained results only apply to these two decades (Figures 9–11).

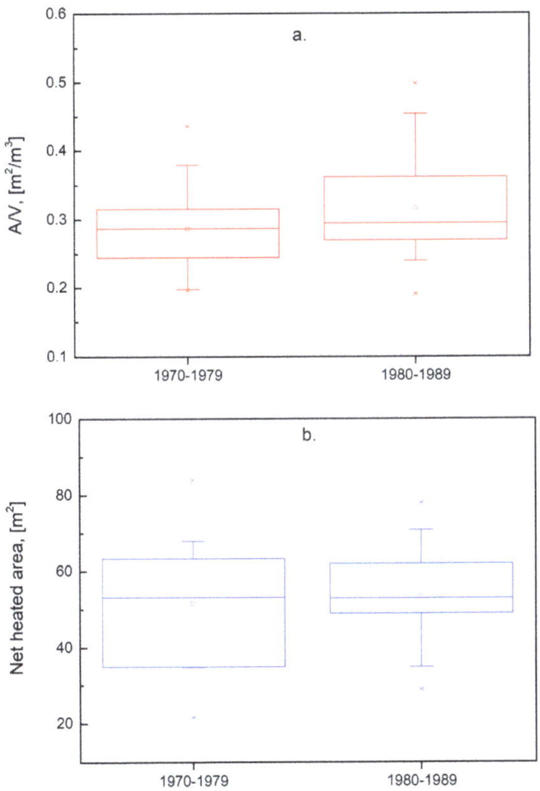

Figure 9. A/V ratio (**a**) and net heated area (**b**) of the analysed panel buildings.

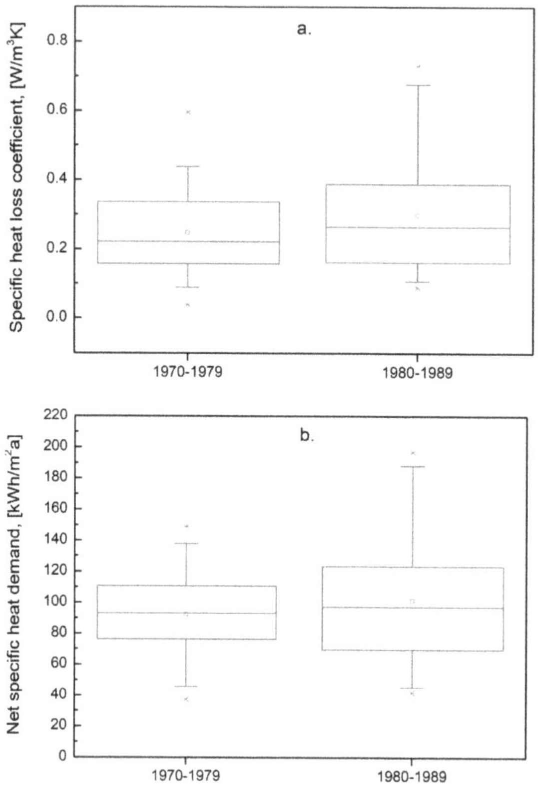

Figure 10. Specific heat loss coefficient (**a**) and net specific heat demand (**b**) of the analysed panel buildings.

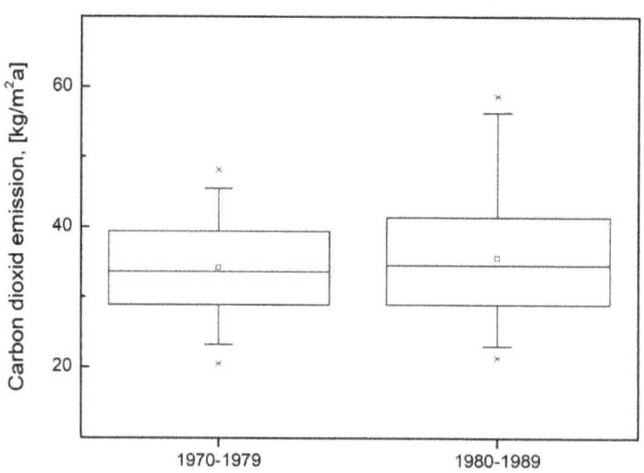

Figure 11. Specific CO_2 emissions of the analysed panel buildings.

Most of the analysed panel buildings have ten or eleven storeys, but some have five levels. The insulation layer inserted between the two steel concrete layers was 6–8 cm thick. Regarding the A/V parameter and the net heated area of a flat, no significant differences can be observed between the buildings built in the two decades. Moreover, the differences are not significant when comparing the net heated area of flats in masonry-structured multifamily buildings and panel buildings. Analysing the specific CO_2 emissions, it can be observed that, despite thermal refurbishments, in the framework of this study, no building was identified with a specific CO_2 emission below 20 kg/m^{2}a.

4.4. Comparison of Urban–Rural Single-Family Houses

To show the variation trends during the analysed 93 years, linear regression curves were determined (Figures 12–14). The goodness of fit is given by the adjusted R-squared values.

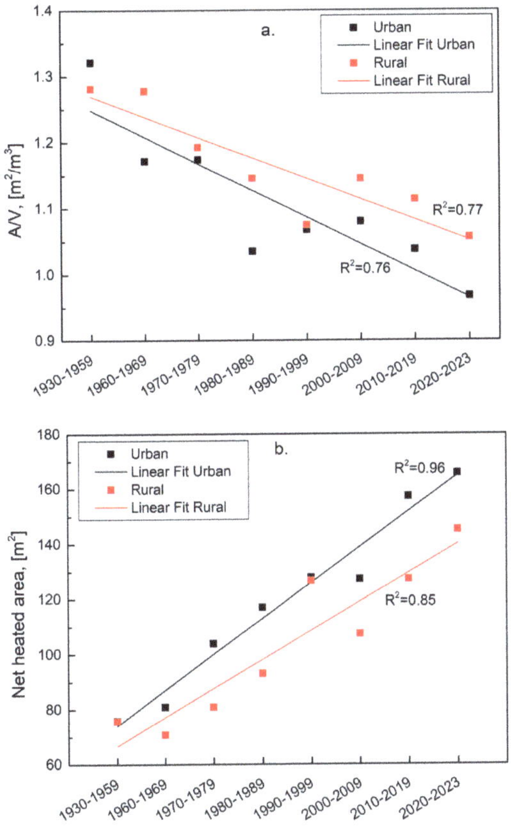

Figure 12. Trends of A/V ratio (**a**) and net heated area (**b**) of the analysed urban–rural single-family buildings (R^2—adjusted R-squared).

The statistical analysis of the differences between groups (means of urban–rural single-family houses, urban single-family houses–masonry-structured multifamily buildings, and rural single-family houses–masonry-structured multifamily buildings) was performed at a $p = 0.05$ significance level using the paired sample t-test.

It can be observed that the trend is similar in the case of each analysed parameter, yet important differences can be identified. While the A/V parameter of rural single-family houses is higher, the net heated area of urban single-family houses exceeds the net heated area of rural houses. This means that urban buildings are larger but energy conscious, being more compact. The differences are significant in both cases. The trend lines of the specific heat loss coefficient are almost parallel. The heat loss coefficients of single-family houses in rural areas exceeds those in urban areas; hence, the heat demand of rural houses is higher than the heat demand of urban buildings. The differences are significant in both cases. However, the differences between CO_2 emissions are not significant. The reason is the fact that in rural buildings, biomass and wood are used extensively, while in urban regions, mainly natural gas is used (the CO_2 emissions of wood and biomass are lower than those of natural gas).

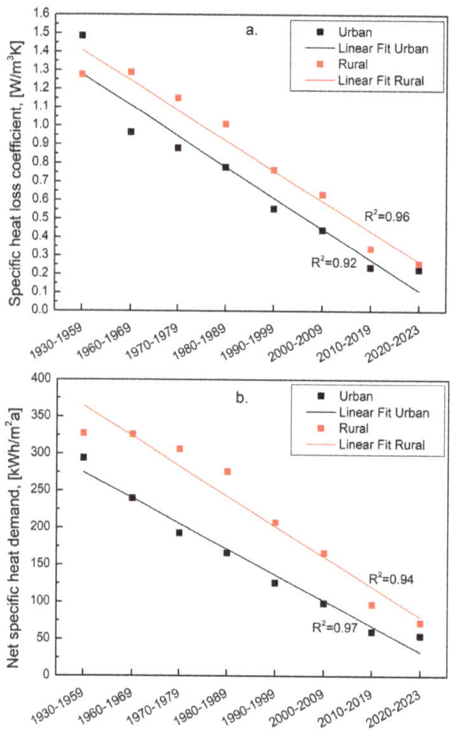

Figure 13. Trends of specific heat loss coefficient (**a**) and net specific heat demand (**b**) of the analysed urban–rural single-family buildings (R^2—adjusted R-squared).

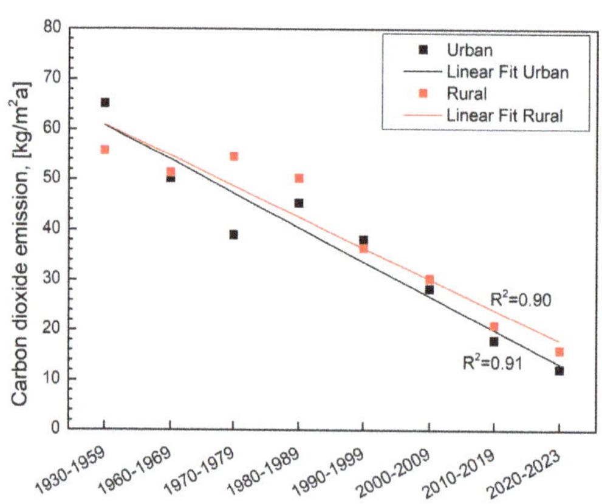

Figure 14. Trends of specific CO_2 emissions of the analysed urban–rural single-family buildings (R^2—adjusted R-squared).

4.5. Comparison of Urban Single-Family Houses and Masonry-Structured Multifamily Buildings

The trend lines for the specific heat loss coefficient, net specific heat demand and CO_2 emissions for urban single-family houses and masonry-structured multifamily buildings are presented in Figures 15 and 16. It can be observed that single-family houses show a steeper trend line in both cases. As a result, the trend lines intersect. It can generally be said that the heat loss coefficient and the heat demand of single-family houses are higher than those of multifamily buildings. However, there are decades when the situation is the inverse. While the differences between the urban and rural specific heat loss coefficients are significant, the differences between the net specific heat demand are not significant.

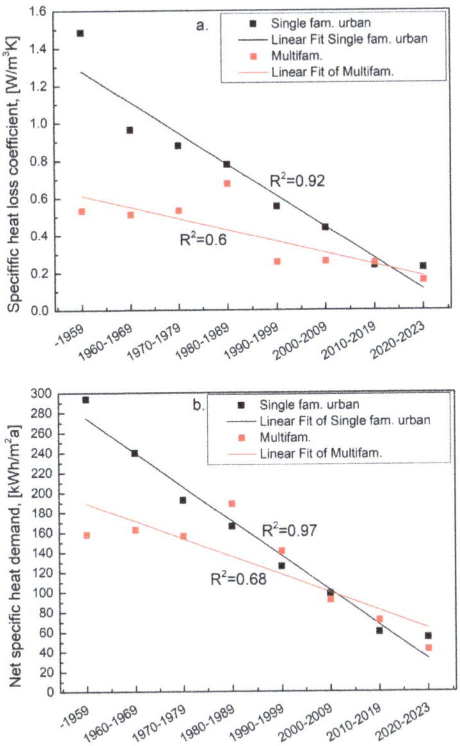

Figure 15. Trends of specific heat loss coefficient (**a**) and net specific heat demand (**b**) of the analysed urban single-family and urban masonry-structured multifamily buildings (R^2—adjusted R-squared).

Similarly, the inclination angle of the specific carbon dioxide emissions trend line in the case of single-family houses is higher than that of multifamily buildings. This means that the rate of carbon dioxide emission reduction from decade to decade was greater for single-family houses than for multifamily buildings. The differences between CO_2 emissions are not significant. The reason can be the fact that until 1990, in multifamily buildings, gas convectors were widely used as heaters. Higher efficiency can be obtained in the case of central heating systems with radiators. After 1990, similar heat sources and heating systems were used, and the differences between the net specific heat demand are not significant.

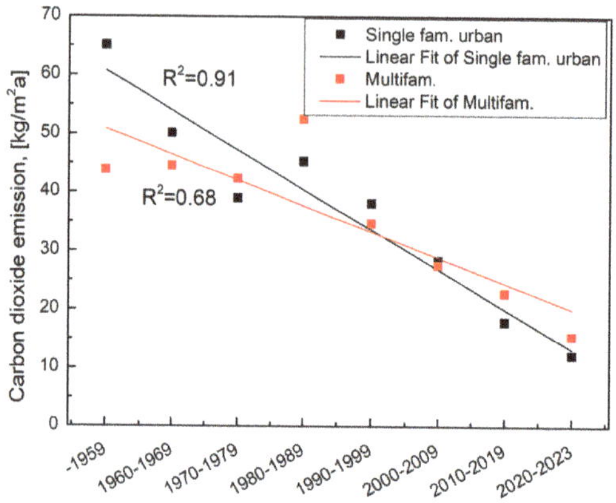

Figure 16. Trends of specific CO_2 emissions of the analysed urban single-family and masonry-structured multifamily buildings (R^2—adjusted R-squared).

4.6. Comparison of Rural Single-Family Houses and Masonry-Structured Multifamily Buildings

Finally, the specific heat loss coefficient, the net specific heat demand and the specific CO_2 emissions were compared in the case of rural single-family houses and multifamily buildings (Figures 17 and 18). There is no point in comparing the A/V coefficient and the net heated area. The results are similar to those presented in the previous subsection. The differences are significant in the case of the specific heat loss coefficient and net specific heat demand, but the differences are not significant in the case of specific CO_2 emissions. The reason can be found in the heat source (natural gas vs. biomass and wood) and heating system type (gas convector vs. central heating system).

In the case of panel buildings, having data only for two decades prevents a comprehensive comparison across all the categories. The mean A/V parameter of panel buildings was 0.286 m^{-1} for the decade 1970–1979 and 0.316 m^{-1} for the decade 1980–1989. The mean specific heat loss values were 0.247 W/m^3K for the decade 1970–1979 and 0.298 W/m^3K for the decade 1980–1989. The mean specific heat demands were 92.4 kWh/m^2a (SD = 27.92) for the decade 1970–1979 and 101.38 kWh/m^2a (SD = 38.27) for the decade 1980–1989. The specific CO_2 emissions were 34.18 kg/m^2a (SD = 7.26) for the decade 1970–1979 and 35.63 kg/m^2a (SD = 9.18) for the decade 1980–1989. It should be noted that almost all panel buildings are connected to the district heating system, where heat is produced primarily using natural gas. In the same decades, masonry-structured multifamily buildings had mean specific CO_2 emissions of 42.35 kg/m^2a (SD = 12.62) and 52.55 kg/m^2a (SD = 18.05), respectively. Urban single-family houses exhibited a mean specific CO_2 emission of 38.97 kg/m^2a (SD = 21.55) for the decade 1970–1979 and 45.31 kg/m^2 a (SD = 23.38) for the decade 1980–1989. In contrast, rural single-family buildings fared worse, with emissions of 54.54.53 kg/m^2a (SD = 30.17) and 50.25 kg/m^2a (SD = 26.73), respectively. It can be stated that panel buildings built during their construction period demonstrate better energy performance, with lower CO_2 emissions than all the other building types. However, all the other building categories have shown substantial improvements in their energy performance over the past few decades, while progress has stalled for panel buildings. The issue is that the cost of thermal and energy refurbishment for these buildings is substantial, and the flats are privately owned, meaning that each owner must contribute to the refurbishment expenses. Unfortunately, many homeowners, such as pensioners, are unable to afford these costs. Nevertheless, there are project programmes for panel buildings that allow the residential community to apply for funding to cover up to 50% of the refurbishment costs. However, this does not guarantee a reduction in CO_2 emissions below the required 20 kg/m^{2}a, as this is heavily influenced by the heating mode and energy source. The district heating company must fully or at least partially transition from natural gas to renewable energy sources (e.g., geothermal).

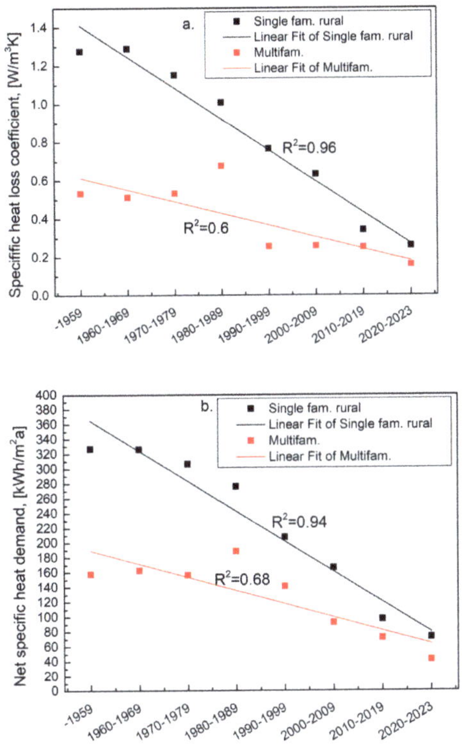

Figure 17. Trends of specific heat loss coefficient (**a**) and net specific heat demand (**b**) of the analysed rural single-family and urban masonry-structured multifamily buildings (R^2—adjusted R-squared).

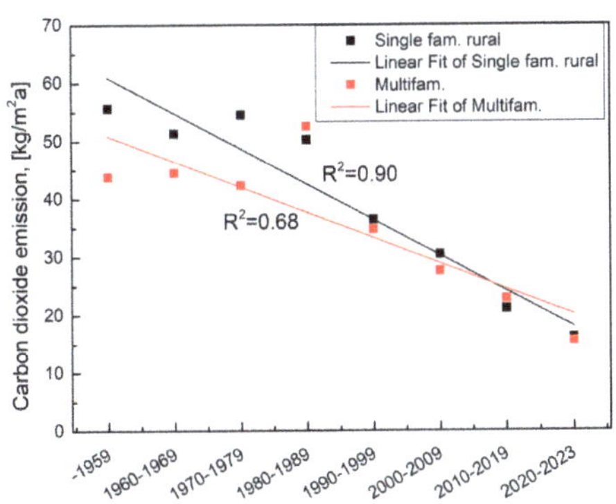

Figure 18. Trends of specific CO_2 emissions of the analysed rural single-family and masonry-structured multifamily buildings (R^2—adjusted R-squared).

4.7. Good Examples of Refurbishments

Between the analysed 605 houses and flats, good examples could be identified and selected which can serve as the model for other buildings.

4.7.1. Single-Family Houses

In the case of single-family houses, the same model can be followed by either urban or rural houses; the only difference is the heat carrier: wood and biomass cannot be recommended in cities.

Wall structures need 15–20 cm of insulation, while slabs under the attic or flat roofs need at least 30 cm of insulation. The floor laid on the floor or the slab above the cellar should have, at minimum, a 10 cm insulation layer. Windows and doors should have an overall heat transfer coefficient lower than 1.1 W/m^2K. With these thermal characteristics of the envelope, the specific heat loss coefficient will be 0.2 ± 0.04 W/m^3K. Having thermostat-controlled central heating with radiators, a buffer tank and wood as the heat carrier, the specific CO_2 emission will be 9.8–10.6 kg/m^2a. By installing a central heating system with an air–water thermal pump as the heat source, the specific CO_2 emission will be 10 ± 0.4 kg/m^2a. By installing a 5.1 kW_p photovoltaic system with the optimal orientation and inclination angle, the specific CO_2 emission can be reduced to 4.0 ± 0.5 kg/m^2a. If the photovoltaic system has a peak output of 15 kW_p, then the specific CO_2 emission will be negative: -11.8 kg/m^2a. It can be observed that in these refurbished model buildings, natural gas is not used at all.

4.7.2. Masonry-Structured Multifamily Buildings

The building taken as an example has five floors and 48 apartments (Figure A1). The walls are constructed from B30 masonry blocks without insulation prior to thermal refurbishment. The flat roof has a thickness of 20 cm of reinforced concrete and 15 cm of perlite concrete (average thickness). The slab above the cellar consists of reinforced concrete without any insulation. The net floor area of the building is 2486.1 m^2, and the A/V parameter is 0.52. The specific heat demand was found to be 0.871 W/m^3K, while the net specific heat demand was 251.67 kWh/m^2a. Gas convectors are utilised for heating, and hot water is provided by gas instantaneous water heaters. The specific carbon dioxide emission was found to be 58.8 kg/m^2a. After the initial step of energy refurbishment, the external walls received an additional 12 cm of EPS thermal insulation, while the flat roof was enhanced with an additional 15 cm of EPS insulation. Meanwhile, the slab above the cellar was equipped with an additional 10 cm of EPS thermal insulation. The specific heat demand decreased to 0.189 W/m^3K, and the net specific heat demand was 34.11 kWh/m^2a. Heating was supplied through a central heating system with a buffer tank, maintaining supply and return temperatures of 55 °C and 45 °C, respectively. The heat was centrally controlled, depending on the outdoor temperature. An air–water heat pump was utilised as the heat source (installed on the flat roof), and hot water was prepared indirectly through a storage tank. Following these interventions, the specific carbon dioxide emission was reduced to 10.82 kg/m^2a. In the second step of refurbishments, solar energy integration took place. The flat roof area measures 702.1 m^2. A 47.3 kWp photovoltaic system was installed on the roof (the output of a photovoltaic panel is 430 W, the distance between the panel rows is 1.71 m, the area required by a 1 kW system is 8.78 m^2, the inclination angle is 20 degrees, and the panels are oriented toward the south). A 100 kWh storage package was installed in accordance with new regulations, with a total area requirement of 450 m^2 for the system. Additionally, a 10 m^2 solar collector system was mounted on the roof, oriented towards the south, with a 40-degree inclination angle. With these two systems, the specific CO_2 emission was reduced to 2.59 kg/m^{2}a. It can be concluded that zero emissions (resulting from the heat demand for heating and hot water preparation) could have been achieved if the external building structures had received thicker thermal insulation layers during the first phase of the renovation (external walls at least 15 cm, flat roof 30 cm).

4.7.3. Panel Buildings

From an energy perspective, the main difference between the panel building and the masonry-structured multifamily building is that the panel building is connected to the district heating system. There are local regulations that recommend or allow district heating in downtowns only. Homeowners have few possibilities to change the heat source or integrate renewable energy sources. These essential decarbonisation steps must be carried out by the district heating company. However, rising energy prices have placed district heating companies in a challenging position. The heat losses from the pipe network reduce the system's efficiency. Furthermore, substantial electricity is consumed by the circulation pumps and control motor valves in the heat centres. This energy consumption is related to the district heating system rather than the building, but through the primary energy, it significantly influences the energy performance and CO_2 emissions of the building.

The building taken as an example has eleven floors, 88 apartments and an advantageous A/V parameter of 0.302 (Figure A2). The net heated area is 5579.37 m^2. Before thermal refurbishment, the layers of sandwich panels were as follows: 15 cm reinforced concrete, 5 cm EPS and 7 cm reinforced concrete. The flat roof had 20 cm of reinforced concrete and 15 cm (average thickness) of perlite concrete. The slab above the cellar was 20 cm uninsulated reinforced concrete with additional covering layers. The heating system was a one-pipe-with-bypass central heating system with control of the supplied heat in the heat centre depending on the outdoor temperature. The radiators were fitted with thermostatic valves. The supply temperature was 90 °C, and the return temperature was 70 °C. The heat was produced in gas boilers and provided by the district heating system, with a primary energy transformation factor of 1.26. The hot water was produced indirectly in storage tanks. The specific heat loss coefficient was 0.509 W/m^{3}K, and the net specific heat demand was 160.6 kWh/m^{2}a. The specific carbon dioxide emission was 44.9 kg/m^{2}a. By adding 12 cm EPS to the external walls, 20 cm to the flat roof and 12 cm to the slab above the cellar, the specific heat loss coefficient was reduced to 0.102 W/m^{3}K, and the specific net heat demand fell to 25.78 kWh/m^{2}a. A 4 kWp photovoltaic system oriented to the south, with an inclination angle of 20°, was installed on the roof to cover the auxiliary electricity demand of the heating and hot water system (8 kWh accumulators are also necessary). Assuming that the heat source is a biomass-based power plant, the primary energy transformation factor can be reduced to 0.5394. This factor currently cannot be reduced to zero, because of the auxiliary energy consumption. The supply and return temperatures are 75 °C and 55 °C, respectively. Consequently, the specific carbon dioxide emission will be reduced to 22.99 kg/m^{2}a.

It can be concluded that for panel buildings connected to the district heating system, the CO_2 emissions have a lower limit, which cannot be reduced further using the current technologies. This limitation arises from the efficiency of the district heating system (heat losses, auxiliary energy use), heat source and heat production.

Nevertheless, new technologies could aid in the mitigation of energy use and CO_2 emissions. Currently, these technologies are not widely used because of the higher investment costs. The energy use and CO_2 emission can be reduced further in all residential buildings using special insulation materials [45]. However, in this case, the changing of the physical parameters has to be taken into account [46]. Currently, in residential buildings, fresh air is provided widely through natural ventilation. The ventilation heat demand can be reduced using special personalised ventilation systems with heat recovery [47].

5. Conclusions

- In this study, based on the energy assessment of 605 single-family houses and flats in multifamily buildings, a heat demand and CO_2 emission survey in urban and rural residential buildings in Hajdú-Bihar County (Hungary) was conducted. Analysing the obtained results statistically, the following conclusions can be drawn:
- In both rural and urban regions, the compactness of single-family houses has improved in recent decades (with a more pronounced decrease in urban regions), and the net floor area of the buildings is increasing in both urban and rural areas (though the increase is more notable in urban regions). Hence, the trends of the A/V parameter and net heated area are similar in urban and rural single-family houses; however, the A/V parameter is significantly higher in rural houses, while the net heated area is significantly greater in urban houses.
- Between 1930 and 2023, the specific heat loss coefficient was reduced by 6.7 times in urban buildings and by 4.9 times in rural buildings. For the net specific heat demand, a reduction of 5.4 for urban houses and 4.5 for rural houses is observed, respectively. The specific heat loss coefficient and net specific heat demand exhibit similar trends in both urban and rural single-family houses. However, these parameters are significantly higher in rural buildings.
- The specific CO_2 emissions show a reduction of 5.3 times in urban houses and 3.5 times in rural houses. At a $p = 0.05$ level, no significant differences can be observed between urban and rural single-family houses.
- It can be stated that urban buildings have significantly better thermal properties, though the differences in CO_2 emissions are not significant. The reason for this lies in the heat carriers used. In old rural buildings, biomass and wood are predominantly used, while natural gas is used in old urban buildings. In new buildings, there are no significant differences between the heat sources; however, wood and biomass are still utilised in rural houses, while rarely utilised in urban houses. Heat pumps are widely used in both regions.
- Comparing masonry-structured multifamily buildings with urban and rural single-family houses, similar results were obtained. The specific heat loss coefficient, the net specific heat demand and the specific CO_2 emissions are higher in single-family houses than in multifamily houses. In multifamily houses, the specific heat loss coefficient was reduced by 3.5 times, the net specific heat demand by 3.7 times and the CO_2 emissions by 2.8 times, respectively. The differences between urban single-family houses and masonry-built multifamily buildings are significant only in the case of specific heat losses, while the differences between rural single-family houses and masonry-structured multifamily buildings are significant for both specific heat losses and the net specific heat demand.
- Panel buildings constructed between 1970 and 1989 show the best thermal and energy performance. Due to the high investment costs and the inability to change the heat source, these buildings now exhibit poor thermal performance.
- Single-family houses in both rural and urban regions can be effectively transformed into positive energy houses by utilising heat pumps and properly designed photovoltaic systems with appropriately chosen accumulators.

- In the case of masonry-structured multifamily buildings, zero CO_2 emissions (from heating and hot water preparation) can be achieved by using air–water heat pumps, photovoltaic systems with accumulators and solar collectors.
- For panel buildings connected to the district heating system, the CO_2 emissions have a lower limit that cannot be reduced further by using the current technologies. This is due to the efficiency of the district heating system (including heat losses and auxiliary energy use), heat sources and heat production.

6. Future Work

The further reduction in CO_2 emissions in panel buildings is extremely important. New, energy-efficient solutions for heat supply must be found for these buildings.

7. Limitations

In this study, 605 buildings and flats were assessed from an energy point of view. By increasing the number of assessments, more accurate results would be obtained.

Author Contributions: T.K.: calculations, investigation, writing original draft; B.B.: data curation, calculations, energy analysis of 605 buildings and flats; F.K.: conceptualisation, writing original draft, review and editing. All authors have read and agreed to the published version of the manuscript.

Funding: The work was supported by Project no. TKP2021-NKTA-34. Project no. TKP2021-NKTA-34 has been implemented with the support provided by the Ministry of Innovation and Technology of Hungary from the National Research, Development and Innovation Fund, financed under the TKP2021-NKTA funding scheme.

Data Availability Statement: The data presented in this study are available on request from the corresponding author due to privacy reasons.

Conflicts of Interest: The authors declare that they have no known competing financial interests or personal relationships that could have appeared to influence the work reported in this paper.

Nomenclature

A/V	ratio of the building envelope area A, [m^2] and heated volume V, [m^3];
p	level of significance;
R^2	adjusted R-squared value (goodness-of-fit measure);
SD	standard deviation (average amount of variability in the dataset);
GHG	greenhouse gases;
EPS	expanded polystyrene;
B30,HB38, PTH	bricks with vertical hole types.

Appendix A

Figure A1. Masonry-structured multifamily building.

Figure A2. Panel building.

References

1. Directive (EU) 2018/844 of the European Parliament and of the Council of 30 May 2018 amending Directive 2010/31/EU on the Energy Performance of Buildings and Directive 2012/27/EU on Energy Efficiency, Official Journal of the European Union, 19.6.2018. Available online: http://data.europa.eu/eli/dir/2018/844/oj (accessed on 21 March 2025).
2. Communication from the Commission to the European Parliament, the Council, the European Economic and Social Committee and the Committee of the Regions, Securing our Future Europe's 2040 Climate Target and Path to Climate Neutrality by 2050 Building a Sustainable, Just and Prosperous Society, Strasbourg. 2024. Available online: https://www.carbonbrief.org/qa-european-commission-calls-for-90-cut-in-eu-emissions-by-2040/ (accessed on 21 March 2025).
3. Ürge-Vorsatz, D.; Cabeza, L.F.; Serrano, S.; Barreneche, C.; Petrichenko, K. Heating and cooling energy trends and drivers in buildings. *Renew. Sustain. Energy Rev.* **2015**, *41*, 85–98. [CrossRef]
4. Berardi, U. A cross-country comparison of the building energy consumptions and their trends, Resources. *Conserv. Recycl.* **2017**, *123*, 230–241. [CrossRef]

5. Parshall, L.; Gurney, K.; Hammera, S.A.; Mendoza, D.; Zhou, Y.; Geethakumar, S. Modeling energy consumption and CO_2 emissions at the urban scale: Methodological challenges and insights from the United States. *Energy Policy* **2010**, *38*, 4765–4782. [CrossRef]
6. Zhou, N.; Khanna, N.; Feng, W.; Ke, J.; Levine, M. Scenarios of energy efficiency and CO_2 emissions reduction potential in the buildings sector in China to year 2050. *Nat. Energy* **2018**, *3*, 978–984. [CrossRef]
7. Akan, A.P.; Akan, A.E. Modeling of CO_2 emissions via optimum insulation thickness of residential buildings. *Clean Technol. Environ. Policy* **2022**, *24*, 949–967. [CrossRef]
8. Canbolat, A.S.; Albak, E.I. Multi-Objective Optimization of Building Design Parameters for Cost Reduction and CO_2 Emission Control Using Four Different Algorithms. *Appl. Sci.* **2024**, *14*, 7668. [CrossRef]
9. Zhu, W.; Feng, W.; Li, X.; Zhang, Z. Analysis of the embodied carbon dioxide in the building sector: A case of China. *J. Clean. Prod.* **2020**, *269*, 122438. [CrossRef]
10. Heinonen, J.; Junnila, S. A Carbon Consumption Comparison of Rural and Urban Lifestyles. *Sustainability* **2011**, *3*, 1234–1249. [CrossRef]
11. Heinonen, J.; Junnila, S. Residential energy consumption patterns and the overall housing energy requirements of urban and rural households in Finland. *Energy Build.* **2014**, *76*, 295–303. [CrossRef]
12. Nematchoua, M.K.; Sadeghi, M.; Reiter, S. Strategies and scenarios to reduce energy consumption and CO_2 emission in the urban, rural and sustainable neighbourhoods. *Sustain. Cities Soc.* **2021**, *72*, 103053.2. [CrossRef]
13. Cai, J.; Jiang, Z. Changing of energy consumption patterns from rural households to urban households in China: An example from Shaanxi Province, China. *Renew. Sustain. Energy Rev.* **2008**, *12*, 1667–1680. [CrossRef]
14. Guan, X.; Guo, S.; Xiong, J.; Jia, G.; Fan, J. Energy-related CO_2 emissions of urban and rural residential buildings in China: A provincial analysis based on end-use activities. *J. Build. Eng.* **2023**, *64*, 105686. [CrossRef]
15. Zhang, M.; Song, Y.; Li, P.; Li, H. Study on affecting factors of residential energy consumption in urban and rural Jiangsu. *Renew. Sustain. Energy Rev.* **2016**, *53*, 330–337. [CrossRef]
16. Cheshmehzangi, A. Evaluating the Nexus between Housing and Energy Sectors: The Comparison of Urban, Peri-Urban and Rural Housing Areas in Zhuhai, China. *Energy Power Eng.* **2020**, *12*, 314–338. [CrossRef]
17. Zhou, T.; Luo, X.; Liu, X.; Zhai, X.; Sun, Y.; Liu, G.; Liu, J.; Gao, Y.; Dang, D.; Li, N.; et al. The green and low-carbon development pathways in the urban and rural building sector in Shaanxi Province. *China Energy Build.* **2024**, *306*, 113952. [CrossRef]
18. Du, Q.; Han, X.; Li, Y.; Li, Z.; Xia, B.; Guo, X. The energy rebound effect of residential buildings: Evidence from urban and rural areas in China. *Energy Policy* **2021**, *153*, 112235. [CrossRef]
19. Wang, J.; You, K.; Qi, L.; Ren, H. Gravity center change of carbon emissions in Chinese residential building sector: Differences between urban and rural area. *Energy Rep.* **2022**, *8*, 10644–10656. [CrossRef]
20. Zhou, Q.; Shi, W. Socio-economic transition and inequality of energy consumption among urban and rural residents in China. *Energy Build.* **2019**, *190*, 15–24. [CrossRef]
21. Zhou, Y.; Liu, Y.; Wu, W.; Li, Y. Effects of rural–urban development transformation on energy consumption and CO_2 emissions: A regional analysis in China. *Renew. Sustain. Energy Rev.* **2015**, *52*, 863–875. [CrossRef]
22. He, B.J.; Yang, L.; Ye, M. Building energy efficiency in China rural areas: Situation, drawbacks, challenges, corresponding measures and policies. *Sustain. Cities Soc.* **2014**, *11*, 7–15. [CrossRef]
23. Fan, J.L.; Ju, H.; Wei, Y.M. Residential energy-related carbon emissions in urban and rural China during 1996–2012: From the perspective of five end-use activities. *Energy Build.* **2015**, *96*, 201–209. [CrossRef]
24. Chen, C.; Liu, G.; Meng, F.; Hao, Y.; Zhang, Y.; Casazza, M. Energy consumption and carbon footprint accounting of urban and rural residents in Beijing through Consumer Lifestyle Approach. *Ecol. Indic.* **2019**, *98*, 575–586. [CrossRef]
25. Hu, S.; Yan, D.; Guo, S.; Cui, Y.; Dong, B. A survey on energy consumption and energy usage behavior of households and residential building in urban China. *Energy Build.* **2017**, *148*, 366–378. [CrossRef]
26. Shahi, D.K.; Rijal, H.B.; Shukuya, M. A study on household energy-use patterns in rural, semi-urban and urban areas of Nepal based on field survey. *Energy Build.* **2020**, *223*, 110095. [CrossRef]
27. Santamouris, M.; Papanikolaou, N.; Livada, I.; Koronakis, I.; Georgakis, C.; Argiriou, A.; Assimakopoulos, D.N. On the impact of urban climate on the energy consumption of buildings. *Sol. Energy* **2001**, *70*, 201–216. [CrossRef]
28. Roxon, J.; Ulm, F.J.; Pellenq, R.J.M. Urban heat island impact on state residential energy cost and CO_2 emissions in the United States. *Urban Clim.* **2020**, *31*, 100546. [CrossRef]
29. Magli, S.; Lodi, C.; Lombroso, L.; Muscio, A.; Teggi, S. Analysis of the urban heat island effects on building energy consumption. *Int. J. Energy Environ. Eng.* **2015**, *6*, 91–99. [CrossRef]
30. Li, X.; Zhou, Y.; Yu, S.; Jia, G.; Li, H.; Li, W. Urban heat island impacts on building energy consumption: A review of approaches and findings. *Energy* **2019**, *174*, 407–419. [CrossRef]

31. Paolini, R.; Zania, A.; MeshkinKiya, M.; Castaldo, V.L.; Pisello, A.L.; Antretter, F.; Poli, T.; Cotana, F. The hygrothermal performance of residential buildings at urban and rural sites: Sensible and latent energy loads and indoor environmental conditions. *Energy Build.* **2017**, *152*, 792–803. [CrossRef]
32. Fan, J.L.; Zeng, B.; Hu, J.W.; Zhang, X.; Wang, H. The impact of climate change on residential energy consumption in urban and rural divided southern and northern China. *Env. Geochem Health* **2020**, *42*, 969–985. [CrossRef]
33. Madlener, R.; Sunak, Y. Impacts of urbanization on urban structures and energy demand: What can we learn for urban energy planning and urbanization management? *Sustain. Cities Soc.* **2011**, *1*, 45–53. [CrossRef]
34. Myszczyszyn, J.; Supron, B. Relationship among Economic Growth, Energy Consumption, CO_2 Emission, and Urbanization: An Econometric Perspective Analysis. *Energies* **2022**, *15*, 9647. [CrossRef]
35. Győri, Á.; Huszár, Á.; Balogh, K. Differences in the Domestic Energy Consumption in Hungary: Trends between 2006–2017. *Energies* **2021**, *14*, 6718. [CrossRef]
36. Hrabovszky-Horváth, S.; Pálvölgyi, T.; Csoknyai, T.; Talamon, A. Generalized residential building typology for urban climate change mitigation and adaptation strategies: The case of Hungary. *Energy Build.* **2013**, *62*, 475–485. [CrossRef]
37. Csoknyai, T.; Hrabovszky-Horváth, S.; Georgiev, Z.; Jovanovic-Popovic, M.; Stankovic, B.; Villatoro, O.; Szendrő, G. Building stock characteristics and energy performance of residential buildings in Eastern-European countries. *Energy Build.* **2016**, *132*, 39–52. [CrossRef]
38. Available online: https://www.originlab.com/doc/en/Origin-Help/LR-Algorithm (accessed on 3 May 2025).
39. Available online: https://www.originlab.com/doc/Origin-Help/tTest-PairSample-Algorithm (accessed on 3 May 2025).
40. Kalmár, F.; Bodó, B.; Li, B.; Kalmár, T. Decarbonization Potential of Energy Used in Detached Houses—Case Study. *Buildings* **2024**, *14*, 1824. [CrossRef]
41. Tóth, G.; Jáger, V.; Kovalszky, Z.S.; Bóday, P.; Ádám, D.; Kincses, Á.; Szép, T. Characteristics of Household Energy Consumption in the Shadow of the Russia-Ukraine War—A Case Study from Hungary. *Int. J. Sustain. Energy Plan. Manag.* **2024**, *40*, 52–70. [CrossRef]
42. Hungarian Central Statistical Office. Available online: https://www.ksh.hu/nepszamlalas/tables_regional_09?lang=en (accessed on 26 March 2025).
43. Bodó, B.; Kalmár, F. Analysis of primary energy use of typical buildings in Hungary. *Environ. Eng. Manag. J.* **2014**, *13*, 2725–2731. [CrossRef]
44. Regulation 7/2006. (V. 24.) TNM, (2006) on the Determination of Energy Performance of Buildings of the Minister Without Portfolio, Budapest (in Hungarian). Available online: http://new.etanusitvany.hu/uploads/editor/files/7_2006.pdf (accessed on 21 March 2025).
45. Szabó, Á.; Lakatos, Á. Thermal analysis of aerogels and their vacuum-formed forms, their potential uses, and their effects on the environment. *Case Stud. Therm. Eng.* **2024**, *56*, 104284. [CrossRef]
46. Lakatos, Á. Investigation of the thermal insulation performance of fibrous aerogel samples under various hygrothermal environment: Laboratory tests completed with calculations and theory. *Energy Build.* **2020**, *214*, 109902. [CrossRef]
47. Csáky, I. Air Terminal Devices Developed for Personal Ventilation Systems. *Energies* **2020**, *13*, 1688. [CrossRef]

Disclaimer/Publisher's Note: The statements, opinions and data contained in all publications are solely those of the individual author(s) and contributor(s) and not of MDPI and/or the editor(s). MDPI and/or the editor(s) disclaim responsibility for any injury to people or property resulting from any ideas, methods, instructions or products referred to in the content.

buildings

MDPI

Article

Key Competencies of Built Environment Professionals for Achieving Net-Zero Carbon Emissions in the Ghanaian Construction Industry

Kofi Agyekum [1,2,*], Kezia Nana Yaa Serwaa Sackey [1], Felix Esahe Addoh [1], Hayford Pittri [3], John Sosu [1] and Frederick Owusu Danso [4]

1 Building Science, Engineering and Materials Research Team, Department of Construction Technology and Management, Kwame Nkrumah University of Science and Technology, Kumasi AK384, Ghana; knyssackey@st.knust.edu.gh (K.N.Y.S.S.); faesahe1@knust.edu.gh (F.E.A.); jsosu1@st.knust.edu.gh (J.S.)

2 The Brew Hammond Energy Centre, Kwame Nkrumah University of Science and Technology, Kumasi AK384, Ghana

3 School of Energy, Geoscience, Infrastructure and Society, Heriot-Watt University, Edinburgh EH14 4AS, UK; hp2037@hw.ac.uk

4 Department of Building Technology, Takoradi Technical University, Takoradi P.O. Box 256, Ghana; frederick.danso@ttu.edu.gh

* Correspondence: kagyekum.cap@knust.edu.gh or agyekum.kofi1@gmail.com

Abstract: The deficiency in competencies among built environment professionals (BEPs) in achieving sustainability goals presents a significant challenge, contributing substantially to the escalation of carbon emissions globally, with pronounced implications in Ghana. Addressing this issue is critical to bridging the existing knowledge gap concerning the role of key professional competencies in mitigating carbon emissions. This study, therefore, seeks to examine and synthesize the essential competencies required by BEPs to support the attainment of net-zero carbon emissions within the Ghanaian construction industry (GCI). A quantitative research approach was employed, utilizing a structured questionnaire survey to examine the opinions of 125 professionals, including architects, engineers, and construction managers. The questions were developed based on a review of the related literature. The data collected was analyzed using one-sample *t*-tests, multiple linear regression, and ANOVA to assess the significance and impact of the identified competencies on sustainability outcomes. The key competencies identified included "value engineering", "stakeholder engagement for low-carbon development", "circular impact assessment", and "reverse logistics for sustainable material use". This research also revealed the key competencies' contributions to attaining environmental sustainability in the Ghanaian construction industry. Some key outcomes are "proper planning and provision of detailed net-zero carbon building specifications for contractors" and "promotion and implementation of net-zero carbon buildings". It was identified that actions towards net-zero carbon emissions are the leading contributor to environmental sustainability, whereas the essential competencies have a greater impact on sustainable resource use. The findings highlight gaps in the current practices and underscore the need for improved professional training and development to meet sustainability goals. This study concludes that while professionals in the GCI are aware of sustainability objectives, significant improvements are needed in the application of sustainable practices.

Keywords: net-zero carbon emissions; sustainability; competencies; built environment professionals; construction; Ghana

1. Introduction

In the past decade, environmental products and services have drawn much public attention. Growing awareness of the harm done to the environment is encouraging consumers to purchase products or engage in activities labelled "green," "eco-friendly," or "environmentally friendly", where the welfare of society and the environment are considered in addition to the consumers' satisfaction [1,2].

Industries whose operations directly contribute to pollution have traditionally been viewed as the primary drivers of environmental concerns. However, with growing awareness of the environmental impact of daily human activities, there is an increasing recognition that responsibility for mitigating resource consumption, carbon emissions, and environmental pollution lies with everyone, including businesses. This shared obligation extends to the construction sector, which plays a significant role in addressing these challenges [3].

The World Commission on Environment and Development [4] (p. 12) states that "Many present development trends leave increasing numbers of people poor and vulnerable, while at the same time degrading the environment", where the construction sector is culpable. According to the United Nations Environment Programme [5], the built environment sector, responsible for delivering constructed assets, is among the largest contributors to climate change. It accounts for over a third (37%) of global energy-related carbon emissions, significantly driving climate change impacts [5]. The traditional approach to waste management in the construction industry, involving the generating, collecting, and disposing of waste in landfills or through incineration, is directly accountable for this global change. To address climate change challenges, it is necessary to shift from assets that contribute to climate change to assets that promote a net-zero economy while protecting vulnerable assets from the impacts of climate change [6,7]. As a result, several organizations, governments, and regions have incorporated the Net-Zero Carbon Buildings (NZCB) approach into their policies and climate goals. For instance, the World Green Building Council, through its "Advancing Net Zero" initiative, aims for all buildings to operate at net-zero carbon between 2030 and 2050 [8]. In Ghana, the drive towards achieving net-zero carbon (NZC) goals in the building sector is still at a formative stage [9]. Several policy measures and initiatives lay the groundwork for its realization. Notably, Ghana has committed to a 15% reduction in greenhouse gas emissions by 2030 under its updated Nationally Determined Contributions (NDCs) as part of the Paris Agreement, which encompasses the building sector as a critical area for action [9,10]. This framework prioritizes renewable energy, low-carbon hydrogen, electric vehicles, and clean cook stoves to reduce emissions from transportation and industry. Furthermore, government programs, such as the National Rooftop Solar Programme, initiated in 2016, aim to boost renewable energy adoption in residential sectors, indirectly supporting NZC objectives [11]. Despite these efforts, the implementation of NZC principles, especially in sectors like construction, remains limited due to barriers such as low awareness among construction professionals, financial constraints, and lack of specific regulatory frameworks tailored to NZC buildings [9,10]. Recent studies emphasize the urgent need for improved competencies among built environment professionals to support the transition towards NZC practices [9], indicating a growing recognition of the regulatory and skills gap that must be addressed for widespread adoption. Therefore, while Ghana's broader climate commitments and renewable energy initiatives offer a supportive backdrop, the establishment of a detailed, sector-specific roadmap for NZC buildings remains a crucial next step [9,10].

Cabral and Dhar [12] postulated that people must possess green competencies, including green knowledge, green awareness, green skills, green abilities, green attitudes, and green behavior, to promote environmental sustainability. BEPs, especially, play a crucial role in environmental sustainability, and their competencies are indispensable. They utilize the skills and knowledge acquired to deliver sustainable results of the built product and infrastructure throughout the design and execution stages of the built asset. Given their role in the execution of construction projects, their knowledge, skills, and competencies are significant for mitigating environmental pollution caused by the construction industry, taking into account the social and economic aspects of sustainable development [13].

The competencies of BEPs are critical to driving this transition. These professionals are responsible for making design, material selection, and construction decisions that directly impact a building's carbon footprint over its lifecycle [14]. However, recent studies reveal a gap in the skills, knowledge, and motivation required to implement sustainable practices effectively [15,16]. In particular, the lack of formal education and training on sustainability and deconstruction principles limits the ability of professionals in Ghana to integrate low-carbon strategies into their projects [16]. Moreover, assessing competencies is essential to identifying the existing gaps and developing targeted capacity-building initiatives [15]. Without a clear understanding of the skills required for sustainable design, construction, and management, efforts to achieve net-zero carbon targets may be wasted [14]. As Agyekum et al. [15] emphasized, designers' expertise in material durability, construction processes, and coordination is fundamental to reducing waste and promoting circular economy principles. Therefore, systematic evaluation of professionals' competencies will provide the empirical foundation needed to inform policy, curriculum development, and professional training, ensuring that the Ghanaian construction industry aligns with global sustainability goals.

Numerous studies have explored sustainable construction from the perspectives of architects, quantity surveyors, and other stakeholders, focusing on their roles in advancing the understanding of sustainable practices. Research has also examined the pricing of sustainable materials and the motivation behind architects designing sustainable projects [13]. However, there is limited discussion on the competencies of BEPs and their impact on achieving sustainable construction, despite their competencies being pivotal for its success [13]. This research endeavors to review these crucial competencies to fill this knowledge gap in an attempt to inform the industry and practice in achieving net-zero carbon emissions in the construction industry. The objectives established to guide this study are (1) to identify the essential competencies required by BEPs in achieving net-zero carbon in construction and (2) to determine the significance of the essential competencies on the various elements of environmental sustainability.

Assessing key competencies' contributions to environmental sustainability in the construction industry is essential for understanding how professional skills directly influence sustainable outcomes. Identifying these competencies provides critical insights into their role in reducing carbon emissions, conserving resources, and minimizing environmental degradation. This assessment will also highlight the existing skill gaps among built environment professionals, guiding capacity-building initiatives, professional training, and curriculum development to better equip practitioners for delivering sustainable projects.

Determining the significance of essential competencies on various elements of environmental sustainability further enables a targeted approach to addressing specific sustainability challenges. It clarifies which competencies have the greatest impact on aspects such as energy efficiency, waste reduction, and resource conservation. This understanding informs strategic decision-making across project phases and fosters the adoption of innovative practices.

2. Literature Review

2.1. Conceptual Review

The concept of environmental sustainability boils down to the cutback of greenhouse gases in the atmosphere [17]. Greenhouse gases (GHGs) are atmospheric gases that trap heat and contribute to the greenhouse effect, a natural process that keeps Earth warm enough to support life [18]. Carbon dioxide (CO_2), one greenhouse gas, is a crucial element of the carbon cycle and plays a vital role in controlling the planet's temperature [19]. The strong capacity of CO_2 to retain energy leads to a climatic system that sustains life [20]. However, the rise in human activities has increased GHGs [21]. These GHGs trap excessive heat within the atmosphere, causing global warming and climate change [18]. Currently, climate change is the paramount concern of the twenty-first century. It is one of the greatest threats facing humanity, thus, immediate measures must be taken to decrease the emission of these gases [22].

The term "net-zero carbon" gained fame after The Intergovernmental Panel on Climate Change [23] emphasized the urgent need to achieve net-zero emissions by around 2050 to limit global warming to 1.5 degrees Celsius. Net-zero carbon, in this context, refers to the goal of ensuring that the overall greenhouse gas emissions produced during the entire life cycle of a project amount to zero or a negative value [24]. This is the final state where all carbon emissions are offset. The pathway toward achieving this goal is termed net-zero carbon reduction. This is an ongoing process to reduce carbon emissions by adopting renewable energy, improving energy efficiency, and changing consumption patterns [25]. Net-zero carbon therefore ensures that climatic conditions are brought down to acceptable levels [21]. This goal can be achieved, as the proficiency and competencies of individuals play a pivotal role in the successful attainment of any goal.

Wong [26] defined competencies as a person's traits that result in or are responsible for optimal or superior performance. Other authors have also presented competency as the readiness to apply or utilize a group of interrelated knowledge, skills, and abilities to accomplish critical or key work functions or tasks in a prescribed workplace environment [27]. Key elements that make up an individual's competency are knowledge, skills, attitudes, and motives. Present-day models connect competencies to the standards of performance through behavioral measures [27].

2.2. Theoretical Frameworks

The study of competencies in low-carbon construction is deeply rooted in theoretical frameworks that address the multifaceted challenges of environmental sustainability, behavioral change, and organizational capabilities. The two primary frameworks discussed are competency-based theory (CBT) and sustainability theory (ST).

The use of competency-based theory in construction for low-carbon buildings is well-suitable given the current transformation of the construction business towards sustainability [28]. In line with Azemikhah's [29] competency-based theory, competency-based practice is defined as a systematic plan of practice when the needs of the practitioner involve the application of certain competencies to accomplish the intended objectives. CBT postulates that competencies are the essential inner attributes of the performer, which, when used, make the performer perform better in that context. This theory also focuses on staff development and the constant acquisition of new competencies that correspond to the new demands of the sphere. Furthermore, it addresses the aspect of innovativeness as much as it pertains to sustainability [28]. In essence, the theory underscores the importance of a proactive approach to competency development, where professionals are encouraged to anticipate future challenges and prepare accordingly.

Sustainability theory is the other key theory that informs competencies in realizing low-carbon construction, as outlined in [30]. The World Commission on Environment and Development [4] states that this particular perspective emphasizes the long-term view that development must satisfy the requirements of the current generation without jeopardizing the satisfaction of the next generation's needs. Rockström et al. [31] highlighted that exceeding Earth's environmental boundaries endangers the health and livelihoods of future generations. The sustainability theory, according to Basiago [32], is quite a general theory that gives structure at every level and aspect of construction and contains more general considerations, such as social and economic feasibility, as well as environmental conservation. In the construction industry, these practices refer to selected strategies aiming at managing and achieving sustainable development through the interaction of the social and physical environment and the economy. A major component of ST is interdependency, where many aspects within the built environment are dependent on other aspects [33]. Sustainability theory also supports global sustainability initiatives, like the United Nations SDG, for which global action is needed soon to address climate change.

2.3. Competencies Required of BEPs in Achieving Net-Zero Carbon

There are various existing practices in Ghana aimed at achieving net-zero carbon emissions. These practices are performed by the various construction teams in the built environment, as well as those who have an impact on achieving net-zero carbon in the construction industry. According to Ustaoglu et al. [34], a highly efficient approach involves the construction of energy-efficient buildings by construction designers, such as architects, engineers, surveyors, project managers, estate officers, and draftsmen, resulting in reduced carbon emissions and decreased energy costs. In addition, policymakers and statutory organizations ensure that the laws and regulations that are enacted to warrant net-zero carbon emission are achievable. Additionally, the government organizes training sessions and programs to promote the adoption of net-zero carbon-building practices and the use of related items [35]. According to Bohari et al. [36], contractors and their team members, including skilled and unskilled workers, partake and advise clients in the selection of construction materials that emit little or no carbon into the environment. Construction procurement teams influence emissions reduction in various ways, including selecting suppliers with low carbon emissions and promoting sustainable practices throughout the supply chain. All these practices have led to a significant change in the emission of carbon and, hence, must be promoted. Thus, to navigate towards the construction industry's net-zero carbon goal, it is necessary to understand several competencies and practices that are inherent and need to be applied to the construction sector. These include a broad spectrum of skills and knowledge areas essential for professionals in the built environment to contribute to net-zero carbon construction effectively. According to the World Business Council for Sustainable Development [37], value engineering and lean design and engineering are significant competencies. Stakeholder interest management and connecting reverse logistics with users are also required competencies [38]. Antwi-Afari et al. [39] stated that designing for multiple-use cycles and circular impact assessment are key competencies required. Additionally, Polder et al. [40] indicated that waste-efficient procurement is one of the key outstanding competencies for realizing net-zero carbon construction. The British Standard Institution [41] identified some required competencies, such as the change in communication, addressing generational differences and understanding team dynamism. Wiek et al. [42], based on their findings, stipulated anticipatory competency as a key competency, while Oluleye et al. [43,44] specified the key competencies as being normative competency and waste management. Amstrong [45] affirmed the need to factor culture into sustainability and its practices. Implementation of circular economy is a relative concept, which is also referred to as circular scope definition and management by authors such as Hossain et al. [46], to enhance net-zero carbon construction. Thus, Bocken et al. [47] identified circular business models as one of the competencies that need to be integrated to ensure a shift towards net-zero carbon. Supply chain collaborative integration is another competency that needs to be attained for carrying out net-zero carbon construction [48]. Hopkins [49] also categorized system thinking as one of the core competencies of delivering net-zero carbon construction challenges. The need to construct low-energy buildings was described as one of the competencies of attaining net-zero carbon construction [50]. Munaro et al. [51] stressed that the development of the specification for components and materials contributes to the implementation of net-zero carbon construction. To get net-zero carbon construction right, BEPs need to embrace and possess the various competencies outlined in Table 1.

Table 1. Key competencies for achieving net-zero carbon emissions.

	Competencies	Literature Source(s)
1	Value engineering	[37]
2	Lean design and construction	
3	Stakeholder engagement for low-carbon development	[38]
4	Reverse logistics for sustainable material use	
5	Designing for multiple-use cycles	[39]
6	Circular impact assessment, e.g., LCA	
7	Sustainable procurement for waste reduction	[40]
8	Effective communication for sustainable change management	[41]
9	Workforce engagement for sustainable knowledge transfer	
10	Leadership in high-performance sustainable teams	
11	Foresight and risk management for climate-resilient design	[42]
12	Ethical and regulatory compliance in net-zero construction	[43,44]
13	Waste management	
14	Cultural competency for global sustainability practices	[45]
15	Circular scope definition and management	[46]
16	Circular business model integration	[47]
17	Supply chain collaborative integration	[48]
18	System thinking	[49]
19	Designing for near-zero energy buildings	[50]
20	Specification writing for components and materials	[51]

These competencies are important in enhancing and improving environmental sustainability, mainly in the construction sector in developing countries, of which Ghana is a part. According to the World Economic Forum [52], the main elements of environmental sustainability in the construction industry are reducing carbon emissions, improving energy efficiency, promoting sustainable resource use, improving waste management, and adopting circular economy principles. Thus, the application of the required competencies directly impacts various aspects of the construction process, from initial planning and design to execution and lifecycle management, and produces various results, which end up contributing to the elements of environmental sustainability and not just carbon-emission reduction. As Yang et al. [53] noted, these competencies lead to specifications-related planning, which enables contractors to understand the nature of the project and, in the process, help to acquire sustainable materials, reduce wastage, and apply energy efficiency. The various competencies also help by providing technical services, producing construction documents, ensuring safety during the development of a net-zero carbon building, satisfying the client's goals in executing a net-zero carbon building, and ensuring effective project logistics, site investigations, and environmental issues analysis [54]. Likewise, Sahid et al. [55] identified the key subject matter of BEPs in terms of promoting and implementing a net-zero carbon policy. According to Latief et al. [56], the right competencies aid in proper cost and time management, as well as making good decisions. Liu et al. [57] emphasized that recruiting technical experts with specialized knowledge and enhancing the reputation and image of the company towards achieving net-zero is a key contribution of the various competencies. In the view of Huang et al. [58], the application of skills and competencies helps in the development of net-zero carbon building to promote sustainability. The various competencies also promote the integration of green and net-zero carbon design elements and strategies into the net-zero carbon design building process [59]. These outcomes are expected to have a significant positive impact on the environmental footprint of the construction industry in Ghana.

While some competencies, such as stakeholder engagement, may traditionally align with management roles, and others, like circular impact assessment, align with design professionals, achieving NZC outcomes demands an integrated, cross-disciplinary approach. In NZC projects, architects, engineers, quantity surveyors, and site managers must all contribute collaboratively towards shared sustainability goals. Therefore, a foundational level of competency in each area is necessary for all BEPs, even if the degree of application varies by role. This ensures that every actor within the project lifecycle can make informed decisions that collectively support NZC targets.

The ultimate aim of the initiative for advancing competency development of BEPs is to enhance environmental sustainability in the construction industry by reducing carbon emissions, improving energy efficiency, improving waste management, promoting sustainable resource use, and adopting circular economy principles. These are the five aspects or elements of environmental sustainability, and they are expected to have a significant positive impact on the environmental footprint of the construction industry in Ghana.

3. Methodology

3.1. Research Method and Approach

To achieve comprehensive evaluations aligned with the research objectives, a quantitative research methodology was employed, utilizing a survey approach. This approach facilitated the collection of data from a large sample of BEPs, ensuring the results were representative of the broader population [14,60]. Data collection was conducted exclusively through a structured questionnaire, which enabled the systematic gathering of numerical data for rigorous statistical analyses. A deductive research approach underpinned this study, involving the development of a theory based on prior research, experiences, and specific observations. This approach progressed through literature review, data collection, and the evaluation of the data gathered to confirm or refute the literature, as outlined by [61].

3.2. Survey Design and Distribution

Following a review of the pertinent literature, a questionnaire was prepared to gather data from BEPs in the GCI. Before the commencement of data collection, the questionnaire was pre-tested to ensure its suitability and to identify any ambiguous questions that could lead to misinterpretation. Initially, the questionnaire was reviewed by an academic supervisor in the built environment, followed by a review from seven experienced built environment professionals within the field. Insights gained from this pilot study were used to address potential issues and refine the questions, ensuring clarity and appropriateness in their wording. The feedback obtained during the pre-testing phase informed the development of the final version of the questionnaire, which was subsequently utilized for data collection in this study.

To implement the survey strategy, questionnaires were distributed via email and WhatsApp to the identified sample, with in-person distribution and collection employed for respondents who were more easily accessible. The targeted group included project managers, structural and civil engineers, architects, quantity surveyors, construction managers, and health and safety officers. Due to the absence of a sampling frame for this expert group in Ghana, it was not possible to determine the total population size or statistically calculate an adequate sample size. Purposive sampling was therefore used to identify professionals with substantial expertise in the subject matter and active involvement in the sustainability efforts of the construction industry, as used in similar studies in the same jurisdiction. The data collection process spanned four weeks, during which the respondents were given sufficient time to complete and return their questionnaires. Follow-up reminders were sent to encourage a high response rate and address any queries or concerns from the participants. As a result, a total of 125 responses were successfully gathered.

The questionnaire used in this survey consisted of closed-ended questions. Each variable in the questions was identified during the literature review. It was divided into three sections. Section (A) captured the demographics of the respondents. Section (B) captured information about the first objective, which is the essential competencies for achieving net-zero carbon in construction, ranked on a five-point Likert scale, where 1 = Not Important, 2 = Slightly Important, 3 = Moderately Important, 4 = Very Important, 5 = Extremely Important. The third (C) section encapsulated information about the key competencies' contributions to environmental sustainability elements in construction on a five-point Likert scale, where 1 = Strongly Disagree, 2 = Disagree, 3 = Neutral, 4 = Agree, 5 = Strongly Agree.

3.3. Data Analyses

The data gathered were reviewed and cleansed using Microsoft Excel (Version 365, Microsoft Corporation, Redmond, WA, USA) to ensure accuracy and comprehensiveness. The data were analyzed using the IBM Statistical Package for the Social Sciences (SPSS) (version 27, IBM Corporation, Armonk, NY, USA). An analysis of the demographic data was carried out using frequencies and percentages. One-way ANOVA, multiple linear regression analysis, and one-sample t-test were the three inferential statistics utilized to identify the variables. Multiple linear regression analysis was performed to assess how various competencies contribute to environmental sustainability in the construction industry, ANOVA was used to evaluate the overall significance of the regression models in predicting different aspects of environmental sustainability, and the one-sample t-test indicated whether the mean differences of the competencies were statistically significant, taking into account the mean score that was used to rank the various practices with a hypothesized mean of 3.5 (moderately important); the highest extreme (5) indicates the highest significance level, and the lowest extreme (1) indicates otherwise. Where two or more criteria had the same mean, the one with the lowest standard deviation was assigned the highest significance ranking [62]. The one-sample t-test was conducted at a confidence level of 95% and a p-value of 0.05, where p-values ≤ 0.05 were defined as statistically significant, and p-values > 0.05 were defined as statistically insignificant. Tables and figures were used to portray the data in a way that made them simpler to study and understand.

4. Results and Discussion

4.1. Demographic Data of Respondents

The following section presents the sociodemographic characteristics of the respondents who participated in this study, as presented in Figures 1–3. Regarding their roles within the construction industry, 24 respondents (19.2%) were architects, 30 (24.0%) were engineers, 29 (23.2%) were construction managers (CM), 18 (14.4%) were quantity surveyors (QS), 23 (18.4%) were project managers (PM), and 1 respondent (0.8%) identified as a site manager (SM) (see Figure 1).

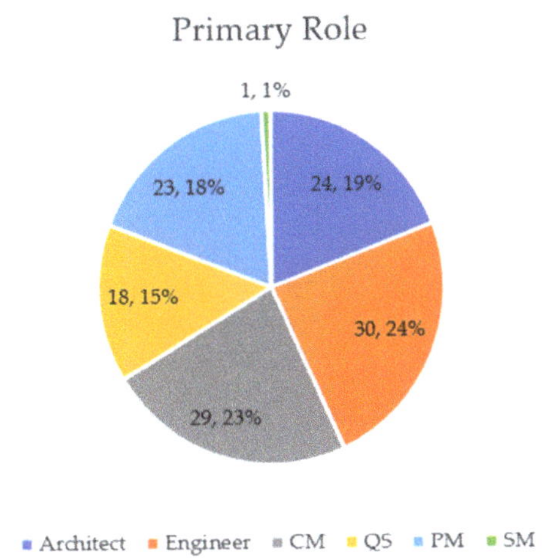

Figure 1. Professional role of respondents.

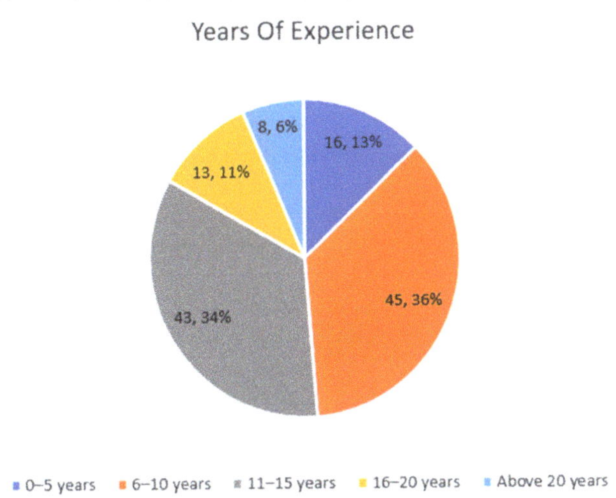

Figure 2. Working experience of respondents.

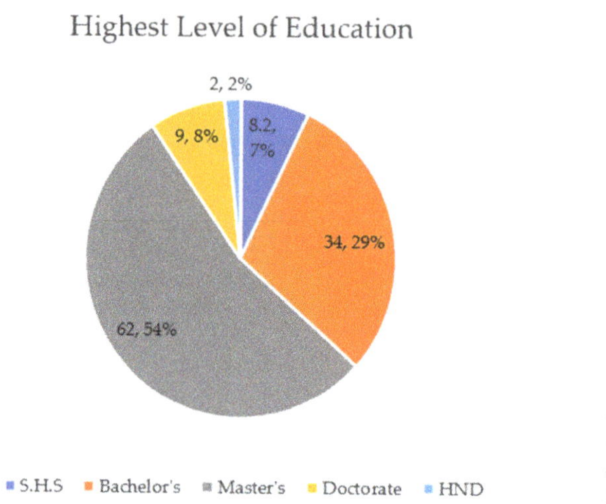

Figure 3. Highest level of education of respondents.

In terms of professional experience (see Figure 2), 16 respondents (12.8%) reported 0–5 years of experience, 45 (36.0%) had 6–10 years of experience, 43 (34.4%) had 11–15 years of experience, 13 (10.4%) had 16–20 years of experience, and 8 respondents (6.4%) had more than 20 years of experience. Regarding educational qualifications (see Figure 3), the majority of respondents (62 individuals, 49.6%) held a master's degree, followed by 34 respondents (27.2%) with a bachelor's degree. A smaller portion of the sample had attained a doctorate (9 respondents, 7.2%), 18 respondents (14.4%) were S.H.S graduates, and 2 respondents (1.6%) reported holding a Higher National Diploma (HND).

4.2. The Essential Competencies Required by BEPs in Achieving Net-Zero Carbon in Construction

Table 2 presents the results of a one-sample *t*-test assessing the essential competencies required by built environment professionals (BEPs) for achieving net-zero carbon emissions in the GCI. The test value was set at 3.5, representing a moderately to very important threshold on the Likert scale, with competencies scoring above this benchmark deemed critical. All the competencies tested returned *p*-values less than 0.05, indicating statistical significance at a 95% confidence level.

The analysis reveals value engineering as the highest-ranked competency, with a mean score of 4.50 and a standard deviation of 0.747. This highlights its crucial role in optimizing resource use and cost-efficiency while minimizing carbon emissions in construction processes. Stakeholder engagement for low-carbon development ranks second (mean = 4.38), emphasizing the importance of collaboration among project stakeholders in fostering sustainable decision-making. Circular impact assessment, such as life cycle assessment (LCA), also scored highly (mean = 4.33), reinforcing the significance of evaluating environmental impacts across a building's lifecycle to support sustainable construction practices.

Additionally, reverse logistics for sustainable material use (mean = 4.29) is positioned as a vital competency, indicating the need for effective systems to manage the reuse and recycling of materials, a critical aspect of achieving circular economy objectives in construction. Other competencies, like cultural competency for global sustainability practices, sustainable procurement for waste reduction, and supply chain collaborative integration, all show strong relevance, with mean scores above 4.20. These findings suggest that BEPs must possess both technical skills and collaborative capabilities to advance sustainability in the sector.

Table 2. One Sample t-test of the essential competencies required by BEPs.

Competencies Required of BEPs in Achieving Net-Zero Carbon Emission	Test Value = 3.5				
	Mean	Standard Deviation	Rank	p-Value	Statistical Significance
Value engineering	4.50	0.747	1st	0.000	Yes
Stakeholder engagement for low-carbon development	4.38	0.657	2nd	0.000	Yes
Circular impact assessment, e.g., LCA	4.33	0.821	3rd	0.000	Yes
Reverse logistics for sustainable material use	4.29	0.760	4th	0.000	Yes
Cultural competency for global sustainability practices	4.26	0.720	5th	0.000	Yes
Sustainable procurement for waste reduction	4.26	0.764	6th	0.000	Yes
Circular scope definition and management	4.26	0.772	7th	0.000	Yes
Effective communication for sustainable change management	4.26	0.844	8th	0.000	Yes
Supply chain collaborative integration	4.25	0.680	9th	0.000	Yes
Designing for multiple-use cycles	4.24	0.865	10th	0.000	Yes
Ethical and regulatory compliance in net-zero construction	4.22	0.694	11th	0.000	Yes
Leadership in high-performance sustainable teams	4.22	0.848	12th	0.000	Yes
Workforce engagement for sustainable knowledge transfer	4.17	0.632	13th	0.000	Yes
Lean design and construction	4.16	0.723	14th	0.000	Yes
System thinking	4.14	0.780	15th	0.000	Yes
Foresight and risk management for climate-resilient design	4.12	0.691	16th	0.000	Yes
Circular business model integration	4.12	0.779	17th	0.000	Yes
Designing for near-zero energy buildings	4.09	0.813	18th	0.000	Yes
Waste management	4.06	0.840	19th	0.000	Yes
Specification writing for components and materials	3.77	1.179	20th	0.012	Yes

Notably, specification writing for components and materials ranks lowest among the competencies (mean = 3.77), though it still demonstrates statistical significance. This indicates room for improvement in this area, as detailed and accurate specification writing is essential for ensuring the procurement and use of sustainable low-carbon materials.

The results underscore that BEPs in Ghana recognize the multidimensional competencies needed to drive net-zero carbon goals, spanning technical, managerial, and collaborative domains. The high ratings for competencies related to stakeholder engagement, circularity, and value engineering suggest a growing awareness of integrated, systems-thinking approaches. However, the relatively lower ranking of specification writing hints at a possible skills gap that requires targeted professional development.

4.2.1. Discussion of Key Competencies

For brevity, the top five competencies are discussed in detail in this section.

Value Engineering

Value engineering (VE) emerged as the most effective competency in this study, aligning with the assertion of the World Business Council for Sustainable Development [37] that VE is a critical skill for built environment professionals (BEPs) in achieving environmental sustainability. VE is increasingly recognized by stakeholders in construction for its capacity to simultaneously reduce costs and enhance performance, particularly in sustainable and low-carbon construction contexts. It promotes design and material choices that minimize resource waste, reduce rework, and optimize lifecycle cost—outcomes that are crucial for advancing NZC goals.

In sustainable construction, VE plays a critical role in identifying where higher initial costs are justified by long-term savings, reduced environmental impact, or enhanced functionality. For example, while a low-carbon material may be more expensive upfront, VE can demonstrate its cost-effectiveness when considering reduced maintenance, energy efficiency, and regulatory compliance benefits over time. In this way, VE enables BEPs to navigate the economic trade-offs inherent in NZC projects, ensuring that sustainability is pursued not at any cost but with financial and functional accountability [63]. The statistical prominence of VE in the findings reinforces its practical significance, particularly in resource-constrained environments like Ghana, where economic efficiency must be balanced with environmental objectives. Projects that incorporate VE principles can lead to fewer design changes during construction, better alignment with NZC performance benchmarks, and enhanced stakeholder satisfaction through cost savings and functional improvements [64].

Given its strategic importance, targeted training on VE, especially, in cost–benefit analysis, sustainable specification, and lifecycle assessment should be embedded into professional development programs and tertiary curricula. Institutions such as the Ghana Institution of Engineers and the Ghana Institute of Architects could partner with sustainability experts to deliver competency-specific workshops and certifications contextualized to local project delivery conditions and regulatory frameworks.

Stakeholder Engagement for Low-Carbon Development

Stakeholder engagement for low-carbon development ranked second among the key competencies, underscoring its central role in aligning diverse project interests to achieve sustainability goals. This result supports prior findings by Ding et al. [38], which emphasize that harmonizing the expectations of clients, contractors, consultants, and communities is essential to delivering successful and sustainable construction outcomes. In the context of NZC projects, where trade-offs between cost, performance, and environmental impact are common, effective stakeholder engagement becomes even more critical. Its high statistical ranking reflects the practical reality that low-carbon initiatives require multi-stakeholder collaboration, early involvement in decision-making, and shared accountability [65].

Practically, poor engagement often results in resistance to sustainable practices, misaligned project objectives, and missed opportunities for innovation. For instance, building users may resist energy-saving measures if their needs are not considered early on, or contractors may overlook low-carbon methods if they are not adequately informed. In Ghana, fragmented communication among stakeholders and weak participatory planning processes remains persistent barriers [60]. These challenges can be mitigated by developing competencies in conflict resolution, participatory design facilitation, and stakeholder mapping which are skills not traditionally emphasized in BEP training.

Circular Impact Assessment

Circular impact assessment is ranked third in significance, reinforcing its growing relevance in supporting sustainable construction decision-making. This finding aligns with the assertion by Antwi-Afari et al. [39] that circular assessment tools enable construction managers and other BEPs to evaluate the implications of material choices and construction activities on carbon emissions throughout a building's lifecycle. LCA, as a core component of circular impact assessment, is widely adopted to measure, reduce, and optimize the environmental footprint of construction materials, design solutions, and operational strategies. It not only informs technical decisions but also strengthens the credibility of sustainability claims, enhances compliance with green building certifications, and supports eco-labeling efforts [66,67].

The high ranking of this competency illustrates a wider global transition from a linear to a circular economy in the construction sector, i.e., one that emphasizes waste minimization, closed-loop resource use, and environmental responsibility. For Ghana's construction industry, this shift is both necessary and challenging. At present, there is limited availability of localized LCA databases, insufficient regulatory mandates for environmental reporting, and a general lack of familiarity among BEPs with circular economy principles [68,69]. As a result, the potential of circular impact assessment remains underutilized despite its clear benefits in reducing embodied carbon and improving resource efficiency.

To address this, developing a circular assessment competency among BEPs in Ghana will require targeted training in LCA tools, integration of circular design thinking into university curricula, and the development of national benchmarks for embodied carbon in building materials. Moreover, partnerships between academia, government agencies, and industry could support pilot projects that demonstrate circular building practices in action. These initiatives would not only build local expertise but also generate data and evidence to inform future policy and market incentives [69].

Reverse Logistics for Sustainable Material Use

Reverse logistics was ranked fourth, highlighting its increasing importance in the pursuit of NZC outcomes. As affirmed by Ding et al. [38], reverse logistics plays a pivotal role in the recycling and reuse of materials, thereby minimizing construction waste and carbon emissions. The responses from the study participants strongly support this perspective, demonstrating that many BEPs recognize the strategic value of this competency. Reverse logistics, broadly defined as the backward flow of products, materials, or components from the point of consumption to the point of origin for recovery or proper disposal, has evolved into a critical aspect of sustainable construction practices.

Its inclusion in the top competencies indicates a growing awareness among BEPs that reclaiming building components at the end of a structure's lifecycle, whether through deconstruction, material reuse, or recycling, can substantially reduce embodied carbon and contribute to circularity. However, practical challenges persist, especially in the Ghanaian context. These include poor material labeling and documentation, lack of infrastructure for material recovery, insufficient regulation on demolition waste handling, and skepticism regarding the quality of reused materials [70]. As noted by Sivasankaran [71], failure to incorporate reverse logistics into project planning can lead to increased waste, environmental degradation, and higher lifecycle costs.

To effectively embed this competency into professional practice, BEPs must be equipped with knowledge of deconstruction principles, waste auditing techniques, and material recovery planning [14,15]. Moreover, construction education and training programs should introduce modules on circular material flows, supported by real-world case studies [14]. At the policy level, developing standards for assessing the quality and safety of reused building materials, as well as creating incentives for contractors who implement reuse strategies, could accelerate industry adoption. Demonstration projects led by public agencies or donor-funded programs could serve as living laboratories for reverse logistics, helping to overcome market and behavioral barriers.

Cultural Competency for Global Sustainability Practices

Cultural competency for global sustainability practices was ranked fifth, reflecting its growing recognition as a foundational skill for achieving contextually appropriate and inclusive sustainability outcomes. As Armstrong [45] argued, cultural competence enables construction professionals, especially managers, to navigate and appreciate the diverse cultural norms, values, and practices that influence sustainability decision-making. This result is significant in the context of Ghana, where traditional practices, community expectations, and local knowledge systems play a critical role in shaping the design, construction, and use of buildings.

A culturally competent BEP is not only aware of these local dynamics but can also integrate them with contemporary sustainability frameworks to produce solutions that are technically sound, socially acceptable, and ethically grounded. As Haq et al. [72] highlighted, this ability helps avoid the imposition of sustainability strategies derived solely from Western models that may be misaligned with the needs, beliefs, or practices of local communities. Ignoring cultural factors in sustainability planning risks resistance, non-compliance, and, in some cases, the reinforcement of existing inequalities, particularly in informal settlements or rural construction contexts.

This competency helps BEPs engage more effectively with communities, tailor communication strategies, and co-create interventions that are more likely to gain local support. It is particularly relevant in public or community-based projects, where success depends on stakeholder buy-in and long-term use of the built environment.

4.3. *The Aspects of Environmental Sustainability and the Significance of Competencies on Each Aspect*

Five aspects of environmental sustainability were identified during the review of the literature, and the respondents were asked to rate their level of agreement on a Likert scale. Their responses were analyzed using the one-sample *t*-test tool, and the results are shown in Table 3. This table shows that all five elements had mean values greater than 3.5 and *p*-values less than 0.05, supporting the role of these aspects in creating environmental sustainability when combined.

Table 3. One-sample *t*-test on the elements of environmental sustainability.

Elements/Aspects of Environmental Sustainability	Test Value = 3.5				
	Mean	**Standard Deviation**	**Rank**	***p*-value**	**Statistical Significance**
Reduced Carbon Emissions	4.53	0.768	1st	0.000	Yes
Improved Waste Management	4.41	0.853	2nd	0.000	Yes
Increased Adoption of Circular Economy Principles	4.34	0.683	3rd	0.000	Yes
Enhanced Energy Efficiency	4.22	0.758	4th	0.000	Yes
Sustainable Resource Use	4.22	0.739	5th	0.000	Yes

From Table 3, the rankings were as follows: reduced carbon emissions (MS = 4.53, SD = 0.768, *p*-value = 0.000), improved waste management (MS = 4.41, SD = 0.853, *p*-value = 0.000), increased adoption of circular economy principles (MS = 4.34, SD = 0.683, *p*-value = 0.000), enhanced energy efficiency (MS = 4.22, SD = 0.758, *p*-value = 0.000), and sustainable resource use (MS = 4.22, SD = 0.739, *p*-value = 0.000).

As stated by Gogg et al. [17], the most impactful way to reach environmental sustainability is the cutback of greenhouse gases in the atmosphere. The results from the analysis were based on the views of the respondents, who ranked reduced carbon emissions first, thereby proving the point of the assertion. The level of agreement was the highest on the fact that reduced carbon emissions is the greatest element that promotes environmental sustainability. According to the World Economic Forum [52], reducing carbon emissions, improving energy efficiency, promoting sustainable resource use, improving waste management, and adopting circular economy principles are all significant in achieving environmental sustainability.

The multiple linear regression model shown in Table 4 presents key statistics, such as R, R-Squared, Adjusted R-Squared, and the Standard Error of the Estimate for evaluating the regression model's fit: R shows the strength of the relationship between independent variables (competencies) and dependent variables (sustainability), R-Squared (coefficient of determination) indicates how much variance in sustainability is explained by the competencies, Adjusted R-Squared accounts for the number of predictors, offering a refined comparison across models, and Standard Error of the Estimate measures prediction accuracy, with lower values indicating better fit. This analysis helps assess how well the model fits the data by comparing R-Squared and Adjusted R-Squared across different models. An ANOVA test was run on the multiple regression analysis to evaluate the overall significance of the models in predicting different aspects of environmental sustainability. All the variables had p-values less than 0.05 and are therefore statistically significant.

Table 4. Multiple regression model summary.

Model	R	R-Squared	Adjusted R-Squared	Std. Error of the Estimate	Rank
1	0.569	0.324	0.209	0.683	2nd
2	0.530	0.281	0.159	0.695	5th
3	0.550	0.303	0.185	0.770	3rd
4	0.604	0.365	0.257	0.637	1st
5	0.540	0.292	0.171	0.622	4th

According to the views of the respondents, it is established that Model 4 provides the best fit among the models tested, offering valuable insights into the key drivers of sustainability in the Ghanaian construction industry. This, however, concludes that the outcomes of the competencies identified have a greater impact on sustainable resource use toward attaining sustainability. Model 2, on the other hand, indicates a weaker fit as compared to the other models, concluding that enhanced energy efficiency is the least impacted by the competencies in achieving sustainability.

Table 5 presents the results of the ANOVA of the multiple regression. Model 4 (*sustainable resource use*) ranked first (Adjusted R-Squared [$A.R^2$] = 0.257, p-value = 0.000), Model 1 (reduced carbon emissions) ranked second ($A.R^2$ = 0.209, p-value = 0.000), Model 3 (improved waste management) ranked third ($A.R^2$ = 0.185, p-value = 0.002), Model 5 (increased adoption of circular economy principles) ranked fourth ($A.R^2$ = 0.171, p-value = 0.003), and Model 2 (enhanced energy efficiency) ranked fifth ($A.R^2$ = 0.159, p-value = 0.004).

The findings of this study reveal that the essential competencies have a major effect on sustainable resource use. As BEPs build on and practice their individual competencies, sustainable resource use is promoted, contributing largely to environmental sustainability. Although reduced carbon emissions is the leading aspect to ensuring environmental sustainability, it is the second most impacted, based on the results. The competencies identified in this study also have positive effects on waste management and the adoption of circular economy principles. Enhanced energy efficiency is the least impacted by the competencies in achieving sustainability.

Table 5. ANOVA of the multiple regression.

Model	Dependent Variable	Description	Sum of Squares	df	Mean Square	F	p-Value	Rank
1	Reduced Carbon Emissions	Regression	23.689	18	1.316	2.820	0.000	2nd
		Residual	49.463	106	0.467			
		Total	73.152	124				
2	Enhanced Energy Efficiency	Regression	20.023	18	1.112	2.305	0.004	5th
		Residual	51.145	106	0.483			
		Total	71.168	124				
3	Improved Waste Management	Regression	27.325	18	1.518	2.560	0.002	3rd
		Residual	62.867	106	0.593			
		Total	90.192	124				
4	Sustainable Resource Use	Regression	24.719	18	1.373	3.385	0.000	1st
		Residual	43.009	106	0.406			
		Total	67.728	124				
5	Increased Adoption of Circular Economy Principles	Regression	16.884	18	0.938	2.425	0.003	4th
		Residual	41.004	106	0.387			
		Total	57.888	124				

5. Conclusions

This study revealed that although the BEPs in Ghana are knowledgeable about sustainability goals, especially as they relate to carbon neutrality, they are still at a very nascent stage in terms of how they operate toward zero-carbon status to achieve net-zero. The first objective helped to identify a list of competencies that BEPs need to play a significant role in the construction industry in achieving net-zero carbon emissions targets. These competencies range from technical competencies to managerial competencies and environmental competencies and are crucial in attaining sustainable construction results. The second objective investigated the extent of the identified competencies in the contribution towards environmental sustainability main elements in the construction industry in Ghana. The five main elements of environmental sustainability were also acknowledged in this study, being reduced carbon emissions, improved waste management, increased adoption of circular economy principles, enhanced energy efficiency, and sustainable resource use. All these elements come together to achieve sustainability in Ghana's construction industry, with reduced carbon emissions being the most significant.

According to this study, the identified competencies required for achieving net-zero carbon were all important and statistically significant. The five highest ranked competencies are value engineering, stakeholder engagement for low-carbon development, circular impact assessment, reverse logistics for sustainable material use, and cultural competency for global sustainability practices. There was a realization that the various outcomes of the competencies had greater influence on sustainable resource use toward sustainability. It was also concluded that reduced carbon emissions is the most significant element of sustainability.

This research significantly contributes to academic knowledge in sustainable construction and competencies for achieving net-zero carbon emissions. It identifies and categorizes essential competencies needed by the Ghanaian construction sector, addressing a crucial gap in sustainability practices, especially in developing nations. By emphasizing the Ghanaian context, this study enhances our understanding of how sustainability objectives can be adopted in similar environments. It also connects theoretical frameworks with practical applications, demonstrating how these competencies can promote net-zero carbon goals in the industry. Ultimately, this research lays the groundwork for future studies on sustainable construction and competency mapping in Ghana and beyond, paving the way for improved construction practices.

Based on the findings of this study, practical recommendations are proposed to enhance the development of competencies. Key strategies include targeted professional development, policy enforcement, collaboration, and incentivization. This includes promoting competency-based training in value management and circular economy, partnering with universities to integrate sustainability into construction education, and ensuring compliance with sustainability laws through regulatory mechanisms. Encouraging industry-wide collaboration, creating associations for green construction, and offering subsidies and tax exemptions for sustainable practices can further support the transition. Additionally, integrating sustainability into project management and ensuring continuous training for project managers are critical to achieving these goals.

This study provides valuable insights into the competencies needed for net-zero carbon emissions in Ghana's construction industry, but it has limitations. These include a small sample size of 125 responses, limited geographic coverage, and reliance on self-reported data. A larger, more geographically diverse sample and longer study period could improve the accuracy and generalizability of the findings.

This study highlights several directions for future research on sustainability competencies in the construction industry. Future research could expand to other regions, track how sustainability competencies develop over time, and explore new practices, like renewable energy and recycling. Including interviews and case studies would provide deeper insights, and more studies should focus on applying these competencies in real projects. Research on the role of government policies in supporting net-zero carbon practices is also recommended.

Author Contributions: Conceptualization, K.A.; Methodology, K.N.Y.S.S.; Formal analysis, H.P.; Investigation, F.E.A.; Resources, F.E.A.; Data curation, K.N.Y.S.S.; Writing—original draft, K.A.; Writing—review & editing, J.S. and F.O.D. All authors have read and agreed to the published version of the manuscript.

Funding: This research received no funding.

Data Availability Statement: Data are contained within the article.

Conflicts of Interest: The authors declare no conflict of interest.

Abbreviations

The following abbreviations are used in this manuscript:

GCI	Ghanaian construction industry
BEPs	Built environment professionals

References

1. Ahn, I.; Kim, S.H.; Kim, M. The relative importance of values, social norms, and enjoyment-based motivation in explaining pro-environmental product purchasing behaviour in the apparel domain. *Sustainability* **2020**, *12*, 6797. [CrossRef]
2. Joshi, Y.; Uniyal, D.P.; Sangroya, D. Investigating consumers' green purchase intention: Examining the role of economic value, emotional value and perceived marketplace influence. *J. Clean. Prod.* **2021**, *328*, 129638. [CrossRef]

3. Udeagha, M.C.; Ngepah, N. The drivers of environmental sustainability in BRICS economies: Do green finance and fintech matter? *World Dev. Sustain.* **2023**, *3*, 100096. [CrossRef]
4. World Commission on Environment and Development (WCED). *Our Common Future*; Oxford University Press: Oxford, UK, 1987.
5. El Sheik, S. *CO_2 Emissions from Building Construction Hit New High, Leaving Sector off Track to Decarbonize by 2050*; United Nations Environment Programme: Nairobi, Kenya, 2022.
6. Colgan, J.D.; Green, J.F.; Hale, T.N. Asset revaluation and the existential politics of climate change. *Int. Organ.* **2021**, *75*, 586–610. [CrossRef]
7. Paterson, M. *Climate Capitalism: Global Warming and the Transformation of the Global Economy*; Cambridge University Press: Cambridge, UK, 2021. [CrossRef]
8. Laski, J.; Burrows, V. *From Thousands to Billions: Coordinated Action towards 100% Net Zero Carbon Buildings by 2050*; World Green Building Council: London, UK, 2017.
9. Pittri, H.; Oduro, S.; Atibila, D.W.; Kwasafo, O.K.; Abdallah, A.; Anteh, E.D. Examining the critical barriers to NZEBs' implementation in the Ghanaian construction industry. *J. Eng. Des. Technol.* **2024**. ahead of print. [CrossRef]
10. Oduro, S.; Pittri, H.; Simons, B.; Baah, B.; Anteh, E.D.; Oduro, J.A. Awareness of net zero energy buildings among construction professionals in the Ghanaian construction industry. *Built Environ. Proj. Asset Manag.* **2024**. ahead of print. [CrossRef]
11. Ohene, E.; Hsu, S.C.; Chan, A.P.C. Feasibility and retrofit guidelines towards net-zero energy buildings in tropical climates: A case of Ghana. *Energy Build.* **2022**, *269*, 112252. [CrossRef]
12. Cabral, C.; Dhar, R.L. Green competencies: Construct development and measurement validation. *J. Clean. Prod.* **2019**, *235*, 887–900. [CrossRef]
13. Donzing, D.; Williams, J.; Kheni, N.A. Evaluating the competencies of built environment consultants in contributing to sustainable construction in Ghana. *Curr. J. Appl. Sci. Technol.* **2023**, *42*, 205–221. [CrossRef]
14. Pittri, H.; Agyekum, K.; Botchway, E.A.; Alencastro, J.; Oladinrin, O.T.; Dompey, A.M.A. Drivers for design for deconstruction (DfD) implementation among design professionals. *Smart Sustain. Built Environ.* **2024**, *13*, 1134–1154. [CrossRef]
15. Agyekum, K.; Pittri, H.; Botchway, E.A.; Dompey, A.M.A. Key attributes of designers' competencies for design for deconstruction (DfD) implementation in Ghana. *Proc. Inst. Civ. Eng. Manag. Procure. Law* **2024**, *177*, 117–126. [CrossRef]
16. Pittri, H.; Agyekum, K.; Ayebeng Botchway, E.; Opoku, A.; Bimpli, I. Design for deconstruction (DfD) implementation among design professionals: Empirical evidence from Ghana. *Int. J. Constr. Manag.* **2023**, *24*, 1387–1397. [CrossRef]
17. Gogg, H.; Smith, J.; Lee, K. *Sustainability and Climate Action in the 21st Century*; Green Earth Publishers: New York, NY, USA, 2015.
18. United Nations. *Greenhouse Gases and the Climate Crisis: Understanding the Greenhouse Effect and Its Impact on Global Warming*; United Nations Framework Convention on Climate Change (UNFCCC): Bonn, Germany, 2024.
19. Coppola, A.I.; Wagner, S.; Lennartz, S.T.; Seidel, M.; Ward, N.D.; Dittmar, T.; Santin, C.; Jones, M.W. The black carbon cycle and its role in the Earth system. *Nat. Rev. Earth Environ.* **2022**, *3*, 516–532. [CrossRef]
20. Schulze-Makuch, D.; Irwin, L.N. *Life in the Universe: Expectations and Constraints*, 3rd ed.; Springer: Cham, Switzerland, 2018. [CrossRef]
21. Kumar, V.; Ranjan, D.; Verma, K. Global climate change: The loop between cause and impact. In *Global Climate Change*; Singh, R.B., Dhiman, R., Eds.; Elsevier: Amsterdam, The Netherlands, 2021; pp. 187–211. [CrossRef]
22. Panda, R.; Maity, M. Global warming and climate change on earth: Duties and challenges of human beings. *Int. J. Res. Eng. Sci. Manag.* **2021**, *4*, 122–125. Available online: https://journal.ijresm.com/index.php/ijresm/article/view/478 (accessed on 4 February 2025).
23. Intergovernmental Panel on Climate Change (IPCC). *Global Warming of 1.5 °C: An IPCC Special Report*; IPCC: Geneva, Switzerland, 2018; Available online: https://www.ipcc.ch/sr15/ (accessed on 27 March 2025).
24. Ohene, E.; Chan, A.P.; Darko, A.; Nani, G. Navigating toward net zero by 2050: Drivers, barriers, and strategies for net zero carbon buildings in an emerging market. *Build. Environ.* **2023**, *242*, 110472. [CrossRef]
25. United Nations. *Paris Agreement*; United Nations Framework Convention on Climate Change (UNFCCC): New York, NY, USA, 2015; Available online: https://unfccc.int/sites/default/files/english_paris_agreement.pdf (accessed on 27 March 2025).
26. Wong, S.C. Competency definitions, development and assessment: A brief review. *Int. J. Acad. Res. Prog. Educ. Dev.* **2020**, *9*, 95–114. [CrossRef]
27. Midhat Ali, M.; Qureshi, S.M.; Memon, M.S.; Mari, S.I.; Ramzan, M.B. Competency framework development for effective human resource management. *SAGE Open* **2021**, *11*, 1–15. [CrossRef]
28. Howaniec, H.; Wróblewski, Ł.; Štverková, H. Competency gaps of employees in the construction sector in terms of the requirements of a low-carbon economy: Polish and Czech case. *Energies* **2021**, *14*, 7868. [CrossRef]

29. Azemikhah, H. The 21st century, the competency era and competency theory. *Open Learn. Inst. TAFE* **2006**, *11*, 1–12. Available online: https://smartcdn.gprod.postmedia.digital/edmontonjournal/wp-content/uploads/2014/03/pa_0058.pdf (accessed on 27 March 2025).
30. Kesidou, S.; Sovacool, B.K. Supply chain integration for low-carbon buildings: A critical interdisciplinary review. *Renew. Sustain. Energy Rev.* **2019**, *113*, 109274. [CrossRef]
31. Rockström, J.; Gupta, J.; Qin, D.; Lade, S.J.; Abrams, J.F.; Andersen, L.S.; Armstrong McKay, D.I.; Bai, X.; Bala, G.; Bunn, S.E.; et al. Safe and just Earth system boundaries. *Nature.* **2023**, *61*, 102–111. [CrossRef]
32. Basiago, A.D. Economic, social, and environmental sustainability in development theory and urban planning practice. *Environmentalist* **1998**, *19*, 145–161. [CrossRef]
33. Pomponi, F.; Moncaster, A. Circular economy for the built environment: A research framework. *J. Clean. Prod.* **2017**, *143*, 710–718. [CrossRef]
34. Ustaoglu, A.; Yaras, A.; Sutcu, M.; Gencel, O. Investigation of the residential building having novel environment-friendly construction materials with enhanced energy performance in diverse climate regions: Cost-efficient, low-energy and low-carbon emission. *J. Build. Eng.* **2021**, *43*, 102617. [CrossRef]
35. Clarke, J.; Littlewood, J.R.; Karani, G. Developing tools to enable the UK construction industry to adopt the active building concept for net zero carbon buildings. *Buildings* **2023**, *13*, 304. [CrossRef]
36. Bohari, A.A.M.; Skitmore, M.; Xia, B.; Teo, M.; Khalil, N. Key stakeholder values in encouraging green orientation of construction procurement. *J. Clean. Prod.* **2020**, *270*, 122246. [CrossRef]
37. WBCSD. *Building for a Sustainable Future: Competencies for Net-Zero Carbon Construction*; World Business Council for Sustainable Development: Geneva, Switzerland, 2020.
38. Ding, Z.; Liu, S.; Luo, L.; Liao, L. A building information modelling-based carbon emission measurement system for prefabricated residential buildings during the materialization phase. *J. Clean. Prod.* **2020**, *264*, 121728. [CrossRef]
39. Antwi-Afari, P.; Ng, S.T.; Chen, J. Developing an integrative method and design guidelines for achieving systemic circularity in the construction industry. *J. Clean. Prod.* **2022**, *354*, 131752. [CrossRef]
40. Polder, J.J.; Bocken, N.M.P.; den Hollander, M.C. Circular procurement in the construction industry: A case study on public procurement of circular building projects in The Netherlands. *J. Clean. Prod.* **2019**, *212*, 550–562.
41. *BS 8903:2010*; Principles and Framework for Designing and Managing Sustainable Procurement. British Standards Institution: London, UK, 2019.
42. Wiek, A.; Withycombe, L.; Redman, C.L. Key competencies in sustainability: A reference framework for academic program development. *Sustain. Sci.* **2011**, *6*, 203–218. [CrossRef]
43. Oluleye, B.I.; Chan, D.W.; Antwi-Afari, P.; Olawumi, T.O. Modeling the principal success factors for attaining systemic circularity in the building construction industry: An international survey of circular economy experts. *Sustain. Prod. Consum.* **2023**, *37*, 268–283. [CrossRef]
44. Oluleye, B.I.; Chan, D.W.; Olawumi, T.O.; Saka, A.B. Assessment of symmetries and asymmetries on barriers to circular economy adoption in the construction industry towards zero waste: A survey of international experts. *Build. Environ.* **2023**, *228*, 109885. [CrossRef]
45. Armstrong, K. Cultivating cultures of sustainability. Harnessing personalities and perspectives to create more sustainable societies. *APS Observer* **2021**, *34*, 43–47. Available online: https://www.psychologicalscience.org/observer/cultivating-cultures-sustainability (accessed on 9 March 2025).
46. Hossain, M.U.; Ng, S.T.; Antwi-Afari, P.; Amor, B. Circular economy and the construction industry: Existing trends, challenges, and prospective framework for sustainable construction. *Renew. Sustain. Energy Rev.* **2020**, *130*, 109948. [CrossRef]
47. Bocken, N.M.P.; Short, S.W.; Rana, P.; Evans, S. A literature and practice review to develop sustainable business model archetypes. *J. Clean. Prod.* **2014**, *65*, 42–56. [CrossRef]
48. Janson, U.; Richter, J.L.; Milios, L.; Johansson, D. Towards a circular building industry. In *Handbook of Sustainability Science in the Future: Policies, Technologies, and Education by 2050*; Springer International Publishing: Cham, Switzerland, 2022; pp. 1–24. [CrossRef]
49. Hopkins, B. Competencies, Sustainability, and the Importance of Systems Thinking. Bryan Hopkins Consulting 2021. Available online: https://www.linkedin.com/pulse/competencies-sustainability-importance-systems-thinking-bryan-hopkins (accessed on 27 March 2025).
50. Ohene, E.; Chan, A.P.C.; Darko, A. Prioritizing barriers and developing mitigation strategies toward net-zero carbon building sector. *Build Environ.* **2022**, *223*, 109437. [CrossRef]
51. Munaro, M.R.; Tavares, S.F.; Bragança, L. Towards circular and more sustainable buildings: A systematic literature review on the circular economy in the built environment. *J. Clean. Prod.* **2020**, *260*, 121134. [CrossRef]

52. World Economic Forum. 3 Areas Sustainable Construction Can Help Build a Greener Future 2022. Available online: https://www.weforum.org/stories/2022/05/3-ways-sustainable-construction-can-forge-a-greener-future/?utm_source=chatgpt.com (accessed on 27 March 2025).
53. Yang, L.; Du, P.; Liao, H.; Zhang, H. Stakeholder behavior analysis of low-carbon housing promotion. *IOP Conf. Ser.: Earth Environ. Sci.* **2021**, *821*, 012012. [CrossRef]
54. Brunklaus, B.; Thormark, C.; Baumann, H. Illustrating limitations of energy studies of buildings with LCA and actor analysis. *Build. Res. Inf.* **2010**, *38*, 265–279. [CrossRef]
55. Sahid; Sumiyati, Y.; Purisari, R. The constrains of green building implementation in Indonesia. *J. Phys. Conf. Ser.* **2020**, *1485*, 012050. [CrossRef]
56. Latief, Y.; Berawi, M.A.; Basten, V.; Budiman, R. Construction performance optimization toward green building premium cost based on Greenship rating tools assessment with value engineering method. *J. Phys. Conf. Ser.* **2017**, *877*, 012041. [CrossRef]
57. Liu, Y.; Lu, Y.; Hong, Z.; Nian, V.; Loi, T.S.A. The "START" framework to evaluate national progress in green buildings and its application in cases of Singapore and China. *Environ. Impact Assess. Rev.* **2019**, *75*, 67–78. [CrossRef]
58. Huang, N.; Bai, L.; Wang, H.; Du, Q.; Shao, L.; Li, J. Social network analysis of factors influencing green building development in China. *Int. J. Environ. Res. Public Health* **2018**, *15*, 2684. [CrossRef]
59. Qin, X.; Mo, Y.; Jing, L. Risk perceptions of the life-cycle of green buildings in China. *J. Clean. Prod.* **2016**, *126*, 148–158. [CrossRef]
60. Agyekum, A.K.; Fugar, F.D.K.; Agyekum, K.; Akomea-Frimpong, I.; Pittri, H. Barriers to stakeholder engagement in sustainable procurement of public works. *Eng. Constr. Archit. Manag.* **2023**, *30*, 3840–3857. [CrossRef]
61. Saunders, M.N.K.; Lewis, P.; Thornhill, A. *Research Methods for Business Students*, 7th ed.; Pearson Education Limited: Harlow, UK, 2016.
62. Ahadzie, D.K.; Proverbs, D.G.; Olomolaiye, P. Towards developing competency-based measures for construction project managers: Should contextual behaviours be distinguished from task behaviours? *Int. J. Proj. Manag.* **2008**, *26*, 631–645. [CrossRef]
63. Wao, J.O. Value engineering evaluation method for sustainable construction. In Proceedings of the AEI 2017, Oklahoma City, OK, USA, 11–13 April 2017; pp. 1091–1102.
64. Gunduz, M.; Aly, A.A.; El Mekkawy, T. Value engineering factors with an impact on design management performance of construction projects. *J. Manag. Eng.* **2022**, *38*, 3. [CrossRef]
65. Rondinel-Oviedo, D.R.; Schreier-Barreto, C. Methodology for selection of sustainability criteria: A case of social housing in Peru. In *The Palgrave Handbook of Sustainability: Case Studies and Practical Solutions*; Brinkmann, R., Garren, S., Eds.; Palgrave Macmillan: Cham, Switzerland, 2018; pp. 385–409. [CrossRef]
66. *ISO 14044:2006*; Environmental Management—Life Cycle Assessment—Principles and Framework. 2nd ed. International Organization for Standardization: Geneva, Switzerland, 2006.
67. Ellen MacArthur Foundation. *Completing the Picture: How the Circular Economy Tackles Climate Change*; Ellen MacArthur Foundation: Cowes, UK, 2019.
68. Amudjie, J.; Agyekum, K.; Adinyira, E.; Amos-Abanyie, S.; Kumah, V.M.A. Awareness and practice of the principles of circular economy among built environment professionals. *Built Environ. Proj. Asset Manag.* **2023**, *13*, 140–156. [CrossRef]
69. Pittri, H.; Godawatte, G.R.; Esangbedo, O.P.; Antwi-Afari, P.; Bao, Z. Exploring barriers to the adoption of digital technologies for circular economy practices in the construction industry in developing countries: A case of Ghana. *Buildings* **2023**, *15*, 1090. [CrossRef]
70. Pittri, H.; Godawatte, G.A.G.R.; Agyekum, K.; Botchway, E.A.; Dompey, A.M.A.; Oduro, S.; Asamoah, E. Examining the barriers to implementing design for deconstruction in the construction industry of a developing country. *Constr. Innov.* **2024**. ahead of print. [CrossRef]
71. Sivasankaran, P. Study on reverse logistics and its significant importance—Review. *Acta Technol.* **2024**, *10*, 131–139. [CrossRef]
72. Haq, S.M.; Pieroni, A.; Bussmann, R.W.; Abd-ElGawad, A.M.; El-Ansary, H.O. Integrating traditional ecological knowledge into habitat restoration: Implications for meeting forest restoration challenges. *J. Ethnobiol. Ethnomed.* **2023**, *19*, 33. [CrossRef] [PubMed]

Disclaimer/Publisher's Note: The statements, opinions and data contained in all publications are solely those of the individual author(s) and contributor(s) and not of MDPI and/or the editor(s). MDPI and/or the editor(s) disclaim responsibility for any injury to people or property resulting from any ideas, methods, instructions or products referred to in the content.

buildings

MDPI

Article

Development of a Solar-Tracking Movable Louver with a PV Module for Building Energy Reduction

Sowon Han [1], Janghoo Seo [2,*] and Heangwoo Lee [3,*]

[1] Department of Design, Graduate School, Sangmyung University, Cheonan-si 31066, Republic of Korea
[2] School of Architecture, Kookmin University, Seoul 136-702, Republic of Korea
[3] College of Design, Sangmyung University, Cheonan-si 31066, Republic of Korea
* Correspondence: seojh@kookmin.ac.kr (J.S.); 2hw@smu.ac.kr (H.L.)

Abstract: In response to rising energy consumption in buildings, this study proposes a solar-tracking movable louver integrated with a photovoltaic (PV) module and evaluates its performance to verify its energy-saving potential. First, the louver system can be configured as either vertical or horizontal by modularizing and rotating its slats. A solar-tracking mechanism for single-axis louver control was also developed and proven effective. Second, for optimal energy-saving performance, the louver operation must respond to external environmental conditions. Its control should account for PV power generation and building energy demands for heating, cooling, and lighting to maintain comfortable indoor and outdoor environments. Third, the proposed louver system achieved a building energy reduction of 4.7–8.8% compared to conventional fixed technologies. However, in winter, the louver may obstruct solar gains, potentially diminishing its effectiveness in reducing energy consumption. While this study demonstrates the potential of the proposed louver technology for energy efficiency, it is limited by the scope of environmental and operational conditions considered in the performance evaluation. Further studies under diverse climatic scenarios are necessary to substantiate its broader applicability.

Keywords: louver; PV module; sun tracking; energy saving; performance evaluation

1. Introduction

According to the Global Energy Trends 2023 report published by Enerdata, global energy consumption in 2022 increased by 1.7% compared to pre-pandemic levels, with this upward trajectory projected to continue [1]. This ongoing increase is expected to double or even triple by the mid-21st century [2]. The sustained rise in energy consumption presents a critical challenge, as it contributes to elevated CO_2 emissions, exacerbating air pollution, and global warming [3]. A significant driver of this trend is the building sector, which has witnessed a marked surge in energy use [4–6]. The 2022 Global Status Report for Buildings and Construction, published by the Global Alliance for Buildings and Construction, reports that, in 2021, the buildings and construction sector accounted for 34% of total global energy consumption. Furthermore, energy use in buildings reached an all-time high in 2021, marking an approximate 4% increase from 2020. This underscores that the energy consumption of the building sector exceeds that of both the industrial and transportation sectors [7]. Consequently, reducing energy consumption in buildings has become a pressing issue, with research actively underway in areas such as building envelopes [8–10], green retrofitting [11–13], and environmentally friendly construction materials [14–16]. Among these strategies, louvers have emerged as a representative building envelope technology. Installed primarily on windows, louvers function as shading devices and are recognized for their simple structure and high efficiency in daylighting and shading. This has led to increasing application and research interest [17,18]. Recent studies have integrated photovoltaic (PV) systems into louver slats, adjusting their angles to enhance energy performance and reduce building energy demand [19–21]. However, these studies have primarily concentrated on maximizing power generation, without fully leveraging the respective advantages and disadvantages of vertical and horizontal louver configurations.

In response, this study proposes a movable louver system composed of vertical slats capable of reconfiguring between vertical and horizontal orientations. It aims to develop a louver-based technology that not only enhances PV power generation through solar tracking but also contributes to building energy reduction and improved indoor lighting environments.

1.1. Types of Louvers and Literature Review

Louvers are common external shading systems installed on windows to regulate the amount of natural light entering indoor spaces [22,23]. By adjusting the slat angles, louvers can provide partial or complete shading while also enabling daylighting, as illustrated in Figure 1 [24]. This capability allows them to reduce cooling energy consumption during summer by limiting the heat gain caused by solar radiation [25].

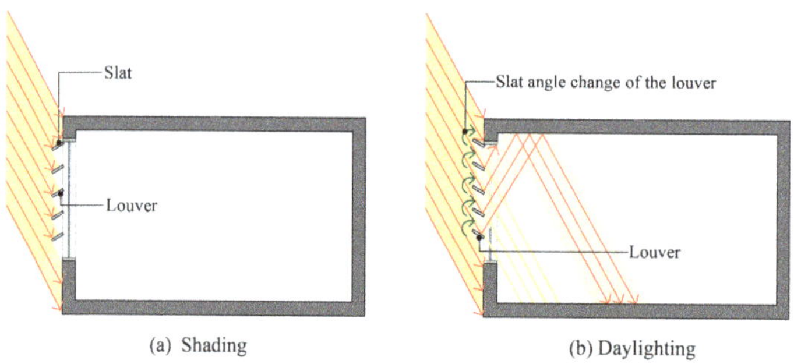

Figure 1. Shading and daylighting effects according to louver slat angles.

Louvers are generally classified as vertical, horizontal, or mixed types based on their installation orientation and shading approach, as summarized in Table 1 [26,27]. Vertical louvers are most effective for blocking sunlight when the sun is at lower angles in the east or west, making them ideal for installation on the east and west facades of buildings [28]. Horizontal louvers, in contrast, are effective in shielding against high-angle sunlight typical of summer and are best suited for south-facing windows [29]. In winter, adjusting the slat angles of horizontal louvers can allow low-angle solar radiation to enter indoor spaces, contributing to passive heating and energy savings [30,31]. Therefore, the choice between vertical and horizontal louvers should be made with careful consideration of solar orientation and environmental context. Mixed-type louvers, which combine elements of both vertical and horizontal configurations, offer the benefits of each. However, they tend to increase the shading surface area, potentially obstructing outdoor views and limiting solar heat gain. This makes them less effective in winter, when increased solar penetration is desirable for heating efficiency [30,31].

This study reviewed previous research on louver systems [19–21,32–37], as summarized in Table 2. Prior studies primarily focus on technological development and performance evaluation based on slat specifications. Among these, several have examined the integration of photovoltaic (PV) modules into louver slats to reduce building energy consumption [19,32–35]. However, research incorporating actual solar-tracking mechanisms into louver systems remains limited. Recent advancements in louver technologies have introduced non-traditional configurations, enhancing performance through novel forms and adaptive behaviors [20,21,36,37]. Nevertheless, such complex forms often entail significant challenges in terms of maintenance and operability. In contrast, the louver system proposed in this study retains the simple and practical structure of conventional vertical and horizontal louvers. It differentiates itself by integrating both PV modules and a solar-tracking mechanism to evaluate the resulting performance improvements. This approach aims to balance technological enhancement with structural simplicity and ease of management.

Table 1. Types and characteristics of louver.

Type		Characteristics	Best Orientation	Image
Horizontal	Overhang	-Prevents the rise of hot air from outside -Potential risk of damage due to wind pressure and snow accumulation	South	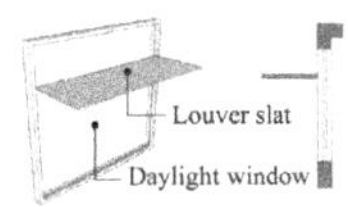
	Overhang with Multiple Blades	-Installed in the form of multiple horizontal slats -Potential risk of damage due to wind pressure and snow accumulation	South	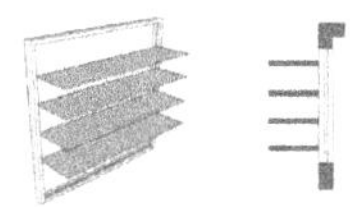
	Overhang Horizontal Louvers	-Horizontal slats divided into uniform sizes -Allows free movement of air -Low snow and wind load	South	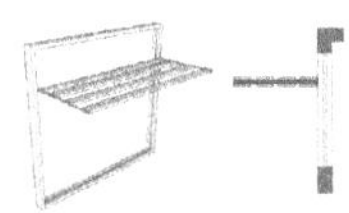
	Overhang Vertical Panel	-Allows free movement of air -Low risk of damage due to snowfall -Causes visual obstruction	South, East, West	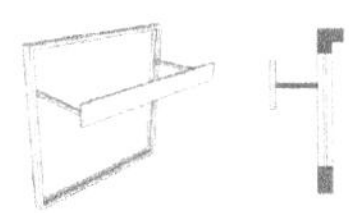
Vertical	Vertical Fin	-Typically used only in hot climates for north-facing windows -May obstruct visibility	East, West	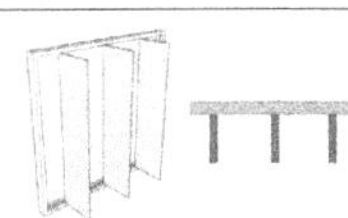
	Slanted Vertical Fin	-Installed at an angle (not perpendicular) to the window -May obstruct visibility	East, West	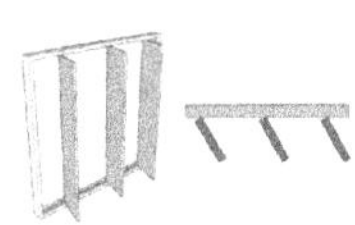
Mixed	Eggcrate	-Used in very hot climates -Tends to trap hot air -Causes high visual obstruction	East, West	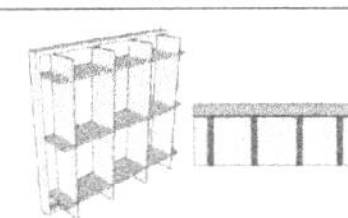
	Eggcrate with Horizontal Louvers	-Applies horizontal slats in the form of overhang horizontal louvers -Partially mitigates the trapping of hot air -Suitable for use in very hot climates -Causes high visual obstruction	East, West	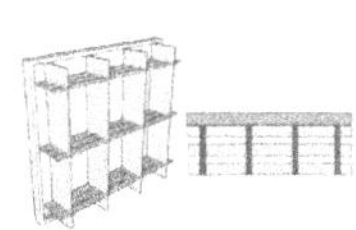

Table 2. Review of previous studies on louver shading systems.

Reference	Objective	Louver			PV Application	Use of Solar-Tracking Technology
		Type (Shape)	Width (m)	Angle		
[19]	Developed optimal control for BIPV to improve energy efficiency and visual comfort	Horizontal	-	−78° to 78°	Applied (PV attached to only half of the slats)	Not applied
[20]	Developed three prototype adaptive facade systems compatible with curvature by applying biomimicry and SMC	A shape that mimics the curvature of a fabric membrane	-	Changes by 20°	Not applied	Not applied
[21]	Presented a responsive envelope using hygroscopic material properties	A form that mimics the shape and function of cactus stomata	-	It closes at 50% relative humidity and opens at 85%	Not applied	Not applied
[32]	Evaluated fixed shading systems with integrated solar power generation in Mediterranean countries	Horizontal	-	-	Applied (Attached to the top surface)	Not applied
[33]	Developed a new ventilated BIPV double skin facade system consisting of transparent amorphous silicon PV modules and inward-opening windows	Horizontal	1.1	Fixed	Applied (Only to the external glass)	Not applied
[34]	Assessed translucent PV ventilated glazing for energy and cost savings	Horizontal	0.88	-	Applied (Attach translucent PV to the window)	Not applied
[35]	Tested prototype PV-integrated louver system for optimal shading, daylighting, and solar energy production	Horizontal	0.41	−90°, −60°, −30°, 0°, 30°, 60°, 90°	Applied (PV attached only to 2/3 of slats)	Not applied
[36]	Hybridized the physical characteristics of thermodynamic SMA springs and PFEA, which balance forces, to propose facade panel module designs	A form that mimics cactus surface	-	The length changes to 31.5, 50.7, and 68.1mm depending on the pressure	Not applied	Not applied
[37]	Designed building envelopes for improved thermal performance through a biomimetic optimization algorithm	A form that mimics the hygroscopic structure of a pine cone	-	Optimal rotation angle 43°, optimal enclosure angle 69°	Not applied	Not applied

1.2. PV Module Concept

PV modules are composed of multiple interconnected PV cells, which serve as the fundamental units of solar power generation [38,39]. As illustrated in Figure 2, a PV cell is typically formed by joining P-type and N-type semiconductors, each possessing distinct electrical properties. To assemble a PV module, multiple PV cells are connected using conductive ribbons that facilitate the flow of current between cells [40]. When photons strike the surface of a PV cell, they are absorbed by the semiconductor material, generating electron–hole pairs. The excited electrons migrate toward the N-type semiconductor, while the corresponding holes move toward the P-type region. These separated charge carriers are driven through an external low-resistance circuit, producing an electric current and thereby generating electrical power. The efficiency of a PV cell is significantly influenced by temperature; as the cell temperature increases, its efficiency tends to decrease [41–43]. In addition, PV cells achieve their highest efficiency when sunlight strikes them perpendicularly. Any deviation from this optimal angle reduces the intensity of light absorption and thereby lowers the energy conversion efficiency [44,45].

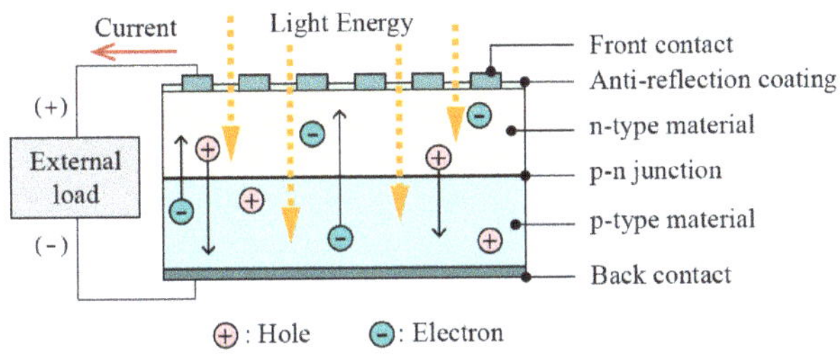

Figure 2. Structure and working principle of PV cell.

PV cells are generally classified into two major categories: crystalline and thin-film types. Crystalline cells are further subdivided into monocrystalline and polycrystalline types based on their fabrication processes. While monocrystalline PV cells offer higher efficiency than polycrystalline variants, they are also associated with higher production costs [46]. In this study, crystalline PV modules were integrated onto the planar slats of the louver system.

1.3. Concept of Solar-Tracking Technology

Solar-tracking technology in photovoltaic (PV) systems is designed to enhance the performance of fixed solar generation systems by rotating PV modules along the sun's trajectory [47]. By adjusting the orientation of solar panels to follow the sun's movement, this technology significantly improves power output and conversion efficiency, making it a critical component in various solar energy applications [48].

Solar-tracking methods are broadly categorized into program-based and sensor-based systems [49–51]. Program-based tracking utilizes astronomical algorithms that define the relative motion of the Earth and the sun. This method calculates the sun's position in real time based on geographic coordinates and time data for a specific location. Since it operates according to predefined algorithms, it is unaffected by weather conditions, offering the advantage of eliminating unnecessary mechanical movements. However, it lacks adaptability to dynamic environmental factors such as cloud cover or shading [52]. In contrast, sensor-based tracking systems use multiple light sensors mounted on the tracking structure to detect the direction of maximum solar irradiance. By comparing the light intensity received by each sensor, the system adjusts the PV module's orientation accordingly. This method enables responsiveness to changing weather and environmental conditions, thereby enhancing overall energy yield. However, it is generally less accurate in determining the exact solar position compared to program-based systems. Solar-tracking devices are also classified by the number of axes involved in the tracking mechanism, as shown in Table 3. These include single-axis and dual-axis systems. A single-axis solar tracker rotates around one axis—either aligned with the solar azimuth (east–west) or the solar altitude (elevation)—to follow the sun's daily movement [53,54]. Typically, the system tilts the PV module toward the north while tracking the sun's position from east to west, thereby improving solar concentration and power generation. Single-axis trackers offer enhanced mechanical stability and increased energy output with relatively simple structure. However, they are limited to tracking the sun's daily movement and do not accommodate seasonal altitude variations, which reduces overall tracking precision and efficiency. Dual-axis trackers use two perpendicular axes to follow both the solar azimuth and altitude angles. This configuration enables comprehensive tracking throughout the day and across seasons, accounting for annual changes in the sun's position [55,56]. Dual-axis systems can employ either passive or active tracking mechanisms. Although they provide superior energy gains and flexibility compared to single-axis systems, they suffer from increased structural complexity. The PV module is typically connected to the tracking frame only at the central rear, which weakens mechanical integrity and increases the risk of damage under adverse weather conditions [57].

Table 3. Operating method for solar tracking.

Method	Characteristics	Image
Single-axis tracking method	-Pivots along a single axis responding to the sun's azimuth or altitude (divided into vertical and horizontal tracking) -Tracks the sun's movement from east to west based on its position -Stability and power generation efficiency can be improved with external support -Does not account for external factors such as weather and shading	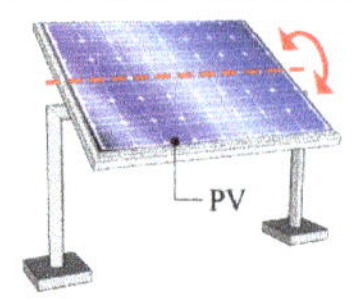 PV
Dual-axis tracking method	-Rotates around two perpendicular axes, simultaneously tracking solar azimuth and altitude angles -Produces energy at a faster rate than single-axis tracking systems -Tracks both the sun's movement from east to west and its altitude -Less affected by clouds and other environmental conditions	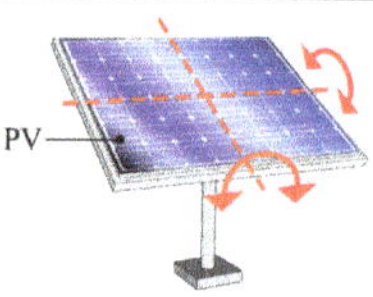 PV

2. Materials and Methods

2.1. Design Suggestion for a PV-Integrated Movable Louver System Utilizing Solar-Tracking Technology

The solar-tracking-based movable louver system proposed in this study is designed to improve photovoltaic (PV) generation efficiency by integrating solar-tracking technology with PV-attached slats. The system also allows for seamless transformation between vertical and horizontal louver configurations, thereby optimizing the shading and daylighting benefits of both types. The system is initially configured as a vertical louver, as shown in Figure 3a. Each slat is divided into modular segments, which are mechanically linked to rotate synchronously at a fixed angle. This design enables the structure to transform into a horizontal louver configuration, thereby combining the benefits of both orientations. The transformation capability allows the louver to be controlled dynamically in response to external environmental conditions, such as sun position and seasonal variations. In addition, as shown in Figure 3b, the slat support frames are designed to shift and overlap in one direction, enabling improved visibility and increased solar access in winter to reduce heating energy demand. Each slat module incorporates a PV panel on one side and a high-reflectivity aluminum surface on the opposite side, which reflects approximately 90% of incident light. By adjusting the slat angle, the system can either generate electricity or enhance indoor lighting, depending on orientation and seasonal need. As shown in Figure 3a, the system includes a solar-tracking device mounted at the top to adjust the slat angles in real time. This tracking mechanism uses a single-axis drive system and features a "⊥"-shaped cross-section with a protruding center, as shown in Figure 4. Light sensors are positioned symmetrically on either side of the central protrusion. As the tracker rotates around a pivot axis, the irradiance values measured by the two sensors vary based on the angle of incidence from the sun. When the difference in sensor readings decreases, the sensor surface becomes more perpendicular to sunlight (Figure 4b). This principle allows the system to track the sun's position accurately and apply the optimal slat angle for both shading and PV performance. Moreover, as shown in Figure 3d, the slat module support frames can move laterally to create an open-louver configuration, further enhancing view access and daylight penetration.

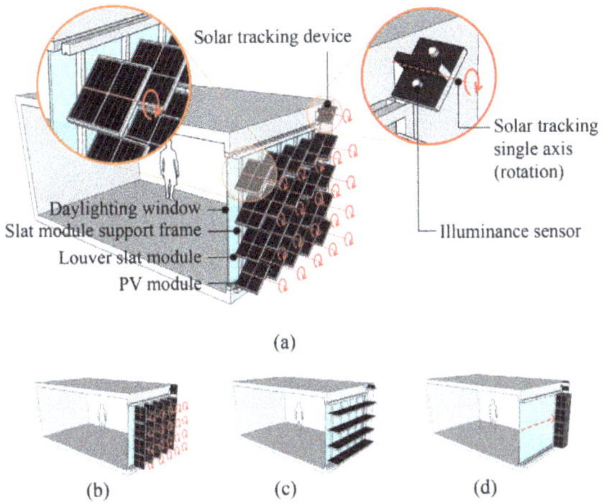

Figure 3. Configuration and operation of the solar-tracking-based movable louver with a PV module: (**a**) System configuration and angle control (slat angle 135°), (**b**) Slat angle 180°, (**c**) Slat angle 90°, (**d**) Open configuration via slat frame movement.

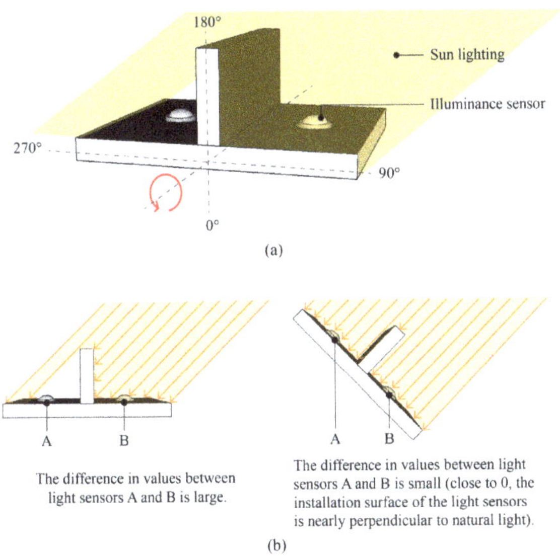

Figure 4. Structure and operation principle of the solar tracking device for the proposed louver system: (**a**) Angle control of the solar tracking device, (**b**) Principle of solar tracking based on the difference in illuminance sensor values.

2.2. Performance Evaluation Environment Setup

To evaluate the performance of the solar-tracking-based movable louver with a PV module, this study constructed a testbed capable of simulating artificial climatic conditions.

As illustrated in Figure 5, the testbed comprises four main components: an indoor space, a daylighting window, an artificial climate chamber, and an energy monitoring system.

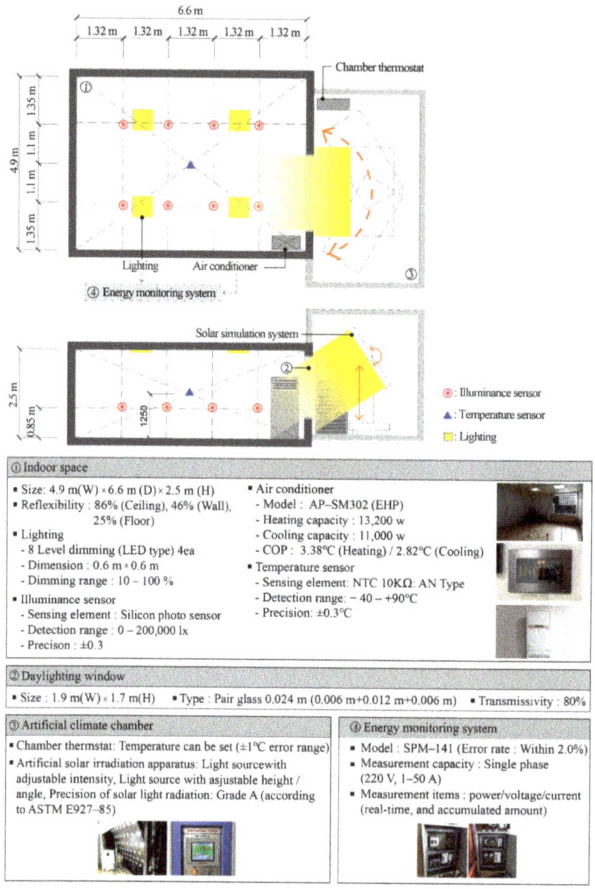

(a) Configuration of the testbed

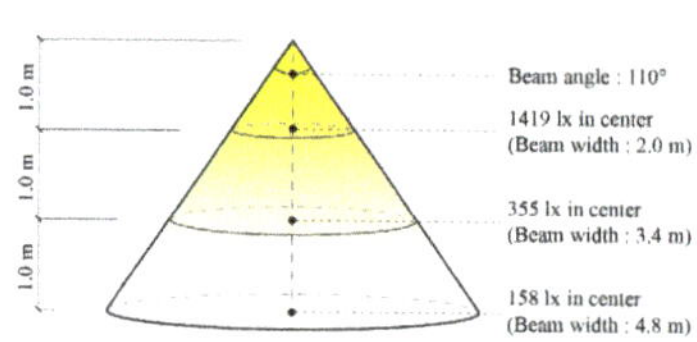

(b) Conical illuminance of lighting

Figure 5. Testbed configuration for performance evaluation.

The setup is detailed as follows. First, the indoor test space has internal dimensions of 6.6 m × 4.9 m × 2.5 m, with walls constructed from 0.1 m thick insulated panels. The surface reflectance values for the floor, walls, and ceiling are 25%, 46%, and 86%, respectively. Second, the indoor space includes four LED-type lighting fixtures and one floor-standing air conditioner, both of which were used to evaluate the energy-saving performance of the louver system. The lighting system supports eight levels of dimming, excluding the OFF state. The installation positions of the lighting fixtures follow the Illuminating Engineering Society (IES) four-point method [58]. Third, eight illuminance sensors were installed within the space, as shown in Figure 5. These were positioned at a height of 0.85 m above the floor to align with the typical working plane height. A temperature sensor was installed at the center of the interior space to monitor ambient thermal conditions. Fourth, the daylighting window into which the proposed louver was installed measures 1.9 m × 1.7 m. The window is composed of double-glazed panes with a light transmittance of 80%. Notably, the window is not centered on the wall but positioned off to one side, as per the testbed configuration. Fifth, the artificial climate chamber is equipped with a solar irradiation apparatus and a temperature control system. The solar simulator can replicate solar azimuth and altitude angles by adjusting both the intensity and elevation of artificial light. This Grade-A simulator, compliant with ASTM E927-85 international standards, ensures high measurement accuracy and reliability [59]. However, due to equipment constraints, it can simulate solar azimuth angles only within the range of 120° to 190°. Finally, the study implemented control of lighting and HVAC systems to maintain appropriate indoor illuminance and thermal comfort levels. Power consumption associated with these operations was continuously recorded using an integrated energy monitoring system.

2.3. Performance Evaluation Method

This study evaluates the shading and power generation performance of the solar-tracking-based movable louver system with an integrated PV module, as proposed in Section 2.1. The evaluation methodology is detailed below.

To verify the effectiveness of the proposed system (Case 4), three reference configurations were established for comparison: vertical louvers (Case 1), horizontal louvers (Case 2), and mixed-type louvers (Case 3). The specifications for each case are presented in Table 4. Slat dimensions were determined based on the physical size of the PV cells used. In Cases 1 and 2, PV cells are affixed to one side of each slat. In Case 3, vertical slats are installed on both lateral sides of the window, while horizontal slats with PV cells are mounted on top. The proposed Case 4 adopts a configuration similar to Case 1 but incorporates modular slats, resulting in a slight reduction in total PV cell count. In addition, Case 4 offers dynamic transformation between vertical and horizontal orientations. As illustrated in Figure 6, the slat angles in Cases 1 and 2 are adjustable within a range of 180°, in 15° increments. For Case 4, the slat modules rotate within a full 360° range, also in 15° increments, allowing for enhanced operational flexibility. PV specifications are summarized in Table 5. Slats or modules without PV cells were treated with aluminum surfaces having a reflectivity of 90% to support daylight reflection.

Table 4. Case setting for performance evaluation.

Case	Image	Louver Type	Movable or Not	Slat				Slat Module		PV Module Area (Number of PV Cells Used)
				Count	Size	Spacing	Angle	Size	Angle	
1	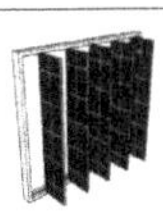	Vertical	Fixed	5	0.03 m (W) × 0.35 m (D) × 1.65 m (H)	0.375 m	Moves within a range of −90° to 90° in 15° increments	-	-	2.43 m² (100)
2		Horizontal	Fixed	4	1.85 m (W) × 0.35 m (D) × 0.03 m (H)	0.41 m	Moves within a range of 0° to 180° in 15° increments	-	-	2.14 m² (88)

Table 4. *Cont.*

Case	Image	Louver Type	Movable or Not	Slat				Slat Module		PV Module Area (Number of PV Cells Used)
				Count	Size	Spacing	Angle	Size	Angle	
3		Mixed	Fixed	9	Combination of Cases 1 and 2	Vertical 0.37 m, horizontal 0.41 m	0	-	-	5.84 m^2 (240)
4		Vertical and horizontal	Movable	5	Each slat is divided into 4 modules	-	0°, 90°	0.03 m (W) × 0.31 m (D) × 0.38 m (H)	Moves within a range of 0° to 360° in 15° increments	1.95 m^2 (80)

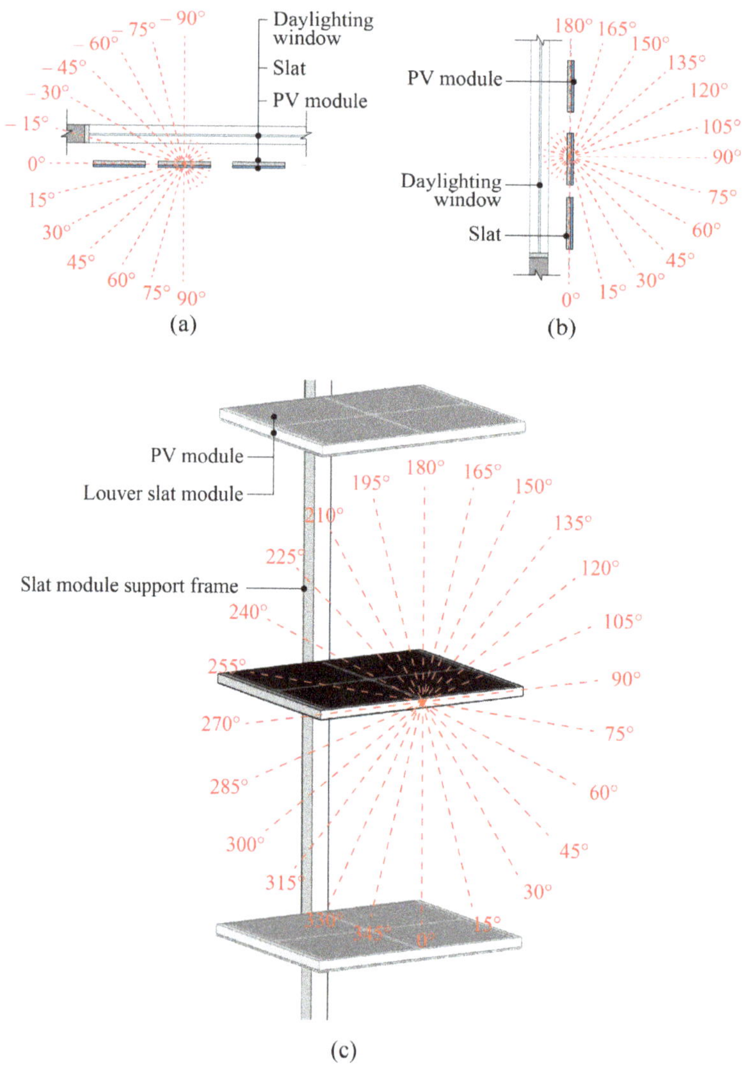

Figure 6. Angle setting for Cases 1, 2, and 4: (**a**) Slat module angle of Case 1, (**b**) Slat module angle of Case 2, (**c**) Slat module angle of Case 4.

Table 5. Specifications of the PV Cell.

Item	Specifications	Item	Specifications
Type	Monocrystalline Silicon	Open-Circuit Voltage	0.678 V
Max. Power (Pmax)	5.35 W	Fill Factor	80.75%
Voltage at Max. Power Point (Vmpp)	0.577 V	Size	0.156 m × 0.156 m
Current at Max. Power Point	9.329 A	Cell Efficiency	21.9%

Second, a mockup was fabricated to evaluate the performance of the proposed louver system. This mockup was constructed using the profiles shown in Figure 7. A solar-tracking device, illustrated in Figure 7e, was mounted on the upper portion of the structure. The specifications of the pyranometer (illuminance sensor) used in the solar-tracking device are presented in Table 6. White-painted aluminum was used for the rear part of the louver system, with a reflectance of 0.6.

(a) (b) (c) (d) (e)

Figure 7. Mockup production for performance evaluation: (**a**) Slat module angle at 0°, (**b**) Slat module angle at 45°, (**c**) Slat module angle at 90°, (**d**) Open configuration via slat movement, (**e**) Solar-tracking device.

Table 6. Specifications of the illuminance sensor used in the solar-tracking device.

Specifications	Values	Specifications	Values
Model	ML-02	Non-linearity at 1000 W/m^2	<0.2%
Wavelength range	400~1100 nm	Directional response at 1000 W/m^2	<10 W/m^2
Response time 95%	<1 ms	Sensitivity (μV/W/m^2)	Approx. 50
Non-stability (change/year)	±2%	Operating temperature range	−30 to +70

Third, the study calculated the electricity generated by the PV modules attached to the louvers. This was performed by recording the voltage at the maximum power point (Vmp) and the corresponding current (Imp) and then calculating the power output as the product of these two values. Since the total PV module area differed across the four test cases, the power generation values were normalized to a per-square-meter basis (W/m^2) to enable fair comparison among cases.

Fourth, lighting energy consumption was evaluated based on the required dimming levels necessary to maintain appropriate indoor illuminance, with the goal of ensuring a visually comfortable environment. A target indoor illuminance of 500 lx was established in accordance with the general office lighting standards specified by the IES of North America [60]. Dimming control was implemented when the minimum illuminance value among the ten installed indoor sensors fell below 500 lx. The dimming level was increased incrementally, starting with the light closest to the sensor registering the lowest value. This process continued until all sensors measured illuminance levels of 500 lx or higher. Lighting energy consumption was derived based on the final dimming configuration for each scenario, providing a basis for evaluating the lighting energy reduction potential of each louver configuration. Lastly, the study determined the optimal slat and slat module angles for minimizing lighting energy consumption. These were defined as the angles at which energy use was lowest. In cases where multiple angles yielded the same minimum energy consumption, the optimal angle was selected based on the highest minimum illuminance value among the sensors—thereby prioritizing uniform lighting distribution while achieving energy efficiency.

Fifth, this study evaluated the operation of air conditioning systems required to maintain appropriate indoor temperatures across different louver configurations. Based on prior research [61], the target indoor temperatures were set to 26 °C for summer and 20 °C for winter. Instead of using the built-in temperature sensor of the air conditioner, this study used a centrally located sensor installed in the middle of the indoor space (as shown in Figure 5). This sensor provided real-time feedback for controlling the air conditioner, enhancing the accuracy and responsiveness of indoor temperature regulation. Heating and cooling energy consumption was monitored to assess the thermal energy-saving performance of each louver configuration. To ensure the validity of the temperature-based evaluation, lighting levels were kept constant at a designated dimming setting during the measurement of heating and cooling energy usage. This control ensured that lighting did not introduce variability in thermal load conditions.

Sixth, the performance evaluation was conducted under controlled seasonal conditions using the artificial climate chamber integrated into the testbed. This study was conducted in Seoul, South Korea, which has distinct seasonal climatic changes. Seoul sits at a latitude of 37.57° N and a longitude of 126.98°; the external environmental conditions of the artificial climate chamber were set based on the climate data of this region. The criteria reflected the climatic characteristics of a typical mid-latitude city and ensured the generalizability of performance evaluations. As presented in Table 7, the environmental parameters were simulated in one-hour intervals from 10:00 a.m. to 2:00 p.m. Due to equipment constraints in the artificial solar radiation system, the 10:00 a.m. condition was partially adjusted to align with the testbed's physical limitations. Each time slot simulated solar azimuth, altitude, external illuminance, and solar radiation levels consistent with seasonal variations. The seasonal temperature settings of the artificial climate chamber were based on the 30-year average climate data for Seoul provided by the Korea Meteorological Administration [61]. Additionally, the solar radiation of the artificial solar radiation system was measured based on illuminance control. The solar radiation information was calculated based on the solar radiation value that reached the vertical surface of the window where the louver was installed (Table 7).

Table 7. Simulated solar altitude, external illuminance, and solar radiation by time for summer and winter.

Season	Temperature of the Artificial Climate Chamber	External Illuminance, Azimuth Angle, Solar Altitude, and Solar Radiation by Time Zone			
		10:00–11:00	11:00–12:00	12:00–13:00	13:00–14:00
Summer	27.1 °C	70 k lx, 120°, 60° 530 W/m^2	75 k lx, 150°, 70° 574 W/m^2	80 k lx, 180°, 76.5° 638 W/ m^2	70 k lx, 203°, 70° 574 W/m^2
Winter	−3.2 °C	20 k lx, 157°, 22.5° 289 W/m^2	25 k lx, 172°, 28° 296 W/m^2	30 k lx, 180°, 29.5° 332 W/m^2	20 k lx, 187°, 26.5° 296 W/m^2

3. Results and Discussion

3.1. Performance Evaluation Results

To validate the effectiveness of the solar-tracking-based movable louver system with a PV module (Case 4), comparative performance evaluations were conducted against existing louver technologies (Cases 1–3). The results are as follows.

First, the functionality of the solar-tracking system was verified by analyzing the ratio of illuminance sensor values recorded between 12:00 and 13:00 in both summer and winter conditions. As shown in Figure 8, the ratio was calculated by dividing the lower illuminance sensor value by the higher one. This ratio was used to determine the directional alignment of the solar-tracking device. In addition, the study evaluated power generation in Case 4 by varying the slat module angle. The resulting power generation curve exhibited a pattern similar to the sensor ratio trend observed in the solar-tracking system. When the maximum power output of the slat module and the maximum sensor ratio were normalized to the same reference, the average difference between the two indicators was found to be 9.9%. This close correlation demonstrates the functional validity and effectiveness of the solar-tracking device developed in this study.

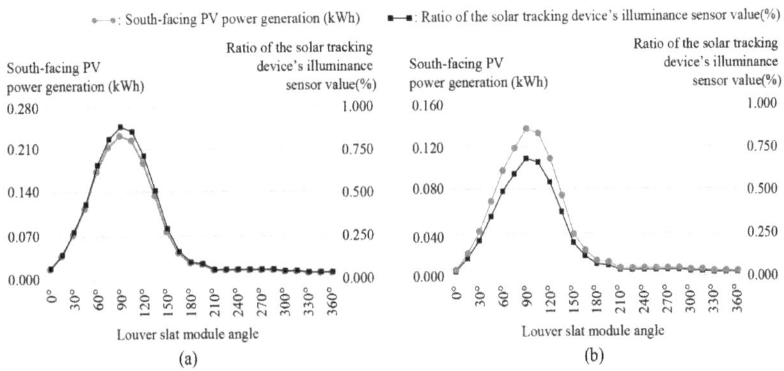

Figure 8. Comparison between the south-facing PV power generation and the illuminance sensor ratio of the solar-tracking device: (**a**) Summer, (**b**) Winter.

Second, Figure 9 presents the power output of the PV modules based on the slat and module angles for each case. For the vertical louver configuration (Case 1), the optimal slat angles to maximize power generation during summer from 10:00 a.m. to 2:00 p.m. were identified as 60°, 30°, 0°, and −30°, respectively. In winter, the corresponding optimal angles were 30°, 15°, 0°, and 0°. The total power generation for Case 1, using these optimal angles, amounted to 1.326 kWh. For the horizontal louver configuration (Case 2), the optimal slat angles during summer were 60°, 75°, 75°, and 75°, while in winter they were 15°, 30°, 30°, and 30°, respectively. The resulting total power generation was 1.366 kWh. The mixed louver configuration (Case 3) was assumed to remain static (non-operational) during the performance evaluation. Despite its larger number of PV modules, the total power generation was 1.539 kWh. For the solar-tracking-based movable louver proposed in this study (Case 4), the optimal slat module angles between 10:00 a.m. and 2:00 p.m. were 60°, 75°, 90°, and 105° in summer and 45°, 75°, 90°, and 105° in winter. The total power generation for Case 4 was 1.387 kWh. These results demonstrate that adapting the slat angle in response to external environmental conditions improves PV generation efficiency. Although Case 4 has fewer PV cells than Cases 1 and 2, its total power generation was 4.4% higher than Case 1 and 1.5% higher than Case 2 when operated at optimal angles. More notably, when normalized by PV module area (1 m^2), Case 4 achieved 23.3% and 10.3% higher power generation compared to Cases 1 and 2, respectively, as shown in Table 8. Although Case 3 incorporated 58.3%, 63.3%, and 66.7% more PV cells than Cases 1, 2, and 4, respectively, its power generation only increased by 13.1%, 11.2%, and 9.9% compared to each of those cases. This limited performance is attributed to the fixed nature of Case 3 and its structural design, which includes PV modules on both sides of vertical slats. Due to direct sunlight conditions, only one side receives effective irradiance, while the other remains shaded. Furthermore, the combination of vertical and horizontal slats causes mutual shading, further reducing overall PV efficiency.

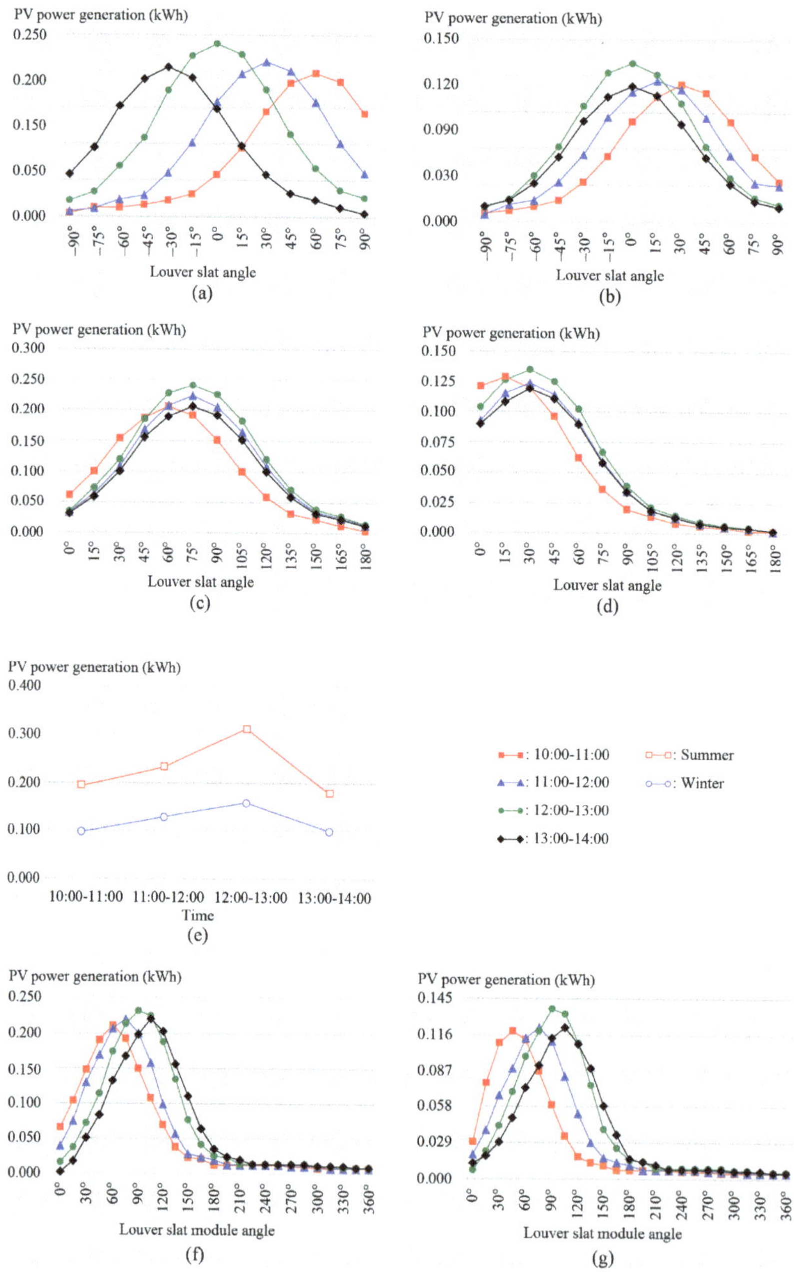

Figure 9. PV power generation by case: (**a**) Summer (Case 1), (**b**) Winter (Case 1), (**c**) Summer (Case 2), (**d**) Winter (Case 2), (**e**) Case 3, (**f**) Summer (Case 4), (**g**) Winter (Case 4).

Third, the lighting energy consumption required to maintain appropriate indoor brightness for each case is illustrated in Figure 10. In Case 1 (vertical louver), the optimal slat angles for reducing lighting energy between 10:00 a.m. and 2:00 p.m. were −90°, −90°, 90°, and 75° during summer and −90°, −90°, 90°, and 90° during winter. Based on these angles, the total lighting energy consumption was 0.831 kWh, as shown in Table 9. For Case 2 (horizontal louver), the optimal slat angles were 150°, 165°, 165°, and 165° during summer and 105°, 120°, 120°, and 120° during winter. The total lighting energy consumption in this case was 0.843 kWh. Case 3 (mixed louver), which was non-operational during evaluation, consumed 0.970 kWh of lighting energy. In Case 4 (the proposed solar-tracking-based movable louver), the optimal slat module angles for reducing lighting energy were 255°, 195°, 240°, and 270° in summer and 195°, 195°, 240°, and 240° in winter. The total lighting energy consumption under these optimal angle conditions was the lowest among all cases at 0.825 kWh. This represents a reduction in lighting energy consumption by 0.7%, 2.1%, and 10.8% compared to Cases 1, 2, and 3, respectively. However, when Case 4 was operated using angles derived from the solar-tracking device—optimized for PV power generation rather than lighting—its total lighting energy consumption was 0.950 kWh. This outcome highlights the trade-off between maximizing power generation and minimizing lighting energy use, as the optimal slat angle for lighting reduction does not always align with that for PV efficiency. During the 10:00–11:00 and 11:00–12:00 time periods, Case 1 demonstrated effective lighting energy reduction by reflecting natural light into parts of the room distant from the daylighting window, as shown in Figure 11. This effect is attributed to the asymmetrical placement of the window in the testbed, which is offset to one side. Similarly, in Case 4, the optimal slat module angles for lighting reduction exceeded 180°, enabling effective reflection of natural light from the rear aluminum panel (opposite the PV side) into the interior space.

Table 8. Optimal angles and total power generation for improving power generation by case.

Case	Optimal Angle of the Slat Module for Improving Power Generation								Total Power Generation (kWh)	Power Generation per 1 m² (kWh)
	Summer				Winter					
	10:00–11:00	11:00–12:00	12:00–13:00	13:00–14:00	10:00–11:00	11:00–12:00	12:00–13:00	13:00–14:00		
1	60°	30°	0°	−30°	30°	15°	0°	0°	1.326	0.546
2	60°	75°	75°	75°	15°	30°	30°	30°	1.366	0.638
3	-	-	-	-	-	-	-	-	1.539	0.264
4	60°	75°	90°	105°	45°	75°	90°	105°	1.387	0.711

Fourth, the heating and cooling energy consumption required to maintain appropriate indoor temperatures for each case is shown in Figure 12. These calculations were performed in conjunction with dimming control for lighting to ensure consistent indoor illuminance levels. During summer, the optimal slat angles for reducing cooling energy from 10:00 a.m. to 2:00 p.m. were 0° for all time intervals in Case 1 and 0° (or 180° due to the horizontal orientation) for all intervals in Case 2. The corresponding cooling energy consumption was 1.574 kWh for Case 1 and 1.619 kWh for Case 2. In both cases, it was beneficial to block direct sunlight to minimize internal heat gain, making a slat angle of 0° preferable. During winter, the optimal slat angles for maximizing solar heat gain and thus reducing heating energy consumption were −45°, −90° (or 90°), −90° (90°), and −90° (90°) in Case 1 and 105°, 120°, 120°, and 120° in Case 2. The calculated heating energy usage was 2.643 kWh for Case 1 and 2.746 kWh for Case 2. These results suggest that maximizing indoor solar exposure through open slat angles is effective in reducing heating demand. In Case 3 (mixed louver), the total heating and cooling energy consumption was 4.864 kWh, which is 1.2% and 3.1% higher than in Cases 1 and 2, respectively. This increase is attributed to its fixed structure, which prevents the system from adapting to varying external environmental conditions. In Case 4 (proposed louver system), the heating and cooling energy consumption based on the optimal slat module angles was 5.0% higher than that of Case 1. This is primarily due to its structural configuration, which slightly reduces its ability to block solar radiation. However, when Case 4 was operated using the solar-tracking system—optimized for PV generation—the total heating and cooling energy dropped to 4.503 kWh. This variation again highlights the performance trade-offs between optimizing for energy savings and for solar energy production.

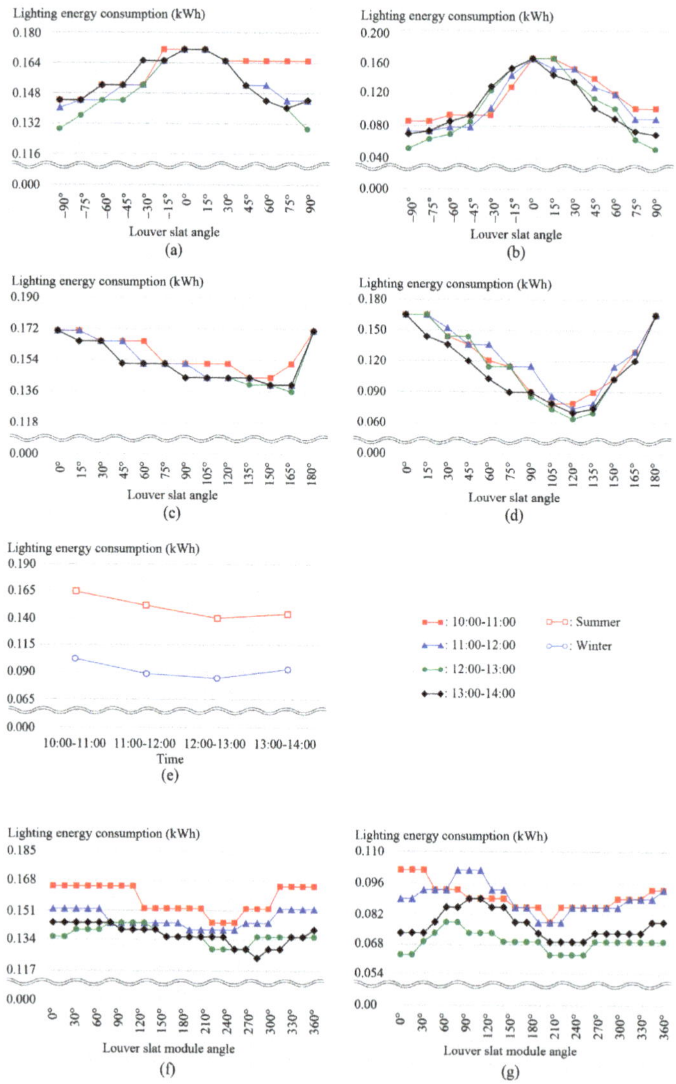

Figure 10. Lighting energy consumption by case: **(a)** Summer–Case 1, **(b)** Winter–Case 1, **(c)** Summer–Case 2, **(d)** Winter–Case 2, **(e)** Case 3, **(f)** Summer–Case 4, **(g)** Winter–Case 4.

Table 9. Optimal angles and total lighting energy usage for reducing lighting energy consumption by case.

Case	Optimal Angle of the Slat Module for Improving Power Generation								Total Lighting Energy Consumption (kWh)
	Summer				Winter				
	10:00–11:00	11:00–12:00	12:00–13:00	13:00–14:00	10:00–11:00	11:00–12:00	12:00–13:00	13:00–14:00	
1	–90°	–90°	90°	75°	–90°	–90°	90°	90°	0.831
2	150°	165°	165°	165°	105°	120°	120°	120°	0.843
3	-	-	-	-	-	-	-	-	0.970
4	* 255°	195°	240°	270°	195°	195°	240°	240°	0.825
	** 60°	75°	90°	105°	45°	75°	90°	105°	0.950

* Slat module angle control to reduce lighting energy consumption in Case 4. ** Operation based on the optimal angle derived from the solar-tracking device of Case 4.

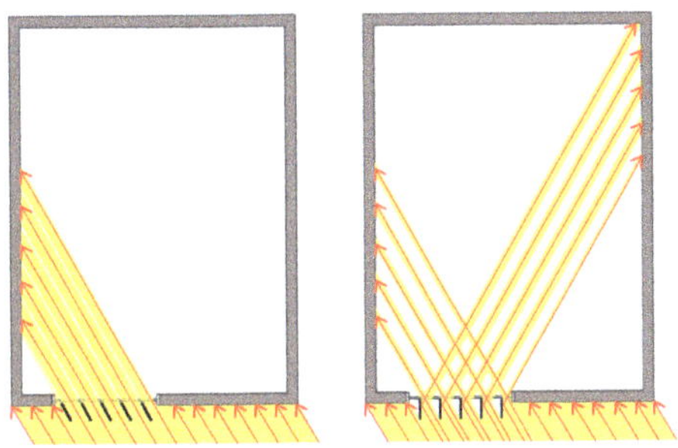

(a) (b)

Figure 11. Inflow process of Case 1 from 11:00 to 12:00 during summer based on slat angle (plan view): (**a**) Slat angle of −60°, (**b**) Slat angle of −90°.

3.2. Discussion

This study compared and analyzed the energy reduction performance of the solar-tracking-based movable louver with a PV module (Case 4) against conventional louver systems. Table 10 presents the optimal slat/module angles for each case when considering three criteria in combination: PV module power generation, lighting energy consumption required to maintain appropriate indoor illuminance, and heating/cooling energy required for thermal comfort. These integrated optimal angles differ from the individually optimized values previously shown in Tables 8, 9 and 11, which were based on single-parameter evaluations. This indicates that controlling louver slats in PV-integrated systems should not rely solely on maximizing PV power generation but must also consider overall building energy usage, including lighting and HVAC loads. The proposed system (Case 4) incorporates solar tracking to autonomously respond to changing environmental conditions and combines the advantages of both vertical and horizontal louver orientations. As a result, Case 4 achieved overall energy reductions of 4.7%, 8.8%, and 5.9% compared to Cases 1 (vertical), 2 (horizontal), and 3 (mixed fixed), respectively. Although Case 4 does not optimize for lighting and heating/cooling energy directly—being driven by solar-tracking logic to enhance PV output—it still demonstrates superior overall performance. Importantly, Case 4 utilizes a smaller PV surface area than the other cases, yet it achieves comparable or improved results, underscoring its efficiency and practicality. In addition, this study examined the benefit of increasing solar access during winter to reduce heating energy consumption. As shown in Figures 3d and 7d, an additional performance evaluation was conducted in which the slat modules were retracted to one side of the daylighting window to maximize solar radiation. Under this configuration, the total energy consumption was measured at 2.259 kWh (Table 12). This represents an average energy reduction of 14.5%

compared to the scenarios where the louver remained in standard operating positions, highlighting the effectiveness of passive solar gain strategies during colder seasons.

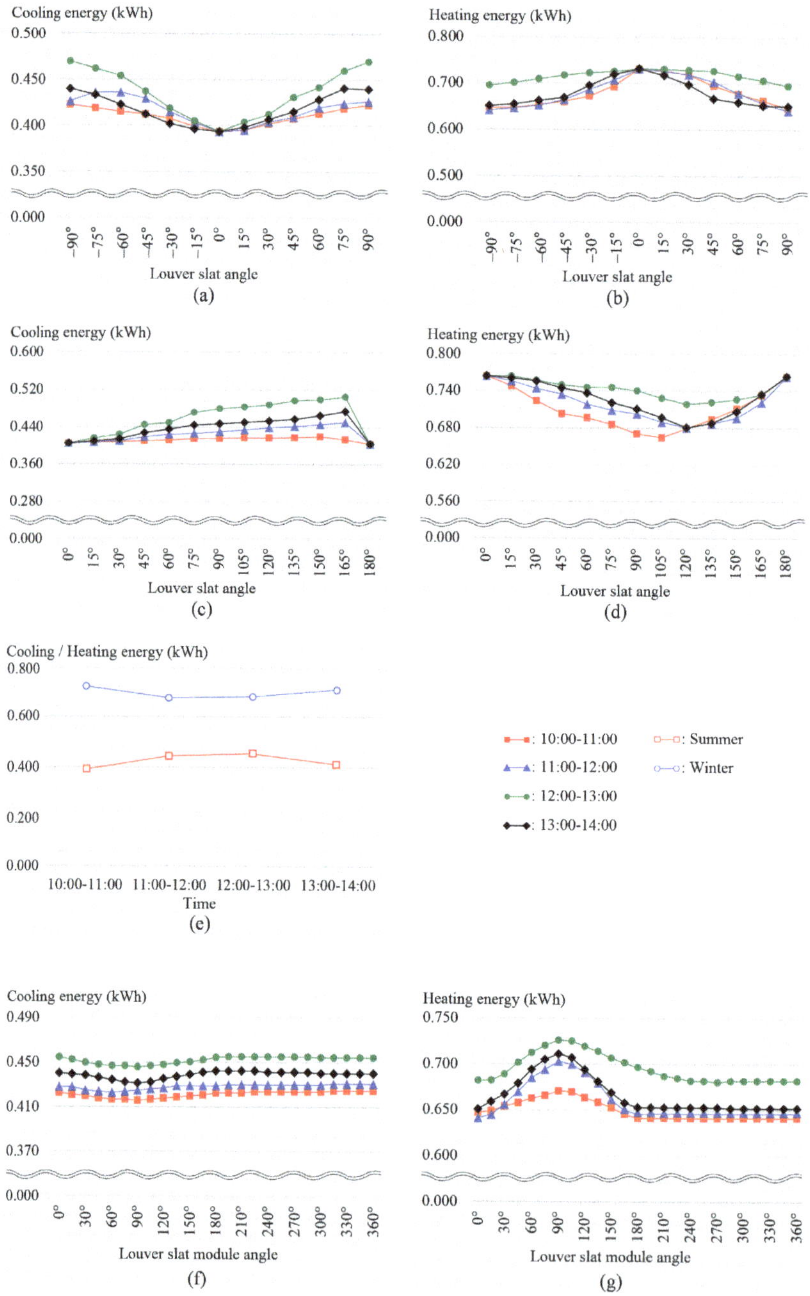

Figure 12. Heating and cooling energy usage by case: (**a**) Summer–Case 1, (**b**) Winter–Case 1, (**c**) Summer–Case 2, (**d**) Winter–Case 2, (**e**) Case 3, (**f**) Summer–Case 4, (**g**) Winter–Case 4.

Table 10. Total energy consumption based on the optimal slat/module angles for integrated energy reduction by case.

Case	Optimal Angle of the Slat Module for Energy Reduction								Total Summer Energy Usage (kWh)	Total Winter Energy Usage (kWh)	* Total Energy Usage (kWh)
	Summer				Winter						
	10:00–11:00	11:00–12:00	12:00–13:00	13:00–14:00	10:00–11:00	11:00–12:00	12:00–13:00	13:00–14:00			
1	60°	30°	0°	−30°	−30°	90°	−90°	−90°	1.424	2.844	4.268
2	60°	75°	60°	75°	105°	120°	60°	120°	1.493	2.965	4.458
3	-	-	-	-	-	-	-	-	1.431	2.889	4.320
4	60°	75°	90°	105°	45°	75°	90°	105°	1.425	2.642	4.067

* Energy consumption total (kWh) = PV module power generation−Lighting energy consumption−Heating and cooling energy consumption.

Table 11. Optimal angles and total heating and cooling energy usage for reducing energy consumption by case.

Case	Optimal Angle of the Slat Module to Improve the Total Heating and Cooling Energy Usage								Total Cooling Energy (kWh)	Total Heating Energy (kWh)	Total Heating and Cooling Energy (kWh)
	Summer				Winter						
	10:00–11:00	11:00–12:00	12:00–13:00	13:00–14:00	10:00–11:00	11:00–12:00	12:00–13:00	13:00–14:00			
1	0°	0°	0°	0°	−45°	−90° (90°)	−90° (90°)	−90° (90°)	1.574	2.643	4.217
2	0° (180°)	0° (180°)	0° (180°)	0° (180°)	105°	120°	120°	120°	1.619	2.746	4.365
3	-	-	-	-	-	-	-	-	1.787	3.077	4.864
4	* 90°	60°	90°	90°	0°	180°	270°	0°	1.715	2.628	4.343
	** 60°	75°	90°	105°	45°	75°	90°	105°	1.718	2.785	4.503

* Controlling the angle of the slat module to reduce heating and cooling energy consumption in Case 4. ** Operation based on the optimal angle derived from the solar-tracking device in Case 4.

Table 12. Total energy consumption of the open structure through slat movement in winter (Case 4).

PV Module Power Generation (kWh)	Total Lighting Energy (kWh)	Total Heating Energy (kWh)	* Total Energy Usage (kWh)
0.014	0.176	2.097	2.259

* Energy consumption total (kWh) = PV module power generation−Lighting energy consumption−Heating and cooling energy consumption.

4. Conclusions

This study proposed a solar-tracking-based movable louver integrated with a photovoltaic (PV) module to enhance the energy performance of conventional louvers. The effectiveness of the system was verified through controlled testbed experiments, and the following conclusions were drawn:

First, the proposed louver system is fundamentally based on a vertical configuration. However, by modularizing the slats and allowing them to rotate, the system can be transformed into either vertical or horizontal configurations. This design enables the benefits of both louver types. A solar tracking device is mounted at the top of the structure, operating on a single-axis rotation mechanism. It features two illuminance sensors positioned on either side of a frame with a cross-sectional shape resembling "⊥". When the difference in sensor readings decreases, the angle of the sensor surface approaches perpendicularity to the sunlight, thus enabling accurate solar tracking based on this principle.

Second, to improve PV power generation, it is essential for sunlight to strike the PV modules perpendicularly. Therefore, the operation of the louver system should be adjusted according to the time of day. The proposed system achieved increases in power generation of 4.4% and 1.5% compared to vertical and horizontal louvers, respectively, when operated at optimal angles. Notably, this performance was achieved despite using fewer PV cells. In contrast, the mixed louver, despite having the largest number of PV cells, showed lower energy efficiency because it was fixed and unable to adapt to changing conditions.

Third, this study considered not only PV power generation but also lighting and heating/cooling energy requirements in evaluating louver performance. The analysis showed that the optimal slat angles for minimizing total energy consumption differ from those that only maximize PV output. This finding highlights the importance of integrated energy management. When accounting for all three energy components—PV generation, lighting, and thermal loads—the proposed louver system reduced total energy consumption by 4.7%, 8.8%, and 5.9% compared to vertical, horizontal, and mixed louver systems, respectively. This demonstrates the proposed system's effectiveness in comprehensive energy savings.

Fourth, in winter, applying PV modules to the louver slats is not as effective for reducing total energy consumption. Instead, allowing natural sunlight to penetrate indoor spaces is more beneficial for minimizing lighting and heating energy. Therefore, the system is designed with an open–close mechanism to allow the slats to be retracted when solar radiation is desirable, as demonstrated in the winter energy-saving scenario. This highlights the importance of adaptable control based on seasonal and environmental conditions.

This study was validated through performance evaluations to verify the effectiveness of technology development aimed at improving the energy reduction performance of louvers. However, the performance evaluations were conducted only under limited conditions, such as specific regions and time periods. Furthermore, the performance evaluations were conducted with an artificial environment rather than a real one. The artificial sunlight used in this study was also limited in its simulation of the azimuth angle of the actual sun. As a result, additional constraints arose from conducting performance evaluations restricted to specific time periods. Therefore, it is necessary to conduct follow-up performance evaluations considering various temporal, climatic, and environmental conditions in the future.

Author Contributions: Conceptualization, S.H. and H.L.; Methodology, S.H.; Writing—original draft preparation, S.H.; Writing—review and editing, J.S. and H.L.; Funding acquisition, H.L.; Visualization, S.H.; Investigation, S.H.; Formal analysis, S.H.; Supervision, H.L. All authors have read and agreed to the published version of the manuscript.

Funding: This research was supported by the National Research Foundation of Korea (NRF) grant funded by the Korean government (MSIT) (RS-2023-00208303).

Data Availability Statement: Data generated or analyzed during the current study are available from the corresponding author on reasonable request.

Conflicts of Interest: The authors declare no conflicts of interest.

References

1. Grobal Energy Trends 2023. Available online: https://www.enerdata.co.kr/publications/reports-presentations/world-energy-trends.html (accessed on 15 June 2023).
2. Barea, G.; Victoria Mercado, M.; Filippín, C.; Monteoliva, J.M.; Villalba, A. New paradigms in bioclimatic design toward climatic change in arid environments. *Energy Build.* **2022**, *266*, 100–112. [CrossRef]
3. Amasyali, K.; El-Gohary, N.M. A review of data-driven building energy consumption prediction studies. *Renew. Sustain. Energ. Rev.* **2018**, *81*, 1192–1205. [CrossRef]

4. Uddin, M.; Ruva, I.; Syed, M.; Hossain, D.; Akter, R.; Tamanna, N.; Rahman, A.; Saka, A. Occupant centric energy renovation strategy for hospital and restaurant building envelop using distinct modelling tools: A case study from low income cultural context. *Energy Build.* **2022**, *272*, 112–338. [CrossRef]
5. Wang, Y.; Dong, Q.; Guo, H.; Yin, L.; Gao, W.; Yao, W.; Sun, L. Indoor thermal comfort evaluation of traditional dwellings in cold region of china: A case study in Guangfu ancient city. *Energy Build.* **2023**, *288*, 113–128. [CrossRef]
6. Gupta, G.; Mathur, S.; Mathur, J.; Nayak, B.K. Comparison of energy-efficiency benchmarking methodologies for residential buildings. *Energy Build.* **2023**, *285*, 112–920. [CrossRef]
7. 2022 Global Status Report for Buildings and Construction. Available online: https://www.unep.org/resources/publication/2022-global-status-report-buildings-and-construction (accessed on 9 November 2022).
8. Moazzeni, M.H.; Ghiabaklou, Z. Investigating the influence of light shelf geometry parameters on daylight performance and visual comfort, a case study of educational space in Tehran, Iran. *Buildings* **2016**, *6*, 26. [CrossRef]
9. Soliman, M.E.; Bo, S. An innovative multifunctional biomimetic adaptive building envelope based on a novel integrated methodology of merging biological mechanisms. *J. Build. Eng.* **2023**, *76*, 106995. [CrossRef]
10. Han, S.; Lee, M.; Lee, H. Effectiveness of internal light shelves to improve daylighting. *Build. Simul.* **2025**, *18*, 747–763. [CrossRef]
11. Kim, J.; Ko, W.; Kim, J.; Byun, J.; Nam, S.; Jeong, S. Analyzing the environmental impacts and empirical limitations of green remodeling with life cycle assessment. *Buildings* **2025**, *15*, 783. [CrossRef]
12. Jaradat, H.; Alshboul, O.A.M.; Obeidat, I.M.; Zoubi, M.K. Green building, carbon emission, and environmental sustainability of construction industry in Jordan: Awareness, actions and barriers. *Ain Shams Eng. J.* **2024**, *15*, 102–441. [CrossRef]
13. Qiang, G.; Tang, S.; Hao, J.; Di Sarno, L.; Wu, G.; Ren, S. Building automation systems for energy and comfort management in green buildings: A critical review and future directions. *Renew. Sust. Energ.* **2023**, *179*, 113–301. [CrossRef]
14. Xu, J.; Sun, J.; Zhao, J.; Zhang, W.; Zhou, J.; Xu, L.; Guo, H.; Liu, Y.; Zhang, D. Eco-friendly wood plastic composites with biomass-activated carbon-based form-stable phase change material for building energy conversion. *Ind. Crops Prod.* **2023**, *197*, 116–573. [CrossRef]
15. Amin, M.S.; Heikal, M.; Negm, H.H.; Abu-Dief, A.M.; Mohamed, O.A. Manufacture of eco-friendly cementitious building materials of high performance from Egyptian industrial solid wastes. *Constr. Build. Mater.* **2023**, *406*, 133–446. [CrossRef]
16. Sierra-Pérez, J.; López-Forniés, I.; Boschmonart-Rives, J.; Gabarrell, X. Introducing eco-ideation and creativity techniques to increase and diversify the applications of eco-materials: The case of cork in the building sector. *J. Clean. Prod.* **2016**, *137*, 606–616. [CrossRef]
17. Hammad, F.; Abu-Hijleh, B. The energy savings potential of using dynamic external louvers in an office building. *Energy Build.* **2010**, *42*, 1888–1895. [CrossRef]
18. Saelens, D.; Parys, W.; Roofthooft, J.; de la Torre, A.T. Assessment of approaches for modeling louver shading devices in building energy simulation programs. *Energy Build.* **2013**, *60*, 286–297. [CrossRef]
19. Jung, S.K.; Kim, Y.; Moon, J.W. Performance evaluation of control methods for PV-integrated shading devices. *Energies* **2020**, *13*, 3171. [CrossRef]
20. Yi, H.; Kim, Y. Prototyping of 4D-printed self-shaping building skin in architecture: Design, fabrication, and investigation of a two-way shape memory composite (TWSMC) façade panel. *J. Build. Eng.* **2021**, *43*, 103–176. [CrossRef]
21. Reichert, S.; Menges, A.; Correa, D. Meteorosensitive architecture: Biomimetic building skins based onmaterially embedded and hygroscopically enabled responsiveness. *Comput. Aid. Des.* **2015**, *60*, 50–69. [CrossRef]
22. Iqbal, W.; Ullah, I.; Hussain, A.; Cho, M.; Park, J.; Lee, K.; Shin, S. Optimizing energy efficiency: Louver systems for sustainable building design. *Buildings* **2025**, *15*, 1183. [CrossRef]
23. Tao, S.; Yu, N.; Jiang, F.; Su, X.; Zhao, K. Correlations for forced convective heat transfer coefficients at the windward building façade with vertical louvers. *Build. Environ.* **2023**, *242*, 110–611. [CrossRef]
24. Luo, Z.; Sun, C.; Dong, Q.; Qi, X. Key control variables affecting interior visual comfort for automated louver control in open-plan office—A study using machine learning. *Build. Environ.* **2022**, *207*, 108–565. [CrossRef]
25. Pourshab, N.; Tehrani, M.D.; Toghraie, D.; Rostami, S. Application of double glazed façades with horizontal and vertical louvers to increase natural air flow in office buildings. *Energy* **2020**, *200*, 117–486. [CrossRef]
26. Rafati, N.; Hazbei, M.; Eicker, U. Louver configuration comparison in three Canadian cities utilizing NSGA-II. *Build. Environ.* **2023**, *229*, 109–939. [CrossRef]
27. Kangazian, A.; Emadian Razavi, S.Z. Multi-criteria evaluation of daylight control systems of office buildings considering daylighting, glare and energy consumption. *Sol. Energy* **2023**, *263*, 111–928. [CrossRef]
28. Palmero-Marrero, A.I.; Oliveira, A.C. Effect of louver shading devices on building energy requirements. *Appl. Energy* **2010**, *87*, 2040–2049. [CrossRef]

29. Luca, F.; Varjas, A. Multi-performance optimization of static shading devices for glare, daylight, view and energy consideration. *Build. Environ.* **2022**, *217*, 109–110. [CrossRef]
30. Fang, J.; Zhao, Y.; Tian, Z.; Lin, P. Analysis of dynamic louver control with prism redirecting fenestrations for office daylighting optimization. *Energy Build.* **2022**, *262*, 112–119. [CrossRef]
31. Alhuwayil, W.K.; Almaziad, F.A.; Abdul Mujeebu, M. Energy performance of passive shading and thermal insulation in multistory hotel building under different outdoor climates and geographic locations. *Case Stud. Therm. Eng.* **2023**, *45*, 102–940. [CrossRef]
32. Mandalaki, M.; Tsoutsos, T.; Papamanolis, N. Integrated PV in shading systems for Mediterranean countries: Balance between energy production and visual comfort. *Energy Build.* **2014**, *77*, 445–456. [CrossRef]
33. Peng, J.; Lu, L.; Yang, H. An experimental study of the thermal performance of a novel photovoltaic double-skin facade in Hong Kong. *Sol. Energy* **2013**, *97*, 293–304. [CrossRef]
34. Chow, T.T.; Pei, G.; Chan, L.S.; Lin, Z.; Fong, K.F. A comparative study of PV glazing performance in warm climate. *Indoor Built Environ.* **2009**, *18*, 32–40. [CrossRef]
35. Kim, J.J.; Jung, S.; Choi, Y.; Kim, J. Optimization of photovoltaic integrated shading devices. *Indoor Built Environ.* **2010**, *19*, 114–122. [CrossRef]
36. Kim, M.; Kim, B.; Koh, J.; Yi, H. Flexural biomimetic responsive building façade using a hybrid soft robot actuator and fabric membrane. *Autom. Constr.* **2023**, *145*, 104–660. [CrossRef]
37. Abdel-Rahman, W.S.M. Thermal performance optimization of parametric building envelopebased on bio-mimetic inspiration. *Ain Shams Eng. J.* **2021**, *12*, 1133–1142. [CrossRef]
38. Alimi, O.A.; Meyer, E.L.; Olayiwola, O.I. Solar photovoltaic modules' performance reliability and degradation analysis—A review. *Energies* **2022**, *15*, 5964. [CrossRef]
39. Li, Z.; Ji, J.; Yuan, W.; Zhao, B.; Zhou, F.; Muin Uddin, M.; Ren, X.; Yu, B.; Song, Z. Experimental & numerical investigation and optimization on a novel flat-plate PV/T system using CdfTe thin-film solar modules of sandwich structure. *Sol. Energy* **2021**, *223*, 261–277. [CrossRef]
40. Han, S.; Lee, H. Evaluation of energy performance and daylight utilization in light shelves integrated with transparent solar panels. *Energy* **2025**, *326*, 136336. [CrossRef]
41. Huld, T.; Amillo, A.M. Estimating PV module performance over large geographical regions: The role of irradiance, air temperature, wind speed and solar spectrum. *Energies* **2015**, *8*, 5159–5181. [CrossRef]
42. Lu, Y.; Li, G. Potential application of electrical performance enhancement methods in PV/T module. *Energy* **2023**, *281*, 128–253. [CrossRef]
43. Han, S.; Seo, J.; Lee, H. Verification of the effectiveness of transparent photovoltaics according to the transmittance level in Seoul, South Korea. *Sol. Energy* **2025**, *294*, 113511. [CrossRef]
44. Hu, Y.; Shen, H.; Yao, Y. A novel sun-tracking and target-aiming method to improve the concentration efficiency of solar central receiver systems. *Renew. Energ.* **2018**, *120*, 98–113. [CrossRef]
45. Lu, J.; Hajimirza, S. Optimizing sun-tracking angle for higher irradiance collection of PV panels using a particle-based dust accumulation model with gravity effect. *Sol. Energy* **2017**, *158*, 71–82. [CrossRef]
46. Fazal, M.A.; Rubaiee, S. Progress of PV cell technology: Feasibility of building materials, cost, performance, and stability. *Sol. Energy* **2023**, *258*, 203–219. [CrossRef]
47. Pirayawaraporn, A.; Sappaniran, S.; Nooraksa, S.; Prommai, C.; Chindakham, N.; Jamroen, C. Innovative sensorless dual-axis solar tracking system using particle filter. *Appl. Energy* **2023**, *338*, 120–946. [CrossRef]
48. Hafez, A.Z.; Yousef, A.M.; Harag, N.M. Solar tracking systems: Technologies and trackers drive types—A review. *Renew. Sustain. Energ. Rev.* **2018**, *91*, 754–782. [CrossRef]
49. Carballo, J.A.; Bonilla, J.; Roca, L.; Berenguel, M. New low-cost solar tracking system based on open source hardware for educational purposes. *Sol. Energy* **2018**, *174*, 826–836. [CrossRef]
50. Boukdir, Y.; Omari, H. Novel high precision low-cost dual axis sun tracker based on three light sensors. *Heliyon* **2022**, *8*, e12412. [CrossRef]
51. Mamodiya, U.; Tiwari, N. Dual-axis solar tracking system with different control strategies for improved energy efficiency. *Comput. Electr. Eng.* **2023**, *111*, 109–920. [CrossRef]
52. Achkari, O.; El Fadar, A.; Amlal, I.; Haddi, A.; Hamidoun, M.; Hamdoune, S. A new sun-tracking approach for energy saving. *Renew. Energ.* **2021**, *169*, 820–835. [CrossRef]
53. Mpodi, E.K.; Tjiparuro, Z.; Matsebe, O. Review of dual axis solar tracking and development of its functional model. *Procedia Manuf.* **2019**, *35*, 580–588. [CrossRef]
54. Zhu, Y.; Liu, J.; Yang, X. Design and performance analysis of a solar tracking system with a novel single-axis tracking structure to maximize energy collection. *Appl. Energy* **2020**, *264*, 114–647. [CrossRef]

55. Saymbetov, A.; Mekhilef, S.; Kuttybay, N.; Nurgaliyev, M.; Tukymbekov, D.; Meiirkhanov, A.; Dosymbetova, G.; Svanbayev, Y. Dual-axis schedule tracker with an adaptive algorithm for a strong scattering of sunbeam. *Sol. Energy* **2021**, *224*, 285–297. [CrossRef]
56. Alomar, O.R.; Ali, O.M.; Ali, B.M.; Qader, V.S.; Ali, O.M. Energy, exergy, economical and environmental analysis of photovoltaic solar panel for fixed, single and dual axis tracking systems: An experimental and theoretical study. *Case Stud. Therm. Eng.* **2023**, *51*, 103–635. [CrossRef]
57. Awasthi, A.; Shukla, A.K.; Murali Manohar, S.R.; Dondariya, C.; Shukla, K.N.; Porwal, D.; Richhariya, G. Review on sun tracking technology in solar PV system. *Energy Rep.* **2020**, *6*, 392–405. [CrossRef]
58. Illuminating Engineering Society. *The Lighting Handbook*, 10th ed.; Illuminating Engineering Society (IES): New York, NY, USA, 2011.
59. Zhao, X.; Seo, J.; Lee, H. Performance evaluation of a light shelf with a folding reflector (LSFR) to improve daylighting performance. *Build. Environ.* **2024**, *255*, 111457. [CrossRef]
60. Lee, H.; Kim, S.; Seo, J. Evaluation of a light shelf based on energy consumption for lighting and air conditioning. *Indoor Built Environ.* **2018**, *27*, 1405–1414. [CrossRef]
61. Korea Meteorological Administration. Available online: https://www.kma.go.kr/eng/weather/climate/worldclimate.jsp (accessed on 8 June 2025).

Disclaimer/Publisher's Note: The statements, opinions and data contained in all publications are solely those of the individual author(s) and contributor(s) and not of MDPI and/or the editor(s). MDPI and/or the editor(s) disclaim responsibility for any injury to people or property resulting from any ideas, methods, instructions or products referred to in the content.

MDPI AG
Grosspeteranlage 5
4052 Basel
Switzerland
Tel.: +41 61 683 77 34

Buildings Editorial Office
E-mail: buildings@mdpi.com
www.mdpi.com/journal/buildings

Disclaimer/Publisher's Note: The title and front matter of this reprint are at the discretion of the Guest Editor. The publisher is not responsible for their content or any associated concerns. The statements, opinions and data contained in all individual articles are solely those of the individual Editor and contributors and not of MDPI. MDPI disclaims responsibility for any injury to people or property resulting from any ideas, methods, instructions or products referred to in the content.

www.ingramcontent.com/pod-product-compliance
Lightning Source LLC
Chambersburg PA
CBHW061609220926
51850CB00005B/463
* 9 7 8 3 7 2 5 8 5 0 5 9 4 *